Signaling and Communication in Plants

Series Editor

František Baluška, IZMB, Department of Plant Cell Biology, University of Bonn, Bonn, Nordrhein-Westfalen, Germany

This book series will be devoted to diverse aspects of signaling and communication at all levels of plant organization, starting from single molecules and ending at ecological communities. The individual volumes will interlink molecular biology with plant physiology and the behavior of individual organisms, right up to the system analysis of whole plant communities and ecosystems.

Plants have developed a robust signaling apparatus with both chemical and physical communication pathways. The chemical communication is based either on vesicular trafficking pathways or accomplished directly through cell-cell channels known as plasmodesmata. Moreover, there are numerous signal molecules generated within cell walls and also diffusible signals, such as nitric oxide, reactive oxygen species and ethylene, penetrating cells from exocellular space. Physical communication on the other hand is based on electrical, hydraulic, and mechanical signals.

The integrative view of this book series will foster our understanding of plant communication throughout the individual plant and with other communicative systems such as fungi, nematodes, bacteria, viruses, insects, other plants, and predatory animals.

Tariq Aftab
Editor

Auxins, Cytokinins and Gibberellins Signaling in Plants

Springer

Editor
Tariq Aftab
Department of Botany
Aligarh Muslim University
Aligarh, India

ISSN 1867-9048 ISSN 1867-9056 (electronic)
Signaling and Communication in Plants
ISBN 978-3-031-05426-6 ISBN 978-3-031-05427-3 (eBook)
https://doi.org/10.1007/978-3-031-05427-3

This Springer imprint is published by the registered company Springer Nature Switzerland AG
The registered company address is: Gewerbestrasse 11, 6330 Cham, Switzerland

Preface

Being multicellular organisms, plant and animal growth have a conspicuous feature in common: both plant and animal growth are regulated by hormones. Plant hormones have pivotal roles in the regulation of plant growth, development, and reproduction. Additionally, they emerged as cellular signaling molecules with key functions in the regulation of responses to various abiotic and biotic stressors. Their signaling pathways are interconnected in a complex network, which provides plants with an enormous regulatory potential to rapidly adapt to their environment and to utilize their limited resources for growth and survival in a cost-efficient manner.

Auxin is a hormone molecule whose activity levels are most important for its regulatory roles during plant cell, organ, and tissue development. Therefore, the precise regulation of auxin levels is an essential mechanism to fine-tune the activity of this powerful hormone during plant growth and development. Extensive genetic and molecular studies have demonstrated that the auxin transport is also involved in plant responses to environmental stimuli, such as low temperature, light, and gravity. In the past several years, there are some evidences pointing to the regulatory role of the auxin signaling in plant response to salinity and drought.

In plants, the cytokinins were defined as substances stimulating cell division (cytokinesis) in tissue cultures. Apart of this effect, cytokinins exhibit a wide range of physiological functions, including regulation of shoot and root apical meristems, stimulation of branching, vascular development, chloroplast differentiation, stabilization of the structure and function of the photosynthetic machinery, delay of senescence, stomata opening, and elevation of the sink strength and nutritional signaling. Targeted elevation of cytokinin levels was found to increase the tolerance of plants to abiotic stresses, at least partially by diminishing the negative stress effects on photosynthesis. Recently, function of cytokinins in biotic stress responses has been also recognized.

Gibberellins are a class of diterpenoid acids that regulate many aspects of plant growth and development including seed germination, stem elongation, leaf expansion, and flower and fruit development. The broad implication of gibberellins in plant development is strictly associated to tight regulation of their metabolism by multiple environmental and endogenous factors, ranging from light and temperature to other hormones including feedback control. Distribution patterns and finely tuned concentration gradients govern plant growth and development. These hormones regulating key processes in plants; many of them are of significant agricultural importance, such as seed germination, root and shoot elongation, flowering, and fruit patterning.

Understanding the significant roles of these phytohormones in plant biology and from agriculture point of view, the current subject has recently attracted the attention of scientists from across the globe. Therefore, I bring forth a comprehensive volume **Auxins, Cytokinins and Gibberellins Signaling in Plants** highlighting the various prospects involved in current scenario. I am hopeful that this comprehensive volume will furnish the requisite of all those who are working or have interest in the proposed topic.

I am highly grateful to all our contributors for accepting our invitation for not only sharing their knowledge and research, but for venerably integrating their expertise in dispersed information from diverse fields in composing the chapters and enduring editorial suggestions to finally produce this venture. I also thank Springer-Nature team for their generous cooperation at every stage of the book production.

Lastly, thanks are also due to well-wishers, research students and editor's family members for their moral support, blessings and inspiration in the compilation of this book.

Aligarh, India Tariq Aftab

Contents

About the Editor

Dr. Tariq Aftab received his Ph.D. in the Department of Botany at Aligarh Muslim University, India, and is currently an Assistant Professor there. He is the recipient of a prestigious Leibniz-DAAD fellowship from Germany, Raman Fellowship from the Government of India, and Young Scientist Awards from the State Government of Uttar Pradesh (India) and Government of India. After completing his doctorate, he has worked as Research Fellow at National Bureau of Plant Genetic Resources, New Delhi and as Post-doctorate Fellow at Jamia Hamdard, New Delhi, India. Dr. Aftab also worked as Visiting Scientist at Leibniz Institute of Plant Genetics and Crop Plant Research (IPK), Gatersleben, Germany, and in the Department of Plant Biology, Michigan State University, USA. He is a member of various scientific associations from India and abroad.

He has edited 12 books with international publishers, including Elsevier Inc., Springer Nature and CRC Press (Taylor & Francis Group), co-authored several book chapters, and published over 80 research papers in peer-reviewed international journals. His research interests include physiological, proteomic, and molecular studies on medicinal and crop plants.

How the Three Organ-Produced Signals: Auxin, Cytokinin and Gibberellin, Induce and Regulate Wood Formation and Adaptation

Roni Aloni

Abstract The chapter clarifies the hormonal mechanisms that regulate wood formation in plants focusing on hardwood trees. Uncovering the specific role of each of the hormones: auxin, gibberellin and cytokinin in controlling vascular differentiation. Explaining the hormonal control of vessels and fibers along the plant axis from leaves to roots, and during the growth season. Clarifying how the environment, by controlling plant growth and dimensions, regulates the internal hormonal mechanisms that shape the rate of vessel widening and their final size. How the environment has modified the sensitivity of the cambium to the auxin signal during the evolution of ring-porous trees, resulting in their typical very wide earlywood vessels, followed by latewood fibers with thick secondary walls.

1 Introduction

This chapter provides an overview of the internal hormonal mechanisms that control and regulate xylem differentiation in plants, focusing on wood formation in trees, clarifying the evolution and specialization of these mechanisms in temperate deciduous hardwood trees. Special attention is paid to explain a major topic in xylogenesis, namely, the control of vessel size within the whole tree as affected by external and internal factors, evolving the ring-porous wood pattern under extreme environmental conditions during the recent 50 million years.

In order to explain the evolution of ring-porous trees, I provide a summary of the three major hormonal signals that regulate wood formation for those unfamiliar with the subject and as a preamble to a discussion on the control of vessels and fibers in forest trees. Then, I focus on three major topics in vascular differentiation and the recent advances made in each, demonstrating the gradual conceptual evolution of ideas as a natural process, by presenting three hypotheses that stem from each other: (i) the control of vessel width and density along the tree axis (*auxin gradient*

R. Aloni (✉)
School of Plant Sciences and Food Security, Tel Aviv University, Tel Aviv, Israel
e-mail: alonir@tauex.tau.ac.il

T. Aftab (ed.), *Auxins, Cytokinins and Gibberellins Signaling in Plants*, Signaling and Communication in Plants,
https://doi.org/10.1007/978-3-031-05427-3_1

hypothesis), (ii) environmental adaptation of the xylem in plants (*vascular adaptation hypothesis*), and (iii) the evolution of ring-porous wood in temperate deciduous hardwood trees (*limited-growth hypothesis*).

2 The Conducting and Supporting Cells in the Wood of Angiosperms

The wood of plants is produced by the meristematic vascular cambium and is termed secondary xylem. Its main function is water transport from roots to leaves and the support of the plant body. Two conductive water conduits are found in wood, the tracheids which are typical to conifers and vessels that are dominant in angiosperms. A *tracheid* is usually a long water-conducting conduit that has no perforations. A *vessel* is a long tube build of vessel elements, that their common walls have perforations. The vessel is therefore a more efficient conduit than a tracheid, since water flow through vessel elements occurs via openings, namely, perforations, rather than diffusion through the primary cell walls, through the bordered pits of tracheids (Tyree and Zimmermann 2002).

A vessel has endings walls in the lower and upper ends (Aloni 2021). Transport of water from vessel to vessel occurs through pits along their vessel elements or their end cell wall. It should be noted, that vessels and not vessel elements are the physiologically operating units of water transport regarding cavitation and embolism (Zimmermann 1983; Zhang et al. 2018). Embolism of a large vessel is usually followed by the outgrowths of the surrounding parenchyma cells into the vessel, a phenomenon known as tyloses (Zimmermann 1983), which forms blockages to penetration and movement of fungi and bacteria into and along the air-filled nonfunctional vessel. There are species that gum may plug the air-filled vessel to prevent possible penetration damage. Tyloses and gum plugs develop naturally in hardwood trees when the functional water transporting sapwood gradually turns into a stable heartwood that is resistant to rot (De Micco et al. 2016).

There is a positive correlation between *conductive efficiency* and *vulnerability* to water stress and freezing inducing embolism. Wide and long vessels that are efficient conduits are more vulnerable to cavitation and embolism induced by freezing and water stresses than narrow vessels and tracheids. The widest earlywood vessels of ring-porous trees (see below) operate for a relatively short duration, usually for only one growth season and become nonfunctional at the end of the season. On the other hand, tracheids and narrow vessels are safe conduits that function for long periods of a few years, but are less efficient in water transport (Tyree and Sperry 1989; Tyree and Ewers 1991).

The vessels are accompanied by xylem parenchyma cells and fibers. A *fiber* in the xylem and phloem is usually a long cell characterized by thick and hard lignified secondary cell walls with simple pits. The latter provides the mechanical strength for supporting the plant body. During evolution, both xylem fibers and vessels, have originated from tracheids of more primitive plants (Bailey and Tupper 1918; Evert and Eichhorn 2013; Aloni 2021).

2.1 The Importance of Vessel Width for Water Conductance

The hydraulic performance of trees is crucially affected by vessel diameter (Tyree and Zimmermann 2002; Lucas et al. 2013; Hacke et al. 2017; Williams et al. 2019; Olson et al. 2021; Aloni 2021), which also affects wood adaptation (Aloni 1987, 2015) and xylem pathology (Aloni and Ullrich 2008; Ullrich et al. 2019). Therefore, it is important to understand and clarify the mechanisms that control the diameter of these vascular conduits in plants.

Vessel diameter has a very important functional significance in water conduction. In ideal capillaries, conductivity is proportional to the fourth power of the radius, or diameter (Zimmermann 1983), which means that at a given pressure gradient the relative volumes of water flowing through capillaries, or vessels, of diameters: 1, 2, 3, 4, and 5, are: 1, 16, 81, 256, and 625, respectively. A cross section of a ring-porous wood demonstrates that most of the water would have flown through the very wide earlywood vessels, whereas the narrow latewood vessels would be inefficient in water conductance (Fig. 1). Yet, the narrow latewood vessels are important for plant survival as they continue to function when the wide earlywood vessels stop functioning, following cavitation and embolism.

2.2 The Problem of Wide Earlywood Vessel Formation in Temperate Deciduous Hardwood Trees

In temperate deciduous broad-leafed trees, the size differences of the vessels in the early- and latewoods are quite marked and two categories of deciduous trees are determined: *diffuse-porous species* and *ring-porous species* (Fig. 1). In diffuse-porous wood the vessels produced along the season are more or less uniform in size, whereas in ring-porous wood the vessels produced at the beginning of the growth season are significantly wider than those produced at the end of the season (Evert 2006; Aloni 2021).

Although Hartig (1853) and Russow (1883) observed long ago the formation of wide vessels in the trunk of *Quercus* trees at the very early stage of bud development, the mechanisms that regulate the differentiation of wide earlywood vessels in temperate deciduous ring-porous trees still need clarifications (Suzuki et al. 1996; Sass-Klaassen et al. 2011; Takahashi et al. 2013; Pérez-de-Lis et al. 2016; Lavrič et al. 2017; Puchałka et al. 2017; Rodriguez-Zaccaro et al. 2019; Zhu et al. 2020). Interestingly, ring-porous species produce diffuse-porous xylem in their young leader, twigs and branches during their first year, while they produce the typical ring-porous xylem

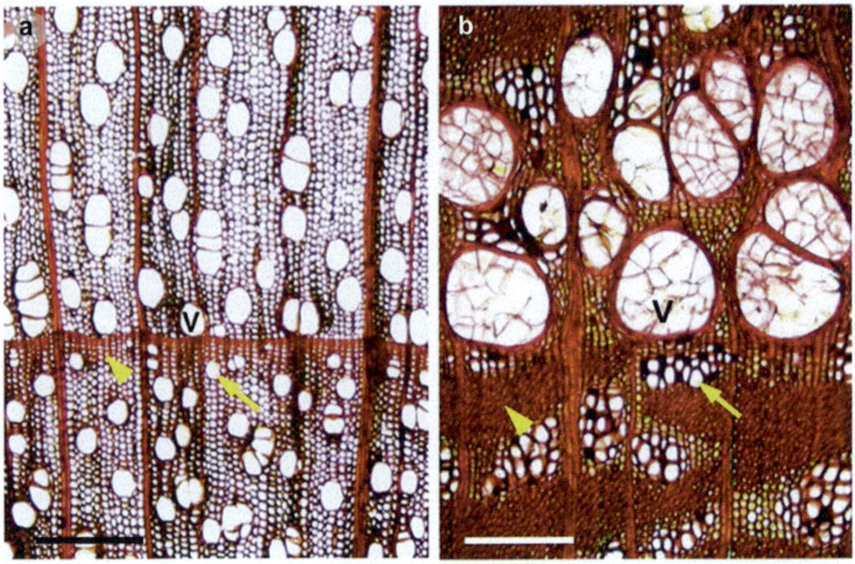

Fig. 1 Transverse sections showing the borderline and transition from the latewood (*down*) to earlywood (*up*), in the diffuse-porous wood of *Acer rubrum* (**a**), in comparison with the ring-porous wood of *Robinia pseudoacacia* (**b**), analyzed during winter dormancy, stained with safranin and fast green. In both photomicrographs, earlywood vessels are marked by V, latewood vessels are marked with an arrow, and latewood fibers with an arrowhead. Both micrographs are at the same orientation and magnification (scale bars = 500 μm). A, the diffuse-porous wood shows functional open vessels with relatively similar width forming a continuous pattern, which was induced by the continuous production of auxin-producing young leaves along the growth season. The vessels of the new year are wider than those of the previous season due to the increased distance of the new year's leaves from the transverse-section site. The earlywood fibers have thin secondary walls, while the latewood fibers (arrowhead) have somewhat thicker secondary cell walls associated with narrow vessels (arrow) due to leaf maturation, which promoted gibberellin production toward the end of the growth season. B, the very wide and vulnerable earlywood vessels (V) in the ring-porous wood are already plugged by tyloses at the end of the growth season. These wide vessels were induced by extremely low-auxin-concentration streams originating in dormant looking buds, in a slow widening process along a few weeks, which started before bud break. Whereas, the latewood of the previous year contains the narrow functioning latewood vessels (arrow) that would transport water for more than one growth season. The ring-porous latewood is also characterized by numerous latewood fibers with very thick lignified secondary cell walls. Both, the safe narrow latewood vessels and thick-wall fibers were induced by the gibberellin-producing mature leaves

along their mature trunk (Cochard & Tyree, 1990; Lo Gullo et al. 1995; Takahashi et al. 2013; Rodriguez-Zaccaro et al. 2019); these differences between the two types of wood porosity produced along the same ring-porous tree were recently elucidated (Aloni 2021) and will be discussed below. Additionally, it is unclear how the regulating mechanisms of wide earlywood vessel formation are influenced by climatic factors in different hardwood trees in temperate forests (Sass-Klaassen et al. 2011; Pérez-de-Lis et al. 2016, 2018; Hacke et al. 2017; Gričar et al. 2018, 2020; Zhu et al. 2020).

In ring-porous species, the first earlywood vessels are very wide and are formed at, or just prior to, the breaking of buds. Conversely, in diffuse-porous trees the earlywood vessels are formed much later and appear in the trunk when the leaves are one-fourth to fully expended (Lodewick 1928). Furthermore, in ring-porous trees the initial wide vessels develop almost simultaneously all along the main stem (Atkinson and Denne 1988), whereas in a diffuse-porous tree they are restricted to the base of the buds, appearing first in the lower branches and later in the upper ones (Atkinson and Denne 1987). The wide earlywood vessels in ring-porous trees are also very long and can extend along the length of the stem itself (Greenidge 1952; Zimmermann and Jeje 1981), whereas the earlywood vessels of diffuse-porous species are narrow and much shorter, usually less than 1 m (Zimmermann and Jeje 1981).

When young trees are completely debudded in late winter, before any bud activity can be observed, new earlywood vessels differentiate in ring-porous trees, but are entirely absent in diffuse-porous species (Wareing 1951; Reines 1959). Young growing leaves are known to produce the auxin hormone that moves downward and induces cambial cell divisions and the differentiation of vessels along its pathway (Snow 1935; Jacobs 1952; Sachs 1981; Aloni 1987; Scarpella and Helariutta 2010). Therefore, it was difficult to explain the cambial reactivation and wide earlywood vessel differentiation in both normal and debudded ring-porous trees. Wareing (1951, 1958) suggested that in the cambium of ring-porous trees there is a high initial reserve of an auxin precursor that enables early cambium reactivation and rapid spread of earlywood vessel formation at an early stage of bud development. Presumably, in ring-porous species this reserve of auxin is accumulated during the previous season, whereas in diffuse-porous species little or no such reserved is formed. Wareing also hypothesized that the very wide earlywood vessels of ring-porous trees are induced by high supply of auxin in spring (Wareing 1951, 1958; Digby and Wareing 1966). However, results that contradict Wareing hypothesis were obtained in experiments on ring-porous trees (Aloni 1991), showing that application of moderate or high auxin concentrations to debudded trees, before bud break, inhibited the formation of wide earlywood vessels and yielded narrow vessels in the earlywood, and will be discussed below. These contradicting results point out a need for a new general hypothesis to solve the problem of how the wide earlywood vessels are induced, and account for the fundamental differences in wood porosity between ring-porous and diffuse-porous species, which will be clarified in this review.

3 The Three Major Hormonal Signals that Regulate Wood Formation

The major signaling molecules that regulate vascular differentiation and plant development are the plant hormones, also called phytohormones (Went and Thimann 1937). The hormones can be produced in any living plant cell at extremely low concentrations. They may act locally or at a distance from the producing cells.

Very few phytohormonal signals enable regulation and adaptation in remarkably simple mechanisms. The developmental process could be carried out by either a single developmental signal, or by very limited number of signals. The use of one or two signals is an economical way for carrying out major integrating roles.

The three primary phytohormonal signals that control vascular differentiation are: auxin, gibberellin, and cytokinin. Additional hormonal signals may be involved in specific responses to the environment, various stresses, wounding, and regulation of specific cell differentiation. The role of the hormonal signals and their molecular mechanisms in vascular differentiation were extensively reviewed in recent years by Caño-Delgado et al. (2010), Scarpella and Helariutta (2010), Lucas et al. (2013), Aloni (2013a, 2015), Furuta et al. (2014), Zhang et al. (2014), De Rybel et al. (2016), Scarpella (2017), Hellmann et al. (2018), Taiz et al. 2018; Fukuda and Ohashi-Ito (2019), Agustí and Blázquez (2020), and Aloni (2021).

All these three primary hormones are moving signals that are transported in specific pathways through the primary (originate from procambium) and secondary (originate from cambium) vascular tissues (Aloni 2010, 2015, 2021). In addition, hormonal movement through young parenchyma cells can induce regenerative differentiation (Jacobs 1952; Sachs 1981; Aloni 2021).

Auxin is the young leaf signal (Jacobs 1952; Sachs 1981; Aloni et al. 2003), gibberellin is the mature leaf signal (Dayan et al. 2012), and cytokinin is the root cup signal (Aloni et al. 2004, 2005). The continuous flow of these hormonal signals enables the plant to continuously respond to changing environmental cues.

The three hormonal signals are mainly produced by different plant organs and thus informing the stem cells of the embryonic vascular cambium, through which they move, on the physiological strength and quantity of the producing organs and their developmental stage. The vascular tissues are induced and regulated accordingly and the produced vascular elements reflect the developmental phase and amount of the plant organs. Thus, for example, during early spring when there are mainly young leaves on the stem of a hardwood tree, the auxin they produce is the main signal flowing through the cambium which, therefore, produces mainly sieve tubes and vessels; while during late summer when there are mainly mature leaves building large foliage biomass, their produced gibberellin becomes the dominant signal resulting in the formation of numerous fibers building stronger wood, which supports the enlarged shoot. Leaf development and biomass are regulated by environmental conditions (i.e., photoperiod, water availability, temperature, and nutrients), which control the production of wood and the type, quantity and patterns of its differentiating vascular cell.

Understanding the role of each hormonal signal is the key to understand how these moving signals design plant development, structure and vascular tissue differentiation under different environmental conditions.

3.1 *Auxin (IAA) from Young Leaves Induces Vessel Differentiation*

Developing buds and young growing leaves synthesize the auxin hormone, namely, indole-3-acetic acid (IAA), which is primarily produced in the hydathodes (Aloni 2001; Aloni et al. 2003; Baylis et al. 2013; Yagi et al. 2021), moves polarly downward to the root tips and induces vessels along the auxin pathways (Jacobs 1952; Sachs 1981; Aloni 2010, 2021). Auxin is a limiting factor for vessel differentiation, in its absence there is no vessel development. The polar auxin movement from the young leaves to the roots, which induces the vessels occurs through the procambium, parenchyma cells and cambium. Auxin stimulates cambial reactivation in spring and induces earlywood vessel formation along the cambium (Aloni 1991). The polar movement of IAA is continuous, ensuring the formation of continuous vessels, which transport water from root to leaves. Wounding that interrupt the auxin flow, results in bypasses of new auxin streams that induce vessel regeneration around the injury (Jacobs 1952; Sachs 1981; Berleth et al. 2000; Scarpella and Helariutta 2010; Aloni 2021).

3.2 *Cytokinin (CK) from Root Tips Promotes Cambium Sensitivity and Vascularization*

Roots do not induce wood formation nor must they be present in order to form xylem in stem tissues. However, the root apices, specifically the root caps, are sources of cytokinin that promotes cambial activity (cell division) and vessel differentiation (Aloni et al. 2005, 2006; Matsumoto-Kitano et al. 2008; Nieminen et al. 2008). Cytokinins from the root tips increase the sensitivity of the cambium to the auxin signal originating in young leaves (Baum et al. 1991; Aloni 1993, 1995; Aloni et al. 2003). Cytokinin prevents the usually rapid occurring IAA conjugation (Coenen and Lomax 1997), therefore, elevated CK concentration enables the transport of extremely low-IAA concentrations via the cambium, which may explain the increased sensitivity of the cambium to the auxin hormone. Experimental evidence from transformed plants (Zhang et al. 1995; Eklöf et al. 1997) supports the idea that reduced auxin concentrations can elevate cytokinin concentration, which would enhance tissue sensitivity to the auxin signal (Trewavas 1983; Aloni 1991; Bradford and Trewavas 1994; Barbez et al. 2012). The experiments demonstrate that auxin or cytokinin modify the content of the other hormone by affecting its rate of synthesis. Reduced IAA concentration increases free CK level (Palni et al. 1988; Zhang et al. 1995; Eklöf et al. 1997). Elevated CK enhances cambium sensitivity to extremely low-concentration-IAA streams originating in swelling buds and creates the special physiological conditions that enable slow vessel widening until secondary wall deposition, resulting in the wide earlywood vessels of ring-porous trees (Aloni 1991, 2001, 2021).

3.3 Gibberellin (GA) from Mature Leaves Induces Fiber Differentiation

Mature leaves are major sources of gibberellin (Hess and Sachs 1972; Aloni 1979; Dayan et al. 2012). The GA is the specific hormonal signal that induces fiber differentiation (Aloni 1979; Dayan et al. 2012). The transport of GA along the plant axis is not polar; therefore, GA induces fiber formation in both above and below the producing leaves (Dayan et al. 2012; Aloni 2021). The bioactive gibberellins (GA_1 and GA_4) were predominantly found in the expansion zone of differentiating secondary xylem cells in *Populus*, suggesting that the role of GA in early stages of wood formation, is promoting cell elongation (Israelsson et al. 2005). GA, which promotes cambial activity, reduces vessel width, resulting in narrow vessels in the latewood (Aloni 2021). Ring-porous trees develop auxin-producing young leaves in the beginning of the growth season; but during most of the season they have mainly mature leaves (Aloni et al. 1997), therefore the wood of ring-porous trees is characterized by numerous hard lignified latewood fibers and narrow vessels, both are regulated by the GA originating in their mature leaves.

4 Control of Vessel Size and Density Along the Tree Axis

A well-documented phenomenon is the downward gradual and continuous increase in vessel size from leaves to roots. This widening in vessel diameter was found along leaves from the tip to the base of the leaf (Colbert and Evert 1982; Russell and Evert 1982; Lechthaler et al. 2019). A continuous gradual increase in vessel diameter and vessel length was demonstrated from twigs to branches, downward along the stem and into the roots of *Acer rubrum* trees (Zimmermann and Potter 1982). Vessels are narrow at the leaves and their diameter increases gradually downward and continuously along the stem (Carlquist 1975; Zimmermann 1983; Aloni and Zimmermann 1983; Sorce et al. 2013; Lazzarin et al. 2016; Williams et al. 2019; Olson et al. 2021; Aloni 2021) and the root (Riedl 1937; Fahn 1964). Whereas vessel density decreases from leaves to roots, as was found in many plant species (Fegel 1941; Carlquist 1976; Aloni and Zimmermann 1983; Leitch 2001; Sorce et al. 2013; Zhao 2015).

Although Olson et al. (2021) suspect that the increase in vessel width from leaves to roots is an adaptation of trees to their environment, the picture that they present is incomplete because they do not consider the biological causing factors, namely, the hormonal mechanisms that control vessel widening from leaves to roots and adjust plants to their environment (Aloni 2013b, 2015, 2021; Agustí and Blázquez 2020). These hormonal mechanisms provide the answers to the questions raised by Olson et al. (2021), and will be clarified below.

To explain the mechanism that controls the general increase in vessel width and decrease in vessel density from leaves to roots, Aloni and Zimmermann (1983) proposed the *auxin gradient hypothesis* (that was first called the *six-point hypothesis*) suggesting that the polar transport of the auxin hormone from leaves to roots is the morphogenetic signal that creates a gradual polar gradient in the vascular cambium along the plant axis providing directional and location information to the differentiating cells (i.e., vessels, tracheids, fibers and sieve tubes) along the morphogenetic field.

The hypothesis proposes that the final size of a conduit is determined by the rate of cell differentiation. Since cell expansion ceases after the secondary wall is deposited, high-auxin concentrations near the young leaves induce narrow vessels, because of their rapid differentiation, allowing only limited time for cell widening. Conversely, further down, low-auxin concentrations result in slow differentiation, which permits more cell expansion before secondary wall deposition and therefore results in wide vessels at the base of the stem.

Vessel density is controlled by, and positively correlates with auxin concentration; consequently high-IAA concentrations (near the auxin producing young leaves) induce great vessel density, while low-IAA concentrations (further down, towards the roots) diminish density. Consequently, vessel density decreases from leaves to roots.

The *auxin gradient hypothesis* was experimentally confirmed by showing that various auxin concentrations applied to decapitated stems induce substantial gradients of increasing vessel diameter and decreasing vessel density from the auxin source towards the roots. High-auxin concentration yielded numerous vessels that remained narrow because of their rapid differentiation; low-auxin concentration resulted in slow differentiation and therefore in fewer and wider vessels (Fig. 2) (Aloni and Zimmermann 1983).

Studies on transgenic plants with altered levels of auxin confirmed the general relations between IAA concentration and vessel size and density. Thus, auxin overproducing plants (i.e., over-expressing the *iaaM* gene) contained many more vessel elements than did control plants, and their vessels were narrow (Klee et al. 1987); conversely, plants with lowered IAA levels (i.e., expressing the *iaaL* gene as an anti-auxin gene) contained fewer vessels of generally larger size (Romano et al. 1991).

A recent study of conduit diameter in the earth's tallest tree species: *Eucalyptus regnans*, *Sequoia sempervirens*, and *Sequoiadendron giganteum*, that were 86–105 m tall and exceeded 85% of the maximum height for each species, showed the typical gradual increase in conduit width along the upper parts of their shoots. However, at the base of their trunks, below about 60 m from the tree tops, vessel and tracheid diameters approached their maximum size, as they did not continue to expand; demonstrating that at the base of these giant trees, there is a limitation to conduit widening (Williams et al. 2019).

In spring, the first very wide earlywood vessels in temperate deciduous ring-porous trees initiate a few weeks before the onset of leaf expansion (Suzuki et al. 1996; Sass-Klaassen et al. 2011; Takahashi et al. 2013; Lavrič et al. 2017; Puchałka et al. 2017; Gričar et al. 2020). The width of these wide earlywood vessels increases slowly along a few weeks, because it is stimulated and induced by low-level streams of auxin produced by dormant looking buds, before swelling. This slow vessel widening

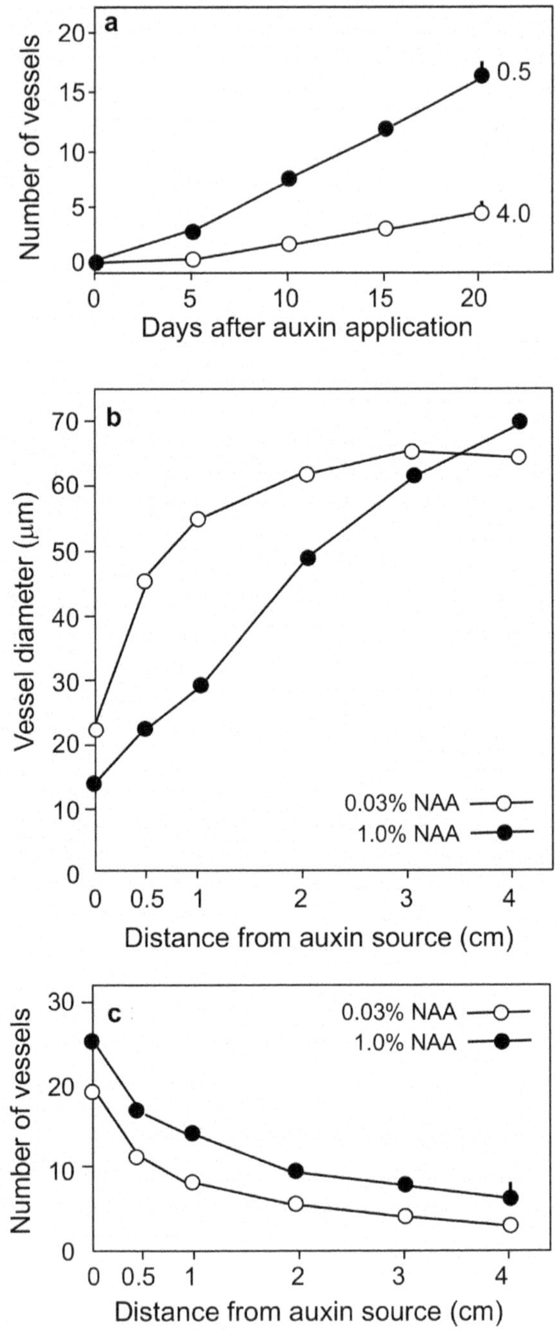

Fig. 2 Effects of applied auxin concentration (0.03% NAA, 0.1% NAA, or 1.0% NAA w/w in lanolin, renewed every 3 days) on secondary vessel differentiation in the second internode above the cotyledons of *Phaseolus vulgaris*, observed after 3 weeks of hormonal applications, on the top of the internode after the shoot above it was excised. a Effect of distance (0.5 and 4.0 cm) from 0.1% NAA application site on the rate of secondary vessel formation, showing intensive vessel differentiation near (0.5 cm) the site of auxin application. b Effect of 0.03% and 1.0% NAA on the radial diameter of the late-formed secondary vessels, along the studied internode, showing the substantial increase in vessel diameter with increasing distance from the applied auxin. c Effect of 0.03% and 1.0% NAA on the number of secondary vessels induced along a xylem radius, as affected by distance from the auxin source. *Vertical bars* indicate standard errors which are comparable at all points (from Aloni and Zimmermann 1983)

can occur only because the cambium of ring-porous trees has become very sensitive to low-IAA concentrations, as will be explained below.

5 Adaptation of the Xylem to the Tree's Environment

Vascular plants grow in different environments, ranging from deserts to rain forests and from arctic regions to the tropics. Comparative anatomical studies (e.g., Baas and Carlquist 1985; De Micco et al. 2008; Wheeler and Baas 2019) reveal similarities in structure of the vascular system in plants grown in extreme habitats *versus* ones grown in favorable environments. Desert (Carlquist and Hoekman 1985; Fahn et al. 1986), arctic, and alpine shrubs (Carlquist 1975) are characterized by very narrow vessels in high density. Such vascular systems are considered adaptive safety mechanisms against drought and freezing (Baas et al. 2004; Lucas et al. 2013). Conversely, forest trees and lianas, which characterize the tropics and rain forests, have low density vessels of very wide diameter at the base of their stems (Carlquist 1975; Zimmermann 1983; Ewers 1985; Tyree and Sperry 1989), which affords maximal efficiency of water conduction (Ellmore and Ewers 1985; Tyree and Ewers 1991; Tyree and Zimmermann 2002; Olson et al. 2021) and is considered to be an adaptation to mesic conditions.

In order to explain the adaptation of plants' vascular systems to the environment, Aloni (1987) proposed the *vascular adaptation hypothesis* suggesting that the environment controls the plant's vascular system through its control of plant's development, height, and shape. Limiting conditions suppress plant growth and shorten the active growth period, which restrict plant development resulting in small plants. Conversely, favorable conditions allow growth activity throughout the year, enabling more growth and consequently well-developed plants and maximal height.

The height of the plant and the degree of its branching determine gradients of auxin along the plant's axis. In small shrubs, which are typical to extreme stressful environmental conditions, the distances from the young leaves to the roots are very short and no substantial decreasing gradient of auxin can be formed. Therefore, the concentrations of IAA along these small plants are relatively high and result in rapid differentiation of numerous very narrow vessels in the greatest densities (as predicted by the *auxin gradient hypothesis*, Aloni and Zimmermann 1983). Conversely, in large trees and in long lianas, the very great distances from the young auxin-producing leaves to the roots enable a substantial decrease in auxin concentrations in their lower parts, leads to slow conduit differentiation that allows more cell expansion before secondary wall deposition, resulting in very wide vessels in low density at their base.

The *vascular adaptation hypothesis* (Aloni 1987) was confirmed experimentally (Aloni 1988, 2021) and by analysing the correlation between plant size and vessel diameter on a large scale of collected species from a wide range of growth conditions (Olson and Rosell 2013). Finally, the hypothesis explains why a tree that grows in very limited conditions will produce numerous narrow vessels in high density, in comparison with a tree of the same species that develops under favorable conditions and will produce wide vessels in low density at its base.

6 Evolution of Ring-Porous Wood in Temperate Deciduous Hardwood Trees

6.1 The Limited-Growth Hypothesis

In temperate deciduous broad-leaved trees, the size differences of vessels in the earlywood and latewood are quite marked and two main xylem categories can be distinguished: *diffuse-porous wood* and *ring-porous wood*. In diffuse-porous wood the vessels are more or less uniform in size (Fig. 1a), whereas in ring-porous wood the vessels produced at the beginning of the growth season are significantly wider (Fig. 1b) than those produced at the end of the season (Evert 2006; Aloni 2021). Earlywood vessels in ring-porous trees can be huge (width of up to 500 μm and length of the entire tree) and therefore are very efficient in water conductance, but their size makes them vulnerable. The wide earlywood vessels usually function during one season and then they become occluded by tyloses, or gum to plug the air-filled vessel and prevent possible penetration damage of pathogenic bacteria and fungi (Aloni et al. 1997; Tyree and Zimmermann 2002; Evert 2006; Aloni 2021). Tyloses formation in earlywood vessels occurs earlier under drier conditions (Pérez-de-Lis et al. 2018). When tree species that have already developed ring-porosity during evolution grow under favourable conditions, they can reach large sizes, although they usually show a slow growth pattern in comparison with faster growing diffuse-porous trees.

The challenge to understand the mechanisms that have shaped earlywood vessel patterns during the evolution of temperate deciduous hardwood trees requires elucidation of the roles of tissue sensitivity to auxin (Trewavas 1983; Bradford and Trewavas 1994; Barbez et al. 2012) and the specific hormonal signalling in these trees (Aloni 1991, 2001, 2013a, 2021). It has been suggested that ring-porous trees have originated from diffuse-porous species (Aloni 1991; Wheeler and Baas 1991). The development of ring porosity has probably arisen independently multiple times during the diversification of angiosperms, and different lineages might therefore have modified mechanisms in different families.

To explain how ring-porous wood has developed during the evolution of temperate deciduous hardwood trees, Aloni (1991) proposed the *limited-growth hypothesis*, suggesting that during the evolution of temperate deciduous hardwood trees, the ring-porous species have developed from diffuse-porous species under selective pressures in limiting environments, which resulted in limited vegetative growth. It was further postulated that under extreme environmental conditions the selection for ring-porous wood has led to a decrease in the intensity of vegetative growth and reduced foliar biomass, causing a decrease in auxin levels. The latter promoted an increase in cytokinin levels, which induced an increase in the sensitivity of the cambium to extremely low-concentration-auxin streams originating in swelling buds. These changes created the unique physiological conditions that enable slow vessel differentiation, which promotes a long widening process, starting before bud break, resulting in wide earlywood vessels at the beginning of the growth season (Aloni 1991).

6.2 Supporting Evidence for the Limited Growth Hypothesis

Evidence that supports the *limited growth hypothesis* comes from observations that a diffuse-porous tree (*Populus euphratica*) and a ring-porous tree (*Quercus ithaburensis*) can change their porosity under opposite environmental conditions (Liphschitz 1995). Thus, under stress conditions when extension growth is suppressed both tree species produced narrow annual rings characterized by ring-porous wood (as predicted by the *limited-growth hypothesis*), whereas under favorable conditions when extensive growth is intensive, both species produce wide annual rings with diffuse-porous wood (Liphschitz 1995).

The increased cambium sensitivity to IAA in ring-porous trees enables early cambium reactivation at the beginning of the growth season before bud break. This was evident in stem diameter measurements of the ring-porous *Zelkova serrata* saplings in their early leafless state, showing stem swelling 2–6 weeks before bud opening. During this developmental stage, actively dividing cambial cells, and immature slowly widening early-wood vessels that derived from them, are very soft, as they have not yet deposited their hard-secondary cell walls (Yoda et al. 2003), which therefore enables and promotes vessel widening.

The substantial increase in cambial sensitivity to auxin in ring-porous trees created the special internal conditions that enable them to respond to initial flows of extremely low-IAA concentrations originating in dormant looking (before swelling) buds a few weeks before bud break (Aloni 1991, 2001; Aloni and Peterson 1997; Aloni et al. 1997), stimulating slow vessel differentiation, which permits more time for cell expansion, promoting the widening of the differentiating earlywood vessels before their secondary wall deposition (as expected by the *auxin gradient hypothesis* of Aloni and Zimmermann 1983), resulting in the formation of very wide earlywood vessels. Therefore, the first wide earlywood vessels of ring-porous trees are initiated six to two weeks before the onset of leaf expansion (Suzuki et al. 1996; Sass-Klaassen et al. 2011; Takahashi et al. 2013; Lavrič et al. 2017; Puchałka et al. 2017; Gričar et al. 2020) and cause stem swelling before bud opening (Yoda et al. 2003). The pattern of ealywood vessel maturation in the ring porous hardwoods, *Quercus serrata* and *Robinia pseudoacacia* progressed downward. The first mature earlywood vessel elements appeared at bud break, first at the top of the stem, and continue downward to the lower parts of the stem (Kudo et al. 2015).

Conversely, in diffuse-porous species, the first earlywood vessels are initiated two to seven weeks after the onset of leaf expansion (Suzuki et al. 1996; Takahashi et al. 2013), and because of the low cambium sensitivity in diffuse-porous trees, their cambium requires high auxin concentrations (from fast growing young leaves) for reactivation. These results explain the old report of Priestley and Scott (1936) who found that in a deciduous ring-porous tree the cambium undergoes extremely fast reactivation before bud break, which occurs almost simultaneously in the branches and along the trunk. This is why the bark of deciduous ring-porous trees may be peeled a few days before any bud swelling can be observed in spring. Conversely, a deciduous diffuse-porous species requires several weeks for a 'wave' of cambial reactivation to extend from the twigs of a large tree downward to the base of its trunk (Priestley and Scott 1936).

An opposite explanation for the differentiation of wide-earlywood-vessel in ring-porous trees was suggested by Wareing (1951) who studied cambial reactivation and wood formation in ring-porous *versus* diffuse-porous trees. Wareing suggested that the characteristic pattern of early rapid spread of cambium reactivation and the development of wide-earlywood vessels in ring-porous species "is due to the presence in the cambium of a reserve of auxin-precursor, which makes possible the rapid spread of wide-vessel formation throughout the tree, at an early stage of development of the buds". Wareing also mentioned the possibility that the dormant-looking buds of ring-porous trees that initiated cambial reactiviation "were no longer 'physiologically' dormant" (Wareing 1951), in other words, it is possible that the dormant-looking buds possibly started to produce the auxin hormone.

On the contrary, experimental results on ring-porous trees (Aloni 1991) that aimed to test the *limited growth hypothesis*, demonstrated that the wide earlywood vessels are induced by very low auxin stimulation. Evidently, an extremely low-auxin concentration (0.003% Naphthaleneacetic acid (NAA) in lanolin w/w) applied to disbudded shoots of *Melia azedarach* trees, induced wide earlywood vessels (Fig. 3b) in the deciduous ring-porous trees (Aloni 1991, 2001), but this low-auxin concentration was not strong enough to stimulate any earlywood vessel differentiation in deciduous diffuse-porous trees (Aloni 2001). This was true also with 0.01% NAA (in lanolin) that induced more medium-size earlywood vessels (Fig. 3c) but was not strong enough to induce earlywood vessels in diffuse-porous trees. On the other hand, a high-auxin concentration (1% NAA in lanolin) induced rapid differentiation of narrow earlywood vessel in the ring-porous trees, because of fast secondary wall deposition that prevented vessel widening, and therefore remained narrow vessels (Fig. 3d). The high-auxin concentration induced earlywood vessel differentiation in diffuse-porous trees (Aloni 2001). These results clearly demonstrate that the wide earlywood vessels of a ring-porous tree are induced by extremely low-auxin stimulation before bud swelling. Whereas in diffuse-porous trees there is a need for high-auxin concentrations produced in fast growing young leaves for inducing their typical regular size earlywood vessels (Aloni 1991, 2001, 2013a, 2021).

The auxin produced by the buds and young leaves induces early vessel differentiation first immediately below the buds. Complete early differentiation of the earlywood vessels occurs first in the upper stem region, and then progresses to the middle and lower regions during bud swelling in the ring-porous *Quercus serrata* seedlings (Kudo et al. 2018). During this downward earlywood vessel differentiation process, the developing buds and young shoot organs are supplied by the functional network

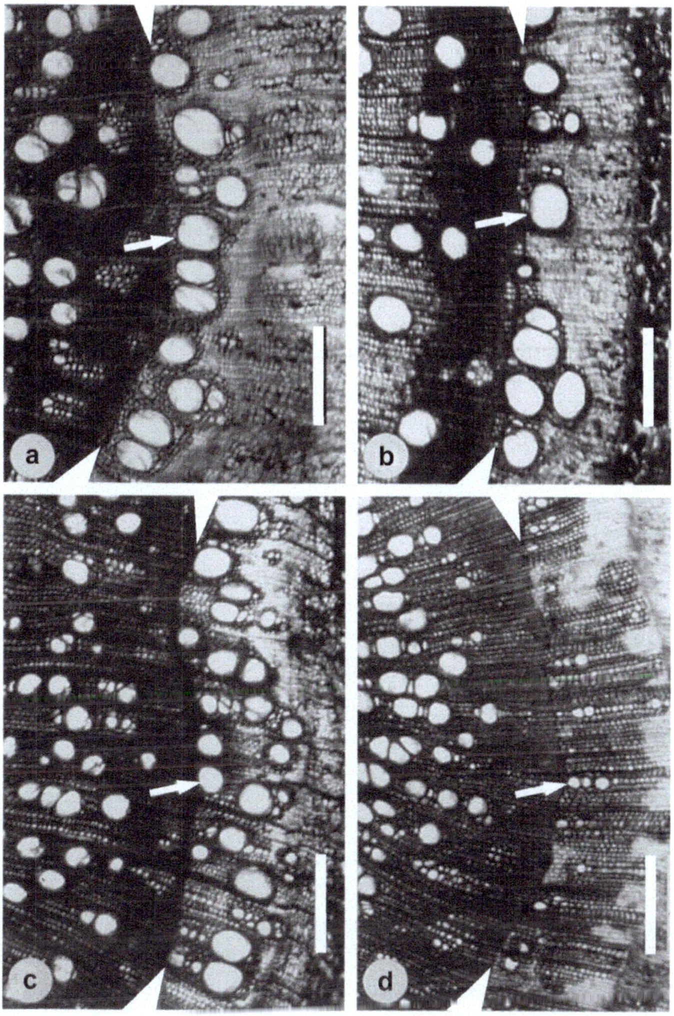

Fig. 3 The effect of auxin (1-Naphthaleneacetic acid (NAA) in lanolin) concentration on the width of earlywood vessel differentiation is shown in transverse sections in stems of the ring-porous tree *Melia azedarach*. All photomicrographs were taken from the same experiment, run in Tel Aviv from February 15 to March 15, 1986, and are presented in the same orientation and magnification (Bars = 250 μm). All the sections were taken 50 mm below the apical bud, which was left intact (**a**), or was replaced by a range of auxin concentrations: low, 0.003% NAA (**b**), medium, 0,01% NAA (**c**), or high, 0.1% NAA (**d**). The auxin was applied in the form of a lanolin paste, which was renewed every 3 days. The photomicrographs show a substantial decrease in the diameter of the earlywood vessels (*white arrows*) with increasing auxin concentration (**b–d**). The low auxin concentration induced wide vessels (**b**). The two higher concentrations induced many more xylem cells (along a radius) with narrower vessels (**c**, **d**). The highest auxin concentration tested (0.1% NAA) resulted in very narrow earlywood vessels (**d**). The borderline between the latewood of 1985 (*left*) and the new earlywood of 1986 (*right*) is marked with *white triangles*. The experiment was repeated three times (in 1984, 1985, and 1986) with 5–10 stems per treatment; yielding the same results (from Aloni 1991)

of previous year's narrow latewood vessels, while the wide earlywood vessels of the current year differentiate and slowly mature (Kudo et al. 2018).

6.3 *Development of Two Wood-Porosity Patterns Along the Same Ring-Porous Tree*

As was clarified above by the *limited-growth hypothesis* (Aloni 1991), for inducing the wide earlywood vessels in ring-porous trees there is the unique requirement for very low-auxin stimulation, originating in dormant-looking buds, early in the growth season along the very sensitive cambium of temperate deciduous ring-porous trees. These special conditions in the trunk of a ring-porous tree allow the early slow and long widening process of earlywood vessels, resulting in wide earlywood vessels before the secondary lignified wall is deposited (Aloni 1991; Suzuki et al. 1996; Sass-Klaassen et al. 2011; Takahashi et al. 2013; Lavrič et al. 2017; Puchałka et al. 2017).

A different pattern of vessel differentiation occurs in the twigs. In the youngest twigs, the secondary wall deposition and lignification of the first-formed vessels, relative to the time of leaf appearance, is faster in the more sensitive ring-porous trees, starting about two weeks earlier than in the twigs of the diffuse-porous trees (Takahashi et al. 2013). In ring-porous trees, the first-formed vessels of the year deposited lignified secondary walls in the twigs around the time of leaf appearance, at the time that the wide earlywood vessels in the trunk continue the slow widening process up to the deposition of their lignified secondary wall, occurring at full leaf expansion (Takahashi et al. 2013). The rapid vessel differentiation and early secondary wall deposition in the twigs does not allow vessel expansion, resulting in narrow vessels in a diffuse-porous pattern in twigs of both the ring- and diffuse-porous trees. Therefore, in the two types of wood porosity trees, the wood produced in the twigs and branches during their first year (when they are only a few weeks/months old) is characterized by a diffuse-porous wood pattern (Cochard and Tyree 1990; Lo Gullo et al. 1995; Takahashi et al. 2013; Rodriguez-Zaccaro et al. 2019).

It should be emphasized that this diffuse-porous wood pattern in the youngest twigs/branches of ring-porous trees is induced by high-auxin concentrations produced by their young leaves, and it is not due to the influence of "cambial age", as suggested by Rodriguez-Zaccaro et al. (2019). Therefore, there is no need for "older cambia" to produce the wide earlywood vessels typical to ring-porous wood pattern (Rodriguez-Zaccaro et al. 2019), but only the requirement for the unique endogenous conditions of sensitive cambium that responds to extremely low-auxin stimulation, from dormant-looking and swelling buds, during early spring, allowing the slow vessel widening process typically forming the wide-earlywood vessels along the stem of temperate deciduous ring-porous trees (Aloni 1991, 2001, 2013a, 2021).

6.4 Leaf Phenology, Earlywood- and Latewood Development in Temperate Deciduous Hardwood Trees

Climate changes influence leaf phenology and tree development, which shape their adaptation and evolution; influencing cambial sensitivity, wood differentiation and vessel patterns. Fossil records indicate that the ring-porous wood pattern has developed under various environmental stresses especially during the past 50 million years, when the global climates have been undergoing active changes (Evert and Eichhorn 2013). The evolution of ring-porous trees has adapted them to survive under shorter and limiting growth seasons.

Diffuse-porous species start the growth season a few weeks earlier than ring-porous trees and have a longer growth season which is characterized by continuous production of auxin-producing young leaves during a few months (Aloni et al. 1997). Conversely, the more specialized ring-porous trees that are well adapted to limiting environments are late-leafing trees (Lechowicz 1984).

Ring-porous trees produce young leaves for only a short period of a few weeks and later they have mainly mature leaves (Aloni et al. 1997; Aloni 2021). Because young diffuse-porous trees possess greater growth intensity they might produce more xylem per year than young ring-porous trees (Aloni et al. 1997). In diffuse-porous trees, the continuous development of new auxin-producing young leaves along the growth season stimulates continuous production of new narrow vessels along the entire growth season with relatively thin-wall fibers. Whereas in ring-porous trees, the dominating mature leaves, which produce gibberellin (Aloni 1979; Dayan et al. 2012), induce the development of numerous well-developed hard lignified fibers during most of the growth season, with only a few narrow vessels limited in their width by gibberellin. These diverse earlywood and latewood properties in ring-porous wood, namely, the soft wide earlywood vessels *versus* the numerous hard thick-wall latewood fibers affect lumber stability and can have major effects on wood and fiber utilization.

7 Conclusions

The environment controls plant development by shaping plant's growth, its rate of development and morphology, which regulate plant's physiology and anatomy. During the process of plants adaptation to their environments, there are gradual and continuous changes in the internal regulating hormonal mechanisms that shape the differentiation and structure of the vascular tissues.

The produced-leaf signal is modified during leaf development. The young growing leaf produced auxin, while during development the maturing leaf becomes a major source of gibberellin. The specific leaf's hormonal signal determines the type of the induced vascular element: auxin from young leaves induces vessels, while gibberellin from mature leaves is the specific signal that induces fibers.

The hormonal signal concentration determines the rate of cell differentiation. While the latter determines the final size of the conduit. Slow vessel differentiation induced by low-auxin stimulation, enables a long period of vessel widening until secondary wall deposition, resulting in the formation of wide vessels; e.g., either (1) at the base of stems of long lianas and tall trees away from the auxin-producing young leaves, or (2) at the earlywood of ring-porous trees characterized by sensitive cambium, where the earlywood vessels along the trunk are induced by extremely low-auxin-concentration streams, originating in dormant looking buds.

During the evolution of ring-porous trees under selective pressures of limiting environments, their natural stress selection has shortened the growth season, decreased vegetative growth and reduced foliar biomass, resulting in the reduction of the auxin hormone produced in young leaves. The latter promoted an increase in cytokinin levels inducing an increase in the sensitivity of the cambium to extremely low-concentration-auxin streams originating in dormant looking buds, before swelling, a few weeks before bud break. These low-auxin streams induce a slow and a long vessel widening process, which permits more cell expansion before secondary wall deposition, therefore resulting in the typical wide earlywood vessels of ring-porous trees.

Contrariwise, rapid vessel differentiation induced by high-auxin stimulation produced in young growing leaves results in narrow vessels in the new twigs and branches during their first growth season in both diffuse- and ring-porous trees; as well as in short trees, or along the upper parts of tall trees.

Although the late-leafing ring-porous trees have a shorter growth season, they have been well adapted to their limiting environment by producing wide and efficient earlywood vessels and numerous hard latewood fibers; the latter are induced by gibberellin produced in mature leaves. This combination makes the ring-porous trees very efficient in water uptake with very strong trunks due to their extremely wide earlywood vessels combined with thick lignified latewood fibers.

References

Agustí J, Blázquez MA (2020) Plant vascular development: mechanisms and environmental regulation. Cell Mol Life Sci 77:3711–3728

Aloni R (1979) Role of auxin and gibberellin in differentiation of primary phloem fibers. Plant Physiol 63:609–614

Aloni R (1987) Differentiation of vascular tissues. Annu Rev Plant Physiol 38:179–204

Aloni R (1988) Vascular differentiation within the plant. In: Roberts LW, Gahan PB, Aloni R (eds) Vascular differentiation and plant growth regulators. Springer, Berlin, Heidelberg, pp 39–62

Aloni R (1991) Wood formation in deciduous hardwood trees. In: Raghavendra AS (ed) Physiology of trees. Wiley, New York, pp 175–197

Aloni R (1993) The role of cytokinin in organised differentiation of vascular tissues. Aust J Plant Physiol 20:601–608

Aloni R (1995) The induction of vascular tissues by auxin and cytokinin. In: Davies PJ (ed) Plant hormones: physiology, biochemistry and molecular biology. Kluwer, Dordrecht, pp 531–546

Aloni R (2001) Foliar and axial aspects of vascular differentiation - hypotheses and evidence. J Plant Growth Regul 20:22–34
Aloni R (2010) The induction of vascular tissues by auxin. In: Davies PJ (ed) Plant hormones: biosynthesis, signal transduction, action! Kluwer Academic Publishers, Dordrecht, pp 485–506
Aloni R (2013a) The role of hormones in controlling vascular differentiation. In: Fromm J (ed) Cellular aspects of wood formation. Springer, Berlin, pp 99–139
Aloni R (2013b) Role of hormones in controlling vascular differentiation and the mechanism of lateral root initiation. Planta 23:819–830
Aloni R (2015) Ecophysiological implications of vascular differentiation and plant evolution. Trees 29:1–16
Aloni R (2021) Vascular differentiation and plant hormones. Springer Nature, Cham, Switzerland
Aloni R, Alexander JD, Tyree MT (1997) Natural and experimentally altered hydraulic architecture of branch junctions in *Acer saccharum* Marsh. and *Quercus velutina* Lam. trees. Trees 11:255–264
Aloni R, Aloni E, Langhans M, Ullrich CI (2006) Role of cytokinin and auxin in shaping root architecture: regulating vascular differentiation, lateral root initiation, root apical dominance and root gravitropism. Ann Bot 97:883–893
Aloni R, Langhans M, Aloni E, Dreieicher E, Ullrich CI (2005) Root-synthesized cytokinin in *Arabidopsis* is distributed in the shoot by the transpiration stream. J Exp Bot 56:1535–1544
Aloni R, Langhans M, Aloni E, Ullrich CI (2004) Role of cytokinin in the regulation of root gravitropism. Planta 220:177–182
Aloni R, Peterson CA (1997) Auxin promotes dormancy callose removal from the phloem of *Magnolia kobus* and callose accumulation and earlywood vessel differentiation in *Quercus robur*. J Plant Res 110:37–44
Aloni R, Schwalm K, Langhans M, Ullrich CI (2003) Gradual shifts in sites of free-auxin production during leaf-primordium development and their role in vascular differentiation and leaf morphogenesis in *Arabidopsis*. Planta 216:841–853
Aloni R, Ullrich CI (2008) Biology of crown gall tumors. In: Tzfira T, Citovsky VH (eds) Agrobacterium, from biology to biotechnology. Springer, New York, pp 565–591
Aloni R, Zimmermann MH (1983) The control of vessel size and density along the plant axis—a new hypothesis. Differentiation 24:203–208
Atkinson CJ, Denne MP (1987) Reactivation of vessel expansion in relation to budbreak in sycamore (*Acer pseudoplantanus*) trees. Can J for Res 17:1166–1174
Atkinson CJ, Denne MP (1988) Reactivation of vessel production in ash (*Fraxinus excelsior* L.) trees. Ann Bot 61:679–688
Baas P, Carlquist S (1985) A comparison of the ecological wood anatomy of the floras of southern California and Israel. IAWA Bull NS 6:349–353
Baas P, Ewers FW, Davis SD, Wheeler EA (2004) Evolution of xylem physiology. In: The evolution of plant physiology, Hemsley AE, Poole I (eds) Linnean Soc Sympm Series, no. 21. Academic Press, Amsterdam, pp 273–295
Bailey IW, Tupper WW (1918) Size variation in tracheary cells: I. A comparison between the secondary xylems of vascular cryptogams, gymnosperms and angiosperms. Proc Am Acad Arts Sci 54:149–204
Baylis T, Cierlik I, Sundberg E, Mattsson J (2013) Short internodes/stylish genes, regulators of auxin biosynthesis, are involved in leaf vein development in *Arabidopsis thaliana*. New Phytol 197:737–750
Barbez E, Kubeš M, Rolčík J, Béziat C, Pěnčík A, Wang B, Rosquete MR, Zhu J, Dobrev PI, Lee Y, Zažímalovà E, Petrášek J, Geisler M, Friml J, Kleine-Vehn J (2012) A novel putative auxin carrier family regulates intracellular auxin homeostasis in plants. Nature 485:119–122
Baum SF, Aloni R, Peterson CA (1991) The role of cytokinin in vessel regeneration in wounded *Coleus* internodes. Ann Bot 67:543–548
Berleth T, Mattsson J, Hardtke CS (2000) Vascular continuity and auxin signals. Trends Plant Sci 5:387–393

Bradford KJ, Trewavas AJ (1994) Sensitivity thresholds and variable time scales in plant hormone action. Plant Physiol 105:1029–1036
Caño-Delgado A, Lee JY, Demura T (2010) Regulatory mechanisms for specification and patterning of plant vascular tissues. Annu Rev Cell Dev Biol 26:605–637
Carlquist S (1975) Ecological strategies of xylem evolution. University of California Press, Berkeley
Carlquist S (1976) Wood anatomy of Roridulaceae: ecological and phylogenetic implications. Am J Bot 63:1003–1008
Carlquist S, Hoekman DA (1985) Ecological wood anatomy of the woody southern Californian flora. IAWA Bull 6:319–346
Cochard H, Tyree MT (1990) Xylem dysfunction in *Quercus*: Vessel sizes, tyloses, cavitation and seasonal changes in embolism. Tree Physiol 6:393–407
Coenen C, Lomax TL (1997) Auxin-cytokinin interactions in higher plants: old problems and new tools. Trends Plant Sci 2:351–356
Colbert JT, Evert RF (1982) Leaf vasculature in sugar cane (*Saccharum officinarum* L.). Planta 156:136–151
Dayan J, Voronin N, Gong F, Sun T-P, Hedden P, Fromm H, Aloni R (2012) Leaf-induced gibberellin signaling is essential for internode elongation, cambial activity, and fiber differentiation in tobacco stems. Plant Cell 24:66–79
De Micco V, Aronne G, Baas P (2008) Wood anatomy and hydraulic architecture of stems and twigs of some Mediterranean trees and shrubs along a mesic-xeric gradient. Trees 22:643–655
De Micco V, Balzano A, Wheeler E, Baas P (2016) Tyloses and gums: a review of structure, function and occurrence of vessel occlusions. IAWA J 37:186–205
De Rybel B, Mähönen AP, Helariutta Y, Weijers D (2016) Plant vascular development: from early specification to differentiation. Nat Rev Mol Cell Biol 17:30–40
Digby J, Wareing PF (1966) The effect of applied growth hormones on cambial division and the differentiation of cambial derivatives. Ann Bot 30:539–548
Eklöf S, Åstot C, Blackwell J, Moritz T, Olsson O, Sandberg G (1997) Auxin-cytokinin interactions in wild-type and transgenic tobacco. Plant Cell Physiol 38:225–235
Ellmore GS, Ewers FW (1985) Hydraulic conductivity in trunk xylem of elm, *Ulmus amaerican*. IAWA Bull NS 6:303–307
Evert RF (2006) Esau's plant anatomy: meristems, cells, and tissues of the plant body: their structure, function, and development. Wiley, New Jersey
Evert RF, Eichhorn SE (2013) Raven biology of plants, 8th edn. Freeman, New York
Ewers FW (1985) Xylem structure and water conduction in conifer trees, dicot trees and liana. IAWA Bull NS 6:309–317
Fahn A (1964) Some anatomical adaptations of desert plants. Phytomorphology 14:93–102
Fahn A, Werker E, Baas P (1986) Wood anatomy and identification of trees and shrubs from israel and adjacent regions. Israel Acad Sci Jerusalem
Fegel CA (1941) Comparative anatomy and varying physical properties of trunk, branch and root wood in certain northeastern trees. Bull NY State College of Forestry at Syracuse Univ, vol 14, no 2b, Tech Publ 55, pp 1–20
Fukuda H, Ohashi-Ito K (2019) Vascular tissue development in plants. Curr Top Dev Biol 131:141–160
Furuta KM, Hellmann E, Helariutta Y (2014) Molecular control of cell specification and cell differentiation during procambial development. Annu Rev Plant Biol 65:604–638
Greenidge KNH (1952) An approach to the study of vessel length in hardwood species. Am J Bot 3:570–574
Gričar J, Vedenik A, Skoberne G, Hafner P, Prislan P (2020) Timeline of leaf and cambial phenology in relation to development of initial conduits in xylem and phloem in three coexisting sub-mediterranean deciduous tree species. Forests 11:1104
Gričar J, Zavadlav S, Jyske T, Lavrič M, Laakso T, Hafner P, Eler K, Vodnik D (2018) Effect of soil water availability on intra-annual xylem and phloem formation and non-structural carbohydrate pools in stem of *Quercus pubescens*. Tree Physiol 39:222–233

Hacke UG, Spicer R, Schreiber SG, Plavcová L (2017) An ecophysiological and developmental perspective on variation in vessel diameter. Plant Cell Environ 40:831–845
Hartig T (1853) Über die Entwicklung des Jahresringes der Holzpflanzen. Bot Z 11(553–560):569–579
Hellmann E, Ko D, Ruonala R, Helariutta Y (2018) Plant vascular tissues-connecting tissue comes in all shapes. Plants (basel) 7:E109
Hess T, Sachs T (1972) The influence of a mature leaf on xylem differentiation. New Phytol 71:903–914
Israelsson M, Sundberg B, Moritz T (2005) Tissue-specific localization of gibberellins and expression of gibberellin-biosynthetic and signaling genes in wood-forming tissues in aspen. Plant J 44:494–504
Jacobs WP (1952) The role of auxin in differentiation of xylem around a wound. Am J Bot 39:301–309
Klee HJ, Horsch RB, Hinchee MA, Hein MB, Hoffmann MB (1987) The effects of overproduction of two *Agrobacterium tumefaciens* T-DNA auxin biosynthetic gene products in transgenic petunia plants. Genes Dev 1:86–96
Kudo K, Utsumi Y, Kuroda K, Yamagishi Y, Nabeshima E, Nakaba S, Yasue K, Takata K, Funada R (2018) Formation of new networks of earlywood vessels in seedlings of the deciduous ring-porous hardwood *Quercus serrata* in springtime. Trees 32:725–734
Kudo K, Yasue K, Hosoo Y, Funada R (2015) Relationship between formation of earlywood vessels and leaf phenology in two ring-porous hardwoods, *Quercus serrata* and *Robinia pseudoacacia,* in early spring. J Wood Sci 61:455–464
Lavrič M, Eler K, Ferlan M, Vodnik D, Gričar J (2017) Chronological sequence of leaf phenology, xylem and phloem formation and sap flow of *Quercus pubescens* from abandoned karst grasslands. Front Plant Sci 8:314
Lazzarin M, Crivellaro A, Mozzi G, Williams C, Dawson T, Anfodillo T (2016) Tracheid and pit anatomy vary in tandem in a tall giant sequoia. IAWA J 37:172–185
Lechowicz MJ (1984) Why do temperate deciduous trees leaf out at different times? Adaptation and ecology of forest communities. Am Natur 124:821–842
Lechthaler S, Colangeli P, Gazzabin M, Anfodillo T (2019) Axial anatomy of the leaf midrib provides new insights into the hydraulic architecture and cavitation patterns of *Acer pseudoplatanus* leaves. J Exp Bot 70:6195–6201
Leitch MA (2001) Vessel-element dimensions and frequency within the most current growth increment along the length of *Eucalyptus globules* stems. Tree 15:353–357
Liphschitz N (1995) Ecological wood anatomy: changes in xylem structure in Israeli trees. In: Shuming W (ed) Wood anatomy research 1995. Proceedings of the international symposium tree anatomy and wood formation, Tianjin, China. International Academic Publishers, Beijing, pp 12–15
Lodewick JE (1928) Seasonal activity of the cambium in some northeastern trees. Bull NY State College Tech Publ 23
Lo Gullo MA, Salleo S, Piaceri EC, Rosso R (1995) Relations between vulnerability to xylem embolism and xylem conduit dimensions in young trees of *Quercus corris*. Plant Cell Environ 18:661–669
Lucas WJ, Groover A, Lichtenberger R, Furuta K, Yadav SR, Helariutta Y, He XQ, Fukuda H, Kang J, Brady SM, Patrick JW, Sperry J, Yoshida A, López-Millán AF, Grusak MA, Kachroo P (2013) The plant vascular system: evolution, development and functions. J Integr Plant Biol 55:294–388
Matsumoto-Kitano M, Kusumoto T, Tarkowski P, Kinoshita-Tsujimura K, Václavíková K, Miyawaki K, Kakimoto T (2008) Cytokinins are central regulators of cambial activity. Proc Natl Acad Sci USA 105:20027–20031
Nieminen K, Immanen J, Laxell M, Kauppinen L, Tarkowski P, Dolezal K, Tähtiharju S, Elo A, Decourteix M, Ljung K, Bhalerao R Keinonen K, Albert VA, Helariutta Y (2008) Cytokinin signaling regulates cambial development in poplar. Proc Natl Acad Sci USA 105:20032–20037

Olson ME, Anfodillo T, Gleason SM, McCulloh KA (2021) Tip-to-base xylem conduit widening as an adaptation: causes, consequences, and empirical priorities. New Phytol 229:1877–1893
Olson ME, Rosell JA (2013) Vessel diameter-stem diameter scaling across woody angiosperms and the ecological causes of xylem vessel diameter variation. New Phytol 197:1204–1213
Palni LM, Burch L, Horgan R (1988) The effect of auxin concentration on cytokinin stability and metabolism. Planta 174:231–234
Pérez-de-Lis G, Rossi S, Vázquez-Ruiz RA, Rozas V, García-González I (2016) Do changes in spring phenology affect earlywood vessels? Perspective from the xylogenesis monitoring of two sympatric ring-porous oaks. New Phytol 209:521–530
Pérez-de-Lis G, Rozas V, Vázquez-Ruiz RA, García-González I (2018) Do ring-porous oaks prioritize earlywood vessel efficiency over safety? Environmental effects on vessel diameter and tyloses formation. Agric Meteorol 248:205–214
Priestley JH, Scott LI (1936) A note upon summer wood production in the tree. Proc. Leeds Phil. Soc. 3:235–248
Puchałka R, Koprowski M, Gričar J, Przybylak R (2017) Does tree-ring formation follow leaf phenology in Pedunculate oak (*Quercus robur* L.)? Eur J Forest Res 136:259–268
Reines M (1959) The initiation of cambial activity in black cherry. For Sci 5:70–73
Riedl H (1937) Bau und Leistungen des Wurzelholzes. Jb Wiss Bot 85:1–75 (English translation available from: National Translations Center, 35 West 33rd St, Chicago, IL 60616)
Rodriguez-Zaccaro FD, Valdovinos-Ayala J, Percolla MI, Venturas MD, Pratt RB, Jacobsen AL (2019) Wood structure and function change with maturity: age of the vascular cambium is associated with xylem changes in current-yėar growth. Plant Cell Environ 42:1816–1831
Romano CP, Hein MB, Klee HJ (1991) Inactivation of auxin in tobacco transformed with the indoleacetic acid-lysine synthetase gene of *Pseudomonas savastanoi*. Genes Dev 5:438–446
Russell SH, Evert RF (1982) Leaf vasculature in maize (*Zea mays* L.) Bot Soc Am Misc Publ 162:23
Russow EI (1883) Ueber Tüpfelbidung und Inhalt der Bastparenchym-und Baststrahlzellen der Dikotylen und Gymnospermen. 2. Ueber den Inhalt der parenchymastrischen Elemente der Rinde vor und während des Knospenaustriebes und Beginns der Cambiumthätigkeit in Stamm und Wurzel der einheimischen Lignosen. Bot Centralblatt 13:271–275
Sachs T (1981) The control of patterned differentiation of vascular tissues. Adv Bot Res 9:151–262
Sass-Klaassen U, Sabajo CR, den Ouden J (2011) Vessel formation in relation to leaf phenology in pedunculate oak and European ash. Dendrochronologia 29:171–175
Scarpella E (2017) The logic of plant vascular patterning. Polarity, continuity and plasticity in the formation of the veins and of their networks. Curr Opin Genet Dev 45:34–43
Scarpella E, Helariutta Y (2010) Vascular pattern formation in plants. Curr Top Dev Biol 91:221–265
Snow R (1935) Activation of cambial growth by pure hormones. New Phytol 34:347–360
Sorce C, Giovannelli A, Sebastiani L, Anfodillo T (2013) Hormonal signals involved in the regulation of cambial activity, xylogenesis and vessel patterning in trees. Plant Cell Rep 32:885–898
Suzuki M, Yoda K, Suzuki H (1996) Phenological comparison of the onset of vessel formation between ring-porous and diffuse-porous deciduous trees in a Japanese temperate forest. IAWA J 17:431–444
Taiz L, Zeiger E, Møller IM, Murphy A (2018) Fundamentals of plant physiology. Sinauer, Sunderland, MA
Takahashi S, Okada N, Nobuchi T (2013) Relationship between the timing of vessel formation and leaf phenology in ten ring-porous and di_use-porous deciduous tree species. Ecol Res 28:615–624
Trewavas AJ (1983) Is plant development regulated by changes in concentration of growth substances or by changes in the sensitivity to growth substances? TIBS 8:354–357
Tyree MT, Ewers FW (1991) The hydraulic architecture of trees and other woody plants. New Phytol 119:345–360
Tyree MT, Sperry JS (1989) Vulnerability of xylem to cavitation and embolism. Annu Rev Plant Physiol 40:19–36

Tyree MT, Zimmermann MH (2002) Xylem structure and the ascent of sap, 2nd edn. Springer, Berlin
Ullrich CI, Aloni R, Saeed MEM, Ullrich W, Efferth T (2019) Comparison between tumors in plants and human beings: mechanisms of tumor development and therapy with secondary plant metabolites. Phytomedicine 64:153081
Wareing PF (1951) Growth studies in woody species. 4. The initiation of cambial activity in ring-porous species. Physiol Plant 4:546–562
Wareing PF (1958) The physiology of cambial activity. J Inst Wood Sci 1:34–42
Went FW, Thimann KV (1937) Phytohormones. Macmillan, New York
Wheeler EA, Baas P (1991) A survey of the fossil record for dicotyledonous wood and its significance for evolutionary and ecological wood anatomy. IAWA Bull NS 12:275–332
Wheeler E, Baas P (2019) Wood evolution: Baileyan trends and functional traits in the fossil record. IAWA J 40:488–529
Williams CB, Anfodillo T, Crivellaro A, Lazzarin M, Dawson TE, Koch GW (2019) Axial variation of xylem conduits in the Earth's tallest trees. Trees 33:1299–1311
Yagi H, Tamura K, Matsushita T, Shimada T (2021) Spatiotemporal relationship between auxin dynamics and hydathode development in *Arabidopsis* leaf teeth. Plant Signal Behav. https://doi.org/10.1080/15592324.2021.1989216
Yoda K, Wagatsuma H, Suzuki M, Suzuki H (2003) Stem diameter changes before bud opening in *Zelkova serrata* saplings. J Plant Res 116:13–18
Zhang W, Feng F, Tyree MT (2018) Seasonality of cavitation and frost fatigue in *Acer mono* Maxim. Plant Cell Environ 41:1278–1286
Zhang J, Nieminen K, Serra JA, Helariutta Y (2014) The formation of wood and its control. Curr Opin Plant Biol 17:56–63
Zhang R, Zhang X, Wang J, Letham DS, McKinney SA, Higgins TJV (1995) The effect of auxin on cytokinin levels and metabolism in transgenic tobacco tissue expressing an *ipt* gene. Planta 196:84–94
Zhao X (2015) Effects of cambial age and flow path-length on vessel characteristics in birch. J Res 20:175–185
Zhu L, Cooper DJ, Yuan D, Li Z, Zhang Y, Liang H, Wang X (2020) Regional scale temperature rather than precipitation determines vessel features in earlywood of Manchurian ash in temperate forests. JGR Biogeosci 125:e2020JG005955.
Zimmermann MH (1983) Xylem structure and the ascent of sap. Springer, Berlin, Heidelberg
Zimmermann MH, Jeje AA (1981) Vessel-length distribution in stems of some American woody plants. Can J Bot 59:1882–1892
Zimmermann MH, Potter D (1982) Vessel-length distribution in branches, stem and roots of *Acer rubrum* L. IAWA Bull 3:103–109

Role of Plant Growth Regulators in the Plant-Environment Interaction and Epigenetic Regulation of Auxin

Clelia De-la-Peña and Víctor M. Loyola-Vargas

Abstract Relationships among plant growth regulators (PGR) occur at all metabolism levels, from the mutual regulation of their biosynthesis, degradation, transport, and signaling, to the control of the gradient distribution of auxins (Aux) by cytokinins (CKs) or vice-versa. Gene regulation in Aux signalling has become an important point for turning on or off certain sets of genes during environmental stress, plant development, and hormone interaction. Despite the considerable amount of literature regarding Aux biosynthesis, transport, and signalling, few reports available have explored its complex role in epigenetic regulation. In this book chapter, we analyze the interaction of Aux with CKs, their mutual interrelation, and the role played by epigenetic regulation in Aux.

1 Introduction

It is a popular saying that the winner takes all. However, in the case of auxin, the major player in plant development, this is not necessarily the case. In many cases, auxins recruit other plant growth regulators (PGR), or these PGR recruit auxins to induce a physiological effect. Ultimately, the interaction among these groups of PGR produces the final response in plant tissues. This fact is particularly essential in the case of the interaction of plants with the environment. To date, interactions of auxins (Aux) with abscisic acid (ABA) (Mohammadi et al. 2021), brassinosteroid (BR) (Casal and Balasubramanian 2019; Zhou et al. 2013), cytokinins (CKs) (Aloni 2021; Großkinsky and Petrášek 2019; Piotrowska-Niczyporuk et al. 2020), ethylene (ET) (Kim et al. 2021; Mohammadi et al. 2021), gibberellic acid (GA) (Jing et al. 2020b;

C. De-la-Peña (✉)
Centro de Investigación Científica de Yucatán, Unidad de Biotecnología, Calle 43 No. 130 x 32 y 34. Col. Chuburná de Hidalgo, 97205 Mérida, Yucatán, México
e-mail: clelia@cicy.mx

V. M. Loyola-Vargas
Centro de Investigación Científica de Yucatán, Unidad de Bioquímica y Biología Molecular de Plantas, Calle 43 No. 130 x 32 y 34. Col. Chuburná de Hidalgo, 97205 Mérida, Yucatán, México

T. Aftab (ed.), *Auxins, Cytokinins and Gibberellins Signaling in Plants*,
Signaling and Communication in Plants,
https://doi.org/10.1007/978-3-031-05427-3_2

Li et al. 2020), jasmonates (JA) (Blázquez et al. 2020; Ghorbel et al. 2021; Sun et al. 2021), salicylic acid (SA) (Salopek-Sondi et al. 2013; Westfall et al. 2016), nitric oxide (NO) (Shiraz et al. 2020), and strigolactones (SL) (Bellegarde and Sakakibara 2021; Blázquez et al. 2020; Ravazzolo et al. 2021) have been documented. However, many of these interactions are specific to root physiology, such as the auxin-ABA crosstalk for the development of lateral roots. Aux also interact in roots with CKs, GA, BR, JA, SA, ET, and ABA to control root growth in general. It may be in the root that the most complex interactions of the Aux take place (Mazzoni-Putman et al. 2021).

Specific relationships among PGR occur at all metabolism levels, from the mutual regulation of their biosynthesis, transport, and signaling, to the control of the gradient distribution of Aux by CKs or vice versa. These interactions have uncovered a highly complex panorama in the interaction between the different PGR on their way to regulating the different metabolic functions that lead to the development and maintenance of plant life. In this book chapter, we have chosen to focus only on the Aux/CKs partnership, understanding that in some physiological processes, not only these two, but multiple PGR participate. Moreover, we have described the importance of epigenetics in the biosynthesis, transport, signaling and conjugation of auxin. We are only at the beginning of understanding the complexity of signaling in plants under epigenetic regulation.

2 Relationship of Aux and CKs

The relationship between Aux and CKs has a long history. Almost from the moment the CKs were discovered, the interaction between Aux and CKs was exposed (Amasino 2005; Skoog and Miller 1957). When the concentration of CKs increases and the concentration of Aux in the callus is kept low, shoots emerge. On the other hand, if the concentration of Aux increases and the concentration of CKs decrease, then roots are produced. In this way, it was discovered that the Aux/CKs ratio has an essential function in cell differentiation. Variation in this relationship has been widely used to induce in vitro callus, shoots, roots, and plantlets, as well as to induce somatic embryogenesis (Loyola-Vargas et al. 2008; Loyola-Vargas and Ochoa-Alejo 2016; Singh et al. 2021; Zhao et al. 2021).

Most of the research on the relationship between Aux and CKs has focused on cell differentiation. An important number of scientific publications show the effect of Aux and CK concentrations on the morphogenic response (Bernula et al. 2020; Singh et al. 2021; Zhao et al. 2021). However, the effect on the Aux/CK ratio goes further. For specific aspects of the interaction between Aux and CKs, one of the following reviews can be consulted (Aloni 2021; Chandler and Werr 2015; El-Showk et al. 2013; Großkinsky and Petrášek 2019; Hwang et al. 2012; Ludwig-Müller et al. 2017; Schaller et al. 2015; Singh and Sinha 2017; Skalický et al. 2018; Su et al. 2011; Vanstraelen and Benková 2012; Wu et al. 2021).

The cross-regulation between Aux and CKs is essential in the control of shoot branching (Shimizu-Sato et al. 2009), the size of the root apical meristem (Dello Ioio et al. 2007; Dello Ioio et al. 2008), and lateral root patterning (Laplaze et al. 2008). However, it should be noted that the crosstalk between Aux and CKs is complex and has multiple players. The interaction between Aux and CKs can vary if it occurs in the leaves, in the root, or some other tissue or organ (Hussain et al. 2021), and if the plant is subjected to one or different environmental stress or senescence (Hussain et al. 2021). In all cases, the mutual regulation between Aux and CKs occurs both antagonistically in the root meristem (Chandler and Werr 2015; Kurepa et al. 2019; Zhao et al. 2010) and synergistically for plant growth and development (Yang et al. 2017) and for the development of the shoot apical meristem (Chandler and Werr 2015; Zhao et al. 2010) and these interactions are usually cell and/or tissue-specific (Danilova et al. 2020; Kieber and Schaller 2018; Wu et al. 2021).

The finely tuned cross-talk between Aux and CKs controls root growth and development, particularly the formation and maintenance of the root apical meristem (Del Bianco et al. 2013; Dello Ioio et al. 2007; Dello Ioio et al. 2008), the vascular pattern in roots (Bishopp et al. 2011b) and lateral root patterning (Laplaze et al. 2008; Schaller et al. 2015), mainly under stress conditions (Argueso et al. 2009; O'Brien and Benkova 2013; Tognetti et al. 2017; Wu et al. 2021). Also this cross-talk participates in the regulation of apical meristems, the development of the gynoecium and female gametophyte (Müller et al. 2017), the organogenesis and phyllotaxy in the shoot (Schaller et al. 2015) and shoot branching (Shimizu-Sato et al. 2009). Even third signals use the Aux/CKs ratio changes to carry out their control. This is the situation of parthenocarpy enhanced by sugars, which occurs by promoting Aux and CKs signaling (Wang et al. 2021a).

How does the cross-talk between Aux and CKs work? In general, it depends on the tissue where the interaction occurs. For instance, in the roots, both Aux (Petersson et al. 2009) and CKs (Antoniadi et al. 2015) form gradients along the root. What is interesting about this is that Aux are mostly absent in the root zones where CKs are most abundant (Antoniadi et al. 2015; Petersson et al. 2009). The CKs are found in a higher concentration in the lateral root cells and are absent in the first eight cells of the epidermis and the cortex. Then they are present in intermediate amounts similar to those present in the vascular bundle and the pericycle. In the case of Aux, its concentration is low in the epidermis cells and much higher in the lateral root cells. In the vascular bundle, the concentration of both PGR is very similar (Antoniadi et al. 2015; Petersson et al. 2009).

The cross-talk between Aux and CKs also includes all aspects of the homeostasis of the two PGR, biosynthesis, transport, and signaling (Bishopp et al. 2011a; Moubayidin et al. 2009; Müller et al. 2017; Pernisová et al. 2011). CKs induce Aux biosynthesis in *Arabidopsis* young root and shoot tissues, and this induction requires CK signaling through histidine phosphotransfer proteins (AHPs) and type-A ARABIDOPSIS RESPONSE REGULATORS (ARRs) (Jones et al. 2010). CKs also regulate local Aux metabolism (Casanova-Sáez et al. 2021; Di et al. 2016; Di Mambro et al. 2017, 2019; Jones et al. 2010; Pierdonati et al. 2019; Yan et al. 2017) and the polar Aux transport (PAT) (Dello Ioio et al. 2008; Liu et al. 2017; Moore et al.

2015; Müller and Sheen 2008; Muraro et al. 2016; Ruzicka et al. 2009; Waldie and Leyser 2018), and modulate CK signaling through the activation of *ARABIDOPSIS RESPONSE REGULATOR 7 (ARR7)* and *ARR15* (El-Showk et al. 2013; Kurepa et al. 2019). The regulation of the transport of Aux by CKs can be at the level of transcription and translational activations of *AUXIN/LIKE AUX* (*Aux/LAX*) genes and *PIN-FORMED* (*PIN*) family genes (Pernisova et al. 2009; Ruzicka et al. 2009) or the modification of PIN proteins. This disruption of the polar auxin transport (PAT) by CKs changes the Aux gradient formation necessary for the lateral root development (Laplaze et al. 2008). However, the interaction between CKs and PIN transporters is more complex. For example, CKs negatively affect *PIN1*, *PIN2*, and *PIN* 3, but act positively in *PIN 7* (Ruzicka et al. 2009). This inhibition of PIN1, PIN2, and PIN3 is mediated by the Aux/IAA SHORT HYPOCOTYL2 (SHY2, IAA3). *SHY2* transcription is induced by CKs, probably through ARR1 and ARR2 (Dello Ioio et al. 2008; Moubayidin et al. 2010).

A recent discovery adds a new layer of complexity to the already complex interaction between Aux and CKs. In the formation of the lateral roots, TRANSPORTER OF IBA1 (TOB1), an indole-3-butyric acid (IBA) transporter that blocks lateral root formation, was discovered. CKs transcriptionally induce this transporter (Michniewicz et al. 2019). Although it is not yet clear what the role of IBA is in plant cells, IBA can be considered a precursor in the biosynthesis of IAA or an Aux by itself. The fact that CKs modify their transport and this modification produces a physiological change that suggests a role for the IBA that was unknown until now. Consideration should also be given to the transport of Aux and CKs (Kramer and Bennett 2006). Aux move throughout the plant and, in particular, to the interior of its different organs (Overvoorde et al. 2010). Aux movement is very relevant for cross-talk with CKs, as it is one of the targets of CK action during the interaction of Aux with CKs (Bernula et al. 2020; Bishopp et al. 2011a). CKs are also transported through the plant, and this transport is biologically relevant (Hluska et al. 2021; Tessi et al. 2021).

On the other hand, the treatment of Aux resulted in reduced CK biosynthesis in *Arabidopsis thaliana* (Nordstrom et al. 2004). In the biosynthesis of CKs, Aux control the expression of the *ISOPENTENYL TRANSFERASE* (*IPT*) gene (Tanaka et al. 2006). Aux positively modulate the CK biosynthesis via direct control of the *IPT* gene transcript accumulations, mediated by the AUXIN RESPONSE FACTOR 19 (ARF19) transcription factor (Cancino-García et al. 2020; Cheng et al. 2013; Miyawaki et al. 2004; Yang et al. 2017). Auxins regulate at least two steps in the biosynthesis of terpenic CKs and one in their degradation. Aux not only participate in the regulation of CK biosynthesis but also in their degradation by regulating the expression of some of the members of the *CYTOKININ OXIDASE/DEHYDROGENASE (CKX)* family, in particular down-regulating the expression of *CKX2*, *CKX4*, and *CKX7*, while the inhibition of Aux transport down-regulates the expression of *CKX1* and *CKX6* (Jones et al. 2010; Werner et al. 2006).

Both Aux (Liu et al. 2019; Ljung 2013; Ludwig-Müller 2011; Ostrowski and Ciarkowska 2021) and CKs (Bajguz and Piotrowska 2009; Lulsdorf et al. 2013; Pokorná et al. 2020) are conjugated with other molecules. In the case of Aux, they are conjugated with different amino acids and glucose (Méndez-Hernández et al. 2021; Wojtaczka et al. 2022), while CKs are also conjugated with various carbohydrates, mainly glucose (Romanov and Schmülling 2022). These conjugates are also transported through the plant (Nguyen et al. 2021; Romanov and Schmülling 2022).

Gene regulation in Aux biosynthetic and signalling machinery has become an important point for dynamic "on" or "off" switches in certain sets of genes during environmental stress (Rhaman et al. 2020; Sonkar et al. 2021), plant development (Casanova-Sáez and Voß 2019; Mateo-Bonmatí et al. 2019), and PGR interaction (Pacurar et al. 2014; Wu et al. 2021) driven by epigenetic processes mainly via the histone repressive mark H3K27me3 (He et al. 2012; Lafos et al. 2011). Epigenetic processes are directly affected by environment and, therefore, when plants are exposed to different environmental conditions, their epigenetics change dramatically, expressing or repressing the transcription of many genes and altering the action of Aux (Campos-Rivero et al. 2017; Huq 2018; Salazar-Iribe and De-la-Peña 2020; Tognetti et al. 2012; Zhou and Luo 2018).

3 Epigenetics

The two major epigenetic mechanisms are DNA methylation and histone modifications (Allis et al. 2015). DNA methylation in plants occurs in three sites or contexts: CG, CHG and CHH (where H could be Adenine, Cytosine or Thymine) (Chan et al. 2005; De Mendoza et al. 2018). These sequences are mainly distributed in repetitive regions of the genome that are part of transposons, centromeres, 5S and 45S ribosomal genes, in promoter regions and in highly expressed gene coding regions (Pikaard and Mittelsten Scheid 2014; Zilberman et al. 2007). In the case of histone, these epigenetic modifications occur in different amino acid residues in any of the five histone tails (H2A, H2B, H3, H4 and H1) and histone variants (e.g. H1.1, H2A.Z, H3.3 and H1.3). The most studied histone modifications are acetylation and methylation (Allis et al. 2015). Acetylation is mediated by the reversible activity of histone acetyltransferases (HATs) and histone deacetylases (HDACs), while methylation is catalysed by histone methyltransferases (HMT) and histone demethylases (DMET).

DNA methylation and histone modifications are part of the epigenetic regulation governed by PGR, in which plants can take advantage of their totipotent nature (Campos-Rivero et al. 2017; Duarte-Aké et al. 2019; Us-Camas et al. 2014; Zhu 2010). There are four principal players in chromatin rearrangements that participate in regulating auxin homeostasis: (1) chromatin remodelers; (2) histone modifiers; (3) polycomb proteins and (4) DNA methylation enzymes. Both DNA methylation and histone modifications operate in several plant processes in which auxin has an important role, such as floral patterning and determinacy, endosperm development, hypocotyl elongation and leaf growth (Mateo-Bonmatí et al. 2019).

4 Epigenetic Players in Plant Hormone Regulation

Both histone acetylation and deacetylation play an important role in gene regulation and have been implicated in PGR´ signalling (Chen and Wu 2010; Chen et al. 2010; Zhou et al. 2005; Zhu 2010). For instance, knocking out HDA6 and HDA19 causes ABA hypersensitivity (Chen and Wu 2010; Chen et al. 2010) and both ethylene as well as JA induce the expression of HDA6 and HDA19 (Zhou et al. 2005). Also, the PIF7 transcription factor recruits an unknown HAT enzyme(s) to promote histone acetylation in H4K5, H3K9, and H3K27 at *YUC8*, resulting in its expression (Peng et al. 2018).

Besides acetylation, the methylation mark H3K4me3, related to active transcription, has been shown to regulate several genes that affect PGR. For instance, ARABIDOPSIS HOMOLOG OF TRITHORAX 1 (ATX1) is a histone methyltransferase that functions as a "writer," incorporating methyl groups in the lysine of histones. This "writer" directly targets the *9-CIS-EPOXYCAROTENOID DIOXYGENASE 3* (*NCED3*). NCED3 plays a key role in the ABA biosynthesis pathway and ATX1 regulates its transcriptional activity. *Atx1* knockout mutants undergoing dehydration stress showed ABA-related phenotypes (Ding et al. 2011).

In mammalian as well as plant cells, POLYCOMB GROUP (PcG) protein complexes perform the formation, perpetuation, and epigenetic inheritance of the H3K27me3 modification (Derkacheva and Hennig 2014; Khan et al. 2015; Lu et al. 2011). H3K27me3 is one of the epigenetic marks most investigated in genes involved in auxin biosynthesis (*YUCs, CYPs, TAA1/TARs, SUR1, NITs*), inactivation (*GH3s, IAMT*), transport (*PINs, AUX/LAXs*), and signalling (*TIR1/AFBs, IAAs, ARFs*), revealing that this histone modification exerts a profound effect on auxin action (He et al. 2012; Lafos et al. 2011). *YUC1*, *NIT2* and *PIN1* are regulated by this transcriptional repressive mark, and it is likely that PRC2 is the "writer" in these genes. For instance, H3K27me3 targets *ARF* expression indirectly, through miRNAs, and directly targets 14 *AUX/ IAA* genes and several gene loci encoding the auxin transporter genes *PINFORMED 1* (*PIN1*), *PIN4*, *PIN7*, and *PIN8*, which are differentially methylated in leaves and meristems (Lafos et al. 2011).

On the other hand, the expression of *YUCCA* (*YUC*) genes, involved in Aux biosynthesis, increases under the wound response, mediating JA signals, and JA increases the transcript levels of *YUC8* and *YUC9* in *Arabidopsis* (Hentrich et al. 2013).

Another histone modification that has been analysed under PGR action is histone ubiquitination, which is a repressive mark. HUB1/2 is also required in circadian rhythms (Himanen et al. 2012), seed dormancy (Liu et al. 2007), photomorphogenesis (Bourbousse et al. 2012), flowering (Cao et al. 2008; Gu et al. 2009; Xu et al. 2009), immune responses (Zhang et al. 2015b; Zou et al. 2014) and plant defense (Dhawan et al. 2009; Wang et al. 2021b). The *Cytokinin induced root waving 2* (*ckrw2*) mutant is auxin deficient and *CKRW2* is identical to histone monoubiquitination1 (HUB1), a gene encoding an E3 ligase required for histone H2B monoubiquitination (H2Bub1) (Zhang et al. 2021), which normally occurs on K143 and K145 in Arabidopsis. Therefore, the participation of this histone modification in the auxin-cytokinin relationship is evident. In fact, the expression of *CKRW2* is induced by cytokinin (Zhang et al. 2021).

In general, Aux affects the state of chromatin in genes related to biosynthesis, transport, conjugation or degradation through the regulation of chromatin remodelers, enzymes involved in DNA methylation or histone modifications (Chung et al. 2019; Hasegawa et al. 2018; Wu et al. 2015) (Table 1).

Table 1 Epigenetic regulation in Aux biosynthesis, transport, signalling and conjugation

Process and proteins involved	*Genes*	Epigenetic regulation	References
Biosynthesis			
Tryptophan synthase β	*TRP2*	DNA methylation and H3K9me2 H2Bub1	Yang et al. (2021) Zhang et al. (2021)
Amidase	*AMI1*	H2Bub1	Zhang et al. (2021)
Tryptophan aminotransferase	*TAA1* *TAR1,2*		Jing et al. (2020a)
Aldehyde oxidase	*AO1*	None reported	
Flavin monooxygenase	*YUC1,4* *YUC7*	H3K27me3 DNA methylation	Do et al. (2019), He et al. (2012), Jing et al. (2020a), Xu et al. (2018), Zhang et al. (2021)
Cytochrome P450	*CYP79B2/3*	None reported	
Nitrilase	*NIT2*	H3K27me3	Jing et al. (2020a)
Transport			
Auxin influx transporter	*AUX1* *LAX*	H3K9ac and H3K18ac	Wang et al. (2016)
Auxin efflux carrier	*PIN1*	H3K27me3	He et al. (2012), Jing et al. (2020a)
Serine threonine kinase	*PID*	None reported	
ABC transporter	*ABC1,19*	None reported	
Signalling			
Aux/IAA transcription factor	*IAA1-25*	H3K27me3	He et al. (2012)
Auxin response factor	*ARF1-23*		Jing et al. (2020a)
F-box	*TIR1* *AFB*	None reported	
Small Auxin Up RNA	*SAUR*	None reported	
Conjugation			
Gretchen Hagen 3	*GH3*	H3K27me3	He et al. (2012)

5 Epigenetics and Auxin Biosynthesis

Since YUCCAs (YUCs) function critically in Aux biosynthesis, their regulation has to be precise and fast under different environmental conditions such as light level, ambient temperature, and stress signals (Do et al. 2019). *YUC1* and *YUC4* promoter regions were found to exhibit a decrease in H3K27me3 that was related to early Aux-mediated de novo root regeneration (Chen et al. 2016). Furthermore, LIKE HETEROCHROMATIN 1 (LHP1), a member of the Polycomb Repressive Complex 2 (PRC2), was found to directly bind to some *YUC* gene promoters, thus negatively regulating their expression (Rizzardi et al. 2011). In fact, the PRC2 accessory protein LIKE HETEROCHROMATIN 1 (LHP1) (Derkacheva et al. 2013) is recruited to *YUC1, YUC2, YUC4, YUC5, YUC6, YUC8, YUC9,* and *YUC10* promoters to control their expression (Rizzardi et al. 2011). On one example, PRC2 is recruited by the C2H2-type zinc-finger transcription factor SUPERMAN (SUP) protein, which physically interacts with *YUC1* and *YUC4* promoters, repressing auxin biosynthesis by the incorporation of H3K27me3 (Xu et al. 2018). On the other hand, the chromatin remodeling factors CHROMATIN REMODELLING 11 (CHR11) and CHR17 are recruited to the *YUC4* promoter during floral primordium formation, promoting its expression (Yamaguchi et al. 2018). However, in floral organ development, SUP actively represses *YUC1* and *YUC4* expression and, thereby, Aux biosynthesis at the boundaries between carpels and stamen primordia (Xu et al. 2018). Moreover, Chen et al. (2016) found that *YUC1* and *YUC4* transcript levels increased in leaf explants at 4, 8, and 12 h after culture and this increase was associated with the reduction of H3K27me3 levels at these gene loci. *YUC1* and *YUC4* are involved in early Aux-mediated de novo root regeneration in a strong correlation with a decrease in H3K27me3 in their promoter regions (Chen et al. 2016). Using chromatin immunoprecipitation-chip (ChIP-chip) assays, which analyzed the wide-genome distribution of an specific epigenetic mark, He et al. (2012) compared the H3K27me3 levels at different Aux-related gene loci such as *YUC4*, *PIN1*, and *IAA2* in the cultured *Arabidopsis* leaf and callus samples. It was found that in the callus H3K27me3 levels decreased first in these genes (He et al. 2012).

YUC8 involved in the Aux biosynthesis in response to shade conditions is induced by the MORF-RELATED GENE 2 (MRG2), a H3K4me3/H3K36me3-binding protein, which interacts with PHYTOCHROME-INTERACTING FACTOR 7 (PIF7). This PIF7 transcription factor recruits HAT enzymes to promote histone acetylation in H4K5, H3K9, and H3K27 at the *YUC8* promoter, facilitating its expression (Peng et al. 2018). On the other hand, at high temperature, acetylation in the histone H2A. Z is removed from the *YUC8* locus by the HISTONE DEACETYLASE 9 (HDA9) providing a looser chromatin conformation that allows PIF4-mediated activation of *YUC8* transcription (Van Der Woude et al. 2019). In the case of *YUC10*, it is repressed in maternal-derived tissues by the FERTILIZATION-INDEPENDENT SEED-PRC2 (FIS-PRC2) complex, which marks the target loci with H3K27me3 during the development of female gametes (Figueiredo et al. 2015; Mateo-Bonmatí et al. 2019; Wolff et al. 2011).

DNA methylation has been shown to alter auxin homeostasis by regulating the transcription of YUCs (Mateo Bonmatí et al. 2019). The *drm1 drm2 cmt3 (ddc)* triple mutant, which is defective in both maintenance of and de novo DNA methylation, has a higher expression of *YUC1*, *YUC2* and *TAA1* genes in leaves than in roots (Forgione et al. 2019). The authors found that the promoter of *YUC2* was hypomethylated in the *ddc* mutant, which means higher expression of the gene and, therefore, more IAA abundance. De novo DNA methylation is normally positioned at the CHH sites promoting gene repression. Plants grown at high temperatures showed severe reductions in CHH methylation, triggering the up-regulation of *YUC2* (Gyula et al. 2018).

6 Epigenetics and Auxin Transport

Auxin transport is mediated by carriers in the cell membrane, such as AUXIN-RESISTANT 1/LIKE AUXIN-RESISTANT (AUX1/LAX), PIN-FORMED (PIN) and ATP-BINDING CASSETTE SUBFAMILY B TRANSPORTER (ABCB)/MDR/PGP. PIN and ABC are responsible for auxin efflux while AUX1/LAX and PIN-like proteins are responsible for auxin influx (Geisler et al. 2017; Prát et al. 2018; Swarup and Bhosale 2019). Comparative studies showed differential H3K27me3 levels in *PIN1*, *PIN4*, *PIN7*, and *PIN8* during leaf differentiation from the apical meristem (Lafos et al. 2011). During pluripotency acquisition, a large decrease of H3K27me3 levels was found at *PIN1* in leaf-derived callus (He et al. 2012). In Arabidopsis, BRAHMA (BRM), an orthologue of the *Saccharomyces cerevisiae* SWI/SNF chromatin remodeling ATPase, was found to directly target the chromatin of the Aux efflux transporter genes *PIN1*, *PIN2*, *PIN3*, *PIN4*, and *PIN7*, in part by acting antagonistically with the H3K27me3-associated chromatin repression mark mediated by the Polycomb group (PcG) proteins (Yang et al. 2015). In the case of acetylation, it was found that in Arabidopsis, the inhibition of histone deacetylation alters Aux distribution due to PIN1 degradation in the root tip (Nguyen et al. 2013).

In addition to histone modifications, DNA methylation also is affected in Aux transport. *PIN1*, *PIN3* and *PIN4* presented low expression in leaves in the *drm1 drm2 cmt3* triple mutant (*ddc*), which is defective in maintenance and de novo DNA methylation in Arabidopsis. When the concentration of IAA was quantified, it was found that IAA was higher in leaves than in roots in the *ddc* mutants (Forgione et al. 2019).

AUX1 is an efflux transporter, and the increased in the acetylation of H3K9 and H3K18 at AUX1 chromatin was observed in the histone deacetylase-binding factors SWI-INDEPENDET3-LIKE (SNL1) and (SNL2) double mutant *snl1 snl2*. This increase in H3K9ac and H3K18ac was related to *AUX1* expression, which regulates radicle emergence (Wang et al. 2016).

7 Epigenetics and Auxin Signalling

The function of different epigenetic mechanisms, mainly histone acetylation, has been investigated in the auxin signalling pathway (Manzano et al. 2012; Nguyen et al. 2013; Weiste and Dröge-Laser 2014; Yamamuro et al. 2016). For instance, Yamamuro et al. (2016) described the involvement between epigenetic modifications and PGR action, highlighting the roles of epigenetic modifications in auxin signalling and distribution. One of the first signals of auxin presence is through the Aux-IAA degradation, liberating ARF; then the free ARF transcription activators can activate the expression of auxin-responsive genes (Lau et al. 2008; Mockaitis and Estelle 2008) by recruiting histone acetyltransferases that facilitate chromatin opening in the target genes (Yamamuro et al. 2016). ARF3 and ARF4 have been found to repress SHOOTMERISTEMLESS (STM) via histone deacetylation, which promotes flower initiation at the reproductive meristem (Chung et al. 2019).

On the other hand, when auxin is absent, the AUX/IAA proteins can directly bind to ARFs and interact with TOPLESS (TPL), forming the complex ARF-AUX/IAA-TPL that can bind to AuxRE (Causier et al. 2012; Leyser 2018; Mockaitis and Estelle 2008; Szemenyei et al. 2008; Woodward and Bartel 2005). This complex recruits a histone deacetylase (HDAC), promoting heterochromatin shift (Kagale and Rozwadowski 2011; Krogan and Long 2009; Long et al. 2006; Yamamuro et al. 2016). Nguyen et al. (2020) investigate the function of HAT in auxin signalling and root morphogenesis. The authors suggest that auxin signalling is controlled by AUX/IAA-HDA and ARF-HAT. This balance in auxin signalling is critical to normal root morphogenesis. Recent studies have explained the functions of different Arabidopsis HDAs (such as HDA6, HDA9, and HDA19) in auxin signalling as well as biosynthesis (Kuhn et al. 2020; Van Der Woude et al. 2019; Yuan et al. 2019).

Nuclear factor-YC homologs (NF-YCs), which redundantly repress light-controlled hypocotyl elongation, interact with HDA15 in the H4 acetylation that is reduced by light, while in the dark this NF-YCs-HDA15 complex decreases, allowing H4 acetylation to increase (Tang et al. 2017). In the *hda15* mutant the expression of both *IAA6* and *IAA19* increased, which strongly suggests that Aux signalling is dependent on histone acetylation (Tang et al. 2017).

Although most of the studies are related to histone deacetylation, a few studies indicate the involvement of histone acetylation in the Aux response (Chandler 2016; Saiga et al. 2012; Weiste and Dröge-Laser 2014). Some studies have implied a positive role for the histone acetyl transferase complex in ARF transcriptional activity (such as ARF5/MONOPTEROS) (Anzola et al. 2010; Chandler 2016; Kornet and Scheres 2009; Saiga et al. 2012; Wu et al. 2015).

8 Epigenetic and Auxin Conjugation

Most endogenous IAA is found as a conjugate form at the carboxyl group. This conjugation can occur with monosaccharides, polysaccharides, myoinositol, glycoproteins, amino acids, peptides, carbohydrates, choline and other molecules (Bajguz and Piotrowska 2009). Each of these conjugations has a specific role, such as IAA storage, transport, catabolism, degradation or homeostasis.

GRETCHEN HAGEN 3 (*GH3*) genes are regulated by *ARF8*, *ARF17* and *ARF19* (De Rybel et al. 2010; Mallory et al. 2005; Tian et al. 2004; Zhang et al. 2015a). *GH3-8* encodes an IAA-amino synthetase that prevents free IAA accumulation and activates disease resistance in SA and JA signalling (Ding et al. 2008). Ding et al. (2008) found that the overexpression of *GH3-8* enhances disease resistance to the rice pathogen *Xanthomonas oryzae* pv *oryzae.* In the presence of auxin, the bZIPs can directly bind to the *GH3.3* promoter, recruiting the HAT and GCN5 via an adapter protein (ADA2b/PROPORZ1) to activate the expression of this *GH3.3* gene (Weiste and Dröge-Laser 2014). During leaf-derived callus formation, consistent correlations were found between a decrease in H3K27me3 levels and upregulation of *GH3.1, GH3.2, GH3.3, GH3.6,* and *GH3.17* (He et al. 2012).

9 Conclusion

Undoubtedly, there are substantial gaps in knowledge related to how auxins interact with other PGR for the smooth functioning and response of plants exposed to different environmental conditions. However, the knowledge that is being revealed about the role that epigenetics plays in the biosynthesis, degradation, transport, signalling and conjugation of auxins has opened paths to understand their rapid and subtle regulation by the repressive mark H3K27me3.

Work should be done to understand how other PGR recruit Aux and vice versa. At the same time, how these mechanisms change from tissue to tissue must be analyzed, particularly how they are epigenetically regulated.

Out of the large number of genes known to participate in auxin homeostasis, there are only a few for which direct epigenetic regulation has been described, revealing the widespread poor understanding of this topic. Further multidisciplinary efforts are imperative to better understand the different epigenetic players for PGR action. This knowledge will be necessary for the eventual manipulation of PGR homeostasis in order to improve plant fitness and adaptation.

We can conclude that the chromatin modifications such as histone acetylation and methylation as well as DNA methylation play an important role in the regulation of auxin biosynthesis, signalling, transport and conjugation genes.

References

Allis CD, Caparros ML, Jenuwein T, Reinberg D (2015) Epigenetics, 2nd edn. Cold Spring Harbor Laboratory Press, Cold Spring Harbor, NY

Aloni R (2021) The hormonal signals that regulate plant vascular differentiation. Springer International Publishing, pp 55–96

Amasino R (2005) 1955: Kinetin arrives. The 50th anniversary of a new plant hormone. Plant Physiol 138(3):1177–1184. https://doi.org/10.1104/pp.104.900160

Antoniadi I, Plačková L, Simonovik B, Doležal K, Turnbull C, Ljung K, Novák O (2015) Cell-type-specific cytokinin distribution within the Arabidopsis primary root apex. Plant Cell 27(7):1955–1967. https://doi.org/10.1105/tpc.15.00176

Anzola JM, Sieberer T, Ortbauer M, Butt H, Korbei B, Weinhofer I, Mâllner AE, Luschnig C (2010) Putative Arabidopsis Transcriptional Adaptor Protein (PROPORZ1) is required to modulate histone acetylation in response to auxin. Proc Natl Acad Sci 107(22):10308–10313

Argueso CT, Ferreira FJ, Kieber JJ (2009) Environmental perception avenues: the interaction of cytokinin and environmental response pathways. Plant, Cell Environ 32(9):1147–1160. https://doi.org/10.1111/j.1365-3040.2009.01940.x

Bajguz A, Piotrowska A (2009) Conjugates of auxin and cytokinin. Phytochemistry 70(8):957–969. https://doi.org/10.1016/j.phytochem.2009.05.006

Bellegarde F, Sakakibara H (2021) Nitrate-dependent modulation of root system architecture in maize: a balance between strigolactone and auxin pathways. Plant Cell Physiol 62(4):541–542. https://doi.org/10.1093/pcp/pcab032

Bernula D, Benkő P, Kaszler N, Domonkos I, Valkai I, Szőllősi R, Ferenc G, Ayaydin F, Fehér A, Gémes K (2020) Timely removal of exogenous cytokinin and the prevention of auxin transport from the shoot to the root affect the regeneration potential of Arabidopsis roots. Plant Cell, Tissue Organ Cult 140(2):327–339. https://doi.org/10.1007/s11240-019-01730-3

Bishopp A, Benková E, Helariutta Y (2011a) Sending mixed messages: auxin-cytokinin crosstalk in roots. Curr Opin Plant Biol 14(1):10–16. https://doi.org/10.1016/j.pbi.2010.08.014

Bishopp A, Help H, El-Showk S, Weijers D, Scheres B, Friml J, Benková E, Ari & Helariutta (2011b) A mutually inhibitory interaction between auxin and cytokinin specifies vascular pattern in roots. Curr Biol 21(11):917–926. https://doi.org/10.1016/j.cub.2011.04.017

Blázquez MA, Nelson DC, Weijers D (2020) Evolution of plant hormone response pathways. Annu Rev Plant Biol 71(1):327–353. https://doi.org/10.1146/annurev-arplant-050718-100309

Bourbousse C, Ahmed I, Roudier F, Zabulon G, Blondet E, Balzergue S, Colot V, Bowler C, Barneche F (2012) Histone H2B nonoubiquitination facilitates the rapid modulation of gene expression during *Arabidopsis* photomorphogenesis. PLoS Genet 8(7):e1002825. https://doi.org/10.1371/journal.pgen.1002825

Campos-Rivero G, Osorio-Montalvo P, Sánchez-Borges R, Us-Camas R, Duarte-Aké F, De-la-Peña C (2017) Plant hormone signaling in flowering: an epigenetic point of view. J Plant Physiol 214:16–27. https://doi.org/10.1016/j.jplph.2017.03.018

Cancino-García VJ, Ramírez-Prado JH, De-la-Peña C (2020) Auxin perception in Agave is dependent on the species' Auxin Response Factors. Sci Rep 10(1):3860. https://doi.org/10.1038/s41598-020-60865-y

Cao Y, Dai Y, Cui S, Ma L (2008) Histone H2B monoubiquitination in the chromatin of *FLOWERING LOCUS C* regulates flowering time in *Arabidopsis* The Plant Cell 20(10):2586–2602

Casal JJ, Balasubramanian S (2019) Thermomorphogenesis. Annu Rev Plant Biol 70(1):321–346. https://doi.org/10.1146/annurev-arplant-050718-095919

Casanova-Sáez R, Mateo-Bonmatí E, Ljung K (2021) Auxin metabolism in plants. Cold Spring Harb Perspect Biol 13(3):a039867. https://doi.org/10.1101/cshperspect.a039867

Casanova-Sáez R, Voß U (2019) Auxin metabolism controls developmental decisions in land plants. Trends Plant Sci 24(8):741–754. https://doi.org/10.1016/j.tplants.2019.05.006

Causier B, Ashworth M, Guo W, Davies B (2012) The TOPLESS interactome: a framework for gene repression in Arabidopsis. Plant Physiol 158(1):423–438. https://doi.org/10.1104/pp.111.186999

Chan SWL, Henderson IR, Jacobsen SE (2005) Gardening the genome: DNA methylation in *Arabidopsis thaliana*. Nat Rev Genet 6(5):351–360. https://doi.org/10.1038/nrg1601

Chandler JW (2016) Auxin response factors. Plant, Cell Environ 39:1014–1028

Chandler JW, Werr W (2015) Cytokinin–auxin crosstalk in cell type specification. Trends Plant Sci 20(5):291–300. https://doi.org/10.1016/j.tplants.2015.02.003

Chen L, Tong J, Xiao L, Ruan Y, Liu J, Zeng M, Huang H, Wang J-W, Xu L (2016) *YUCCA*-mediated auxin biogenesis is required for cell fate transition occurring duringde *novo* root organogenesis in Arabidopsis. J Exp Bot 67(14):4273–4284. https://doi.org/10.1093/jxb/erw213

Chen L-T, Wu K (2010) Role of histone deacetylases HDA6 and HDA19 in ABA and abiotic stress response. Plant Signal Behav 5(10):1318–1320

Chen LT, Luo M, Wang YY, Wu K (2010) Involvement of Arabidopsis histone deacetylase HDA6 in ABA and salt stress response. J Exp Bot 61(12):3345–3353

Cheng ZJ, Wang L, Sun W, Zhang Y, Zhou C, Su YH, Li W, Sun TT, Zhao XY, Li XG, Cheng Y, Zhao Y, Xie Q, Zhang XS (2013) Pattern of auxin and cytokinin responses for shoot meristem induction results from the regulation of cytokinin biosynthesis by AUXIN RESPONSE FACTOR3. Plant Physiol 161(1):240–251. https://doi.org/10.1104/pp.112.203166

Chung Y, Zhu Y, Wu M-F, Simonini S, Kuhn A, Armenta-Medina A, Jin R, Østergaard L, Gillmor CS, Wagner D (2019) Auxin response factors promote organogenesis by chromatin-mediated repression of the pluripotency gene *SHOOTMERISTEMLESS*. Nat Commun 10(1). https://doi.org/10.1038/s41467-019-08861-3

Danilova MN, Doroshenko AS, Kudryakova NV, Klepikova AV, Shtratnikova VY, Kusnetsov VV (2020) The crosstalk between cytokinin and auxin signaling pathways in the control of natural senescence of *Arabidopsis thaliana* leaves. Russ J Plant Physiol 67(6):1028–1035. https://doi.org/10.1134/s1021443720060035

De Mendoza A, Bonnet A, Vargas-Landin DB, Ji N, Li H, Yang F, Li L, Hori K, Pflueger J, Buckberry S, Ohta H, Rosic N, Lesage P, Lin S, Lister R (2018) Recurrent acquisition of cytosine methyltransferases into eukaryotic retrotransposons. Nat Commun 9(1). https://doi.org/10.1038/s41467-018-03724-9

De Rybel B, Vassileva V, Parizot B, Demeulenaere M, Grunewald W, Audenaert D, Van Campenhout J, Overvoorde P, Jansen L, Vanneste S, Möller B, Wilson M, Holman T, Van Isterdael G, Brunoud G, Vuylsteke M, Vernoux T, De Veylder L, Inzé D, Weijers D, Bennett MJ, Beeckman T (2010) A novel Aux/IAA28 signaling cascade activates GATA23-dependent specification of lateral root founder cell identity. Curr Biol 20(19):1697–1706. https://doi.org/10.1016/j.cub.2010.09.007

Del Bianco M, Giustini L, Sabatini S (2013) Spatiotemporal changes in the role of cytokinin during root development. New Phytol 199(2):324–338. https://doi.org/10.1111/nph.12338

Dello Ioio R, Linhares FS, Scacchi E, Casamitjana-Martinez E, Heidstra R, Costantino P, Sabatini S (2007) Cytokinins determine Arabidopsis root-meristem size by controlling cell differentiation. Curr Biol 17(8):678–682. https://doi.org/10.1016/j.cub.2007.02.047

Dello Ioio R, Nakamura K, Moubayidin L, Perilli S, Taniguchi M, Morita MT, Aoyama T, Costantino P, Sabatini S (2008) A genetic framework for the control of cell division and differentiation in the root meristem. Science 322(5906):1380–1384. https://doi.org/10.1126/science.1164147

Derkacheva M, Hennig L (2014) Variations on a theme: polycomb group proteins in plants. J Exp Bot 65(10):2769–2784. https://doi.org/10.1093/jxb/ert410

Derkacheva M, Steinbach Y, Wildhaber T, Mozgová I, Mahrez W, Nanni P, Bischof S, Gruissem W, Hennig L (2013) *Arabidopsis* MSI1 connects LHP1 to PRC2 complexes. EMBO J 32(14):2073–2085. https://doi.org/10.1038/emboj.2013.145

Dhawan R, Luo H, Foerster AM, Abuqamar S, Du H-N, Briggs SD, Scheid OM, Mengiste T (2009) HISTONE MONOUBIQUITINATION1 interacts with a subunit of the mediator complex and regulates defense against necrotrophic fungal pathogens in Arabidopsis. Plant Cell 21(3):1000–1019. https://doi.org/10.1105/tpc.108.062364

Di D-W, Wu L, Zhang L, An C-W, Zhang T-Z, Luo P, Gao H-H, Kriechbaumer V, Guo G-Q (2016) Functional roles of Arabidopsis *CKRC2/YUCCA8* gene and the involvement of PIF4 in the regulation of auxin biosynthesis by cytokinin. Sci Rep 6(1):36866. https://doi.org/10.1038/srep36866

Di Mambro R, De Ruvo M, Pacifici E, Salvi E, Sozzani R, Benfey PN, Busch W, Novak O, Ljung K, Di Paola L, Marée AFM, Costantino P, Grieneisen VA, Sabatini S (2017) Auxin minimum triggers the developmental switch from cell division to cell differentiation in the Arabidopsis root. Proc Natl Acad Sci 114(36):E7641–E7649. https://doi.org/10.1073/pnas.1705833114

Di Mambro R, Svolacchia N, Dello Ioio R, Pierdonati E, Salvi E, Pedrazzini E, Vitale A, Perilli S, Sozzani R, Benfey PN, Busch W, Costantino P, Sabatini S (2019) The lateral root cap acts as an auxin sink that controls meristem size. Curr Biol 29(7):1199-1205.e4. https://doi.org/10.1016/j.cub.2019.02.022

Ding X, Cao Y, Huang L, Zhao J, Xu C, Li X, Wang S (2008) Activation of the indole-3-Acetic Acid-Amido Synthetase GH3-8 suppresses expansin expression and promotes salicylate- and jasmonate-independent basal immunity in rice. Plant Cell 20(1):228–240. https://doi.org/10.1105/tpc.107.055657

Ding Y, Avramova Z, Fromm M (2011) The Arabidopsis trithorax-like factor ATX1 functions in dehydration stress responses via ABA-dependent and ABA-independent pathways. Plant J 66(5):735–744. https://doi.org/10.1111/j.1365-313X.2011.04534.x

Do BH, Phuong VTB, Tran G-B, Nguyen NH (2019) Emerging functions of chromatin modifications in auxin biosynthesis in response to environmental alterations. Plant Growth Regul 87(1):165–174. https://doi.org/10.1007/s10725-018-0453-x

Duarte-Aké F, Us-Camas R, Cancino-García VJ, De-la-Peña C (2019) Epigenetic changes and photosynthetic plasticity in response to environment. Environ Exp Bot 159:108–120. https://doi.org/10.1016/j.envexpbot.2018.12.010

El-Showk S, Ruonala R, Helariutta Y (2013) Crossing paths: cytokinin signalling and crosstalk. Development 140(7):1373–1383. https://doi.org/10.1242/dev.086371

Figueiredo DD, Batista RA, Roszak PJ, Köhler C (2015) Auxin production couples endosperm development to fertilization. Nat Plants 1(12):15184. https://doi.org/10.1038/nplants.2015.184

Forgione I, Wołoszyńska M, Pacenza M, Chiappetta A, Greco M, Araniti F, Abenavoli MR, Van Lijsebettens M, Bitonti MB, Bruno L (2019) Hypomethylated *drm1 drm2 cmt3* mutant phenotype of *Arabidopsis thaliana* is related to auxin pathway impairment. Plant Sci 280:383–396. https://doi.org/10.1016/j.plantsci.2018.12.029

Geisler M, Aryal B, Di Donato M, Hao P (2017) A critical view on ABC transporters and their interacting partners in auxin transport. Plant Cell Physiol 58(10):1601–1614. https://doi.org/10.1093/pcp/pcx104

Ghorbel M, Brini F, Sharma A, Landi M (2021) Role of jasmonic acid in plants: the molecular point of view. Plant Cell Rep 40(8):1471–1494. https://doi.org/10.1007/s00299-021-02687-4

Großkinsky DK, Petrášek J (2019) Auxins and cytokinins—the dynamic duo of growth-regulating phytohormones heading for new shores. New Phytol 221(3):1187–1190. https://doi.org/10.1111/nph.15556

Gu X, Jiang D, Wang Y, Bachmair A, He Y (2009) Repression of the floral transition via histone H2B monoubiquitination. Plant J 57(3):522–533. https://doi.org/10.1111/j.1365-313x.2008.03709.x

Gyula P, Baksa I, Tóth T, Mohorianu I, Dalmay T, Szittya G (2018) Ambient temperature regulates the expression of a small set of sRNAs influencing plant development through *NF-YA2* and *YUC2*. Plant, Cell Environ 41(10):2404–2417. https://doi.org/10.1111/pce.13355

Hasegawa J, Sakamoto T, Fujimoto S, Yamashita T, Suzuki T, Matsunaga S (2018) Auxin decreases chromatin accessibility through the TIR1/AFBs auxin signaling pathway in proliferative cells. Sci Rep 8(1). https://doi.org/10.1038/s41598-018-25963-y

He C, Chen X, Huang H, Xu L (2012) Reprogramming of H3K27me3 is critical for acquisition of pluripotency from cultured *Arabidopsis* tissues. PLoS Genet 8(8):e1002911. https://doi.org/10.1371/journal.pgen.1002911

Hentrich M, Böttcher C, Düchting P, Cheng Y, Zhao Y, Berkowitz O, Masle J, Medina J, Pollmann S (2013) The jasmonic acid signaling pathway is linked to auxin homeostasis through the modulation of *YUCCA8* and *YUCCA9* gene expression. Plant J 74(4):626–637. https://doi.org/10.1111/tpj.12152

Himanen K, Woloszynska M, Boccardi TM, De Groeve S, Nelissen H, Bruno L, Vuylsteke M, Van Lijsebettens M (2012) Histone H2B monoubiquitination is required to reach maximal transcript levels of circadian clock genes in Arabidopsis. Plant J 72(2):249–260. https://doi.org/10.1111/j.1365-313x.2012.05071.x

Hluska T, Hlusková L, Emery RJN (2021) The hulks and the deadpools of the cytokinin universe: a dual strategy for cytokinin production, translocation, and signal transduction. Biomolecules 11(2):209. https://doi.org/10.3390/biom11020209

Huq E (2018) Direct convergence of light and auxin signaling pathways in Arabidopsis. Mol Plant 11(4):515–517. https://doi.org/10.1016/j.molp.2018.02.002

Hussain S, Nanda S, Zhang J, Rehmani MIA, Suleman M, Li G, Hou H (2021) Auxin and cytokinin interplay during leaf morphogenesis and phyllotaxy. Plants 10(8):1732. https://doi.org/10.3390/plants10081732

Hwang I, Sheen J, Müller B (2012) Cytokinin signaling networks. Annu Rev Plant Biol 63(1):353–380. https://doi.org/10.1146/annurev-arplant-042811-105503

Jing T, Ardiansyah R, Xu Q, Xing Q & Müller-Xing R (2020a) Reprogramming of cell fate during root regeneration by transcriptional and epigenetic networks. Front Plant Sci 11(317). https://doi.org/10.3389/fpls.2020a.00317

Jing W, Zhang S, Fan Y, Deng Y, Wang C, Lu J, Sun X, Ma N, Shahid MO, Li Y, Zhou X (2020b) Molecular evidences for the interactions of auxin, gibberellin, and cytokinin in bent peduncle phenomenon in rose (*Rosa* sp.). Int J Mol Sci 21(4):1360. https://doi.org/10.3390/ijms21041360

Jones B, Gunnerås SA, Petersson SV, Tarkowski P, Graham N, May S, Dolezal K, Sandberg G, Ljung K (2010) Cytokinin regulation of auxin synthesis in *Arabidopsis* involves a homeostatic feedback loop regulated via auxin and cytokinin signal transduction. Plant Cell 22(9):2956–2969. https://doi.org/10.1105/tpc.110.074856

Kagale S, Rozwadowski K (2011) EAR motif-mediated transcriptional repression in plants. Epigenetics 6(2):141–146. https://doi.org/10.4161/epi.6.2.13627

Khan AA, Lee AJ, Roh T-Y (2015) Polycomb group protein-mediated histone modifications during cell differentiation. Epigenomics 7(1):75–84. https://doi.org/10.2217/epi.14.61

Kieber JJ, Schaller GE (2018) Cytokinin signaling in plant development. Development 145(4). https://doi.org/10.1242/dev.149344

Kim JY, Park Y-J, Lee J-H, Kim ZH, Park C-M (2021) EIN3-mediated ethylene signaling attenuates auxin response during hypocotyl thermomorphogenesis. Plant Cell Physiol 62(4):708–720. https://doi.org/10.1093/pcp/pcab028

Kornet N, Scheres B (2009) Members of the GCN5 histone acetyltransferase complex regulate PLETHORA-mediated root stem cell niche maintenance and transit amplifying cell proliferation in Arabidopsis. Plant Cell 21(4):1070–1079

Kramer EM, Bennett MJ (2006) Auxin transport: a field in flux. Trends Plant Sci 11(8):382–386. https://doi.org/10.1016/j.tplants.2006.06.002

Krogan NT, Long JA (2009) Why so repressed? Turning off transcription during plant growth and development. Curr Opin Plant Biol 12(5):628–636. https://doi.org/10.1016/j.pbi.2009.07.011

Kuhn A, Ramans Harborough S, McLaughlin HM, Natarajan B, Verstraeten I, Friml J, Kepinski S, Østergaard L (2020) Direct ETTIN-auxin interaction controls chromatin states in gynoecium development. eLife 9. https://doi.org/10.7554/elife.51787

Kurepa J, Shull TE, Smalle JA (2019) Antagonistic activity of auxin and cytokinin in shoot and root organs. Plant Direct 3(2):e00121. https://doi.org/10.1002/pld3.121

Lafos M, Kroll P, Hohenstatt ML, Thorpe FL, Clarenz O, Schubert D (2011) Dynamic regulation of H3K27 trimethylation during *Arabidopsis* differentiation. PLoS Genet 7(4):e1002040. https://doi.org/10.1371/journal.pgen.1002040

Laplaze L, Benkova E, Casimiro I, Maes L, Vanneste S, Swarup R, Weijers D, Calvo V, Parizot B, Herrera-Rodriguez MBA, Offringa R, Graham N, Doumas P, Friml J, Bogusz D, Beeckman T, Bennett M (2008) Cytokinins act directly on lateral root founder cells to inhibit root initiation. Plant Cell 19(12):3889–3900. https://doi.org/10.1105/tpc.107.055863

Lau S, Jürgens G, De Smet I (2008) The evolving complexity of the auxin pathway. Plant Cell 20(7):1738–1746. https://doi.org/10.1105/tpc.108.060418

Leyser O (2018) Auxin signaling. Plant Physiol 176(1):465–479. https://doi.org/10.1104/pp.17.00765

Li J, Yang Y, Chai M, Ren M, Yuan J, Yang W, Dong Y, Liu B, Jian Q, Wang S, Peng B, Yuan H, Fan H (2020) Gibberellins modulate local auxin biosynthesis and polar auxin transport by negatively affecting flavonoid biosynthesis in the root tips of rice. Plant Sci 298:110545. https://doi.org/10.1016/j.plantsci.2020.110545

Liu J, Moore S, Chen C, Lindsey K (2017) Crosstalk complexities between auxin, cytokinin, and ethylene in Arabidopsis root development: from experiments to systems modeling, and back again. Mol Plant 10(12):1480–1496. https://doi.org/10.1016/j.molp.2017.11.002

Liu Q, Chen T-T, Xiao D-W, Zhao S-M, Lin J-S, Wang T, Li Y-J, Hou B-K (2019) *OsIAGT1* is a glucosyltransferase gene involved in the glucose conjugation of auxins in rice. Rice 12(1). https://doi.org/10.1186/s12284-019-0357-z

Liu Y, Koornneef M, Soppe WJJ (2007) The absence of Histone H2B monoubiquitination in the *Arabidopsis hub1* (*rdo4*) mutant reveals a role for chromatin remodeling in seed dormancy. Plant Cell 19(2):433–444. https://doi.org/10.1105/tpc.106.049221

Ljung K (2013) Auxin metabolism and homeostasis during plant development. Development 140(5):943–950. https://doi.org/10.1242/dev.086363

Long JA, Ohno C, Smith ZR, Meyerowitz EM (2006) TOPLESS regulates apical embryonic fate in *Arabidopsis*. Science 312(5779):1520–1523. https://doi.org/10.1126/science.1123841

Loyola-Vargas VM, De-la-Peña C, Galaz-Ávalos R, Quiroz-Figueroa F (2008) Plant tissue culture molecular biomethods handbook. Springer, pp 875–904

Loyola-Vargas VM, Ochoa-Alejo N (2016) Somatic embryogenesis. an overview. In: Loyola-Vargas VM, Ochoa-Alejo N (eds) Somatic embryogenesis: fundamental aspects and applications. Springer International Publishing, Cham, pp 1–8

Lu F, Cui X, Zhang S, Jenuwein T, Cao X (2011) Arabidopsis REF6 is a histone H3 lysine 27 demethylase. Nat Genet 43(7):715–719. https://doi.org/10.1038/ng.854

Ludwig-Müller J (2011) Auxin conjugates: their role for plant development and in the evolution of land plants. J Exp Bot 62(6):1757–1773. https://doi.org/10.1093/jxb/erq412

Ludwig-Müller J, Auer S, Jülke S, Marschollek S (2017) Manipulation of auxin and cytokinin balance during the *Plasmodiophora brassicae–Arabidopsis thaliana* interaction. In: Dandekar T, Naseem M (eds) Auxins and Cytokinins in plant biology: methods and protocols. Springer, New York, pp 41–60

Lulsdorf MM, Yuan HY, Slater SMH, Vandenberg A, Han X, Zaharia LI, Abrams SR (2013) Endogenous hormone profiles during early seed development of *C. arietinum* and *C. anatolicum*. Plant Growth Regul 71(2):191–198. https://doi.org/10.1007/s10725-013-9819-2

Mallory AC, Bartel DP, Bartel B (2005) MicroRNA-directed regulation of Arabidopsis *AUXIN RESPONSE FACTOR17* is essential for proper development and modulates expression of early auxin response genes. Plant Cell 17(5):1360–1375. https://doi.org/10.1105/tpc.105.031716

Manzano C, Ramirez-Parra E, Casimiro I, Otero S, Desvoyes B, De Rybel B, Beeckman T, Casero P, Gutierrez C, del Pozo C, J, (2012) Auxin and epigenetic regulation of SKP2B, an F-Box that represses lateral root formation. Plant Physiol 160(2):749–762. https://doi.org/10.1104/pp.112.198341

Mateo-Bonmatí E, Casanova-Sáez R, Ljung K (2019) Epigenetic regulation of auxin homeostasis. Biomolecules 9(10):623

Mazzoni-Putman SM, Brumos J, Zhao C, Alonso JM, Stepanova AN (2021) Auxin interactions with other hormones in plant development. Cold Spring Harb Perspect Biol. https://doi.org/10.1101/cshperspect.a039990

Méndez-Hernández HA, Quintana-Escobar AO, Uc-Chuc MA, Loyola-Vargas VM (2021) Genome-wide analysis, modeling, and identification of amino acid binding motifs suggest the involvement of *GH3* genes during somatic embryogenesis of *Coffea canephora*. Plants 10(10):2034. https://doi.org/10.3390/plants10102034

Michniewicz M, Ho C-H, Enders TA, Floro E, Damodaran S, Gunther LK, Powers SK, Frick EM, Topp CN, Frommer WB, Strader LC (2019) TRANSPORTER OF IBA1 links auxin and cytokinin to influence root architecture. Dev Cell 50(5):599-609.e4. https://doi.org/10.1016/j.devcel.2019.06.010

Miyawaki K, Matsumoto-Kitano M, Kakimoto T (2004) Expression of cytokinin biosynthetic isopentenyltransferase genes in *Arabidopsis*: tissue specificity and regulation by auxin, cytokinin, and nitrate. Plant J 37(1):128–138. https://doi.org/10.1046/j.1365-313x.2003.01945.x

Mockaitis K, Estelle M (2008) Auxin receptors and plant development: a new signaling paradigm. Annu Rev Cell Dev Biol 24(1):55–80. https://doi.org/10.1146/annurev.cellbio.23.090506.123214

Mohammadi F, Naghavi MR, Peighambari SA, Khosravi Dehaghi N, Khaldari I, Bravi E, Marconi O, Perretti G (2021) Abscisic acid crosstalk with auxin and ethylene in biosynthesis and degradation of inulin-type fructans in chicory. Plant Biol 23(4):636–642. https://doi.org/10.1111/plb.13252

Moore S, Zhang X, Mudge A, Rowe JH, Topping JF, Liu J, Lindsey K (2015) Spatiotemporal modelling of hormonal crosstalk explains the level and patterning of hormones and gene expression in *Arabidopsis thaliana* wild-type and mutant roots. New Phytol 207(4):1110–1122. https://doi.org/10.1111/nph.13421

Moubayidin L, Di Mambro R, Sabatini S (2009) Cytokinin–auxin crosstalk. Trends Plant Sci 14(10):557–562. https://doi.org/10.1016/j.tplants.2009.06.010

Moubayidin L, Perilli S, Dello Ioio R, Di Mambro R, Costantino P, Sabatini S (2010) The rate of cell differentiation controls the Arabidopsis root meristem growth phase. Curr Biol 20(12):1138–1143. https://doi.org/10.1016/j.cub.2010.05.035

Müller B, Sheen J (2008) Cytokinin and auxin interaction in root stem-cell specification during early embryogenesis. Nature 453(7198):1094–1097. https://doi.org/10.1038/nature06943

Müller CJ, Larsson E, Spíchal L, Sundberg E (2017) Cytokinin-auxin crosstalk in the gynoecial primordium ensures correct domain patterning. Plant Physiol 175(3):1144–1157. https://doi.org/10.1104/pp.17.00805

Muraro D, Larrieu A, Lucas M, Chopard J, Byrne H, Godin C, King J (2016) A multi-scale model of the interplay between cell signalling and hormone transport in specifying the root meristem of *Arabidopsis thaliana*. J Theor Biol 404:182–205. https://doi.org/10.1016/j.jtbi.2016.04.036

Nguyen CT, Tran G-B, Nguyen NH (2020) Homeostasis of histone acetylation is critical for auxin signaling and root morphogenesis. Plant Mol Biol 103(1–2):1–7. https://doi.org/10.1007/s11103-020-00985-1

Nguyen H, Kim J, Jeong C, Hong S-W, Lee H (2013) Inhibition of histone deacetylation alters Arabidopsis root growth in response to auxin via PIN1 degradation. Plant Cell Rep 32(10):1625–1636. https://doi.org/10.1007/s00299-013-1474-6

Nguyen HN, Nguyen TQ, Kisiala AB, Emery RJN (2021) Beyond transport: cytokinin ribosides are translocated and active in regulating the development and environmental responses of plants. Planta 254(3). https://doi.org/10.1007/s00425-021-03693-2

Nordstrom A, Tarkowski P, Tarkowska D, Norbaek R, Astot C, Dolezal K, Sandberg G (2004) Auxin regulation of cytokinin biosynthesis in *Arabidopsis thaliana*: A factor of potential importance for auxin-cytokinin-regulated development. Proc Natl Acad Sci 101(21):8039–8044. https://doi.org/10.1073/pnas.0402504101

O'Brien J, Benkova E (2013) Cytokinin cross-talking during biotic and abiotic stress responses. Front Plant Sci 4. https://doi.org/10.3389/fpls.2013.00451

Ostrowski M, Ciarkowska A (2021) Pea GH3 acyl acid amidosynthetase conjugates IAA to proteins in immature seeds of *Pisum sativum* L.—A new perspective on formation of high-molecular weight conjugates of auxin. J Plant Physiol 256:153312. https://doi.org/10.1016/j.jplph.2020.153312

Overvoorde P, Fukaki H, Beeckman T (2010) Auxin control of root development. Cold Spring Harb Perspect Biol 2(6):a001537–a001537. https://doi.org/10.1101/cshperspect.a001537

Pacurar DI, Perrone I, Bellini C (2014) Auxin is a central player in the hormone cross-talks that control adventitious rooting. Physiol Plant 151(1):83–96. https://doi.org/10.1111/ppl.12171

Peng M, Li Z, Zhou N, Ma M, Jiang Y, Dong A, Shen W-H, Li L (2018) Linking PHYTOCHROME-INTERACTING FACTOR to histone modification in plant shade avoidance. Plant Physiol 176(2):1341–1351. https://doi.org/10.1104/pp.17.01189

Pernisova M, Klima P, Horak J, Valkova M, Malbeck J, Soucek P, Reichman P, Hoyerova K, Dubova J, Friml J, Za ZE, Hejatko J (2009) Cytokinins modulate auxin-induced organogenesis in plants via regulation of the auxin efflux. Proc Natl Acad Sci 106(9):3609–3614. https://doi.org/10.1073/pnas.0811539106

Pernisová M, Kuderová A, Hejátko J (2011) Cytokinin and auxin interactions in plant development: metabolism, signalling, transport and gene expression. Curr Protein Pept Sci 12(2):137–147

Petersson SV, Johansson AI, Kowalczyk M, Makoveychuk A, Wang JY, Moritz T, Grebe M, Benfey PN, Sandberg G, Ljung K (2009) An auxin gradient and maximum in the *Arabidopsis* root apex shown by high-resolution cell-specific analysis of IAA distribution and synthesis. Plant Cell 21(6):1659–1668. https://doi.org/10.1105/tpc.109.066480

Pierdonati E, Unterholzner SJ, Salvi E, Svolacchia N, Bertolotti G, Dello Ioio R, Sabatini S, Di Mambro R (2019) Cytokinin-dependent control of *GH3* group II family genes in the *Arabidopsis* root. Plants 8(4):94. https://doi.org/10.3390/plants8040094

Pikaard CS, Mittelsten Scheid O (2014) Epigenetic regulation in plants. Cold Spring Harb Perspect Biol 6(12):a019315–a019315. https://doi.org/10.1101/cshperspect.a019315

Piotrowska-Niczyporuk A, Bajguz A, Kotowska U, Zambrzycka-Szelewa E & Sienkiewicz A (2020) Auxins and cytokinins regulate phytohormone homeostasis and thiol-mediated detoxification in the green alga *Acutodesmus obliquus* exposed to lead stress. Sci Rep 10(1). https://doi.org/10.1038/s41598-020-67085-4

Pokorná E, Hluska T, Galuszka P, Hallmark HT, Dobrev PI, Záveská Drábková L, Filipi T, Holubová K, Plíhal O, Rashotte AM, Filepová R, Malbeck J, Novák O, Spíchal L, Brzobohatý B, Mazura P, Zahajská L, Motyka V (2020) Cytokinin N-glucosides: Occurrence, metabolism and biological activities in plants. Biomolecules 11(1):24. https://doi.org/10.3390/biom11010024

Prát T, Hajný J, Grunewald W, Vasileva M, Molnár G, Tejos R, Schmid M, Sauer M, Friml J (2018) WRKY23 is a component of the transcriptional network mediating auxin feedback on PIN polarity. PLoS Genet 14(1):e1007177. https://doi.org/10.1371/journal.pgen.1007177

Ravazzolo L, Boutet-Mercey S, Perreau F, Forestan C, Varotto S, Ruperti B, Quaggiotti S (2021) Strigolactones and auxin cooperate to regulate maize root development and response to nitrate. Plant Cell Physiol. https://doi.org/10.1093/pcp/pcab014

Rhaman MS, Imran S, Rauf F, Khatun M, Baskin CC, Murata Y, Hasanuzzaman M (2020) Seed priming with phytohormones: An effective approach for the mitigation of abiotic stress. Plants 10(1):37. https://doi.org/10.3390/plants10010037

Rizzardi K, Landberg K, Nilsson L, Ljung K, Sundås-Larsson A (2011) *TFL2/LHP1* is involved in auxin biosynthesis through positive regulation of *YUCCA* genes. Plant J 65(6):897–906. https://doi.org/10.1111/j.1365-313x.2010.04470.x

Romanov GA, Schmülling T (2022) On the biological activity of cytokinin free bases and their ribosides. Planta 255(1). https://doi.org/10.1007/s00425-021-03810-1

Ruzicka K, Simaskova M, Duclercq J, Petrasek J, Zazimalova E, Simon S, Friml J, Van Montagu MCE, Benkova E (2009) Cytokinin regulates root meristem activity via modulation of the polar auxin transport. Proc Natl Acad Sci 106(11):4284–4289. https://doi.org/10.1073/pnas.0900060106

Saiga S, Möller B, Watanabe-Taneda A, Abe M, Weijers D, Komeda Y (2012) Control of embryonic meristem initiation in Arabidopsis by PHD-finger protein complexes. Development 139(8):1391–1398. https://doi.org/10.1242/dev.074492

Salazar-Iribe A, De-la-Peña C (2020) Auxins, the hidden player in chloroplast development. Plant Cell Rep 39(12):1595–1608. https://doi.org/10.1007/s00299-020-02596-y

Salopek-Sondi B, Šamec D, Mihaljević S, Smolko A, Pavlović I, Janković I, Ludwig-Müller J (2013) Influence of stress hormones on the auxin homeostasis in *Brassica rapa* seedlings. Plant Cell Rep 32(7):1031–1042. https://doi.org/10.1007/s00299-013-1412-7

Schaller GE, Bishopp A, Kieber JJ (2015) The yin-yang of hormones: cytokinin and auxin interactions in plant development. Plant Cell 27(1):44–63. https://doi.org/10.1105/tpc.114.133595

Shimizu-Sato S, Tanaka M, Mori H (2009) Auxin–cytokinin interactions in the control of shoot branching. Plant Mol Biol 69(4):429–435. https://doi.org/10.1007/s11103-008-9416-3

Shiraz M, Sami F, Siddiqui H, Yusuf M, Hayat S (2020) Interaction of auxin and nitric oxide improved photosynthetic efficiency and antioxidant system of *Brassica juncea* plants under salt stress. J Plant Growth Regul. https://doi.org/10.1007/s00344-020-10268-0

Singh P, Sinha AK (2017) Interplay between auxin and cytokinin and its impact on mitogen activated protein kinase (MAPK). In: Dandekar T, Naseem M (eds) Auxins and cytokinin's in plant biology: methods and protocols. Springer, Springer, pp 93–100

Singh V, Patel R, Kumar S, Sahu MP, Ahirwal A (2021) Plant growth regulators and their use in plant growth and development. Agric Environ 2:28–28

Skalický V, Kubeš M, Napier R, Novák O (2018) Auxins and Cytokinins—the role of subcellular organization on homeostasis. Int J Mol Sci 19(10):3115

Skoog F, Miller C, Chemical regulation of growth and organ formation in plant tissue cultured. In, (1957) In vitro. Symp Soc Exp Biol 11:118–131

Sonkar S, Sharma L, Singh RK, Pandey B, Rathore SS, Singh AK, Porwal P, Singh SP (2021) Plant stress hormones nanobiotechnology. In: Al-Khayri JM, Israil Ansari M, Kumar Sing A (eds) Nanobiotechnology. Springer International Publishing, Switzerland, pp 349–373

Su Y-H, Liu Y-B, Zhang X-S (2011) Auxin–cytokinin interaction regulates meristem development. Mol Plant 4(4):616–625. https://doi.org/10.1093/mp/ssr007

Sun D, Zhang L, Yu Q, Zhang J, Li P, Zhang Y, Xing X, Ding L, Fang W, Chen F, Song A (2021) Integrated signals of jasmonates, sugars, cytokinins and auxin influence the initial growth of the second buds of chrysanthemum after decapitation. Biology 10(5):440. https://doi.org/10.3390/biology10050440

Swarup R, Bhosale R (2019) Developmental roles of AUX1/LAX Auxin influx carriers in plants. Front Plant Sci 10(1306). https://doi.org/10.3389/fpls.2019.01306

Szemenyei H, Hannon GJ, Long J (2008) TOPLESS mediates auxin-dependent transcriptional repression during *Arabidopsis* embryogenesis. Science 319(5868):1384–1386. https://doi.org/10.1126/science.1151461

Tanaka M, Takei K, Kojima M, Sakakibara H, Mori H (2006) Auxin controls local cytokinin biosynthesis in the nodal stem in apical dominance. Plant J 45(6):1028–1036. https://doi.org/10.1111/j.1365-313x.2006.02656.x

Tang Y, Liu X, Liu X, Li Y, Wu K, Hou X (2017) *Arabidopsis* NF-YCs mediate the light-controlled hypocotyl elongation via modulating histone acetylation. Mol Plant 10(2):260–273. https://doi.org/10.1016/j.molp.2016.11.007

Tessi TM, Brumm S, Winklbauer E, Schumacher B, Pettinari G, Lescano I, González CA, Wanke D, Maurino VG, Harter K, Desimone M (2021) Arabidopsis AZG2 transports cytokinins *in vivo* and regulates lateral root emergence. New Phytol 229(2):979–993. https://doi.org/10.1111/nph.16943

Tian C-E, Muto H, Higuchi K, Matamura T, Tatematsu K, Koshiba T, Yamamoto KT (2004) Disruption and overexpression of auxin response factor 8 gene of Arabidopsis affect hypocotyl elongation and root growth habit, indicating its possible involvement in auxin homeostasis in light condition. Plant J 40(3):333–343. https://doi.org/10.1111/j.1365-313X.2004.02220.x

Tognetti VB, Bielach A, Hrtyan M (2017) Redox regulation at the site of primary growth: auxin, cytokinin and ROS crosstalk. Plant, Cell Environ 40(11):2586–2605. https://doi.org/10.1111/pce.13021

Tognetti VB, Mühlenbock P, Van Breusegem F (2012) Stress homeostasis—the redox and auxin perspective. Plant, Cell Environ 35(2):321–333. https://doi.org/10.1111/j.1365-3040.2011.02324.x

Us-Camas R, Rivera-Solís G, Duarte-Aké F, De-la-Peña C (2014) In vitro culture: an epigenetic challenge for plants. Plant Cell, Tissue Organ Cult 118(2):187–201. https://doi.org/10.1007/s11240-014-0482-8

Van Der Woude LC, Perrella G, Snoek BL, Van Hoogdalem M, Novák O, Van Verk MC, Van Kooten HN, Zorn LE, Tonckens R, Dongus JA, Praat M, Stouten EA, Proveniers MCG, Vellutini E, Patitaki E, Shapulatov U, Kohlen W, Balasubramanian S, Ljung K, Van Der Krol AR, Smeekens S, Kaiserli E, Van Zanten M (2019) HISTONE DEACETYLASE 9 stimulates auxin-dependent thermomorphogenesis in *Arabidopsis thaliana* by mediating H2A.Z depletion. Proceed Natl Acad Sci 116(50):25343–25354. https://doi.org/10.1073/pnas.1911694116

Vanstraelen M, Benková E (2012) Hormonal interactions in the regulation of plant development. Annu Rev Cell Dev Biol 28(1):463–487. https://doi.org/10.1146/annurev-cellbio-101011-155741

Waldie T, Leyser O (2018) Cytokinin targets auxin transport to promote shoot branching. Plant Physiol 177(2):803–818. https://doi.org/10.1104/pp.17.01691

Wang M, Su L, Cong Y, Chen J, Geng Y, Qian C, Xu Q, Chen X, Qi X (2021a) Sugars enhance parthenocarpic fruit formation in cucumber by promoting auxin and cytokinin signaling. Sci Hortic 283:110061. https://doi.org/10.1016/j.scienta.2021.110061

Wang P, Jiang H, Boeren S, Dings H, Kulikova O, Bisseling T, Limpens E (2021b) A nuclear-targeted effector of *Rhizophagus* irregularis interferes with histone 2B mono-ubiquitination to promote arbuscular mycorrhization. New Phytol 230(3):1142–1155. https://doi.org/10.1111/nph.17236

Wang Z, Chen F, Li X, Cao H, Ding M, Zhang C, Zuo J, Xu C, Xu J, Deng X, Xiang Y, Soppe WJJ, Liu Y (2016) *Arabidopsis* seed germination speed is controlled by SNL histone deacetylase-binding factor-mediated regulation of *AUX1*. Nat Commun 7(1):13412. https://doi.org/10.1038/ncomms13412

Weiste C & Dröge-Laser W (2014) The *Arabidopsis* transcription factor bZIP11 activates auxin-mediated transcription by recruiting the histone acetylation machinery. Nat Commun 5(1). https://doi.org/10.1038/ncomms4883

Werner T, Köllmer I, Bartrina I, Holst K, Schmülling T (2006) New insights into the biology of cytokinin degradation. Plant Biol 8(03):371–381

Westfall CS, Sherp AM, Zubieta C, Alvarez S, Schraft E, Marcellin R, Ramirez L, Jez JM (2016) *Arabidopsis thaliana* GH3.5 acyl acid amido synthetase mediates metabolic crosstalk in auxin and salicylic acid homeostasis. Proceed Natl Acad Sci 113(48):13917–13922. https://doi.org/10.1073/pnas.1612635113

Wojtaczka P, Ciarkowska A, Starzynska E, Ostrowski M (2022) The GH3 amidosynthetases family and their role in metabolic crosstalk modulation of plant signaling compounds. Phytochemistry 194:113039. https://doi.org/10.1016/j.phytochem.2021.113039

Wolff P, Weinhofer I, Seguin J, Roszak P, Beisel C, Donoghue MTA, Spillane C, Nordborg M, Rehmsmeier M, Köhler C (2011) High-resolution analysis of parent-of-origin allelic expression in the Arabidopsis endosperm. PLoS Genet 7(6):e1002126. https://doi.org/10.1371/journal.pgen.1002126

Woodward AW, Bartel B (2005) A receptor for Auxin. Plant Cell 17(9):2425–2429. https://doi.org/10.1105/tpc.105.036236

Wu L, Wang J-L, Li X-F, Guo G-Q (2021) Cytokinin-controlled gradient distribution of auxin in Arabidopsis root tip. Int J Mol Sci 22(8):3874. https://doi.org/10.3390/ijms22083874

Wu M-F, Yamaguchi N, Xiao J, Bargmann B, Estelle M, Sang Y & Wagner D (2015) Auxin-regulated chromatin switch directs acquisition of flower primordium founder fate. eLife 4:e09269. https://doi.org/10.7554/eLife.09269

Xu L, Ménard R, Berr A, Fuchs J, Cognat V, Meyer D, Shen W-H (2009) The E2 ubiquitin-conjugating enzymes, AtUBC1 and AtUBC2, play redundant roles and are involved in activation of *FLC* expression and repression of flowering in *Arabidopsis thaliana*. Plant J 57(2):279–288, https://doi.org/10.1111/j.1365-313x.2008.03684.x

Xu Y, Prunet N, Gan ES, Wang Y, Stewart D, Wellmer F, Huang J, Yamaguchi N, Tatsumi Y, Kojima M, Kiba T, Sakakibara H, Jack TP, Meyerowitz EM, Ito T (2018) SUPERMAN regulates floral whorl boundaries through control of auxin biosynthesis. EMBO J 37(11):e97499. https://doi.org/10.15252/embj.201797499

Yamaguchi N, Huang J, Tatsumi Y, Abe M, Sugano SS, Kojima M, Takebayashi Y, Kiba T, Yokoyama R, Nishitani K, Sakakibara H, Ito T (2018) Chromatin-mediated feed-forward auxin biosynthesis in floral meristem determinacy. Nature Commun 9(1). https://doi.org/10.1038/s41467-018-07763-0

Yamamuro C, Zhu J-K, Yang Z (2016) Epigenetic modifications and plant hormone action. Mol Plant 9(1):57–70. https://doi.org/10.1016/j.molp.2015.10.008

Yan Z, Liu X, Ljung K, Li S, Zhao W, Yang F, Wang M, Tao Y (2017) Type B response regulators act as central integrators in transcriptional control of the auxin biosynthesis enzyme TAA1. Plant Physiol 175(3):1438–1454. https://doi.org/10.1104/pp.17.00878

Yang J, Zhou X, Wu S, Gu D, Zeng L, Yang Z (2021) Involvement of DNA methylation in regulating the accumulation of the aroma compound indole in tea (*Camellia sinensis*) leaves during postharvest processing. Food Res Int 142:110183. https://doi.org/10.1016/j.foodres.2021.110183

Yang S, Li C, Zhao L, Gao S, Lu J, Zhao M, Chen C-Y, Liu X, Luo M, Cui Y, Yang C, Wu K (2015) The Arabidopsis SWI2/SNF2 chromatin remodeling ATPase BRAHMA targets directly to *PINs* and is required for root stem cell niche maintenance. Plant Cell 27(6):1670–1680. https://doi.org/10.1105/tpc.15.00091

Yang ZB, Liu G, Liu J, Zhang B, Meng W, Müller B, Hayashi KI, Zhang X, Zhao Z, De Smet I, Ding Z (2017) Synergistic action of auxin and cytokinin mediates aluminum-induced root growth inhibition in Arabidopsis. EMBO Rep 18(7):1213–1230. https://doi.org/10.15252/embr.201643806

Yuan L, Chen X, Chen H, Wu K, Huang S (2019) Histone deacetylases HDA6 and HDA9 coordinately regulate valve cell elongation through affecting auxin signaling in Arabidopsis. Biochem Biophys Res Commun 508(3):695–700. https://doi.org/10.1016/j.bbrc.2018.11.082

Zhang L, Luo P, Bai J, Wu L, Di D-W, Liu H-Q, Li J-J, Liu Y-L, Khaskheli AJ, Zhao C-M & Guo G-Q (2021) Function of histone H2B monoubiquitination in transcriptional regulation of auxin biosynthesis in Arabidopsis. Commun Biol 4(1). https://doi.org/10.1038/s42003-021-01733-x

Zhang S, Wang S, Xu Y, Yu C, Shen C, Qian Q, Geisler M, Jiang DA, Qi Y (2015a) The auxin response factor, OsARF19, controls rice leaf angles through positively regulating *OsGH3-5* and *OsBRI1*. Plant, Cell Environ 38(4):638–654. https://doi.org/10.1111/pce.12397

Zhang Y, Li D, Zhang H, Hong Y, Huang L, Liu S, Li X, Ouyang Z, Song F (2015b) Tomato histone H2B monoubiquitination enzymes SlHUB1 and SlHUB2 contribute to disease resistance against *Botrytis cinerea* through modulating the balance between SA- and JA/ET-mediated signaling pathways. BMC Plant Biol 15(1). https://doi.org/10.1186/s12870-015-0614-2

Zhao X, Song J, Zeng Q, Ma Y, Fang H, Yang L, Deng B, Liu J, Fang J, Zuo L, Yue J (2021) Auxin and cytokinin mediated regulation involved in vitro organogenesis of papaya. J Plant Physiol 260:153405. https://doi.org/10.1016/j.jplph.2021.153405

Zhao Z, Andersen SU, Ljung K, Dolezal K, Miotk A, Schultheiss SJ, Lohmann JU (2010) Hormonal control of the shoot stem-cell niche. Nature 465(7301):1089–1092. https://doi.org/10.1038/nature09126

Zhou C, Zhang L, Duan J, Miki B, Wu K (2005) *HISTONE DEACETYLASE19* is involved in jasmonic acid and ethylene signaling of pathogen response in Arabidopsis. Plant Cell 17(4):1196–1204

Zhou J-J, Luo J (2018) The PIN-FORMED auxin efflux carriers in plants. Int J Mol Sci 19(9):2759. https://doi.org/10.3390/ijms19092759

Zhou X-Y, Song L, Xue H-W (2013) Brassinosteroids regulate the differential growth of Arabidopsis hypocotyls through auxin signaling components IAA19 and ARF7. Mol Plant 6(3):887–904. https://doi.org/10.1093/mp/sss123

Zhu YX (2010) The epigenetic involvement in plant hormone signaling. Chin Sci Bull 55(21):2198–2203

Zilberman D, Gehring M, Tran RK, Ballinger T, Henikoff S (2007) Genome-wide analysis of *Arabidopsis thaliana* DNA methylation uncovers an interdependence between methylation and transcription. Nat Genet 39(1):61–69

Zou B, Yang D-L, Shi Z, Dong H, Hua J (2014) Monoubiquitination of Histone 2B at the disease resistance gene locus regulates its expression and impacts immune responses in Arabidopsis. Plant Physiol 165(1):309–318. https://doi.org/10.1104/pp.113.227801

The Role of Auxin and Cytokinin Signaling Components in *de novo* Shoot Organogenesis

Tatjana Ćosić and Martin Raspor

Abstract De novo shoot organogenesis (DNSO) is a widely used procedure for obtaining regenerated plant shoots for biotechnological, industrial and conservation purposes. Particular morphogenic events leading to shoot regeneration are induced by plant hormones auxin and cytokinin that are exogenously supplemented to the regenerating explants through the nutrient media. It was shown, in *Arabidopsis* and in other plant species, that the early stages of DNSO are crucially dependent on auxin signaling and include the development of pluripotent primordia that are developmentally identical to lateral root primordia, whereas in the later stages, these primordia acquire the shoot identity and develop a shoot apical meristem (SAM) in a process that is governed mostly by cytokinin, but is also influenced by auxin signaling. In this chapter, we discuss the current state of knowledge about the particular components of auxin and cytokinin signaling that are specifically involved in the phytohormonal regulation of DNSO. Throughout DNSO, the auxin signaling multicellular domains are generated through the action of auxin influx and efflux carriers, whereas AUXIN RESPONSE FACTORs (ARFs), that are modulated by the Aux/IAA repressors, regulate gene expression during various stages of DNSO. Conversely, the cytokinin signals relevant for DNSO are perceived overwhelmingly through the receptor ARABIDOPSIS HISTIDINE KINASE4 (AHK4) and effected through the type-B ARABIDOPSIS RESPONSE REGULATORs (ARRs), whereas type-A ARRs act as attenuators of the cytokinin signals. Throughout DNSO, auxin and cytokinin signals enter mutual crosstalk, and crosstalk with other phytohormones, sugars and other developmental signals, for a complex and fine-tuned regulation of shoot regeneration.

T. Ćosić · M. Raspor (✉)
Department of Plant Physiology, Institute for Biological Research "Siniša Stanković" – National Institute of Republic of Serbia, University of Belgrade, Belgrade, Serbia
e-mail: martin@ibiss.bg.ac.rs

T. Ćosić
e-mail: tatjana@ibiss.bg.ac.rs

T. Aftab (ed.), *Auxins, Cytokinins and Gibberellins Signaling in Plants*,
Signaling and Communication in Plants,
https://doi.org/10.1007/978-3-031-05427-3_3

1 Introduction

Auxin and cytokinin are major phytohormones regulating plant growth and development. Because of their complementary roles, and sometimes even antagonistic effects on the growth and development of particular plant tissues, they have been dubbed "the yin and yang of phytohormones" (Schaller et al. 2015).

Skoog and Miller (1957) noted that the ratio of cytokinin to auxin in the plant tissue culture media determines the developmental fate of plant tissues, whereby high auxin and low cytokinin promote the differentiation of roots, whereas high exogenous cytokinin and low auxin stimulate the development of shoot tissue. Ever since this discovery, carefully optimized protocols, consisting of sequences of regeneration media, have been used to induce various morphogenic responses of plant tissues (Ikeuchi et al. 2019). The endogenous phytohormonal profiles of in vitro cultured plants, that already differ from their soil-grown counterparts due to an array of differences in environmental conditions and physiological status (Raspor et al. 2020), are further altered by the exogenously supplied auxin and cytokinin, resulting in different morphogenic responses such as proliferation of callus tissue (Ikeuchi et al. 2013), regeneration of adventitious shoots (Valvekens et al. 1988) or roots (Efroni et al. 2016; Xu 2018).

De novo shoot organogenesis (DNSO) is a process in which regeneration of adventitious shoots is induced from explants of plant tissues of various origin through the application of auxin and cytokinin to the nutrient media where the explants are incubated (Valvekens et al. 1988). Classical, two-step protocols rely on a sequence of two media, with the first, callus induction medium (CIM) containing high auxin and low cytokinin concentrations, whereas the second, shoot induction medium (SIM) contains high cytokinin and low auxin (Valvekens et al. 1988; Che et al. 2002; Cary et al. 2002). However, in certain cases, the regeneration of shoots does not require a pre-incubation step on CIM, thus, one-step shoot regeneration on SIM is possible (Ćosić et al. 2015; Alvarez et al. 2020; Lee et al. 2020).

Because DNSO is widely applied in genetic engineering, biotechnology, conservation and fundamental research, it is being extensively studied. The molecular mechanisms of DNSO have been thoroughly characterized and they were proven to critically rely on auxin and cytokinin signaling (Motte et al. 2014; Raspor et al. 2021). In this chapter, we provide an overview of the known roles for the individual elements of the auxin and cytokinin signaling machineries, in regulating DNSO. We mostly focus on knowledge obtained from DNSO in *Arabidopsis thaliana* as the most studied model for plant development; additionally, we present particular results obtained on other model species such as tomato or kohlrabi, when relevant data are available.

2 CIM: Ensuring Sufficient Endogenous Auxin

In the two-step shoot regeneration protocol, CIM is applied to induce the disorganized proliferation of callus tissue, a necessary step in indirect shoot organogenesis (Ikeuchi et al. 2013; Fehér 2019). Direct shoot organogenesis can be however induced without the proliferation of callus tissue (Kareem et al. 2016; Alvarez et al. 2020; Pasternak et al. 2020). In both direct and indirect shoot organogenesis, the pluripotency of plant tissues depends on their ability to respond to auxin by forming pluripotent primordia that are developmentally identical to the primordia of lateral roots (Atta et al. 2009; Sugimoto et al. 2010) and contain a population of stem cells, corresponding to the root stem cell niche (Muñoz et al. 2017; Rosspopoff et al. 2017). Upon transfer to SIM, cytokinin from the nutrient media induces the transdifferentiation of these pluripotent primordia into shoot primordia, that subsequently develop a shoot apical meristem (SAM) (Fig. 1).

Synthetic auxins used in regeneration media can exhibit phytotoxic effects (Peterson et al. 2016) and induce unfavorable somaclonal variations in regenerated

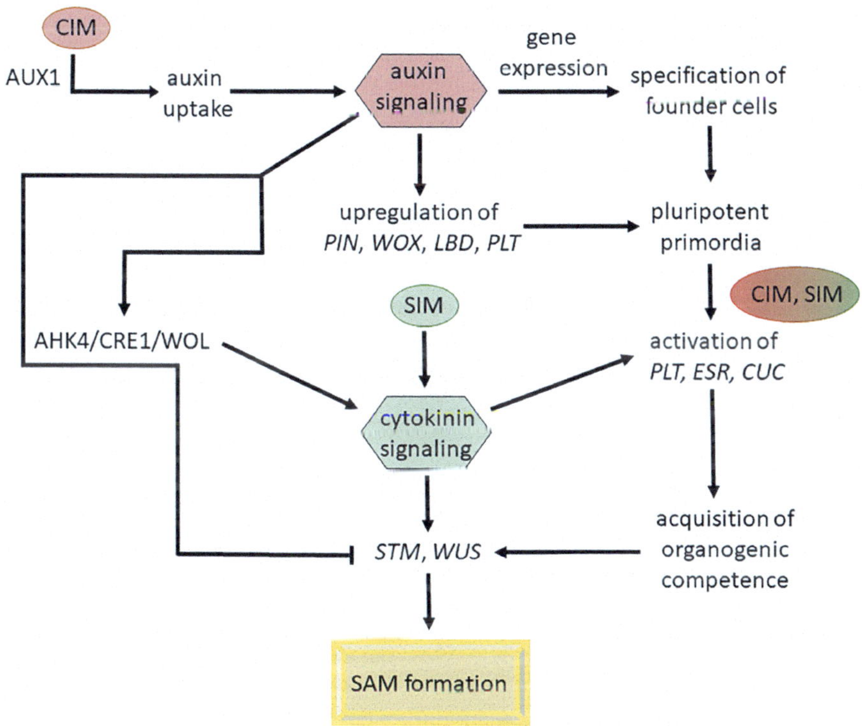

Fig. 1 Regulation of de novo shoot organogenesis (DNSO) by auxin and cytokinin supplied through the callus induction medium (CIM) and shoot induction medium (SIM). The particular steps of DNSO, eventually leading to the formation of the shoot apical meristem (SAM) are explained in the text

plants (Neelakandan and Wang 2012). Genetic analyses for clonal fidelity are advisable in order to validate the safety of particular regeneration protocols for the genetic integrity of each species (Stanišić et al. 2015; Thakur et al. 2016; Niazian et al. 2017; Tikendra et al. 2019; Das et al. 2020). To minimize the risk from possible adverse effects of high auxin concentrations from CIM on the regenerated plants, one-step shoot regeneration protocols are studied as an alternative to classical two-step regeneration (Ćosić et al. 2015; Kaur et al. 2017). Valvekens et al. (1988) noted that it is possible to regenerate shoots from *Arabidopsis* root explants without pre-incubation on CIM, with the limitation that shoots are regenerated only at the proximal end of the explant. A one-step shoot regeneration protocol was developed for DNSO of kohlrabi (*Brassica oleracea* var. *gongylodes*), whereby seedlings cultivated on different SIM formulae without preincubation on CIM, were used as explants to obtain callus formation and subsequent shoot regeneration (Ćosić et al. 2015). Histological and gene expression analyses indicated that in one-step shoot regeneration, organogenesis in kohlrabi followed the same sequence of morphogenic events as in the two-step regeneration of *Arabidopsis*, suggesting that one-step and two-step shoot regeneration differ in environmental requirements for initiation (the presence of auxin in the nutrient media), but not in the mechanism or dynamics of the process (Ćosić et al. 2015, 2019). Analyses of endogenous phytohormone content revealed that even in the absence of CIM, high endogenous auxin levels could be sufficient to induce the callus formation, indicating that endogenous phytohormone content is more relevant to the induction of morphogenic events, than the composition of nutrient media (Ćosić et al. 2015). Additionally, it was shown that even during SAM formation, a process which is critically dependent on cytokinin-rich SIM, regenerating shoots may contain high endogenous levels of auxin, which is in certain cases even critical for the efficiency of shoot regeneration (Kakani et al. 2009; Ćosić et al. 2015; Koike et al. 2017).

Thus, the role of CIM in DNSO is to ensure sufficient auxin supply to plant tissues in case their endogenous auxin is not sufficient to induce callus formation and the development of pluripotent primordia. Conversely, when endogenous auxin (or auxin-to-cytokinin ratio) in the regenerating explants is high, incubation on CIM is not necessary.

3 The Roles of Auxin Signaling Components in DNSO

From the specification of the founder cells that will give rise to the pluripotent primordia, to the establishment of the SAM—the entire process of DNSO critically depends on auxin signaling (Fig. 2).

The first phases of DNSO, that occur during the incubation of explants on CIM, are directly relying on auxin signaling. Auxin influx carriers such as AUXIN-RESISTANT1 (AUX1) are responsible of auxin uptake from CIM (Kakani et al. 2009) and its local accumulation in the xylem pole of the pericycle, where local auxin maxima induce the expression of genes leading to the specification of founder

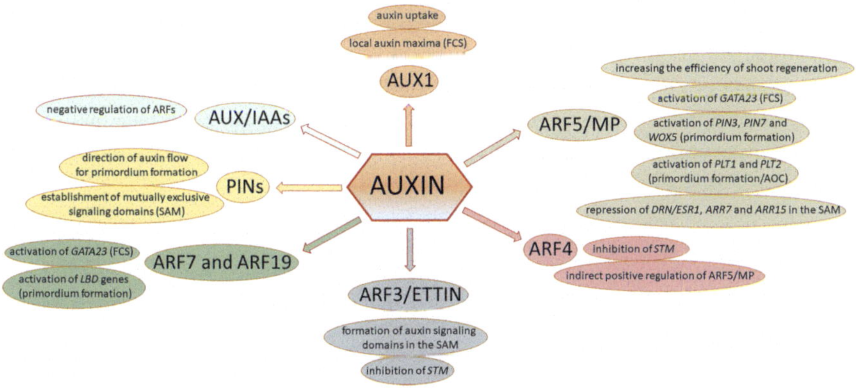

Fig. 2 The roles of auxin signaling components in de novo shoot organogenesis (DNSO), as determined in the model species *Arabidopsis thaliana*. The regulation of DNSO by auxin is dependent on auxin transporters (AUX1, PINs), auxin response factors (ARFs), and the negative regulators of auxin signaling (Aux/IAAs). Founder cell specification (FCS) and primordium initiation are identically regulated in DNSO, and in lateral root formation, as the pluripotent primordia in DNSO are developmentally identical to lateral root primordia. The full names of the genes are given in the text. AOC = acquisition of organogenic competence; SAM = shoot apical meristem

cells that will give rise to the pluripotent primordia (Sugimoto et al. 2010; De Rybel et al. 2010; Fig. 1).

Subsequently, the auxin signaling cascade induces the upregulation of *PIN-FORMED* (*PIN*) genes, encoding auxin efflux carriers, as well as an array of transcription factors belonging to the families *WUSCHEL-RELATED HOMEOBOX* (*WOX*), *LATERAL ORGAN BOUNDARIES DOMAIN* (*LBD*), and *PLETHORA* (*PLT*), which regulate the development of pluripotent primordia (Benková et al. 2003; Sugimoto et al. 2010; Fan et al. 2012; Kareem et al. 2015).

Next, the pluripotent primordia acquire the organogenic competence for transdifferentiation into shoot primordia. The acquisition of organogenic competence is mediated by the gene families *PLT*, *ENHANCER OF SHOOT REGENERATION* (*ESR*) and *CUP-SHAPED COTYLEDON* (*CUC*). This process is initiated on CIM, but subsequent incubation on SIM is necessary, as these genes are regulated by both auxin and cytokinin signaling cascades (Banno et al. 2001; Daimon et al. 2003; Kareem et al. 2015; Rosspopoff et al. 2017; Luo et al. 2018).

In the last phase of DNSO, the SAM formation is regulated by the morphogenic genes *SHOOTMERISTEMLESS* (*STM*) and *WUSCHEL* (*WUS*), that are cytokinin-induced; however, both the morphogenic effects of STM, and the expression of *WUS* depend on cytokinin signaling through the receptor *ARABIDOPSIS HISTIDINE KINASE4/CYTOKININ RESPONSE1/WOODEN LEG* (*AHK4/CRE1/WOL*), which is previously induced by the auxin signaling cascade during the incubation on CIM (Jasinski et al. 2005; Gordon et al. 2009; Adibi et al. 2016). Additionally, *WUS* is indirectly regulated by the auxin signaling cascade (Zhao et al. 2010; Luo et al. 2018) and *STM* can also be subject to negative regulation by auxin (Chung et al. 2019).

Conversely, WUS was shown to repress auxin signaling through the tissue-specific histone deacetylation of loci containing the *ARF* genes, within the central zone of the SAM (Ma et al. 2019). Thus, even though the effects of auxin are traditionally associated with the first phases of DNSO during CIM incubation, and the incubation on SIM is interpreted in relation to cytokinin signaling, the auxin signaling cascade actually regulates all the phases of DNSO (Fig. 1).

3.1 Setting the Stage for Auxin Signaling: Auxin Transporters

The auxin transporters are important for DNSO in two distinct ways: they mediate both the uptake of exogenous auxin into plant tissues, and they regulate the morphogenic course of DNSO by modulating local endogenous auxin concentrations (Raspor et al. 2021). For the explants incubated on CIM, the uptake of auxin and its transport into plant tissues represents the necessary prerequisite for the hormonal regulation of DNSO. Auxin uptake and transport into plant cells is likely to primarily rely on the auxin influx carrier AUXIN-RESISTANT1 (AUX1) (Fig. 2), as the *Arabidopsis aux1* mutants are unable to form calli on regular CIM; however, additional influx carriers and/or passive diffusion likely also account for auxin uptake from the media to a lesser extent, as callus formation and successful shoot regeneration are possible even for *aux1* mutants when CIM is supplied with higher auxin concentrations (Kakani et al. 2009).

The redistribution of auxin between the individual cells of the explant tissues critically affects the morphogenic course of DNSO. The morphogenic effects of auxin are shaped by auxin gradients between cells, and the initiation of an array of morphogenic events depends on the establishment of local multicellular domains containing lower or higher auxin levels compared to the surrounding tissue—local auxin minima and maxima (Adamowski and Friml 2015). These local auxin minima and maxima are established through temporal and spatial changes in expression of the genes coding for polar auxin transporters of the *PIN-FORMED* (*PIN*) family. The PIN transporters are auxin efflux carriers, whose polar pattern of distribution on the plasma membrane is regulated on the transcriptional and post-translational level, as well as through endocytic trafficking; thus, differential expression of *PIN* genes between neighboring cells and the redistribution of PIN proteins on the plasma membrane, can affect the polar flow of auxin into the tissue or away from it, enabling the establishment of a local auxin maximum or minimum (Zhou and Luo 2018). The establishment of local auxin maxima or minima was proven to be crucial at several points in various phases of DNSO, thus making the precise temporal and spatial patterns of the expression of *PIN* genes critical for the correct course of DNSO.

Such local auxin maxima, dependent on the expression of the auxin influx carrier gene *AUX1*, are necessary for the initiation of the entire process of DNSO, through the specification of founder cells that will subsequently give rise to a pluripotent primordium (Fig. 2). During the formation of lateral root primordia, which are developmentally identical to the pluripotent primordia in the process of DNSO, *AUX1* is expressed in the xylem pole of the pericycle before the first periclinal division of the pre-founder cells (Marchant et al. 2002), enabling the formation of local auxin maxima that are necessary for founder cell specification, a process dependent on intense auxin signals (De Rybel et al. 2010). The dependence of founder cell specification on local auxin maxima provides a regulatory mechanism of lateral inhibition, whereby established auxin maxima function as sinks that deplete auxin from the surrounding groups of cells, resulting in a discrete number of distinct foci of founder cell specification (Laskowski et al. 2008). At this early stage, PIN-mediated auxin efflux needs to be downregulated because it would impair the maintenance of auxin maxima (Motte et al. 2014). Subsequent upregulation of *PIN3* and *PIN7* (Marhavý et al. 2013) and redistribution of PIN1 to enable acropetal movement of auxin towards the tip of the growing primordium (Benková et al. 2003), will take place during primordium formation.

Another point in the process of DNSO that is critically relying on the establishment of PIN-dependent auxin maxima, is the formation of *CUP-SHAPED COTYLEDON2* (*CUC2*)-expressing domains which will give rise to the shoot progenitors during the acquisition of shoot identity. When the expression of the SAM-related gene *WUS* is initiated by cytokinin signaling in the shoot promeristem, groups of cells within local, PIN1-dependent auxin maxima, continue expressing *CUC2*, but not *WUS*. Thus, PIN1 plays an important role in local microaccumulation of auxin that leads to the assignment of shoot progenitor identity to the *CUC2*-expressing cells within larger *WUS*-expressing domains (Gordon et al. 2007; Motte et al. 2014). The expression of *PIN1* within the shoot promeristem is dependent on cytokinin from the SIM, but requires cytokinin perception through AHK4/CRE1/WOL, which is previously induced during the incubation on CIM (Gordon et al. 2009). In a later phase, auxin efflux through PIN1 is crucial for the establishment of localized *WUS*-expressing domains (Cheng et al. 2013). The expression of *PIN1*, and the correct establishment of *CUC2*- and *WUS*-expressing domains are functionally required for shoot regeneration, since shoot formation is severely impaired in *pin1* mutants (Gordon et al. 2007; Cheng et al. 2013).

3.2 IAA28, IAA14, ARF7 and ARF19: Regulators of Early Stages of DNSO

Transcription of auxin-responsive genes is facilitated by a class of transcription factors, called AUXIN RESPONSE FACTORs (ARFs). In the absence of auxin, canonical ARFs are subject to regulatory repression by the Auxin/INDOLE-3-ACETIC ACID (Aux/IAA) proteins, the negative regulators of auxin signaling (Fig. 2). Conversely, in the presence of auxin, ARFs are released from the repressive interaction with Aux/IAAs and they become transcriptionally active, whereas

Aux/IAAs are targeted for ubiquitin-mediated degradation (Roosjen et al. 2018). Not all ARFs interact with all Aux/IAA repressors; rather, they are organized into regulatory modules that rely on specific interactions between particular members of the ARF and Aux/IAA families, with distinct roles in plant development (Weijers et al. 2005).

At least two distinct auxin response modules were shown to participate in the regulation of lateral root primordia (LRP) formation in *Arabidopsis*: one consists of the Aux/IAA repressor IAA14/SOLITARY ROOT and the partially redundant auxin response factors ARF7 and ARF19 (Okushima et al. 2005), whereas the other one consists of the repressor IAA12/BODENLOS and its target ARF5/MONOPTEROS (De Smet et al. 2010; De Rybel et al. 2010). Additionally, both modules are affected by the repressive influence of IAA28 during the specification of founder cells of LRP (De Rybel et al. 2010). Although initially described using lateral root formation as model system, the identical regulatory interactions occur during founder cell specification and the formation of pluripotent primordia in DNSO, as has been reported later (Fan et al. 2012).

The earliest auxin-dependent signaling events during founder cell specification involve the induction of *GATA23* in the xylem pole of the pericycle, before the first periclinal division that marks the onset of the primordium formation (De Rybel et al. 2010). GATA23 belongs to the family of GATA motif-binding transcription factors, that are regulators of cell fate specification in multicellular eukaryotes (Muro-Pastor et al. 1999). Expression of *GATA23* is initiated by auxin primarily through the activity of ARF7 and ARF19, as *arf7arf19* double mutants are almost completely deficient in its induction by auxin. However, ARF5, ARF6 and ARF8 are also involved in the induction of *GATA23*. The induction of *GATA23* by ARFs is negatively regulated by IAA28 (De Rybel et al. 2010).

Beside *GATA23*, other important regulatory targets of ARF7 and ARF19 include members of the gene family *LATERAL ORGAN BOUNDARIES DOMAIN* (*LBD*), which are involved in the activation of a broad array of developmental responses (including cell cycle activation, cell wall remodeling, lipid and ROS metabolism, epigenetic regulation, etc.) that lead to the development of LRP (Okushima et al. 2007; Lee et al. 2009; Berckmans et al. 2011). The identical, partially redundant activation of *LBD16*, *LBD17*, *LBD18*, and *LBD29* by ARF7 and ARF19 was shown to underlie the formation of pluripotent primordia in the organogenic calli of *Arabidopsis* during DNSO; moreover, transgenic overexpression of each of these four *LBD* genes could induce the formation of organogenic calli on nutrient media without auxin. These findings are commonly regarded as a definite proof that the pluripotent primordia that are formed in DNSO, are developmentally identical to the LRP (Fan et al. 2012).

The roles of *ARF7* and *ARF19* in the regulation of *GATA23* and the *LBD* genes during pluripotent primordium formation (Fig. 2) were found to be partially redundant: single *arf7* and *arf19* mutants failed to display the severe symptoms of auxin dysfunction that were observed in the double *arf7arf19* mutants (Okushima et al. 2005; De Rybel et al. 2010). The *Arabidopsis* genes *AtARF7* and *AtARF19* are closely related paralogs and originate from a single recent gene duplication event.

Their next closest paralogs are *AtARF5*, *AtARF6* and *AtARF8* (Okushima et al. 2005). Together, these five ARFs are known as "canonical", or "class A ARFs", as opposed to the "class B" and "class C" ARFs (Roosjen et al. 2018).

The repressor IAA14 regulates the activity of ARF7 and ARF19 both in lateral root (Fukaki et al. 2005) and pluripotent primordium formation in DNSO (Fan et al. 2012), given the developmental equivalency of these processes. The gene *IAA14* was identified as the target of a dominant gain-of-function mutation *solitary root* (*slr*), which rendered the mutants completely unable to initiate lateral root primordia, although their unbranched primary root was otherwise normal (Fukaki et al. 2002). A similar, barely milder phenotype was observed in double loss-of-function *arf7arf19* mutants, confirming the role of IAA14 in regulating ARF7 and ARF19 in primordium initiation (Okushima et al. 2005). The IAA14-dependent repression of ARF7 and ARF19 was shown to rely on the activity of the chromatin remodeling factor PICKLE (PKL) (Fukaki et al. 2006). Additionally, beside the binding of auxin to the repressor IAA14 which is a mechanism typical of the auxin signaling cascade, an additional layer of regulation by auxin is provided through posttranscriptional modification of *IAA14*, *ARF7* and *ARF19* transcripts, involving the activation of alternative polyadenylation sites, that affect transcript stability (Hong et al. 2018). Taken together, these regulatory mechanisms are relevant to the auxin-induced process of pluripotent primordium formation during DNSO, through the actions of the regulatory module IAA14-ARF7-ARF19. Additionally, similarly to *IAA14*, a semi-dominant gain-of-function mutation in *IAA28* was also deficient in lateral root development, suggesting the involvement of similar mechanisms (Rogg et al. 2001).

Beside IAA14 and IAA28, other Aux/IAA repressors were suggested to regulate the ARF7-ARF19 signaling module in the early phases of DNSO. Recently, IAA15 was shown to regulate ARF7 and ARF19 during primordium formation in a similar manner as IAA14 does, including the repression of transcriptional activation of *LBD16* and *LBD29* by ARF7 and ARF19 (Kim et al. 2020). Additionally, the auxin response repressor SHORT HYPOCOTYL2 (SHY2/IAA3), which represents a point of auxin-cytokinin crosstalk because it is also regulated by cytokinin-responsive ARABIDOPSIS RESPONSE REGULATOR1 (ARR1) and ARR12 (Moubayidin et al. 2010), has been suggested as a possible repressor of ARF7 and ARF19. However, its interference with ARF7 and ARF19 in primordium formation is unlikely or of low importance, because the lateral root phenotype of *shy2/iaa3* mutants is different from that of the double *arf7arf19* mutants (Goh et al. 2012).

3.3 *IAA12/BODENLOS and ARF5/MONOPTEROS: The Regulators Involved in All Stages of DNSO*

Differently from the regulatory module IAA14-ARF7-ARF19 whose effects are limited to the early phases of DNSO including founder cell specification and primordium formation, the roles for AUXIN RESPONSE

FACTOR5/MONOPTEROS (ARF5/MP) and its repressor IAA12/BODENLOS (BDL) have been confirmed throughout all the stages of DNSO, from founder cell specification to the development of the SAM (Fig. 2). Accordingly, the genotype *MPΔ*, containing a dominant gain-of-function deletion of the ARF5/MP protein domain responsible for negative regulation by IAA12/BDL (Krogan et al. 2012), was shown to be highly efficient in shoot regeneration (Ckurshumova et al. 2014), and the transgenic expression of *MPΔ* was even proposed as a novel strategy for enhancing the efficiency of shoot regeneration in recalcitrant plant species (Ckurshumova and Berleth 2015).

Mechanistic research and mathematical modeling revealed the nature of the regulatory action of the IAA12-ARF5 regulatory module (Lau et al. 2011; Farcot et al. 2015). When present above a certain concentration threshold, depending on the local auxin concentration, ARF5/MP was shown to generate a feedforward signal through the induction of its own transcription; this signal is modulated by the simultaneous transcriptional activation of *IAA12/BDL*. IAA12/BDL acts as a repressor of the ARF5/MP protein; however, as long as auxin is present above a threshold level, the concentration of IAA12/BDL is subject to a dynamic balance between ARF5/MP-stimulated gene transcription and auxin-induced proteolysis. Such dynamic regulation of auxin signaling through ARF5/MP ensures the maintenance of the auxin signals required for auxin-dependent developmental fate decisions (Lau et al. 2011).

The earliest involvement of ARF5/MP in the regulation of DNSO occurs during founder cell specification, whereby ARF5/MP participates in the induction of *GATA23* expression along with ARF6, ARF7, ARF8 and ARF19; this process is negatively regulated by the repressor IAA28. It is known that the activity of ARF7-ARF19 is essential for the auxin-induced expression of *GATA23* (De Rybel et al. 2010). However, it is not sure whether the activity of ARF5/MP is also essential independently of ARF7-ARF19, or the role of ARF5/MP in this process is minor—since the behavior of *arf5/mp* mutants in the induction of *GATA23* was not reported.

The next point of involvement of ARF5/MP in DNSO is at primordium initiation, whereby the IAA12-ARF5 regulatory module is switched on later than IAA14-ARF7-ARF19, as suggested by the development of abnormal primordia in the gain-of-function *iaa12/bdl* and loss-of-function *arf5/mp* mutants (De Smet et al. 2010). However, the specific effects of the IAA12-ARF5 module at this stage of primordium formation are unclear as of today, as no reports of the underlying mechanism have been made over the last decade (Vangheluwe and Beeckman 2021). A possible role for ARF5/MP in primordium initiation might consist in the transcriptional activation of the auxin efflux carrier genes *PIN3* and *PIN7*, since these genes are considered important markers of primordium initiation (Marhavý et al. 2013), and ARF5/MP has been shown to activate their transcription during lateral root initiation (Krogan et al. 2016).

Another possible regulatory target for the IAA12-ARF5 module during primordium initiation is *WUSCHEL-RELATED HOMEOBOX5* (*WOX5*), as such regulation is in place in the root apical meristem (RAM) stem cell niche (Sarkar et al. 2007; García-Gómez et al. 2017). WOX5 is a transcription factor crucial for the establishment and maintenance of the stem cell niche in the pluripotent primordia during DNSO (Sugimoto et al. 2010). It is upregulated upon callus induction (Kim et al. 2018) and remains expressed in the stem cell niche of the pluripotent primordia, determining their lateral root-like identity (Sugimoto et al. 2010; Rosspopoff et al. 2017). Upon cytokinin treatment, the auxin-dependent *WOX5* expression domain in the stem cell niche is replaced by cytokinin-dependent *WUS* expression, changing the identity of the stem cell niche into shoot-like (Rosspopoff et al. 2017). Taken together, the maintenance of the stem cell niche of the pluripotent primordia, regulated by auxin until their transdifferentiation into shoot primordia, likely relies on signaling through ARF5/MP.

The involvement of ARF5/MP at numerous points throughout the later stages of DNSO confirms its essential importance for mediating the developmental regulation of this morphogenic process by auxin (Fig. 2). Gene regulatory network analyses have revealed complicated regulatory interactions throughout the course of DNSO, whereby ARF5/MP both regulates the expression of transcription factors involved in morphogenic processes, and undergoes feedforward or feedback regulation by some of them. For instance, the transcription factor LBD29, involved in callus/primordium formation during DNSO, is a transcriptional regulator of *ARF5/MP* (Ikeuchi et al. 2018). Interestingly, *LBD29* is itself a regulatory target of ARF7 and ARF19 (Fan et al. 2012).

Furthermore, the PLETHORA (PLT) transcription factors, which play a central role within the gene regulatory network underlying primordium formation in DNSO, are auxin-inducible and depend on auxin signaling through ARF5/MP (Horstman et al. 2014; Kareem et al. 2015). In turn, PLT2 provides a regulatory loop through the upregulation of *ARF5/MP*, the auxin efflux carrier gene *PIN4*, and the auxin biosynthetic gene *YUCCA3* in the quiescent center of the LRP, suggesting similar interactions in the pluripotent primordium during DNSO (Santuari et al. 2016).

A major point in the auxin-mediated regulation of SAM formation is the transcriptional repression of one of the main regulators of the acquisition of organogenic competence and shoot identity, *DORNRÖSCHEN/ENHANCER OF SHOOT REGENERATION1* (*DRN/ESR1*) by ARF5/MP within the SAM stem cell niche (Luo et al. 2018; Fig. 2). Since DRN/ESR1 is an important positive regulator of *CLAVATA3* (*CLV3*) which restricts the expression of the shoot identity regulator *WUSCHEL* (*WUS*), ARF5/MP thus indirectly acts as a negative regulator of *CLV3* and a positive regulator of *WUS*. A role for IAA12/BDL as a universal repressor of ARF5/MP was confirmed in this process, as transgenic plants bearing a dominant gain-of-function *iaa12/bdl* mutation showed substantial upregulation of *CLV3* and repression of *WUS* within the SAM (Luo et al. 2018). Taken together, the role of ARF5/MP as a positive regulator of *WUS* is in accordance with the reports on its enhancing effect on the efficiency of shoot regeneration (Ckurshumova et al. 2014; Zhang et al. 2021), and corroborates its importance as one of the central regulators of DNSO.

3.4 *ARF3 and ARF4: The Two Class B ARFs that Regulate the Development of the SAM*

As opposed to ARF5, ARF6, ARF7, ARF8, and ARF19 that unequivocally act as transcriptional activators, the remaining *Arabidopsis* ARF proteins do not activate gene transcription from canonical auxin promoters in heterologous expression systems and hence were historically hypothesized to act as transcriptional repressors (Guilfoyle and Hagen 2007). However, the categorization of these "class B" and "C" ARFs as transcriptional repressors is currently challenged, as they have been shown to mediate an array of auxin-dependent responses in plants (Roosjen et al. 2018). Furthermore, with the exception of ARF4, the class B and C ARFs do not form important protein–protein interactions with the Aux/IAA repressors (Piya et al. 2014). Thus, the class B and class C ARFs have properties that distinguish them from the "canonical" class A ARF7, ARF19, and ARF5/MP that regulate the early phases of DNSO. Two class B ARFs—ARF3/ETTIN and ARF4, have been reported to participate in the regulation of SAM formation and maintenance, in the late stages of DNSO (Fig. 2).

Recently, a role for ARF4 has been shown in enhancing the efficiency of shoot regeneration (Zhang et al. 2021). Instead of competing with ARF5/MP for binding to its target promoters and decreasing ARF5/MP-dependent gene expression as was previously suggested to be the mode of action of the "repressor ARFs" (Guilfoyle and Hagen 2007), ARF4 was shown to actually saturate the ARF-binding sites on the regulatory protein IAA12/BDL, decreasing its repressive effect on ARF5/MP (Zhang et al. 2021). The promotive effect of ARF4 on shoot organogenesis was dependent on the function of ARF5/MP: although the overexpression of *ARF4* was able to rescue the shoot organogenesis efficiency in *IAA12/BDL*-overexpressing *Arabidopsis*, this effect was missing in the *ARF5/MP* weak allele *mp-S319* background (Zhang et al. 2021). Furthermore, both *ARF3/ETTIN* and *ARF4* were found to be upregulated, along with *WUS*, in DNA-methylation-deficient *met1 Arabidopsis* mutants with enhanced efficiency of shoot regeneration, indicating that the methylation of specific chromatin regions may affect the efficiency of cellular reprogramming events underlying successful shoot regeneration (Li et al. 2011).

Additionally, ARF3/ETTIN was also shown to mediate, together with PIN1, the formation of mutually exclusive multicellular domains influenced by auxin or cytokinin signaling during SAM formation. ARF3/ETTIN inhibited *ISOPENTENYL TRANSFERASE5 (IPT5)*-mediated cytokinin biosynthesis, whereas PIN1 facilitated the establishment of *WUS*-expressing domains to which the cytokinin biosynthesis was confined. This spatial pattern was critically important for shoot regeneration, as its disruption resulted in the formation of tissue resembling uninduced callus (Cheng et al. 2013).

Finally, the regulatory interactions occurring upon the initiation of floral meristems point at possibilities for direct negative regulation of the SAM master regulator *SHOOTMERISTEMLESS* (*STM*) by auxin in the SAM. *STM* was shown to be subject to negative regulation by ARF3/ETTIN, ARF4 and ARF5/MP during floral meristem determinacy (Chung et al. 2019). It is currently unknown whether similar interactions take place during DNSO as well, but their existence in the SAM points out the importance of the balance between auxin and cytokinin signals, and the main SAM master regulator genes, for the maintenance of the SAM and for efficient shoot regeneration.

4 SIM: The Trigger for Primordium Transdifferentiation

While one-step shoot regeneration protocols are applicable, for which the use of CIM is dispensable, DNSO is impossible without a cytokinin-rich SIM. The main reason is that the transdifferentiation of the pluripotent primordium, the process in which its lateral root-like identity is converted to a shoot-like identity, can be achieved only through the supply of substantial quantities of exogenous cytokinin (Rosspopoff et al. 2017). Also, in one-step regeneration protocols, exogenous cytokinin from the regeneration medium can induce auxin biosynthesis, which leads to the increase in endogenous auxin, and consequently to callus formation preceding shoot organogenesis (Ćosić et al. 2015). Thus, cytokinin present in the shoot regeneration media plays multiple roles that are important for shoot organogenesis.

After the explants are incubated on SIM, cytokinin from the media needs to penetrate into the plant tissues and reach the cytokinin receptors, for the signaling cascade to start (Raspor et al. 2021). Cytokinin perception from the endoplasmic reticulum and from the plasma membrane are both likely to be biologically relevant (Romanov et al. 2018; Antoniadi et al. 2020), thus both the intracellular uptake of cytokinin and its presence in the apoplast are potentially important for the induction of the processes of cellular reprogramming that are part of DNSO. Upon uptake by the plant tissues, exogenous cytokinin can undergo metabolic modifications and/or affect the biosynthesis of cytokinin and other plant hormones, thus exogenous supply of different isoprenoid, aromatic or diphenylurea-type cytokinins has diverse effects on endogenous phytohormone composition (Ćosić et al. 2015, 2021). Furthermore, the alterations in endogenous cytokinins affect gene expression in the regenerating explants. Various genes that are involved in the regulation of DNSO are responsive to cytokinin, and they display complex time-course expression patterns on SIM corresponding not only to their specific roles in DNSO, but possibly also to nonspecific induction by cytokinin (Ćosić et al. 2019). Taken together, the effects of SIM incubation on the regenerating explants are diverse, and encompass more processes than just the gene expression and developmental events that are commonly described as part of DNSO.

5 The Roles of Cytokinin Signaling Components in DNSO

Without the addition of cytokinin to the regeneration media, the calli present on CIM remain in an unorganized state, or can acquire a root identity (Rosspopoff et al. 2017; Liu et al. 2018). Although the pluripotent, lateral root-like primordia start acquiring their organogenic competence under the influence of auxin signals, exposure to cytokinin signals is needed for them to complete this process and to subsequently acquire a shoot identity and develop a SAM (Gordon et al. 2009; Rosspopoff et al. 2017; Pernisova et al. 2018).

Additionally, de novo cytokinin biosynthesis is involved in the alternative, CIM-independent pathway of callus formation, which is developmentally distinct from the lateral root-like callus formation. The CIM-independent pathway of callus formation is induced by wounding, and dependent on cytokinin instead of auxin signaling (Iwase et al. 2011; Ikeuchi et al. 2017).

The roles of particular components of cytokinin signaling in DNSO are shown in Fig. 3.

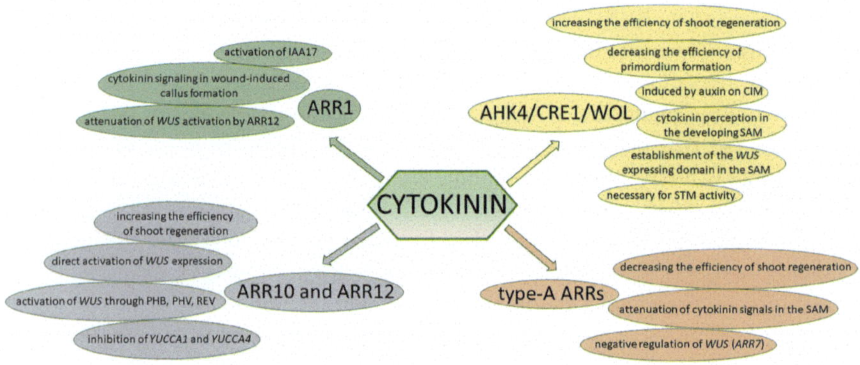

Fig. 3 The roles of cytokinin signaling components in de novo shoot organogenesis (DNSO), as determined in the model species *Arabidopsis thaliana*. The regulation of DNSO by cytokinin is dependent on the cytokinin receptor AHK4/CRE1/WOL, the type-B ARABIDOPSIS RESPONSE REGULATORs (ARRs) that act as transcription factors (ARR1, ARR10, ARR12), and the type-A ARRs, that act as negative regulators of cytokinin signaling. The full names of the genes are given in the text. SAM = shoot apical meristem

5.1 *Cytokinin Perception Through AHK4/CRE1/WOL During SAM Formation: Dependent on Auxin, but Necessary for Cytokinin Action*

Cytokinin perception in *Arabidopsis* is mediated by the three receptors of the ARABIDOPSIS HISTIDINE KINASE (AHK) family: AHK2, AHK3 and AHK4 (Higuchi et al. 2004; Nishimura et al. 2004). Phylogenetic relationships between these *Arabidopsis* HKs, and cytokinin receptors in other groups of plants are currently extensively investigated (Lomin et al. 2018, 2021). Early analyses of *Arabidopsis ahk2*, *ahk3*, and *ahk4* single, double, and triple mutants revealed partially redundant roles in the regulation of developmental processes dependent on cytokinin (Higuchi et al. 2004; Nishimura et al. 2004; Riefler et al. 2006). However, despite partial overlaps in their functions, AHK2, AHK3 and AHK4 have been shown to play distinct roles in plant growth and development, determined both by the differences in their ligand-binding specificities, and development- and tissue-specific expression patterns (Romanov et al. 2006; Stolz et al. 2011).

The cytokinin receptors are located mainly on the endoplasmic reticulum (Wulfetange et al. 2011). However, compelling evidence argues in favor of cytokinin signaling both from the endoplasmic reticulum (Romanov and Schmülling 2021) and from the plasma membrane (Zürcher et al. 2016; Antoniadi et al. 2020), thus it is likely that both signaling pathways play distinct roles in plant growth and development (Nedvěd et al. 2021). It is, though, currently unknown whether cytokinin perception from the endoplasmic reticulum, or from the plasma membrane, is more relevant to the cytokinin responses during DNSO (Raspor et al. 2021).

Available evidence argues for a comprehensive involvement of AHK4/CYTOKININ RESPONSE1/WOODEN LEG (AHK4/CRE1/WOL), but not AHK2 or AHK3 throughout the process of DNSO (Gordon et al. 2009; Pernisova et al. 2018; Fig. 3). Analysis of double *ahk2ahk3*, *ahk2ahk4*, and *ahk3ahk4* mutants revealed that the two *ahk4* DNSO-related phenotypes differed from wild type much more importantly than *ahk2ahk3*. The mutation *ahk4* facilitated the development of pluripotent primordia, but strongly impaired shoot regeneration; thus, it was concluded that cytokinin perception through AHK4/CRE1/WOL negatively regulates primordium formation, but is critically important for SAM formation (Pernisova et al. 2018).

The negative regulation of primordium formation by AHK4/CRE1/WOL is in line with previous reports about the inhibitory effects of cytokinin signals on lateral root development, whereby cytokinin acts antagonistically to the promotive effects of auxin (Chang et al. 2013). Thus, primordium formation is negatively regulated by AHK4/CRE1/WOL in the organogenic callus in a similar way.

Conversely, the role of AHK4/CRE1/WOL is pivotal to cytokinin perception in the developing SAM. The expression of *AHK4/CRE1/WOL* was upregulated in the organogenic callus of *Arabidopsis* during the incubation on CIM and was necessary for the induction of genes by cytokinin during SIM incubation (Gordon et al. 2009). Accordingly, it was shown that cytokinin perception through AHK4/CRE1/WOL underlies the correct positioning of the *WUS*-expressing domain in the developing SAM (Chickarmane et al. 2012; Adibi et al. 2016). The perception of cytokinin through AHK4/CRE1/WOL is supposed to be important also for the function of the other SAM-related morphogenic gene, *STM*, as it was reported that the *wol* mutation abolishes SAM formation in the weak *STM* mutant background, *shootmeristemless-bumpershoot1* (*stm-bum1*) (Jasinski et al. 2005).

5.2 Type-B Cytokinin Response Regulators: A Complex Interplay in the Induction of the SAM Stem Cell Niche

The binding of cytokinin to its receptors occurs either in the lumen of the endoplasmic reticulum, or on the outer surface of the plasma membrane. Cytokinin receptors are transmembrane proteins, whose C-terminal domains are located in the cytosol, where the initial steps of cytokinin signaling occur. Cytokinin binding activates the receptor, leading to the phosphorylation of a conserved histidine (His) residue on its protein kinase domain. Subsequently, a phosphorylation cascade starts, whereby the activator phosphate is transferred first to a conserved aspartate (Asp) residue on the C-terminal domain of the receptor, then to a His residue on the ARABIDOPSIS HISTIDINE PHOSPHOTRANSFER PROTEINs (AHPs). The activated AHPs transmit the activator phosphate to an Asp residue on the receiver domain of type-B ARABIDOPSIS RESPONSE REGULATORs (type-B ARRs), that act as transcriptional activators of responses to cytokinin, through the activity of their DNA-binding Myb-like domain (Kieber and Schaller 2018).

The role of the type-B ARRs has been subject of the most recent additions to the knowledge on the regulation of *Arabidopsis* SAM formation by cytokinin (Zhang et al. 2017; Liu et al. 2020). Four type-B ARRs (ARR1, ARR2, ARR10, and ARR12) have been shown to activate the transcription of *WUS* in the SAM, both directly, and through the activation of HD-ZIP III transcription factors PHABULOSA (PHB), PHAVOLUTA (PHV), and REVOLUTA (REV) (Zhang et al. 2017; Meng et al. 2017; Zubo et al. 2017; Dai et al. 2017; Fig. 3). In addition, ARR1, ARR10 and ARR12 were shown to inhibit the auxin biosynthesis genes *YUCCA1* and *YUCCA4* in the SAM, which is supposed to further enhance *WUS* expression (Meng et al. 2017). Further analysis of the individual roles of ARR1, ARR10 and ARR12 confirmed ARR10 and ARR12 as positive regulators of SAM formation (Zubo et al. 2017; Dai et al. 2017), whereas ARR1 was revealed as a competitive inhibitor of ARR12 for binding to the promoter regions of *WUS* and *CLV3*, actually having an attenuating role in the regulation of DNSO (Liu et al. 2020). On the other hand, the activity of ARR1 was more important than that of ARR10 or ARR12 in the regulation of the CIM-independent (wound-induced) callus formation pathway (Ikeuchi et al. 2017).

The functions of the 11 type-B ARRs in the development of *Arabidopsis* have been widely studied. Analyses of multiple-order mutants of type-B ARRs revealed redundancy in the regulation of cytokinin-controlled developmental responses (Mason et al. 2005). However, microarray analysis revealed that the overwhelming majority of cytokinin-regulated developmental responses required the activity of ARR1, ARR10 and ARR12 (Argyros et al. 2008). Accordingly, the triple *arr1arr10arr12* mutants displayed severe cytokinin deficiency phenotypes, that were comparable to those of the triple *ahk2ahk3ahk4* mutants for cytokinin receptors (Ishida et al. 2008). ARR1, ARR10 and ARR12 were shown to jointly regulate a number of developmental responses, such as the differentiation of protoxylem in the root (Yokoyama et al. 2007), anthocyanin biosynthesis (Das et al. 2012), chloroplast deetiolation (Cortleven et al. 2016), abiotic stress responses (Nguyen et al. 2016; Yan et al. 2021) and gynoecium development (Gómez-Felipe et al. 2021). In this context, the role of ARR1 in attenuating the effects of ARR12 on shoot regeneration during DNSO represents a novel concept—a mechanism for fine-tuning the seemingly redundant action of signaling components of a hormone response pathway (Liu et al. 2020). Taken together, in light of the currently available knowledge, we can conclude that the central signaling components mediating the cytokinin regulation of SAM formation are ARR10 and ARR12, whereas ARR1 acts only as a modulator of that response, but plays an important role in wound-induced callus formation.

5.3 Type-A Cytokinin Response Regulators: The Negative Regulators of SAM Formation

In the *Arabidopsis* cytokinin signaling cascade, beside type-B ARRs, another downstream phosphotransfer target of the AHPs are the negative regulators of cytokinin signaling: type-A ARABIDOPSIS RESPONSE REGULATORs (type-A ARRs). Type-A ARRs are structurally similar to the type-B ARRs in the way that they contain a similar receiver domain with a conserved Asp residue that undergoes phosphorylation by the AHPs. However, they lack a DNA-binding domain, and instead, the activated form of type-A ARRs inhibits the upstream cytokinin signal through an unknown mechanism, thus reducing the sensitivity of the cells to cytokinin (Kieber and Schaller 2018). Similarly to the type-B ARRs, redundant roles in plant development have been determined for various members of the type-A *ARR* gene family through the analysis of multiple-order mutants (To et al. 2004) and overexpressing lines (Ren et al. 2009).

In accordance with their roles in suppressing the cytokinin responses, type-A ARRs have been shown to act as negative regulators of shoot regeneration (Fig. 3). Both *ARR7*- and *ARR15*-overexpressing *Arabidopsis* plants suffered suppressed shoot regeneration, while regeneration was enhanced in both *arr7* and *arr15* loss-of-function mutants (Buechel et al. 2010). Furthermore, WUS was shown to maintain the high sensitivity of the SAM to cytokinin, through the transcriptional repression of four type-A ARRs: *ARR5*, *ARR6*, *ARR7*, and *ARR15*. ARR7 was also shown to negatively regulate *WUS* (Leibfried et al. 2005). One of the type-A ARR genes, *ARR5*, was shown to be transcriptionally activated by the other master regulator of the SAM, STM; however, STM concomitantly upregulated cytokinin biosynthesis genes *ISOPENTENYL TRANSFERASE5* (*IPT5*) and *IPT7*, thus overall reinforcing the cytokinin signal in the SAM (Jasinski et al. 2005; Yanai et al. 2005). Finally, *ARR7* and *ARR15* are negatively regulated by ARF5/MP, providing an additional layer to the positive regulation of shoot regeneration by ARF5/MP, and to the auxin-cytokinin crosstalk in the SAM (Zhao et al. 2010).

During two-step DNSO, the genes encoding type-A ARRs are expressed before the calli are transferred to SIM. For instance, *ARR5* is already expressed in the calli during the incubation on CIM, but its expression is further upregulated upon the transfer of calli from CIM to SIM, which is attributable to its transcriptional responsiveness to the abundant supply of cytokinin. Subsequently, *ARR5* is downregulated upon *WUS* induction (Che et al. 2002). Additionally, *ARR5* was already upregulated during callus formation in the one-step shoot regeneration of kohlrabi (Ćosić et al. 2019).

Comparison between the patterns of expression of *ARR5* and *ARR15* revealed some intriguing differences. Both *ARR5* and *ARR15* are cytokinin-response genes involved in the regulation of shoot regeneration; it has been shown that they are both upregulated upon the transfer to SIM through direct transcriptional activation by two type-B ARRs, *ARR1* (Taniguchi et al. 2007) and *ARR2* (Che et al. 2008). However, the upregulation of *ARR15* on SIM is dependent on the CIM preincubation step, while the upregulation of *ARR5* is not. It has been proposed that the acquisition of organogenic competence proceeds through a progression of steps that critically depend on different duration of exposure to auxin signals. Thus, the expression of *ARR5* on SIM accompanies a wider scale of developmental events than the expression of *ARR15*, which occurs only if a certain level of commitment to organogenic competence has been acquired. *ARR15* was thus designated as a molecular marker for the competence of explants to form green calli on SIM, due to the same level of requirement for CIM preincubation. Additionally, both the *ARR15* expression and the competence to form green calli are cell cycle-independent (Che et al. 2007). It is important to note that the designation of *ARR5* and *ARR15* expression as markers of certain stages of DNSO does not necessarily imply their involvement in the regulation of corresponding processes, but simply the temporal co-occurrence between phenomena. An interpretation for the differential requirements of CIM incubation for the expression of *ARR5* or *ARR15* can be expected to emerge when more details about the dynamics of DNSO and the underlying crosstalk between auxin and cytokinin in its regulation, become available.

6 Crosstalk: The Complexity of the DNSO Signaling Network

In the intricate regulatory pathways that direct plant growth and development, it is impossible to isolate the effects of a particular growth regulator without putting it into a broader developmental context that includes interactions with other regulatory elements. Thus, in the case of DNSO, the effects of auxin and cytokinin cannot be considered separately.

One of the most striking examples of phytohormonal crosstalk is the role of the cytokinin receptor AHK4/CRE1/WOL in shoot regeneration. Cytokinin perception in the developing SAM almost exclusively relies on AHK4/CRE1/WOL (Pernisova et al. 2018); however, the expression of *AHK4/CRE1/WOL* is likely auxin-dependent, as in the two-step shoot regeneration of *Arabidopsis* it required the incubation on CIM (Gordon et al. 2009). Furthermore, although cytokinin signals are essential for shoot regeneration because they ensure the acquisition of shoot identity of the pluripotent primordia, SAM development requires the existence of both cytokinin- and auxin-signaling-dominated domains, that are mutually exclusive, and rely on the auxin efflux carrier PIN1 (Gordon et al. 2007; Cheng et al. 2013). Additionally, the action of WUS, the master regulator of the SAM that critically depends on cytokinin signals, is at the same time indirectly regulated by the major auxin response regulator ARF5/MP, through the downregulation of both *DRN/ESR1* (Luo et al. 2018), and type-A cytokinin response regulator genes *ARR7* and *ARR15* (Zhao et al. 2010). Numerous additional points of auxin-cytokinin crosstalk are in place during SAM formation, such as the inhibition of the cytokinin biosynthesis gene *IPT5* by the auxin signaling component ARF3/ETTIN in the auxin-expressing domains; the inhibition of auxin biosynthesis genes *YUCCA1* and *YUCCA4* by the cytokinin signaling components ARR1, ARR10, and ARR12 inside the organizing center of the SAM, and the upregulation of the auxin signaling repressor IAA17 by the cytokinin response regulator ARR1 (Cheng et al. 2013; Meng et al. 2017; Liu et al. 2020). Other regulator genes of DNSO, such as the genes of the families *PLT*, *ESR*, and *CUC*, are also regulated by both auxin and cytokinin signals.

Beside auxin and cytokinin, other signals contribute to the regulation of DNSO. Gibberellin is a negative regulator of shoot regeneration, and its negative regulation by STM importantly affects shoot regeneration efficiency (Jasinski et al. 2005). Accordingly, *procera*, a mutation in a gene encoding a DELLA protein (negative regulator of gibberellin signaling), dramatically reduced the shoot regeneration efficiency in tomato cv. Micro-Tom (Lombardi-Crestana et al. 2012). Endogenous levels of abscisic, jasmonic and salicylic acid were all reported to negatively affect shoot regeneration in barley (Hisano et al. 2016). Brassinosteroids are likely optimal at physiological concentrations, since exogenous addition of brassinolide negatively affected one-step shoot regeneration in tobacco (Kim et al. 2008), as did *dwarf7-1*, a mutation in a brassinosteroid biosynthesis gene in *Arabidopsis* (Cheon et al. 2010).

Recently, growing attention is being given to the signaling roles of sugars in plant growth and development, and their crosstalk with other developmental cues (Sakr et al. 2018; Wang et al. 2021). Sucrose is particularly relevant for DNSO, as it is commonly supplied to plant tissue culture media, including media used for callus induction and shoot regeneration (Ćosić et al. 2020, 2021). A growing number of reports have revealed the effects of sucrose on shoot regeneration through its crosstalk with auxin and cytokinin. For instance, both sucrose and glucose affect the primordium initiation through the activity of *WOX7* in *Arabidopsis* (Kong et al.

2016). The TARGET OF RAPAMYCIN (TOR) kinase, as a component of sugar signal transduction, has been shown to interact with other developmental signals during DNSO. During primordium initiation, the TOR kinase activates the cell cycle regulator E2Fa jointly with the *LBD* genes (Lee and Seo 2017), whereas later, during SAM formation, TOR mediates the activation of *WUS* in both a light-dependent, and a sucrose-dependent manner (Pfeiffer et al. 2016). Finally, the sucrose in the regeneration media, particularly when present at high concentrations, has been shown to affect the cytokinin uptake and/or endogenous homeostasis, and the expression of organogenesis-related genes in kohlrabi (Ćosić et al. 2021). Thus, the sugar signaling pathways and their crosstalk with auxin and cytokinin signals can be expected to emerge in the next years as important elements in the developmental regulation of DNSO.

7 Conclusion

The molecular mechanisms of DNSO rely on well characterized, multifaceted auxin and cytokinin signaling, strictly regulated both temporally and spatially. Distinct regeneration media are applied to plant tissues, to provide an abundant source of auxin for callus formation and the development of pluripotent primordia on one side, and cytokinin, for the transdifferentiation of the pluripotent primordia and further shoot development, on the other side.

Taken together, the complex regulation of DNSO comprises more processes than mere gene expression along with corresponding morphogenic events, and is achieved not only through auxin and cytokinin signaling, but also through crosstalk with a multitude of other developmental signals, which are being given growing attention. It is hard to say whether all the pieces of the complex puzzle of the regulation of DNSO will ever be put together, but the recent advances in the understanding of the regulatory processes of shoot regeneration are starting to reveal a clearer picture.

Acknowledgements We are thankful to our colleague Dr. Slavica Ninković for the critical reading of the manuscript. The work of the authors is funded through the contract No. 451-03-9/2021-14/200007 of the Ministry of Education, Science and Technological Development of Republic of Serbia.

References

Adamowski M, Friml J (2015) PIN-dependent auxin transport: action, regulation, and evolution. Plant Cell 27:20–32. https://doi.org/10.1105/tpc.114.134874

Adibi M, Yoshida S, Weijers D, Fleck C (2016) Centering the organizing center in the *Arabidopsis thaliana* shoot apical meristem by a combination of cytokinin signaling and self-organization. PLoS ONE 11:e0147830. https://doi.org/10.1371/journal.pone.0147830

Alvarez JM, Bueno N, Cuesta C, Feito I, Ordás RJ (2020) Hormonal and gene dynamics in de novo shoot meristem formation during adventitious caulogenesis in cotyledons of *Pinus pinea*. Plant Cell Rep 39:527–541. https://doi.org/10.1007/s00299-020-02508-0

Antoniadi I, Novák O, Gelová Z, Johnson A, Plíhal O, Simerský R, Mik V, Vain T, Mateo-Bonmatí E, Karady M, Pernisová M, Plačková L, Opassathian K, Hejátko J, Robert S, Friml J, Doležal K, Ljung K, Turnbull C (2020) Cell-surface receptors enable perception of extracellular cytokinins. Nat Commun 11:4284. https://doi.org/10.1038/s41467-020-17700-9

Argyros RD, Mathews DE, Chiang YH, Palmer CM, Thibault DM, Etheridge N, Argyros DA, Mason MG, Kieber JJ, Schaller GE (2008) Type B response regulators of *Arabidopsis* play key roles in cytokinin signaling and plant development. Plant Cell 20:2102–2116. https://doi.org/10.1105/tpc.108.059584

Atta R, Laurens L, Boucheron-Dubuisson E, Guivarc'h A, Carnero E, Giraudat-Pautot V, Rech P, Chriqui D (2009) Pluripotency of Arabidopsis xylem pericycle underlies shoot regeneration from root and hypocotyl explants grown *in vitro*. Plant J 57:626–644. https://doi.org/10.1111/j.1365-313X.2008.03715.x

Banno H, Ikeda Y, Niu QW, Chua NH (2001) Overexpression of Arabidopsis *ESR1* induces initiation of shoot regeneration. Plant Cell 13:2609–2618. https://doi.org/10.1105/tpc.010234

Benková E, Michniewicz M, Sauer M, Teichmann T, Seifertová D, Jürgens G, Friml J (2003) Local, efflux-dependent auxin gradients as a common module for plant organ formation. Cell 115:591–602. https://doi.org/10.1016/S0092-8674(03)00924-3

Berckmans B, Vassileva V, Schmid SPC, Maes S, Parizot B, Naramoto S, Magyar Z, Kamei CLA, Koncz C, Bögre L, Persiau G, De Jaeger G, Friml J, Simon R, Beeckman T, De Veylder L (2011) Auxin-dependent cell cycle reactivation through transcriptional regulation of *Arabidopsis E2Fa* by LATERAL ORGAN BOUNDARY proteins. Plant Cell 23:3671–3683. https://doi.org/10.1105/tpc.111.088377

Buechel S, Leibfried A, To JPC, Zhao Z, Andersen SU, Kieber JJ, Lohmann JU (2010) Role of A-type *ARABIDOPSIS RESPONSE REGULATORS* in meristem maintenance and regeneration. Eur J Cell Biol 89:279–284. https://doi.org/10.1016/j.ejcb.2009.11.016

Cary AJ, Che P, Howell SH (2002) Developmental events and shoot apical meristem gene expression patterns during shoot development in *Arabidopsis thaliana*. Plant J 32:867–877. https://doi.org/10.1046/j.1365-313X.2002.01479.x

Chang L, Ramireddy E, Schmülling T (2013) Lateral root formation and growth of *Arabidopsis* is redundantly regulated by cytokinin metabolism and signalling genes. J Exp Bot 64:5021–5032. https://doi.org/10.1093/jxb/ert291

Che P, Gingerich DJ, Lall S, Howell SH (2002) Global and hormone-induced gene expression changes during shoot development in Arabidopsis. Plant Cell 14:2771–2785. https://doi.org/10.1105/tpc.006668

Che P, Lall S, Howell SH (2007) Developmental steps in acquiring competence for shoot development in *Arabidopsis* tissue culture. Planta 226:1183–1194. https://doi.org/10.1007/s00425-007-0565-4

Che P, Lall S, Howell SH (2008) Acquiring competence for shoot development in Arabidopsis: *ARR2* directly targets A-type ARR genes that are differentially activated by CIM preincubation. Plant Signal Behav 3:99–101. https://doi.org/10.4161/psb.3.2.4958

Cheng ZJ, Wang L, Sun W, Zhang Y, Zhou C, Su YH, Li W, Sun TT, Zhao XY, Li XG, Cheng Y, Zhao Y, Xie Q, Zhang XS (2013) Pattern of auxin and cytokinin responses for shoot meristem induction results from the regulation of cytokinin biosynthesis by AUXIN RESPONSE FACTOR3. Plant Physiol 161:240–251. https://doi.org/10.1104/pp.112.203166

Cheon J, Park SY, Schulz B, Choe S (2010) Arabidopsis brassinosteroid biosynthetic mutant *dwarf7-1* exhibits slower rates of cell division and shoot induction. BMC Plant Biol 10:270. https://doi.org/10.1186/1471-2229-10-270

Chickarmane VS, Gordon SP, Tarr PT, Heisler MG, Meyerowitz EM (2012) Cytokinin signaling as a positional cue for patterning the apical-basal axis of the growing *Arabidopsis* shoot meristem. Proc Natl Acad Sci USA 109:4002–4007. https://doi.org/10.1073/pnas.1200636109

Chung Y, Zhu Y, Wu MF, Simonini S, Kuhn A, Armenta-Medina A, Jin R, Østergaard L, Gillmor CS, Wagner D (2019) Auxin response factors promote organogenesis by chromatin-mediated repression of the pluripotency gene *SHOOTMERISTEMLESS*. Nat Commun 10:886. https://doi.org/10.1038/s41467-019-08861-3

Ckurshumova W, Berleth T (2015) Overcoming recalcitrance—auxin response factor functions in plant regeneration. Plant Signal Behav 10:e993293. https://doi.org/10.4161/15592324.2014.993293

Ckurshumova W, Smirnova T, Marcos D, Zayed Y, Berleth T (2014) Irrepressible *MONOPTEROS/ARF5* promotes *de novo* shoot formation. New Phytol 204:556–566. https://doi.org/10.1111/nph.13014

Cortleven A, Marg I, Yamburenko MV, Schlicke H, Hill K, Grimm B, Schaller GE, Schmülling T (2016) Cytokinin regulates the etioplast-chloroplast transition through the two-component signaling system and activation of chloroplast-related genes. Plant Physiol 172:464–478. https://doi.org/10.1104/pp.16.00640

Ćosić T, Motyka V, Raspor M, Savić J, Cingel A, Vinterhalter B, Vinterhalter D, Trávníčková A, Dobrev PI, Bohanec B, Ninković S (2015) In vitro shoot organogenesis and comparative analysis of endogenous phytohormones in kohlrabi (*Brassica oleracea* var. *gongylodes*): effects of genotype, explant type and applied cytokinins. Plant Cell Tiss Organ Cult 121:741–760. https://doi.org/10.1007/s11240-015-0742-2

Ćosić T, Raspor M, Savić J, Cingel A, Matekalo D, Zdravković-Korać S, Ninković S (2019) Expression profiles of organogenesis-related genes over the time course of one-step *de novo* shoot organogenesis from intact seedlings of kohlrabi. J Plant Physiol 232:257–269. https://doi.org/10.1016/j.jplph.2018.11.004

Ćosić T, Savić J, Raspor M, Cingel A, Ghalawnji N, Vinterhalter B, Ninković S (2020) Effects of different types of sugars and plant growth regulators on kohlrabi seedling growth and development *in vitro*. Arch Biol Sci 72:349–357. https://doi.org/10.2298/ABS200622029C

Ćosić T, Motyka V, Savić J, Raspor M, Marković M, Dobrev PI, Ninković S (2021) Sucrose interferes with endogenous cytokinin homeostasis and expression of organogenesis-related genes during de novo shoot organogenesis in kohlrabi. Sci Rep 11:6494. https://doi.org/10.1038/s41598-021-85932-w

Dai X, Liu Z, Qiao M, Li J, Li S, Xiang F (2017) ARR12 promotes *de novo* shoot regeneration in *Arabidopsis thaliana* via activation of *WUSCHEL* expression. J Integr Plant Biol 59:747–758. https://doi.org/10.1111/jipb.12567

Daimon Y, Takabe K, Tasaka M (2003) The *CUP-SHAPED COTYLEDON* genes promote adventitious shoot formation on calli. Plant Cell Physiol 44:113–121. https://doi.org/10.1093/pcp/pcg038

Das A, Sarkar S, Bhattacharyya S, Gantait S (2020) Biotechnological advancements in *Catharanthus roseus* (L.) G. Don. Appl Microbiol Biotechnol 104:4811–4835. https://doi.org/10.1007/s00253-020-10592-1

Das PK, Shin DH, Choi SB, Yoo SD, Choi G, Park YI (2012) Cytokinins enhance sugar-induced anthocyanin biosynthesis in Arabidopsis. Mol Cells 34:93–101. https://doi.org/10.1007/s10059-012-0114-2

De Rybel B, Vassileva V, Parizot B, Demeulenaere M, Grunewald W, Audenaert D, Van Campenhout J, Overvoorde P, Jansen L, Vanneste S, Möller B, Wilson M, Holman T, Van Isterdael G, Brunoud G, Vuylsteke M, Vernoux T, De Veylder L, Inzé D, Weijers D, Bennett MJ, Beeckman T (2010) A novel Aux/IAA28 signaling cascade activates GATA23-dependent specification of lateral root founder cell identity. Curr Biol 20:1697–1706. https://doi.org/10.1016/j.cub.2010.09.007

De Smet I, Lau S, Voβ U, Vanneste S, Benjamins R, Rademacher EH, Schlereth A, De Rybel B, Vassileva V, Grunewald W, Naudts M, Levesque MP, Ehrismann JS, Inzé D, Luschnig C, Benfey PN, Weijers D, Van Montagu MCE, Bennett MJ, Jürgens G, Beeckman T (2010) Bimodular auxin response controls organogenesis in *Arabidopsis*. Proc Natl Acad Sci USA 107:2705–2710. https://doi.org/10.1073/pnas.0915001107

Efroni I, Mello A, Nawy T, Ip PL, Rahni R, DelRose N, Powers A, Satija R, Birnbaum KD (2016) Root regeneration triggers an embryo like sequence guided by hormonal interactions. Cell 165:1721–1733. https://doi.org/10.1016/j.cell.2016.04.046

Fan M, Xu C, Xu K, Hu Y (2012) LATERAL ORGAN BOUNDARIES DOMAIN transcription factors direct callus formation in *Arabidopsis* regeneration. Cell Res 22:1169–1180. https://doi.org/10.1038/cr.2012.63

Farcot E, Lavedrine C, Vernoux T (2015) A modular analysis of the auxin signalling network. PLoS ONE 10:e0122231. https://doi.org/10.1371/journal.pone.0122231

Fehér A (2019) Callus, dedifferentiation, totipotency, somatic embryogenesis: what these terms mean in the era of molecular plant biology? Front Plant Sci 10:536. https://doi.org/10.3389/fpls.2019.00536

Fukaki H, Tameda S, Masuda H, Tasaka M (2002) Lateral root formation is blocked by a gain-of-function mutation in the *SOLITARY ROOT/IAA14* gene of *Arabidopsis*. Plant J 29:153–168. https://doi.org/10.1046/j.0960-7412.2001.01201.x

Fukaki H, Nakao Y, Okushima Y, Theologis A, Tasaka M (2005) Tissue-specific expression of stabilized SOLITARY-ROOT/IAA14 alters lateral root development in Arabidopsis. Plant J 44:382–395. https://doi.org/10.1111/j.1365-313X.2005.02537.x

Fukaki H, Taniguchi N, Tasaka M (2006) PICKLE is required for SOLITARY-ROOT/IAA14-mediated repression of ARF7 and ARF19 activity during Arabidopsis lateral root initiation. Plant J 48:380–389. https://doi.org/10.1111/j.1365-313X.2006.02882.x

García-Gómez ML, Azpeitia E, Álvarez-Buylla ER (2017) A dynamic gene-hormonal regulatory network model explains multiple cellular behaviors of the root apical meristem of *Arabidopsis thaliana*. PLoS Comput Biol 13:e1005488. https://doi.org/10.1371/journal.pcbi.1005488

Goh T, Kasahara H, Mimura T, Kamiya Y, Fukaki H (2012) Multiple AUX-IAA/ARF modules regulate lateral root formation: the role of *Arabidopsis* SHY2/IAA3-mediated auxin signalling. Phil Trans R Soc B 367:1461–1468. https://doi.org/10.1098/rstb.2011.0232

Gómez-Felipe A, Kierzkowski D, de Folter S (2021) The relationship between AGAMOUS and cytokinin signaling in the establishment of carpeloid features. Plants 10:827. https://doi.org/10.3390/plants10050827

Gordon SP, Heisler MG, Reddy GV, Ohno C, Das P, Meyerowitz EM (2007) Pattern formation during de novo assembly of the *Arabidopsis* shoot meristem. Development 134:3539–3548. https://doi.org/10.1242/dev.010298

Gordon SP, Chickarmane VS, Ohno C, Meyerowitz EM (2009) Multiple feedback loops through cytokinin signaling control stem cell number within the *Arabidopsis* shoot meristem. Proc Natl Acad Sci USA 106:16529–16534. https://doi.org/10.1073/pnas.0908122106

Guilfoyle TJ, Hagen G (2007) Auxin response factors. Curr Opin Plant Biol 10:453–460. https://doi.org/10.1016/j.pbi.2007.08.014

Higuchi M, Pischke MS, Mähönen AP, Miyawaki K, Hashimoto Y, Seki M, Kobayashi M, Shinozaki K, Kato T, Tabata S, Helariutta Y, Sussman MR, Kakimoto T (2004) *In planta* functions of the *Arabidopsis* cytokinin receptor family. Proc Natl Acad Sci USA 101:8821–8826. https://doi.org/10.1073/pnas.0402887101

Hisano H, Matsuura T, Mori IC, Yamane M, Sato K (2016) Endogenous hormone levels affect the regeneration ability of callus derived from different organs in barley. Plant Physiol Biochem 99:66–72. https://doi.org/10.1016/j.plaphy.2015.12.005

Hong L, Ye C, Lin J, Fu H, Wu X, Li QQ (2018) Alternative polyadenylation is involved in auxin-based plant growth and development. Plant J 93:246–258. https://doi.org/10.1111/tpj.13771

Horstman A, Willemsen V, Boutilier K, Heidstra R (2014) AINTEGUMENTA-LIKE proteins: hubs in a plethora of networks. Trends Plant Sci 19:146–157. https://doi.org/10.1016/j.tplants.2013.10.010

Ikeuchi M, Sugimoto K, Iwase A (2013) Plant callus: mechanisms of induction and repression. Plant Cell 25:3159–3173. https://doi.org/10.1105/tpc.113.116053

Ikeuchi M, Iwase A, Rymen B, Lambolez A, Kojima M, Takebayashi Y, Heyman J, Watanabe S, Seo M, De Veylder L, Sakakibara H, Sugimoto K (2017) Wounding triggers callus formation via

dynamic hormonal and transcriptional changes. Plant Physiol 175:1158–1174. https://doi.org/10.1104/pp.17.01035
Ikeuchi M, Shibata M, Rymen B, Iwase A, Bågman AM, Watt L, Coleman D, Favero DS, Takahashi T, Ahnert SE, Brady SM, Sugimoto K (2018) A gene regulatory network for cellular reprogramming in plant regeneration. Plant Cell Physiol 59:770–782. https://doi.org/10.1093/pcp/pcy013
Ikeuchi M, Favero DS, Sakamoto Y, Iwase A, Coleman D, Rymen B, Sugimoto K (2019) Molecular mechanisms of plant regeneration. Annu Rev Plant Biol 70:377–406. https://doi.org/10.1146/annurev-arplant-050718-100434
Ishida K, Yamashino T, Yokoyama A, Mizuno T (2008) Three type-B response regulators, ARR1, ARR10 and ARR12, play essential but redundant roles in cytokinin signal transduction throughout the life cycle of *Arabidopsis thaliana*. Plant Cell Physiol 49:47–57. https://doi.org/10.1093/pcp/pcm165
Iwase A, Mitsuda N, Koyama T, Hiratsu K, Kojima M, Arai T, Inoue Y, Seki M, Sakakibara H, Sugimoto K, Ohme-Takagi M (2011) The AP2/ERF transcription factor WIND1 controls cell dedifferentiation in *Arabidopsis*. Curr Biol 21:508–514. https://doi.org/10.1016/j.cub.2011.02.020
Jasinski S, Piazza P, Craft J, Hay A, Woolley L, Rieu I, Phillips A, Hedden P, Tsiantis M (2005) KNOX action in *Arabidopsis* is mediated by coordinate regulation of cytokinin and gibberellin activities. Curr Biol 15:1560–1565. https://doi.org/10.1016/j.cub.2005.07.023
Kakani A, Li G, Peng Z (2009) Role of AUX1 in the control of organ identity during in vitro organogenesis and in mediating tissue specific auxin and cytokinin interaction in *Arabidopsis*. Planta 229:645–657. https://doi.org/10.1007/s00425-008-0846-6
Kareem A, Durgaprasad K, Sugimoto K, Du Y, Pulianmackal AJ, Trivedi ZB, Abhayadev PV, Pinon V, Meyerowitz EM, Scheres B, Prasad K (2015) *PLETHORA* genes control regeneration by a two-step mechanism. Curr Biol 25:1017–1030. https://doi.org/10.1016/j.cub.2015.02.022
Kareem A, Radhakrishnan D, Wang X, Bagavathiappan S, Trivedi ZB, Sugimoto K, Xu J, Mähönen AP, Prasad K (2016) Protocol: a method to study the direct reprogramming of lateral primordia to fertile shoots. Plant Methods 12:27. https://doi.org/10.1186/s13007-016-0127-5
Kaur A, Reddy MS, Kumar A (2017) Efficient, one step and cultivar independent shoot organogenesis of potato. Physiol Mol Biol Plants 23:461–469. https://doi.org/10.1007/s12298-017-0418-y
Kieber JJ, Schaller GE (2018) Cytokinin signaling in plant development. Development 145:dev149344. https://doi.org/10.1242/dev.149344
Kim JY, Yang W, Forner J, Lohmann JU, Noh B, Noh YS (2018) Epigenetic reprogramming by histone acetyltransferase HAG1/AtGCN5 is required for pluripotency acquisition in *Arabidopsis*. EMBO J 37:e98726. https://doi.org/10.15252/embj.201798726
Kim SH, Bahk S, An J, Hussain S, Nguyen NT, Do HL, Kim JY, Hong JC, Chung WS (2020) A gain-of-function mutant of IAA15 inhibits lateral root development by transcriptional repression of *LBD* genes in Arabidopsis. Front Plant Sci 11:1239. https://doi.org/10.3389/fpls.2020.01239
Kim SL, Lee Y, Lee SH, Kim SH, Han TJ, Kim SK (2008) Brassinolide influences the regeneration of adventitious shoots from cultured leaf discs of tobacco. J Plant Biol 51:221–226. https://doi.org/10.1007/BF03030702
Koike I, Taniguchi K, Shimomura K, Umehara M (2017) Dynamics of endogenous indole-3-acetic acid and cytokinins during adventitious shoot formation in ipecac. J Plant Growth Regul 36:805–813. https://doi.org/10.1007/s00344-017-9684-8
Kong D, Hao Y, Cui H (2016) The WUSCHEL RELATED HOMEOBOX protein WOX7 regulates the sugar response of lateral root development in *Arabidopsis thaliana*. Mol Plant 9:261–270. https://doi.org/10.1016/j.molp.2015.11.006
Krogan NT, Ckurshumova W, Marcos D, Caragea AE, Berleth T (2012) Deletion of *MP/ARF5* domains III and IV reveals a requirement for *Aux/IAA* regulation in *Arabidopsis* leaf vascular patterning. New Phytol 194:391–401. https://doi.org/10.1111/j.1469-8137.2012.04064.x

Krogan NT, Marcos D, Weiner AI, Berleth T (2016) The auxin response factor MONOPTEROS controls meristem function and organogenesis in both the shoot and root through the direct regulation of *PIN* genes. New Phytol 212:42–50. https://doi.org/10.1111/nph.14107

Laskowski M, Grieneisen VA, Hofhuis H, ten Hove CA, Hogeweg P, Marée AFM, Scheres B (2008) Root system architecture from coupling cell shape to auxin transport. PLoS Biol 6:e307. https://doi.org/10.1371/journal.pbio.0060307

Lau S, De Smet I, Kolb M, Meinhardt H, Jürgens G (2011) Auxin triggers a genetic switch. Nat Cell Biol 13:611–615. https://doi.org/10.1038/ncb2212

Lee HW, Kim NY, Lee DJ, Kim J (2009) *LBD18/ASL20* regulates lateral root formation in combination with *LBD16/ASL18* downstream of *ARF7* and *ARF19* in *Arabidopsis*. Plant Physiol 151:1377–1389. https://doi.org/10.1104/pp.109.143685

Lee K, Seo PJ (2017) *Arabidopsis* TOR signaling is essential for sugar-regulated callus formation. J Integr Plant Biol 59:742–746. https://doi.org/10.1111/jipb.12560

Lee MH, Lee J, Jie EY, Choi SH, Jiang L, Ahn WS, Kim CY, Kim SW (2020) Temporal and spatial expression analysis of shoot-regeneration regulatory genes during the adventitious shoot formation in hypocotyl and cotyledon explants of tomato (cv. Micro-Tom). Int J Mol Sci 21:5309. https://doi.org/10.3390/ijms21155309

Leibfried A, To JPC, Busch W, Stehling S, Kehle A, Demar M, Kieber JJ, Lohmann JU (2005) WUSCHEL controls meristem function by direct regulation of cytokinin-inducible response regulators. Nature 438:1172–1175. https://doi.org/10.1038/nature04270

Li W, Liu H, Cheng ZJ, Su YH, Han HN, Zhang Y, Zhang XS (2011) DNA methylation and histone modifications regulate *de novo* shoot regeneration in *Arabidopsis* by modulating *WUSCHEL* expression and auxin signaling. PLoS Genet 7:e1002243. https://doi.org/10.1371/journal.pgen.1002243

Liu J, Hu X, Qin P, Prasad K, Hu Y, Xu L (2018) The *WOX11-LBD16* pathway promotes pluripotency acquisition in callus cells during de novo shoot regeneration in tissue culture. Plant Cell Physiol 59:739–748. https://doi.org/10.1093/pcp/pcy010

Liu Z, Dai X, Li J, Liu N, Liu X, Li S, Xiang F (2020) The type-B cytokinin response regulator ARR1 inhibits shoot regeneration in an ARR12-dependent manner in Arabidopsis. Plant Cell 32:2271–2291. https://doi.org/10.1105/tpc.19.00022

Lombardi-Crestana S, da Silva Azevedo M, Ferreira e Silva GF, Pino LE, Appezzato-da-Glória B, Figueira A, Nogueira FTS, Peres LEP (2012) The tomato (*Solanum lycopersicum* cv Micro-Tom) natural genetic variation *Rg1* and the DELLA mutant *procera* control the competence necessary to form adventitious roots and shoots. J Exp Bot 63:5689–5703. https://doi.org/10.1093/jxb/ers221

Lomin SN, Myakushina YA, Kolachevskaya OO, Getman IA, Arkhipov DV, Savelieva EM, Osolodkin DI, Romanov GA (2018) Cytokinin perception in potato: new features of canonical players. J Exp Bot 69:3839–3853. https://doi.org/10.1093/jxb/ery199

Lomin SN, Savelieva EM, Arkhipov DV, Pashkovskiy PP, Myakushina YA, Heyl A, Romanov GA (2021) Cytokinin perception in ancient plants beyond Angiospermae. Int J Mol Sci 22:13077. https://doi.org/10.3390/ijms222313077

Luo L, Zeng J, Wu H, Tian Z, Zhao Z (2018) A molecular framework for auxin-controlled homeostasis of shoot stem cells in *Arabidopsis*. Mol Plant 11:899–913. https://doi.org/10.1016/j.molp.2018.04.006

Ma Y, Miotk A, Šutiković Z, Ermakova O, Wenzl C, Medzihradszky A, Gaillochet C, Forner J, Utan G, Brackmann K, Galván-Ampudia CS, Vernoux T, Greb T, Lohmann JU (2019) WUSCHEL acts as an auxin response rheostat to maintain apical stem cells in *Arabidopsis*. Nat Commun 10:5093. https://doi.org/10.1038/s41467-019-13074-9

Marchant A, Bhalerao R, Casimiro I, Eklöf J, Casero PJ, Bennett M, Sandberg G (2002) AUX1 promotes lateral root formation by facilitating indole-3-acetic acid distribution between sink and source tissues in the Arabidopsis seedling. Plant Cell 14:589–597. https://doi.org/10.1105/tpc.010354

Marhavý P, Vanstraelen M, De Rybel B, Zhaojun D, Bennett MJ, Beeckman T, Benková E (2013) Auxin reflux between the endodermis and pericycle promotes lateral root initiation. EMBO J 32:149–158. https://doi.org/10.1038/emboj.2012.303

Mason MG, Mathews DE, Argyros DA, Maxwell BB, Kieber JJ, Alonso JM, Ecker JR, Schaller GE (2005) Multiple type-B response regulators mediate cytokinin signal transduction in *Arabidopsis*. Plant Cell 17:3007–3018. https://doi.org/10.1105/tpc.105.035451

Meng WJ, Cheng ZJ, Sang YL, Zhang MM, Rong XF, Wang ZW, Tang YY, Zhang XS (2017) Type-B ARABIDOPSIS RESPONSE REGULATORs specify the shoot stem cell niche by dual regulation of *WUSCHEL*. Plant Cell 29:1357–1372. https://doi.org/10.1105/tpc.16.00640

Motte H, Vereecke D, Geelen D, Werbrouck S (2014) The molecular path to in vitro shoot regeneration. Biotechnol Adv 32:107–121. https://doi.org/10.1016/j.biotechadv.2013.12.002

Moubayidin L, Perilli S, Dello Ioio R, Di Mambro R, Costantino P, Sabatini S (2010) The rate of cell differentiation controls the *Arabidopsis* root meristem growth phase. Curr Biol 20:1138–1143. https://doi.org/10.1016/j.cub.2010.05.035

Muñoz A, Mangano S, González-García MP, Contreras R, Sauer M, De Rybel B, Weijers D, Sánchez-Serrano JJ, Sanmartín M, Rojo E (2017) RIMA-dependent nuclear accumulation of IYO triggers auxin-irreversible cell differentiation in Arabidopsis. Plant Cell 29:575–588. https://doi.org/10.1105/tpc.16.00791

Muro-Pastor MI, Gonzalez R, Strauss J, Narendja F, Scazzocchio C (1999) The GATA factor AreA is essential for chromatin remodelling in a eukaryotic bidirectional promoter. EMBO J 18:1584–1597. https://doi.org/10.1093/emboj/18.6.1584

Nedvěd D, Hošek P, Klíma P, Hoyerová K (2021) Differential subcellular distribution of cytokinins: how does membrane transport fit into the big picture? Int J Mol Sci 22:3428. https://doi.org/10.3390/ijms22073428

Neelakandan AK, Wang K (2012) Recent progress in the understanding of tissue culture-induced genome level changes in plants and potential applications. Plant Cell Rep 31:597–620. https://doi.org/10.1007/s00299-011-1202-z

Nguyen KH, Ha CV, Nishiyama R, Watanabe Y, Leyva-González MA, Fujita Y, Tran UT, Li W, Tanaka M, Seki M, Schaller GE, Herrera-Estrella L, Tran LSP (2016) *Arabidopsis* type B cytokinin response regulators ARR1, ARR10, and ARR12 negatively regulate plant responses to drought. Proc Natl Acad Sci USA 113:3090–3095. https://doi.org/10.1073/pnas.1600399113

Niazian M, Noori SAS, Galuszka P, Tohidfar M, Mortazavian SMM (2017) Genetic stability of regenerated plants *via* indirect somatic embryogenesis and indirect shoot regeneration of *Carum copticum* L. Ind Crop Prod 97:330–337. https://doi.org/10.1016/j.indcrop.2016.12.044

Nishimura C, Ohashi Y, Sato S, Kato T, Tabata S, Ueguchi C (2004) Histidine kinase homologs that act as cytokinin receptors possess overlapping functions in the regulation of shoot and root growth in Arabidopsis. Plant Cell 16:1365–1377. https://doi.org/10.1105/tpc.021477

Okushima Y, Overvoorde PJ, Arima K, Alonso JM, Chan A, Chang C, Ecker JR, Hughes B, Lui A, Nguyen D, Onodera C, Quach H, Smith A, Yu G, Theologis A (2005) Functional genomic analysis of the *AUXIN RESPONSE FACTOR* gene family members in *Arabidopsis thaliana*: unique and overlapping functions of *ARF7* and *ARF19*. Plant Cell 17:444–463. https://doi.org/10.1105/tpc.104.028316

Okushima Y, Fukaki H, Onoda M, Theologis A, Tasaka M (2007) ARF7 and ARF19 regulate lateral root formation via direct activation of *LBD/ASL* genes in *Arabidopsis*. Plant Cell 19:118–130. https://doi.org/10.1105/tpc.106.047761

Pasternak T, Lystvan K, Betekhtin A, Hasterok R (2020) From single cell to plants: mesophyll protoplasts as a versatile system for investigating plant cell reprogramming. Int J Mol Sci 21:4195. https://doi.org/10.3390/ijms21124195

Pernisova M, Grochova M, Konecny T, Plackova L, Harustiakova D, Kakimoto T, Heisler MG, Novak O, Hejatko J (2018) Cytokinin signalling regulates organ identity via the AHK4 receptor in *Arabidopsis*. Development 145:dev163907. https://doi.org/10.1242/dev.163907

Peterson MA, McMaster SA, Riechers DE, Skelton J, Stahlman PW (2016) 2,4-D: past, present, and future: a review. Weed Technol 30:303–345. https://doi.org/10.1614/WT-D-15-00131.1

Pfeiffer A, Janocha D, Dong Y, Medzihradszky A, Schöne S, Daum G, Suzaki T, Forner J, Langenecker T, Rempel E, Schmid M, Wirtz M, Hell R, Lohmann JU (2016) Integration of light and metabolic signals for stem cell activation at the shoot apical meristem. eLife 5:e17023. https://doi.org/10.7554/eLife.17023.001

Piya S, Shrestha SK, Binder B, Stewart CN Jr, Hewezi T (2014) Protein-protein interaction and gene co-expression maps of ARFs and Aux/IAAs in Arabidopsis. Front Plant Sci 5:744. https://doi.org/10.3389/fpls.2014.00744

Raspor M, Motyka V, Ninković S, Dobrev PI, Malbeck J, Ćosić T, Cingel A, Savić J, Tadić V, Dragićević IČ (2020) Endogenous levels of cytokinins, indole-3-acetic acid and abscisic acid in *in vitro* grown potato: a contribution to potato hormonomics. Sci Rep 10:3437. https://doi.org/10.1038/s41598-020-60412-9

Raspor M, Motyka V, Kaleri AR, Ninković S, Lj T, Cingel A, Ćosić T (2021) Integrating the roles for cytokinin and auxin in de novo shoot organogenesis: from hormone uptake to signaling outputs. Int J Mol Sci 22:8554. https://doi.org/10.3390/ijms22168554

Ren B, Liang Y, Deng Y, Chen Q, Zhang J, Yang X, Zuo J (2009) Genome-wide comparative analysis of type-A *ARABIDOPSIS RESPONSE REGULATOR* genes by overexpression studies reveals their diverse roles and regulatory mechanisms in cytokinin signaling. Cell Res 19:1178–1190. https://doi.org/10.1038/cr.2009.88

Riefler M, Novak O, Strnad M, Schmülling T (2006) *Arabidopsis* cytokinin receptor mutants reveal functions in shoot growth, leaf senescence, seed size, germination, root development, and cytokinin metabolism. Plant Cell 18:40–54. https://doi.org/10.1105/tpc.105.037796

Rogg LE, Lasswell J, Bartel B (2001) A gain-of-function mutation in *IAA28* suppresses lateral root development. Plant Cell 13:465–480. https://doi.org/10.1105/tpc.13.3.465

Romanov GA, Schmülling T (2021) Opening doors for cytokinin trafficking at the ER membrane. Trends Plant Sci 26:305–308. https://doi.org/10.1016/j.tplants.2021.02.006

Romanov GA, Lomin SN, Schmülling T (2006) Biochemical characteristics and ligand-binding properties of *Arabidopsis* cytokinin receptor AHK3 compared to CRE1/AHK4 as revealed by a direct binding assay. J Exp Bot 57:4051–4058. https://doi.org/10.1093/jxb/erl179

Romanov GA, Lomin SN, Schmülling T (2018) Cytokinin signaling: from the ER or from the PM? That is the question! New Phytol 218:41–53. https://doi.org/10.1111/nph.14991

Roosjen M, Paque S, Weijers D (2018) Auxin response factors: output control in auxin biology. J Exp Bot 69:179–188. https://doi.org/10.1093/jxb/erx237

Rosspopoff O, Chelysheva L, Saffar J, Lecorgne L, Gey D, Caillieux E, Colot V, Roudier F, Hilson P, Berthomé R, Da Costa M, Rech P (2017) Direct conversion of root primordium into shoot meristem relies on timing of stem cell niche development. Development 144:1187–1200. https://doi.org/10.1242/dev.142570

Sakr S, Wang M, Dédaldéchamp F, Perez-Garcia MD, Ogé L, Hamama L, Atanassova R (2018) The sugar-signaling hub: overview of regulators and interaction with the hormonal and metabolic network. Int J Mol Sci 19:2506. https://doi.org/10.3390/ijms19092506

Santuari L, Sanchez-Perez GF, Luijten M, Rutjens B, Terpstra I, Berke L, Gorte M, Prasad K, Bao D, Timmermans-Hereijgers JLPM, Maeo K, Nakamura K, Shimotohno A, Pencik A, Novak O, Ljung K, van Heesch S, de Bruijn E, Cuppen E, Willemsen V, Mähönen AP, Lukowitz W, Snel B, de Ridder D, Scheres B, Heidstra R (2016) The PLETHORA gene regulatory network guides growth and cell differentiation in Arabidopsis roots. Plant Cell 28:2937–2951. https://doi.org/10.1105/tpc.16.00656

Sarkar AK, Luijten M, Miyashima S, Lenhard M, Hashimoto T, Nakajima K, Scheres B, Heidstra R, Laux T (2007) Conserved factors regulate signalling in *Arabidopsis thaliana* shoot and root stem cell organizers. Nature 446:811–814. https://doi.org/10.1038/nature05703

Schaller GE, Bishopp A, Kieber JJ (2015) The yin-yang of hormones: cytokinin and auxin interactions in plant development. Plant Cell 27:44–63. https://doi.org/10.1105/tpc.114.133595

Skoog F, Miller CO (1957) Chemical regulation of growth and organ formation in plant tissues cultured in vitro. Symp Soc Exp Biol 11:118–130. https://pubmed.ncbi.nlm.nih.gov/13486467/

Stanišić M, Raspor M, Ninković S, Milošević S, Ćalić D, Bohanec B, Trifunović M, Petrić M, Subotić A, Jevremović S (2015) Clonal fidelity of *Iris sibirica* plants regenerated by somatic embryogenesis and organogenesis in leaf-base culture—RAPD and flow cytometer analyses. S Afr J Bot 96:42–52. https://doi.org/10.1016/j.sajb.2014.10.014

Stolz A, Riefler M, Lomin SN, Achazi K, Romanov GA, Schmülling T (2011) The specificity of cytokinin signalling in *Arabidopsis thaliana* is mediated by differing ligand affinities and expression profiles of the receptors. Plant J 67:157–168. https://doi.org/10.1111/j.1365-313X.2011.04584.x

Sugimoto K, Jiao Y, Meyerowitz EM (2010) *Arabidopsis* regeneration from multiple tissues occurs via a root development pathway. Dev Cell 18:463–471. https://doi.org/10.1016/j.devcel.2010.02.004

Taniguchi M, Sasaki N, Tsuge T, Aoyama T, Oka A (2007) ARR1 directly activates cytokinin response genes that encode proteins with diverse regulatory functions. Plant Cell Physiol 48:263–277. https://doi.org/10.1093/pcp/pcl063

Thakur J, Dwivedi MD, Sourabh P, Uniyal PL, Pandey AK (2016) Genetic homogeneity revealed using SCoT, ISSR and RAPD markers in micropropagated *Pittosporum eriocarpum* Royle—an endemic and endangered medicinal plant. PLoS ONE 11:e0159050. https://doi.org/10.1371/journal.pone.0159050

Tikendra L, Koijam AS, Nongdam P (2019) Molecular markers based genetic fidelity assessment of micropropagated *Dendrobium chrysotoxum* Lindl. Meta Gene 20:100562. https://doi.org/10.1016/j.mgene.2019.100562

To JPC, Haberer G, Ferreira FJ, Deruère J, Mason MG, Schaller GE, Alonso JM, Ecker JR, Kieber JJ (2004) Type-A Arabidopsis response regulators are partially redundant negative regulators of cytokinin signaling. Plant Cell 16:658–671. https://doi.org/10.1105/tpc.018978

Valvekens D, Van Montagu M, Van Lijsebettens M (1988) *Agrobacterium tumefaciens*-mediated transformation of *Arabidopsis thaliana* root explants by using kanamycin selection. Proc Natl Acad Sci USA 85:5536–5540. https://doi.org/10.1073/pnas.85.15.5536

Vangheluwe N, Beeckman T (2021) Lateral root initiation and the analysis of gene function using genome editing with CRISPR in *Arabidopsis*. Genes 12:884. https://doi.org/10.3390/genes12060884

Wang M, Le Gourrierec J, Jiao F, Demotes-Mainard S, Perez-Garcia MD, Ogé L, Hamama L, Crespel L, Bertheloot J, Chen J, Grappin P, Sakr S (2021) Convergence and divergence of sugar and cytokinin signaling in plant development. Int J Mol Sci 22:1282. https://doi.org/10.3390/ijms22031282

Weijers D, Benkova E, Jäger KE, Schlereth A, Hamann T, Kientz M, Wilmoth JC, Reed JW, Jürgens G (2005) Developmental specificity of auxin response by pairs of ARF and Aux/IAA transcriptional regulators. EMBO J 24:1874–1885. https://doi.org/10.1038/sj.emboj.7600659

Wulfetange K, Lomin SN, Romanov GA, Stolz A, Heyl A, Schmülling T (2011) The cytokinin receptors of Arabidopsis are located mainly to the endoplasmic reticulum. Plant Physiol 156:1808–1818. https://doi.org/10.1104/pp.111.180539

Xu L (2018) *De novo* root regeneration from leaf explants: wounding, auxin, and cell fate transition. Curr Opin Plant Biol 41:39–45. https://doi.org/10.1016/j.pbi.2017.08.004

Yan Z, Wang J, Wang F, Xie C, Lv B, Yu Z, Dai S, Liu X, Xia G, Tian H, Li C, Ding Z (2021) MPK3/6-induced degradation of ARR1/10/12 promotes salt tolerance in *Arabidopsis*. EMBO Rep 22:e52457. https://doi.org/10.15252/embr.202152457

Yanai O, Shani E, Dolezal K, Tarkowski P, Sablowski R, Sandberg G, Samach A, Ori N (2005) *Arabidopsis* KNOXI proteins activate cytokinin biosynthesis. Curr Biol 15:1566–1571. https://doi.org/10.1016/j.cub.2005.07.060

Yokoyama A, Yamashino T, Amano YI, Tajima Y, Imamura A, Sakakibara H, Mizuno T (2007) Type-B ARR transcription factors, ARR10 and ARR12, are implicated in cytokinin-mediated regulation of protoxylem differentiation in roots of *Arabidopsis thaliana*. Plant Cell Physiol 48:84–96. https://doi.org/10.1093/pcp/pcl040

Zhang MM, Zhang HK, Zhai JF, Zhang XS, Sang YL, Cheng ZJ (2021) ARF4 regulates shoot regeneration through coordination with ARF5 and IAA12. Plant Cell Rep 40:315–325. https://doi.org/10.1007/s00299-020-02633-w

Zhang TQ, Lian H, Zhou CM, Xu L, Jiao Y, Wang JW (2017) A two-step model for de novo activation of *WUSCHEL* during plant shoot regeneration. Plant Cell 29:1073–1087. https://doi.org/10.1105/tpc.16.00863

Zhao Z, Andersen SU, Ljung K, Dolezal K, Miotk A, Schultheiss SJ, Lohmann JU (2010) Hormonal control of the shoot stem-cell niche. Nature 465:1089–1092. https://doi.org/10.1038/nature09126

Zhou JJ, Luo J (2018) The PIN-FORMED auxin efflux carriers in plants. Int J Mol Sci 19:2759. https://doi.org/10.3390/ijms19092759

Zubo YO, Blakley IC, Yamburenko MV, Worthen JM, Street IH, Franco-Zorrilla JM, Zhang W, Hill K, Raines T, Solano R, Kieber JJ, Loraine AE, Schaller GE (2017) Cytokinin induces genome-wide binding of the type-B response regulator ARR10 to regulate growth and development in *Arabidopsis*. Proc Natl Acad Sci USA 114:5995–6004. https://doi.org/10.1073/pnas.1620749114

Zürcher E, Liu J, di Donato M, Geisler M, Müller B (2016) Plant development regulated by cytokinin sinks. Science 353:1027–1030. https://doi.org/10.1126/science.aaf7254

Mechanism of Crosstalk Between Cytokinin and Gibberellin

Ankur Singh and Aryadeep Roychoudhury

Abstract Phytohormones play an integral role in various plant biological processes and regulates the signaling pathways that helps to maintain the growth and development of plants in unheralded environmental conditions. Cytokinin and gibberellin are two such major phytohormones that regulate the growth of the plants; however, various studies have demonstrated that an antagonistic interaction occurs between cytokinin and gibberellin during several physiological processes such as shoot and root elongation, shoot regeneration in culture, cell differentiation and meristem activity. This delicate balance between cytokinin and gibberellin in plants is maintained by various proteins such as KNOX, SPY and SEC. KNOX proteins enhance the expression of cytokinin-biosynthesis gene *Isopentenyl Transferase 7* that accumulates cytokinin in meristems, whereas SPY and SEC are two Serine and Threonine O-linked N-acetylglucosamine (O-GlcNAc) transferase that inhibits the gibberellin response and enhances cytokinin signaling in plants. The development of plants requires a dynamic balance between these two hormones. Thus, the main objective of this book chapter is to present all the recent works that was done focusing the crosstalk between cytokinin and gibberellin. We also tried to explain the role of major components (SPY, SEC and KNOX) involved in this complex network and effects of their mutation in plants.

1 Introduction

Plants have extraordinary capability of being potentially 'immortal'. Some plants survived several years and their death mostly occurred due to external factors such as environmental stress, pathogen infection and other diseases. This long life expectancy of the plants is mainly due to presence of long lasting stem cells that continuously produces new organs and tissues. According to Heidstra and Sabatini (2014), this

A. Singh · A. Roychoudhury (✉)
Department of Biotechnology, St. Xavier's College (Autonomous), 30 Mother Teresa Sarani, Kolkata 700016, West Bengal, India
e-mail: aryadeep.rc@gmail.com

T. Aftab (ed.), *Auxins, Cytokinins and Gibberellins Signaling in Plants*, Signaling and Communication in Plants,
https://doi.org/10.1007/978-3-031-05427-3_4

indeterminate growth of plants mostly occurs after post-embryonic phase in meristematic tissues where stem cells are present. Several chemical messengers contribute in lifelong growth of plants. Phytohormones are one such messenger that tightly regulates the growth and development of plants along with other major processes such as stress response, signaling, metabolism and even death (Weyers and Paterson 2001). With the advent of science, various reports have extensively demonstrated the pivotal role of phytohormones in growth and development of plants. Additionally, phytohormones are also linked with higher stress tolerance (abiotic and biotic stress) in plants that make them a suitable target for consideration during engineering of stress endurance in agronomically important crops (Salvi et al. 2021). Abscisic acid (ABA), cytokinin (CK), gibberellin (GA), salicylic acid (SA), auxin, jasmonic acid (JA), ethylene (ET) and strigolactone (SL) are some of the major phytohormones synthesised in plants. These phytohormones control a wide plethora of dynamic yet highly regulated molecular processes in plants during their entire lifecycle and specially during environmental stresses by regulating the gene expression of protective machineries that confer higher tolerance level in plants exposed to unheralded situations (Kaur et al. 2015).

Previously, it was reported that each hormone has specific functions in plants but recent studies have shown that different hormones have overlapping functions such that a specific output of plant behaviour depends on a specific combination of hormones rather that their individual activity. In last three decades, numerous components of signaling pathways have been identified leading to partial or complete elucidation of involvement of hormones in these complex signaling networks along with their crosstalk with other hormones. Crosstalk between hormones can occur during their biosynthesis or at response level which creates a delicate network of signaling pathway. In this book chapter, our focus will be on CK and GB. We first summarize the current knowledge on the molecular mechanism of biosynthesis, transport and signaling pathway of CK and GB and then finally discuss some recent works demonstrating interaction between these hormones.

2 Cytokinin (CK)

Cytokinins are one of the major phytohormones synthesised in plants that play a pivotal role throughout the life of plants. These N6-prenylated adenine derivatives were initially discovered in *Zea mays* and were ultimately reported in various plants species along with their wide range of functions in growth and development of plants (Zalabák et al. 2013). Initially it was believed that CKs only control cell cycle and cell division, but recent studies has shown the involvement of CKs in other major process of plants such as maintenance of apical dominance, inhibition of root growth, growth of lateral buds, nitrogen signaling pathway, formation of shoot meristem and senescence and expansion of leaves (Fig. 1) (Frébortet al. 2011; Miyawaki et al. 2004; Giulini et al. 2004). According to Del Bianco et al. (2013), isoprenoid CKs (*cis*-zeatin, *trans*-zeatin, isopentenyladenine and dihydrozeatin) are

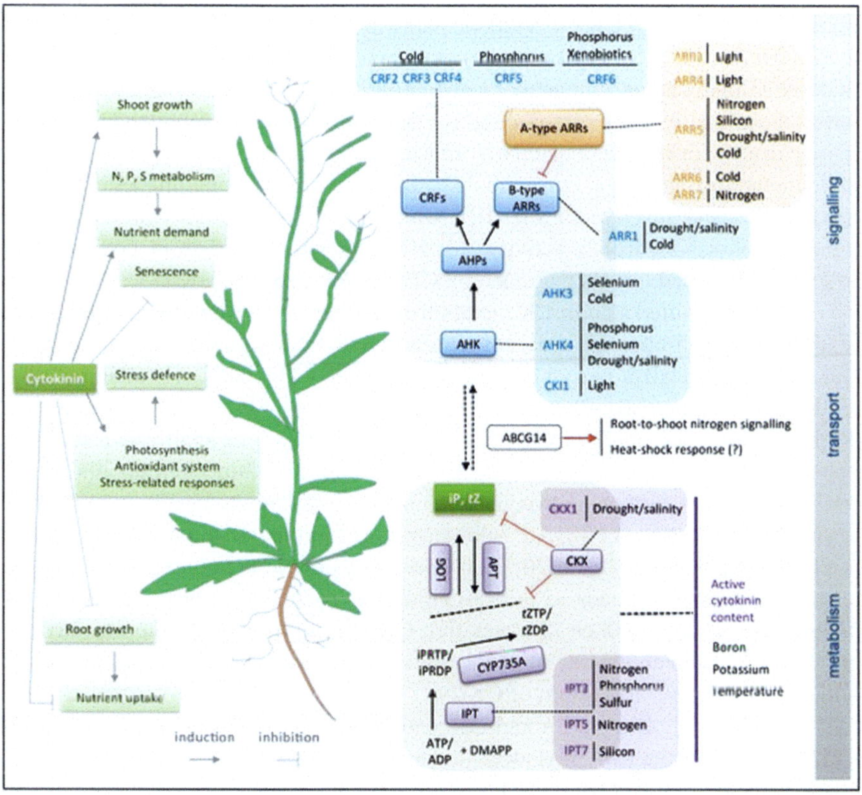

Fig. 1 Crosstalk between abiotic stress signals and cytokinin. Summary of interactive points between cytokinin metabolism and signalling pathways (as currently modelled in Arabidopsis) and abiotic stress response pathways (CC BY: Pavlů et al. 2018)

bioactive CKs synthesised in higher plants. Additionally, Faiss et al. (1997) also reported the presence of N6-(meta hydroxy benzyl) adenine as an aromatic CKs found in lower quantity in plants. Cytokinin plays a major role in variety of growth processes such as chloroplast differentiation, cell division, photosynthesis, nutrient metabolism and maintenance of meristem function (Cortleven et al. 2019; Kurakawa et al. 2007). Additionally, the importance of CKs in conferring stress tolerance in plants has been widely reported.

2.1 Metabolism and Transportation of Cytokinins

Regulation of CK homeostasis is a multistep process which is controlled by the balance between its biosynthesis and degradation. According to Sakakibara (2006), isopentenyladenine and *trans*-zeatin are the most dominant CK found in Arabidopsis.

The rate limiting step of CK biosynthetic pathway is controlled by ATP/ADP Isopentenyltransferase (IPT). According to Sakakibara (2006), inactive CKs are synthesised via de novo pathway and the activity of Lonely Guy (LOG) enzyme is required to convert these inactive forms into free active bases. However, application of adenosine kinase can also convert these inactive forms into active CKs (Schoor et al. 2011). The whole metabolic pathways, i.e., anabolism and catabolism can be divided into three parts: synthesis of inactive CKs by IPT, which is again converted into active form by the phosphoribohydrolase activity of LOG enzyme in second phase and finally in third phase, the used CKs is degraded by CK dehydrogenases (CKX).

The first committed step in CK metabolism can be achieved by two pathways. In the first pathway, 5-phosphate adenosine reacts with hydroxymethylbutenyl disphosphate or dimethylallyl pyrophosphate to yield N6-(2-isopentenyl) adenine riboside 5′-diphosphate (iPRDP) or N6-(2-isopentenyl) adenineriboside 5′-triphosphate (iPRTP) in presence of IPT (Krall et al. 2002). Eventually iPRDP and iPRTP are dephosphorylated to yield N6-(2-isopentenyl) adenine riboside 5′-monophosphate (iPRMP) which is further hydroxylated to *trans*-zeatin by the catalytic activity of enzymes CYP735A1/CYP735A2 belonging to the cytochrome P450 monooxygenase (P450) family (Li et al. 2020). In the second pathway, tRNA-specific adenylate isopentenyltransferase leads to the formation of *cis*-zeatin by isopentenylation of tRNA, degradation of which liberates free CKs (Yevdakova and von Schwartzenberg 2007). The metabolic rate of tRNA is less and thus it is considered as secondary pathway due to lower formation of CKs.

Along with IPT/*trans*-zeatin and *cis*-zeatin CK synthesis pathway, a more efficient CK synthesis pathway was identified in higher plants. In *Arabidopsis thaliana,* dimethylallyl pyrophosphate and adenosine monophosphate is directly converted into iPRMP which is further catalysed by endogenous hydroxylase into zeatin riboside-5′-monophosphate (Li et al. 2020). After their formation, an additional step is required for the activation of CKs which is catalysed by LOG enzyme (Kuroha et al. 2009). They further reported that similar to that of IPT, LOG enzyme is also expressed throughout the plants during development. To ensure the optimum level of CKs in plants a final step occurs where catabolism of excess CKs occurs via *CKX* gene family (Schmullinget al. 2003). According to Frébort et al. (2011), CKX causes irreversible inactivation of CKs via oxidative cleavage at N6 side chain, resulting in the formation of side chain derived aldehyde and adenine.

Transportation of CKs in plants is still not completely known. Initially it was thought that CKs was formed in the roots which are further transported upward to the shoots. In a recent work, Bishopp et al. (2011) reported that along with long distance transport of CKs via phloem through root to shoot, diffusion of CKs also takes place in plants. Bürkle et al. (2003) and Gillissen et al. (2000) identified Equilibrative Nucleoside Transport and Purine Permease as putative transport of CKs via phloem. Recently Zhang et al. (2014) and Ko et al. (2014) identified ATP-binding cassette transporter G14 (AtABCG14) in *Arabidopsis* that plays a pivotal role in transportation of CKs via xylem.

2.2 *Cytokinin Mediated Signaling in Plants*

Similar to that of bacterial two-component system, cytokinin signaling is also mediated by multistep phosphotransfer cascade (El-Showk et al. 2013). According to Hai et al. (2020), CK signal in cells is perceived by a receptor, i.e., histidine kinase (HK) which is followed by transfer of signal via Histidine phosphotransfer (HP). Finally, the transferred signal is received by response regulator (RR) that accordingly regulates the expression pattern of genes or formation of metabolites enabling the plants to respond appropriately to the external stimuli. On receiving any external signals, plants induce the formation of CKs which binds to the cyclases/histidine kinases associated sensory extracellular (CHASE) domain of HK which further transfer the phosphoryl group to the aspartate reside from histidine domain of HK. The histidine residue present on HP eventually receives the phosphoryl group from aspartate domain of HK and further donates it to the RR which ultimately regulates the expression of concerned genes upon receiving the phosphoryl moiety (Hwang et al. 2012).

3 Gibberellin (GA)

Gibberellin is another major growth regulator that controls the growth and development of plants by stimulating cell elongation and division (Colebrook et al. 2014). Gibberellin belongs to large group of tetracyclic diterpenoid carboxylic acid. According to MacMillan (2002), 136 GAs has been extracted from plants produced by fungi or bacteria. Of all the GA synthesised in plants, GA1 and GA4 are the most predominant bioactive form (Sponsel and Hedden 2010). The major functions of GA include stem elongation, reproductive development, seed germination, leaf expansion, flower and seed development and stress tolerance (Yamaguchi 2008). Additionally many non-bioactive forms of GAs exist as a precursor of bioactive form of GAs or as inactive metabolites (Yamaguchi 2008).

3.1 *Metabolism and Transportation of Gibberellin*

The level of GAs in plants is maintained by the synthesis of non-bioactive form of GAs which is further converted in active form followed by degradation of excess active form. Biosynthetic pathway of GAs can be divided into three stages: formation of *ent-kaurene* in proplastids from geranyl geranyl diphosphate (GGDP), formation of C_{20} from *ent-kaurene* by the catalytic activity of cytochrome P450 monooxygenases followed by formation of C_{20} and C_{19}-GAs in cytoplasm. Geranyl geranyl diphosphate acts as a common precursor of GAs, chlorophyll and carotenoids. The first committed step in the formation of GAs is the conversion of GGDP to *ent-kaurene* via a two-step cyclization reaction catalysed by ent-copalyl diphosphate synthase (CPS) and ent-kaurene synthase (KS). In the second stage, *ent-kaurene* undergoes oxidation followed by ring contraction via *ent-kaurene* oxidase (KO) and ent-kaurenoic acid oxidase (KAO) to yield GA_{12}. In the third and final stage, GA_{12} is converted to GA_{53} by 13-hydroxylation. A series of oxidation reaction catalysed by GA3-oxidases (GA3ox), 20-oxidases (GA20ox) and 2-oxoglutarate-dependent dehydrogenase lead to the formation of active form of GAs including GA_1 and GA_4 from GA_{12} and GA_{53}. The four major active GAs (GA_1, GA_3, GA_4 and GA_7) in plants contain a 3β-hydroxyl group (Sun 2008). The rate of activation of GAs can be affected by the rate of their synthesis and deactivation. In recent studies, Zhu et al. (2006) and Varbanova et al. (2007) stated that deactivation of GAs in rice and Arabidopsis is executed by epoxidation (by the catalytic activity of elongated uppermost internode (EUI) which encodes a P450 enzyme (CYP714D1)) of non-13-hydroxylated GAs and methylation (by the catalytic activity of GA methyltransferases 1 and 2) of GAs, respectively (Fig. 2). However, these studies are in early stage and a wide range of work needs to be done to fully decipher the above mentioned deactivation pathways of GAs in other plant species.

Transportation of GAs in plants occurs in both basipetal and acropetal directions (Björklund et al. 2007). Similar to that of auxin, GAs are also subjected to ion-trap mechanism that limits their ability to move out of the cells (Kramer 2006) which lead to the prediction of the existence of GA efflux transporters that effectively translocate GA locally at both cellular and tissue levels (Kramer 2006); however, no GA efflux transporter has been identified in plants till date. In contrast, a number of GA influx transporter has been identified in Arabidopsis (Lacombe and Achard 2016). Kanno et al. (2012) identified a nitrate transporter 1/peptide transporter family (NPF) which was involved in the import of GA. They further stated that different GA transporters might have different affinity for different forms of GA; however, no transporter was identified for transportation of inactive forms of GA. David et al. (2016) further reported that the expression of NPF3 was downregulated in presence of high GA, whereas its expression was upregulated by high ABA level or lack of nitrogen. Along with NPF transporter, SWEET13 and SWEET14 have been linked with the transportation of GA (Chen et al. 2015). The SWEET transporters are mainly linked with the transportation of sucrose in plants (Klemens et al. 2013). Interestingly, the activity of all the above mentioned transporters, i.e., NPF3, SWEET13 and SWEET14 is not only restricted to transportation of active form or a specific intermediate as both the active and intermediate forms of GA were equally transported by the above mentioned transporter in the experiments carried on mutant yeast and oocytes (Kanno et al. 2016). Till date, relatively a large number of proteins have been linked with the transportation of GAs which suggests that GA is transported via complex pathway in cells and tissues and thus further works need to be done to completely characterize the function of these proteins.

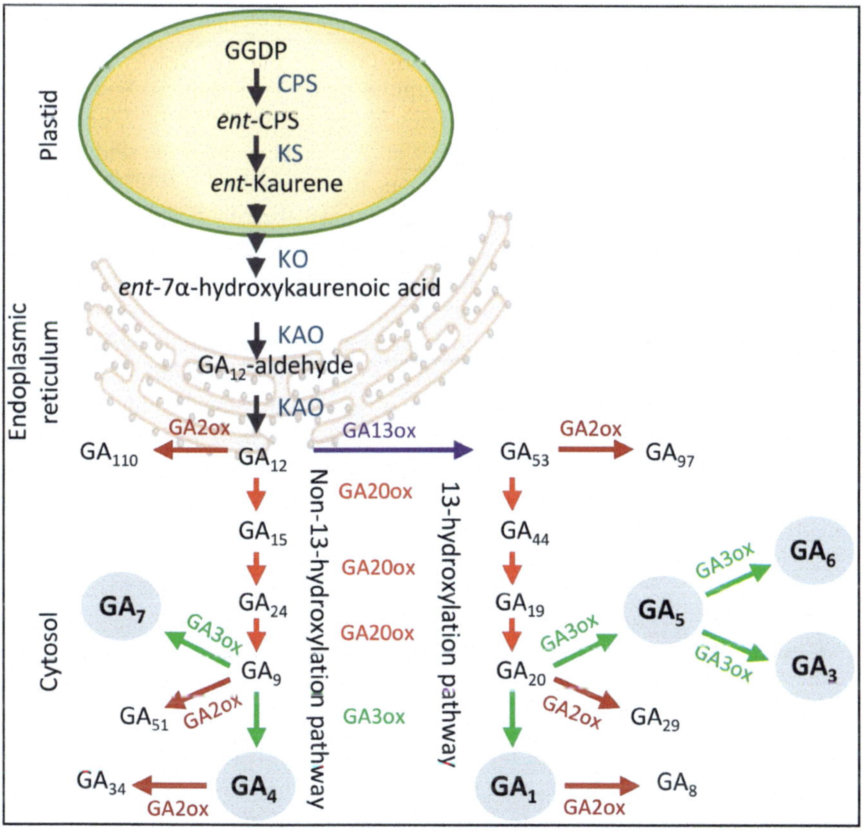

Fig. 2 Generalized scheme of GA biosynthesis and deactivation in higher plants. GA biosynthesis starts in the plastids and is followed by the production of GA_{12} in the endoplasmic reticulum. In the cytoplasm, GA_{12} is processed by GA20ox and GA3ox enzymes in two separate branches to produce bioactive GAs (gray circles). The non-13-hydroxylation yields GA_7 and GA_4, whereas the 13-hydroxylation yields GA_5, GA_6, GA_3, and GA_1. The GA2ox enzymes deactivate the precursors and bioactive GAs (CC BY: Katyayini et al. 2020)

3.2 *Gibberellin Mediated Signaling in Plants*

Genetic studies have led to the identification of both positive and negative signaling components of GA signaling pathway. Of all the components of GA signaling pathway, DELLA proteins are the most extensively studied. DELLA protein is a nuclear protein that belongs to GRAS family of transcriptional regulators and acts as a suppressor of GA signaling. However, till date, the molecular mechanism of GA signal suppression by DELLA protein is still unclear. Itoh et al. (2002) reported the presence of five DELLA proteins (GA Insensitive (GAI), Repressor of *ga1-3* (RGA), RGA Like (RGL) 1, 2 and 3) in *Arabidopsis thaliana* and one (Slender 1) in rice genome. Another important component of GA signaling pathway is GA Insensitive

Dwarf 1 (GID1). According to Griffith et al. (2007), binding of GA with GID1 triggers the interaction between GID1 and DELLA proteins which further stimulates the binding of DELLA protein to SCF E3 ubiquitin ligase via specific F-box proteins (GID2/SLY) that lead to polyubiquitination and degradation of DELLA protein by 26S proteasome (Dill et al. 2004; Sasaki et al. 2003). Thus it can be said that GA signaling pathway involves three major components, i.e., a receptor, a DELLA protein and a F-box protein. Additionally, several studies have also reported involvement of other factors that affect GA responses. Filardo and Swain (2003) reported the presence of SPINDLY (SPY) that acts as inhibitors of GA signaling cascade. The SPY protein resembles mammalian enzymes that modify proteins post-translationally by a specific type of glycosylation, termed tetratricopeptide repeat-containing Serine and Threonine-O-linked GlcNAc transferases (OGT). According to Wells et al. (2001), post-translational modification by SPY affects the stability, localization, phosphorylation and interaction of GAs with other proteins. There is no direct evidence of interaction of DELLA proteins with SPY; however, genetic studies have shown that SPY proteins are required for GA response suppression activity of DELLA (Silverstone et al. 2007).

4 Crosstalk Between Cytokinin and Gibberellin

Cytokinin and gibberellin both play a pivotal role in the regulation of growth and development of plants; however, they show antagonistic behaviour during several developmental processes (Weiss and Ori 2007). Both the hormones interact at various levels including metabolism and signaling pathway like CK promotes and GA inhibits nodule formation in legumes (Maekawa et al. 2009), flower formation in grapes (Srinivasan and Mullins 1981) and tuberization in potato (Rodriguez-Falcon et al. 2006). Yanai et al. (2005) showed that high CK and low GA level in plants is required for normal functions of shoot apical meristem. They further reported that KNOTTED1-LIKE HOMEOBOX (KNOX) transcription factors is involved in the maintenance of regulators of shoot apical meristem, i.e., they upregulate the gene expression of *IPT7* gene which eventually enhances the formation of CKs. Along with this, earlier studies conducted by Hay et al. (2002) and Chen et al. (2004) showed that KNOX transcription factors directly suppress the activities of GA biosynthetic enzymes such as GA-20 oxidases. In further study, Bolduc and Hake (2009) reported that the level of GA deactivating enymes (GA2ox) was enhanced at the base of shoot apical meristem by CK and KNOX1 that blocks the transportation of biologically active forms of GA into shoot apical meristem from nearby tissues. Thus it can be concluded that KNOX transcription factors directly control the balance between CK and GA in shoot apical meristems by enhancing the formation of CK and reducing the level of GA by downregulating and upregulating its formation and degradation in shoot apical meristems, respectively. Brenner et al. (2005) reported that genome profiling of CK-treated Arabidopsis seedlings revealed that CK lowers the expression of GA3ox and GA20ox and promotes the expression of GAI and RGA which further strengthen the negative interaction between these two hormones.

In contrast to shoot apical meristem, later stage of cell elongation and maturation requires high GA signals and low CKs. Greenboim-Wainberg et al. (2005) showed that the accumulation of GA or a mutation in SPY proteins (GA signaling repressors) inhibited the response of CK in Arabidopsis plants. Several other studies have also shown that the CK response in plants is suppressed by cellular GA via inhibiting the activity of SPY, whereas SPY directly upregulates CK signaling pathway in plants. Thus it can be said that, in absence of GA, SPY inhibits GA signaling pathways and induces the CK response in cells; however, higher level of GA in cells suppresses the activity of SPY which eventually results in lower CK signaling in plant cells. However, till date, the mechanism by which SPY inhibits GA signaling in cells is still not clear. Weiss and Ori (2007) assumed that GA can inhibit the activity of a component that directly interacts with SPY which in turn reduces the SPY activity along with CK response in cells. Earlier Ferreira and Kieber (2005) reported that the induction of Type-A RR was inhibited by GA and *spy* mutants which suggests that SPY interacts with, and perhaps modifies (via O-GlcNAc modification), the elements of the cytokinin phosphorelay cascade.

Along with SPY, Hartweck et al. (2002) identified a second OGT gene, i.e., SECRET AGENT (SEC) in Arabidopsis plants which have high similarity to that of animal OGTs. Higuchi et al. (2004) reported that triple mutant of CK receptor shows more severe response in plants as compared to that of single *spy* mutants. Single *sec* mutant do not show any significant alteration in plant phenotype, but double mutation i.e., *sec* and *spy* is lethal for plants (Hartweck et al. 2002). This could be explained by the fact that *sec* mutant contains high GA level or signal which do not show any lethal effects on plants; however, double *sec* and *spy* mutant causes an unregulated alteration in the level of both GA and CK pathways which is lethal for the plants.

Additionally, another interesting level of interaction between CK and GA was explained by Hay et al. (2002). They demonstrated that the alteration in the phenotype caused by overexpression of KNOX was again revived in GA and *spy* mutants. They explained their finding by stating that higher level of GA in cells or *spy* mutant might restore the formation of GA synthesis which was hampered due to overexpression of KNOX. SPY directly controls the KNOX activity which further enhances the biosynthesis of CKs which represent another possible level of interaction between CK and GA. However, this was a very preliminary study and further work needs to been done in coming time to completely decipher the interaction between GA and CK mediated by GA, SPY and KNOX.

5 Conclusion

In this book chapter, we tried to provide an overview of two major plant hormones, i.e., CK and GA and their crosstalk in plants. Based on previous work and data, it can be concluded that an antagonistic relation exists between CK and GA signaling and metabolic pathways. Both the hormones play an integral role in growth and development of plants; however, they act antagonist to each other in several other major processes of plants. This negative relation between the hormones is controlled by various regulators like KNOX, SPY and SEC. SPY and SEC are two OGTs recently identified in Arabidopsis plants exhibiting a high similarity to that of mammalians OGTs. *spy* mutant plants exhibited short hypocotyls, smaller leaves, and deviant phyllotaxy, *sec* mutant do not show any significant alteration in the phenotype of the plants; however, double mutants, i.e., plants having both *sec* and *spy* mutation dies prematurely which led to a hypothesis that SPY has an unidentified function(s) in processes unrelated to GA signaling. Another protein, i.e., KNOX suppress the formation of GA in plants irrespective to that of the level of CK by directly suppressing the expression of GA biosynthetic gene GA20ox. Again the antagonistic effects of CK and GA has been identified in many plant species which led to a conclusion that apart from the minor difference, the major pathways is mainly conserved in most of the plant species. Thus it can be assumed that this complex interaction has been evolved to efficiently maintain the delicate balance between CK and GA in plant tissues as and when required such as in shoot apical meristem tissues where higher CK and lower GA levels are required, whereas cell maturation and elongation requires higher GA and lower CK levels. In spite of the presence of such an important signaling network, their lies a huge gap in our knowledge of crosstalk occurring between CK and GA. Additionally, the biochemical and molecular changes that occurred in *spy* and *sec* double mutants that lead to plant death is not completely known and further work needs to be done to answer these questions. Thus, it is clear that our current knowledge is just the tip of the iceberg of a complex network of interactions that occurs between two major plant hormones, i.e., CK and GA.

Acknowledgements Financial assistance from Science and Engineering Research Board, Government of India through the grant [EMR/2016/004799] and Department of Higher Education, Science and Technology and Biotechnology, Government of West Bengal, through the grant [264(Sanc.)/ST/P/S&T/1G-80/2017] to Dr. Aryadeep Roychoudhury is gratefully acknowledged.

References

Bishopp A, Lehesranta S, Vatén A, Help H, El-Showk S, Scheres B, Helariutta K, Mähönen AP, Sakakibara H, Helariutta Y (2011) Phloem transported cytokinin regulates polar auxin transport and maintains vascular pattern in the root meristem. Curr Biol 21:927–932

Björklund S, Antti H, Uddestrand I, Moritz T, Sundberg B (2007) Cross-talk between gibberellin and auxin in development of Populus wood: gibberellin stimulates polar auxin transport and has a common transcriptome with auxin. Plant J 52:499–511

Bolduc N, Hake S (2009) The maize transcription factor KNOTTED1 directly regulates the gibberellin catabolism gene ga2ox1. Plant Cell 21:1647–1658

Brenner WG, Romanov GA, Kollmer I, Burkle L, Schmulling T (2005) Immediate-early and delayed cytokinin response genes of *Arabidopsis thaliana* identified by genome-wide expression profiling reveal novel cytokinin-sensitive processes and suggest cytokinin action through transcriptional cascades. Plant J 44:314–333

Bürkle L, Cedzich A, Döpke C, Stransky H, Okumoto S, Gillissen B, Kühn C, Frommer WB (2003) Transport of cytokinins mediated by purine transporters of the PUP family expressed in phloem, hydathodes, and pollen of Arabidopsis. Plant J 34:13–26

Chen H, Banerjee AK, Hannapel DJ (2004) The tandem complex of BEL and KNOX partners is required for transcriptional repression of ga20ox1. Plant J 38:276–284

Chen HY, Huh JH, Yu YC, Ho LH, Chen LQ, Tholl D, Frommer WB, Guo WJ (2015) The Arabidopsis vacuolar sugar transporter SWEET2 limits carbon sequestration from roots and restricts *Pythium* infection. Plant J 83:1046–1058

Colebrook EH, Thomas SG, Phillips AL, Hedden P (2014) The role of gibberellin signaling in plant responses to abiotic stress. J Exp Biol 217:67–75

Cortleven A, Leuendorf JE, Frank M, Pezzetta D, Bolt S, Schmülling T (2019) Cytokinin action in response to abiotic and biotic stresses in plants. Plant Cell Environ 42:998–1018

David LC, Berquin P, Kanno Y, Seo M, Daniel-Vedele F, Ferrario-Méry S (2016) N availability modulates the role of NPF3.1, a gibberellin transporter, in GA-mediated phenotypes in Arabidopsis. Planta 244:1315–1328

Del Bianco M, Giustini L, Sabatini S (2013) Spatiotemporal changes in the role of cytokinin during root development. New Phytol 199:324–338

Dill A, Thomas SG, Hu J, Steber CM, Sun T-P (2004) The Arabidopsis F-box protein SLEEPY1 targets GA signaling repressors for GA-induced degradation. Plant Cell 16:1392–1405

El-Showk S, Ruonala R, Helariutta Y (2013) Crossing paths: cytokinin signalling and crosstalk. Development 140:1373–1383

Faiss M, Zalubìlová J, Strnad M, Schmülling T (1997) Conditional transgenic expression of the ipt gene indicates a function for cytokinins in paracrine signaling in whole tobacco plants. Plant J 12:401–415

Ferreira FJ, Kieber JJ (2005) Cytokinin signaling. Curr Opin. Plant Biol 8:518–525

Filardo FF, Swain SM (2003) SPYing on GA signaling and plant development. J Plant Growth Regul 22:163–175

Frébort I, Kowalska M, Hluska T, Frébortová J, Galuszka P (2011) Evolution of cytokinin biosynthesis and degradation. J Exp Bot 62:2431–2452

Gillissen B, Bürkle L, André B, Kühn C, Rentsch D, Brandl B, Frommer WB (2000) A new family of high-affinity transporters for adenine, cytosine, and purine derivatives in Arabidopsis. Plant Cell 12:291–300

Giulini A, Wang J, Jackson D (2004) Control of phyllotaxy by the cytokinin-inducible response regulator homologue ABPHYL1. Nature 430:1031–1034

Greenboim-Wainberg Y, Maymon I, Borochov R, Alvarez J, Olszewski N, Ori N, Eshed Y, Weiss D (2005) Cross talk between gibberellin and cytokinin: the Arabidopsis GA-response inhibitor SPINDLY plays a positive role in cytokinin signaling. Plant Cell 17:92–102

Griffiths J, Murase K, Rieu I, Zentella R, Zhang ZL, Powers SJ, Gong F, Phillips AL, Hedden P, Sun TP, Thomas SG (2007) Genetic characterization and functional analysis of the GID1 gibberellin receptors in Arabidopsis. Plant Cell 18:3399–3414

Hai NN, Chuong NN, Tu NHC, Kisiala A, Hoang XLT, Thao NP (2020) Role and regulation of cytokinins in plant response to drought stress. Plants 9:422

Hartweck LM, Scott CL, Olszewski NE (2002) Two O-linked N-acetylglucosamine transferase genes of *Arabidopsis thaliana* L. Heynh. have overlapping functions necessary for gamete and seed development. Genetics 161:1279–1291

Hay A, Kaur H, Phillips A, Hedden P, Hake S, Tsiantis M (2002) The gibberellin pathway mediates KNOTTED1-type homeobox function in plants with different body plans. Curr Biol 12:1557–1565

Heidstra R, Sabatini S (2014) Plant and animal stem cells: similar yet different. Nat Rev Mol Cell Biol 15:301–312

Higuchi M, Pischke MS, Mahonen AP, Miyawaki K, Hashimoto Y, Seki M, Kobayashi M, Shinozaki K, Kato T, Tabata S et al (2004) In planta functions of the Arabidopsis cytokinin receptor family. Proc Natl Acad Sci USA 101:8821–8826

Hwang I, Sheen J, Muller B (2012) Cytokinin signaling networks. Annu Rev Plant Biol 63:353–380

Itoh H, Ueguchi-Tanaka M, Sato Y, Ashikari M, Matsuoka M (2002) The gibberellin signaling pathway is regulated by the appearance and disappearance of SLENDER RICE1 in nuclei. Plant Cell 14:57–70

Kanno Y, Hanada A, Chiba Y, Ichikawa T, Nakazawa M, Matsui M, Koshiba T, Kamiya Y, Seo M (2012) Identification of an abscisic acid transporter by functional screening using the receptor complex as a sensor. Proc Natl Acad Sci USA 109:9653–9658

Kanno Y, Oikawa T, Chiba Y, Ishimaru Y, Shimizu T, Sano N, Koshiba T, Kamiya Y, Ueda M, Seo M (2016) AtSWEET13 and AtSWEET14 regulate gibberellin-mediated physiological processes. Nat Commun 7:13245

Katyayini NU, Rinne PLH, Tarkowská D, Strnad M, van der Schoot C (2020) Dual role of gibberellin in perennial shoot branching: Inhibition and activation. Front Plant Sci 11:736

Kaur H, Petla BP, Kamble NU, Singh A, Rao V, Salvi P, Ghosh S, Majee M (2015) Differentially expressed seed aging responsive heat shock protein OsHSP18.2 implicates in seed vigor, longevity and improves germination and seedling establishment under abiotic stress. Front Plant Sci 6:713

Klemens PA, Patzke K, Deitmer J, Spinner L, Le Hir R, Bellini C, Bedu M, Chardon F, Krapp A, Neuhaus HE (2013) Overexpression of the vacuolar sugar carrier AtSWEET16 modifies germination, growth, and stress tolerance in Arabidopsis. Plant Physiol 163:1338–1352

Ko D, Kang J, Kiba T, Park J, Kojima M, Do J, Kim KY, Kwon M, Endler A, Song WY, Martinoia E, Sakakibara H, Lee Y (2014) Arabidopsis ABCG14 is essential for the root-to-shoot translocation of cytokinin. Proc Natl Acad Sci USA 111:7150–7155

Krall L, Raschke M, Zenk MH, Baron C (2002) The Tzs protein from *Agrobacterium tumefaciens* C58 produces zeatin riboside 5'-phosphate from 4-hydroxy-3-methyl-2-(E)-butenyl diphosphate and AMP. FEBS Lett 527:315–318

Kramer EM (2006) How far can a molecule of weak acid travel in the apoplast or xylem? Plant Physiol 141:1233–1236

Kurakawa T, Ueda N, Maekawa M, Kobayashi K, Kojima M, Nagato Y, Sakakibara H, Kyozuka J (2007) Direct control of shoot meristem activity by a cytokinin-activating enzyme. Nature 445:652–655

Kuroha T, Tokunaga H, Kojima M, Ueda N, Ishida T, Nagawa S, Fukuda H, Sugimoto K, Sakakibara H (2009) Functional analyses of LONELY GUY cytokinin-activating enzymes reveal the importance of the direct activation pathway in Arabidopsis. Plant Cell 21:3152–3169

Lacombe B, Achard P (2016) Long-distance transport of phytohormones through the plant vascular system. Curr Opin Plant Biol 34:1–8

Li S-M, Zheng H-X, Zhang X-S, Sui N (2021) Cytokinins as central regulators during plant growth and stress response. Plant Cell Rep 40:271–282

MacMillan J (2002) Occurrence of gibberellins in vascular plants, fungi and bacteria. J Plant Growth Regul 20:387–442

Maekawa T, Maekawa-Yoshikawa M, Takeda N, Imaizumi-Anraku H, Murooka Y, Hayashi M (2009) Gibberellin controls the nodulation signaling pathway in *Lotus japonicus*. Plant J 58:183–194

Miyawaki K, Matsumoto-Kitano M, Kakimoto T (2004) Expression of cytokinin biosynthetic isopentenyltransferase genes in Arabidopsis: tissue specificity and regulation by auxin, cytokinin, and nitrate. Plant J 37:128–138

Pavlů J, Novák J, Koukalová V, Luklová M, Brzobohatý B, Černý M (2018) Cytokinin at the crossroads of abiotic stress signalling pathways. Int J Mol Sci 19:2450

Rodriguez-Falcon M, Bou J, Prat S (2006) Seasonal control of tuberization in potato: conserved elements with the flowering response. Annu Rev Plant Biol 57:151–180

Sakakibara H (2006) Cytokinins: activity, biosynthesis, and translocation. Annu Rev Plant Biol 57:431–449

Sakakibara H (2010) Cytokinin biosynthesis and metabolism. In: Davies P (ed) Plant hormones. Springer Netherlands

Salvi P, Manna M, Kaur H, Thakur T, Gandass N, Bhatt D, Muthamilarasan M (2021) Phytohormone signaling and crosstalk in regulating drought stress response in plants. Plant Cell Rep 40:1305–1329

Sasaki A, Itoh H, Gomi K, Ueguchi-Tanaka M, Ishiyama K, KobayashiM JDH, An G, Kitano J, Ashikari M et al (2003) Accumulation of phosphorylated repressor for gibberellin signaling in an F-box mutant. Science 299:1896–1898

Schmulling T, Werner T, Riefler M, Krupkova E, Bartrina y Manns I (2003) Structure and function of cytokinin oxidase/dehydrogenase genes of maize, rice, Arabidopsis and other species. J Plant Res 116:241–252

Schoor S, Farrow S, Blaschke H, Lee S, Perry G, von Schwartzenberg K, Emery N, Moffatt B (2011) Adenosine kinase contributes to cytokinin interconversion in Arabidopsis. Plant Physiol 157:659–672

Silverstone AL, Tseng TS, Swain SM, Dill A, Jeong SY, Olszewski NE, Sun TP (2007) Functional analysis of SPINDLY in gibberellin signaling in Arabidopsis. Plant Physiol 143:987–1000

Sponsel VM, Hedden P (2010) Gibberellin biosynthesis and inactivation. In: Plant hormones: biosynthesis, signal transduction, action! Springer Netherlands, pp 63–94

Srinivasan C, Mullins MG (1981) Physiology of flowering in the grapevine. Am J Enol Vitic 32:47–63

Sun T (2008) Gibberellin metabolism, perception and signaling pathways in Arabidopsis. Arabidopsis Book 6:e0103

Varbanova M, Yamaguchi S, Yang Y, McKelvey K, Hanada A, Borochov R, Yu F, Jikumaru Y, Ross J, Cortes D, Je Ma C, Noel JP, Mander L, Shulaev V, Kamiya Y, Rodermel S, Weiss D, Pichersky E (2007) Methylation of gibberellins by Arabidopsis GAMT1 and GAMT2. Plant Cell 19:32–45

Weiss D, Ori N (2007) Mechanisms of cross talk between gibberellin and other hormones. Plant Physiol 144:1240–1246

Wells L, Vosseller K, Hart GW (2001) Glycosylation of nucleocytoplasmic proteins: signal transduction and O-GlcNac. Science 291:2376–2378

Weyers JDB, Paterson NW (2001) Plant hormones and the control of physiological processes. New Phytol 152:375–407

Yamaguchi S (2008) Gibberellin metabolism and its regulation. Annu Rev Plant Biol 59:225–251

Yanai O, Shani E, Dolezal K, Tarkowski P, Sablowski R, Sandberg G, Samach A, Ori N (2005) Arabidopsis KNOXI proteins activate cytokinin biosynthesis. Curr Biol 15:1566–1571

Yevdakova NA, von Schwartzenberg K (2007) Characterisation of a prokaryote-type tRNA-isopentenyltransferase gene from the moss *Physcomitrella patens*. Planta 226:683–695

Zalabák D, Pospíšilová H, Šmehilová M, Mrízová K, Frébort I, Galuszka P (2013) Genetic engineering of cytokinin metabolism: prospective way to improve agricultural traits of crop plants. Biotechnol Adv 31:97–117

Zhang K, Novak O, Wei Z, Gou M, Zhang X, Yu Y, Yang H, Cai Y, Strnad M, Liu CJ (2014) Arabidopsis ABCG14 protein controls the acropetal translocation of root-synthesized cytokinins. Nat Commun 5:3274

Zhu Y, Nomura T, Xu Y, Zhang Y, Peng Y, Mao B, Hanada A, Zhou H, Wang R, Li P, Zhu X, Mander LN, Kamiya Y, Yamaguchi S, He Z (2006) Elongated uppermost internode encodes a cytochrome P450 monooxygenase that epoxidizes gibberellins in a novel deactivation reaction in rice. Plant Cell 18:442–456

In Vitro Responses of Some Mediterranean Fruit Crops to Auxin, Cytokinin and Gibberellin Treatments

Mouaad Amine Mazri, Meriyem Koufan, Rabha Abdelwahd, and Ilham Belkoura

Abstract Plant growth regulators (PGRs) are organic compounds widely used in tissue culture. PGRs have always been considered as key components of the culture medium, inducing different morphogenetic responses such as caulogenesis, rhizogenesis and somatic embryogenesis. Auxins, cytokinins and gibberellins are the most important PGR groups used in plant tissue culture. They are used either singly or in combination and are added at different concentrations depending on the species, genotype and explant source. Since the beginning of their use in tissue culture, auxins, cytokinins and gibberellins have been greatly involved in the development of efficient micropropagation systems for Mediterranean fruit species. The present chapter reports and discusses the main effects of auxins, cytokinins and gibberellins in the micropropagation of some economically important fruit crops of the Mediterranean region, namely olive (*Olea europaea* L.), cactus pear (*Opuntia* spp.), date palm (*Phoenix dactylifera* L.), argan (*Argania spinosa* L.), fig (*Ficus carica* L.), pomegranate (*Punica granatum* L.), carob (*Ceratonia siliqua* L.) and caper (*Capparis spinosa* L.).

M. A. Mazri (✉)
Unité de Recherche en Agro-Biotechnologie, CRRA de Marrakech, Institut National de la Recherche Agronomique, Avenue Ennasr, BP 415 Rabat Principale, 10090 Rabat, Morocco
e-mail: mouaadamine.mazri@inra.ma; m.a.mazri@gmail.com

M. Koufan
Unité de Recherche en Ressources Naturelles et Produits de Terroir, CRRA d'Agadir, Institut National de la Recherche Agronomique, Avenue Ennasr, BP 415 Rabat Principale, 10090 Rabat, Morocco

R. Abdelwahd
Unité de Recherche en Biotechnologie, CRRA de Rabat, Institut National de la Recherche Agronomique, Avenue Ennasr, BP 415 Rabat Principale, 10090 Rabat, Morocco

I. Belkoura
Laboratoire de Culture In Vitro, Département des Sciences de Base, École Nationale d'Agriculture, BP S/40, 50001 Meknes, Morocco

T. Aftab (ed.), *Auxins, Cytokinins and Gibberellins Signaling in Plants*, Signaling and Communication in Plants,
https://doi.org/10.1007/978-3-031-05427-3_5

1 Introduction

Since their discovery, plant growth regulators (PGRs) have played key roles in plant micropropagation. Indeed, exogenous PGRs interact with the endogenous phytohormones, thus affecting their concentrations and provoking cell division, differentiation and morphogenesis (Gaspar et al. 1996). However, the morphogenetic responses observed in vitro depend on the species, genotype, explant, as well as the type and concentration of added PGRs since the concentrations of endogenous phytohormones varies among genotypes and explants (Gaspar et al. 1996; Mazri 2014; Shen et al. 2020).

In plant cell and tissue culture, auxins and cytokinins are the most commonly used groups of PGRs. Auxins are known for their regulatory role of different aspects of plant growth and development at synthesis and distal sites (Blakeslee and Murphy 2016). They are involved in plant signaling systems and DNA methylation, as well as in basic cell processes such as division and elongation (Libbenga and Mennes 1995). Besides, auxins play a key role in maintaining the polarity and apical dominance of plants (Machakova et al. 2008). In tissue culture, auxins promote different regeneration processes such as callogenesis, somatic embryogenesis and organogenesis (mainly root formation).

Cytokinins represent the other important group of phytohormones involved in plant growth and development (Mazri 2015). In situ, cytokinins have various biological activities and are involved in protein synthesis, chloroplast maturation and cell-cycle control (Van Staden et al. 2008). In tissue culture, cytokinins are used to stimulate cell division and control morphogenesis. Cytokinins are most known for their potent activity in inducing shoots (Van Staden et al. 2008).

Auxins and cytokinins are generally used in combination to achieve different morphogenetic responses such as caulogenesis, rhizogenesis and somatic embryogenesis. The synthetic auxins and cytokinins have similar or superior biological activities to the naturally occurring ones, and the auxin/cytokinin ratio controls the process of differentiation and morphogenesis (Machakova et al. 2008).

Although the mode of action of auxins and cytokinins has not been fully elucidated, auxins are believed to regulate gene expression, whereas cytokinins activate RNA synthesis and stimulate the synthesis and activity of some proteins and enzymes (Machakova et al. 2008; Van Staden et al. 2008).

Regarding gibberellins, they represent a PGR group with more than 100 members widely used in plant tissue culture to stimulate stem and shoot elongation, and for dormancy release and embryo (zygotic and somatic) germination. They are involved in activating the intercalary meristem and reducing the content of endogenous abscisic acid (ABA) (Moshkov et al. 2008). For in vitro germination, gibberellins are used either as pretreatment or incorporated into the culture medium. Among the various gibberellins, the gibberellic acid (GA_3) is the most commonly used in plant tissue culture.

The present chapter is a review of the most common uses of auxins, cytokinins and gibberellins for the micropropagation of some Mediterranean fruit crops

(Table 1), namely olive (*Olea europaea* L.), cactus pear (*Opuntia* spp.), date palm (*Phoenix dactylifera* L.), argan (*Argania spinosa* L.), fig (*Ficus carica* L.), pomegranate (*Punica granatum* L.), carob (*Ceratonia siliqua* L.) and caper (*Capparis spinosa* L.). It highlights the morphogenetic responses depending on the species, genotype/cultivar, explant, as well as the type and concentration of PGRs.

2 Olive (*Olea Europaea* L.)

Olive (*Olea europaea* L.) is a fruit species that belongs to the Oleaceae family. It is native to the Mediterranean region where it plays important economic roles, mainly through the production of olive oil (Lambardi and Rugini 2003; Bajoub et al. 2018). Olive is a species difficult to propagate in vitro. However, during the last forty years, experiments have been undertaken to develop in vitro regeneration systems for this species, which can be used in large-scale propagation and genetic improvement (Mazri et al. 2020). Exogenous auxins, cytokinins and gibberellins have played key roles in the micropropagation of olive through somatic embryogenesis, organogenesis and microcuttings.

2.1 Axillary Bud Culture and Shoot Development

Many authors reported successful axillary shoot growth and proliferation from olive microcuttings. Generally, a cytokinin was added to culture medium. Based on our experiments, zeatin (3 mg L^{-1}) is the most effective cytokinin for axillary bud development in olive (Fig. 1). Chaari Rkhis et al. (2011) also suggested to use zeatin (2–4 mg L^{-1}) for shoot proliferation of cv. Oueslati. In cv. Moraiolo, Ali et al. (2009) recommended the combination of 3 mg L^{-1} zeatin and 0.5 mg L^{-1} 6-benzylaminopurine (BAP). Zeatin is a high active naturally occurring cytokinin that was first identified in *Zea mays* (L.), and that is involved in plant cell proliferation and differentiation (Letham 1963; Wang et al. 2018). Zeatin riboside was also suggested for axillary bud culture of olive. Indeed, Roussos and Pontikis (2002) and Sghir et al. (2005) compared the impact of different cytokinins on axillary shoot development of cvs. Salonenque, Amellau, Lucques, Haouzia, Dahbia, Picholine Marocaine, Picholine du Languedoc, ZDH4 and Koroneiki and noticed the superior effect of zeatin riboside (1–5 mg L^{-1}).

For root induction from microshoots, indole-3-butyric acid (IBA) has been widely used by researchers. IBA is a naturally occurring auxin commonly used in tissue culture to induce rhizogenesis (Machakova et al. 2008). In olive, IBA was either incorporated into the culture medium at 1–6 mg L^{-1} (Sghir et al. 2005; Rostami and Shahsavar 2012), or used as a pretreatment agent, by dipping shoots in an IBA solution for a few seconds (Peixe et al. 2009; Chaari Rkhis et al. 2011). In all cases, IBA promoted root induction from olive microshoots.

Table 1 Examples of in vitro responses of some Mediterranean fruit species to auxins, cytokinins and gibberellins

PGR group	PGR type	Species	Explant	PGR concentration (mg L^{-1})	Morphogenesis/In vitro response	Observations	References
Auxins	2,4-D	*Phoenix dactylifera* (L.)	Shoot tips, adventitious buds	5–100	Embryogenic callus induction	In combination with 0.15–3 mg L^{-1} 2iP	Tisserat (1979); Eke et al. (2005); Eshraghi et al. (2005); Al-Khayri (2011); El Hadrami et al. (1995); Mazri et al. (2017)
	NAA	*Punica granatum* (L.)	Shoots	0.25–1	Root induction	–	El-Agamy et al. (2009); ValizadehKaji et al. (2013)
	Picloram	*Phoenix dactylifera* (L.)	Shoot tips, adventitious buds	10–100	Embryogenic callus induction	In combination with 0.9–3 mg L^{-1} 2iP	Khierallah et al. (2015); Mazri et al. (2017, 2018b)
	Picloram	*Opuntia ficus indica*	Shoot apices devoid of leaf primordia	1–4	Somatic embryogenesis	–	Gomes et al. (2006)
	IBA	*Olea europaea* (L.)	Shoots	4–6	Root induction	–	Sghir et al. (2005); Rostami and Shahsavar (2012)
	IBA	*Olea europaea* (L.)	Zygotic explants	5	Callus induction	In combination with 0.5 mg L^{-1} 2iP	Orinos and Mitrakos (1991); Cererzo et al. (2011); Mazri et al. (2011, 2012); Oulbi et al. (2021)

(continued)

Table 1 (continued)

PGR group	PGR type	Species	Explant	PGR concentration (mg L^{-1})	Morphogenesis/In vitro response	Observations	References
	IBA	*Olea europaea* (L.)	Zygotic explants	0.5	Somatic embryo differentiation	–	Orinos and Mitrakos (1991); Cererzo et al (2011); Oulbi et al. (2021)
	IBA	*Opuntia amyclaea*; *O. ficus indica*	Shoots	1–10	Root induction	–	Escobar et al (1986); Zoghlami et al. (2012)
	IBA	*Ficus carica* (L.)	Shoots	1–2	Root induction	–	Soliman et al (2010); Dhage et al. (2015); Sahraroo et al. (2019)
	IBA	*Punica granatum* (L.)	Shoots	0.25–1	Root induction	–	El-Agamy et al. (2009); ValizadehKaji et al. (2013); Mulaei et al. (2019)
	IBA	*Argania spinosa* (L.) Skeels	Shoots	5	Root induction	In combination with 1–5 mg L^{-1} NAA	Bousselmame et al. (2001); Lamaoui et al. (2019)
	IBA	*Argania spinosa* (L.) Skeels	Shoots	1.5	Root induction	In combination with 0.5 mg L^{-1} NAA	Amghar et al. (2021b)

(continued)

Table 1 (continued)

PGR group	PGR type	Species	Explant	PGR concentration (mg L^{-1})	Morphogenesis/In vitro response	Observations	References
	IBA	*Ceratonia siliqua* (L.)	Shoots	1–2	Root induction and elongation	–	Naghmouchi et al. (2008); Radi et al. (2013); Zouari and El Mtili (2020)
Cytokinins	BAP	*Opuntia amyclaea*; *O. lanigera*; *O. ficus indica*	Cladode segments containing areoles	0.5–2.5	Shoot growth (areole activation)	–	Escobar et al. (1986); Estrada-Luna et al. (2008); Zoghlami et al. (2012)
	BAP	*Opuntia amyclaea*; *O. lanigera*; *O. ficus indica*	Segments of shoots established in vitro	0.1–7.5	Multiple shoot proliferation	–	Escobar et al. (1986); Estrada-Luna et al. (2008); Zoghlami et al. (2012)
	BAP	*Ficus carica* (L.)	Shoot tip explants	0.5–3	Multiple shoot induction and proliferation	Either alone or in combination with GA_3	Mustafa et al. (2013); Danial et al. (2014); Darwesh et al. (2014); Dhage et al. (2015)
	BAP	*Argania spinosa* (L.) Skeels	Epicotyl segments	2	Adventitious shoot induction	–	Amghar et al. (2021a)
	TDZ	*Olea europaea* (L.)	Petioles	6.6	Adventitious shoot induction	In combination with 0.1 mg L^{-1} NAA	Rugini and Caricato (1995); Mazri et al. (2013)

(continued)

Table 1 (continued)

PGR group	PGR type	Species	Explant	PGR concentration (mg L^{-1})	Morphogenesis/In vitro response	Observations	References
	TDZ	*Ficus carica* (L.)	Leaf explants; stem thin cell layers	1–7	Adventitious shoot induction	In combination with other PGRs	Dhage et al. (2015); Abdolinejad et al. (2020)
	Zeatin	*Olea europaea* (L.)	Microcuttings	2–4	Shoot growth	Alone or in combination with 0.5 mg L^{-1} BAP	Ali et al. (2009); Chaari Rkhis et al. (2011)
	Zeatin riboside	*Olea europaea* (L.)	Microcuttings	1–5	Shoot growth	–	Roussos and Pontikis (2002); Sghir et al. (2005)
	Kinetin	*Punica granatum* (L.)	Shoot tips and nodal segments	1–2	Shoot proliferation	In combination with 0.1 mg L^{-1} NAA	ValizadehKaji et al. (2013)
	Meta-topolin	*Capparis spinosa* (L.)	Nodal stem segments	0.6	Shoot proliferation	–	Kereša et al. (2019)
Gibberellins	GA_3	*Olea europaea* (L.)	Somatic embryos	0.1	Somatic embryo germination	In combination with 0.1 mg L^{-1} NAA	Mazri et al. (2020)
	GA_3	*Phoenix dactylifera* (L.)	Somatic embryos	0.5–1	Somatic embryo germination	–	Mazri et al. (2019a)
	GA_3	*Opuntia ficus indica* (L.)	Ovules	1	Callogenesis, somatic embryogenesis	–	Jedidi et al. (2015)
	GA_3	*Opuntia ficus indica* (L.)	Somatic embryos	0.1	Somatic embryo germination	–	Jedidi et al. (2015)
	GA_3	*Argania spinosa* (L.) Skeels	Axillary shoots	1	Shoot elongation	–	Koufan et al. (2018, 2020b)

(continued)

Table 1 (continued)

PGR group	PGR type	Species	Explant	PGR concentration (mg L^{-1})	Morphogenesis/In vitro response	Observations	References
	GA_3	*Argania spinosa* (L.) Skeels	Epicotyl segments	2	Shoot bud multiplication	Either alone or in combination with 1 mg L^{-1} BAP	Amghar et al. (2021a)

Fig. 1 Zeatin promotes shoot elongation in olive (*Olea europaea* L.)

2.2 Adventitious Regeneration

In olive, somatic embryogenesis has been reported by many authors. This regeneration process was achieved in two steps: embryogenic callus formation and embryo differentiation. Generally, an auxin-cytokinin combination was suggested to induce embryogenic calli. In embryo-derived explants of cvs. Arbequina, Dahbia, Picual, Picholine Marcaine as well as wild olive (*Olea europaea* var. *sylvestris*), the combination of 5 mg L^{-1} IBA and 0.5 mg L^{-1} N6-[2-isopentenyl] adenine (2iP) was used for embryogenic callus induction while embryo differentiation was achieved either on PGR-free medium or on a medium containing 0.5 mg L^{-1} IBA (Orinos and Mitrakos 1991; Cererzo et al. 2011; Mazri et al. 2011, 2012; Oulbi et al. 2021).

In shoot apexes and leaf segments derived from shoots maintained in vitro, a high thidiazuron (TDZ) concentration was recommended to stimulate the embryogenesis competence of explants. Indeed, in cvs. Dahbia and Picual, and the wild genotypes StopVert and Ac-18, the combination of 6.6–7 mg L^{-1} TDZ and 0.1 mg L^{-1} 1-naphthaleneacetic acid (NAA) was used. Besides, a medium containing 0.05 mg L^{-1} IBA, 0.1 mg L^{-1} 2iP and 0.1 mg L^{-1} BAP allowed for embryo differentiation (Mazri et al. 2013; Toufik et al. 2014; Narváez et al. 2019).

TDZ has been also used to induce adventitious shoot buds from olive explants. Mencuccini and Rugini (1993) evaluated the effects of different culture media, PGRs, cultivars and explants and found that adding 1.1–2.2 mg L^{-1} TDZ to Murashige and Skoog (MS; Murashige and Skoog 1962) medium promoted adventitious shoot induction from leaf petioles of cvs. Moraiolo, Dolce Agogia and Halidikis. In cvs. Canino and Moraiolo, Rugini and Caricato (1995) observed that the use of 6.6 mg L^{-1} TDZ in combination with 0.1 mg L^{-1} NAA resulted in the formation of single shoots from petiole explants, and somtimes a group of 2 to 3 shoots. Similar results were observed in cv. Dahbia (Mazri et al. 2013). These findings clearly highlight the key role of TDZ in inducing organogenesis in olive. TDZ is a synthetic phenylurea derivative with a high cytokinin-like activity. It was initially used as a cotton defoliant but is now widely used in tissue culture. In fact, TDZ has shown to be a potent compound for shoot bud induction and proliferation, and sometimes was found more effective than adenine-derivative cytokinins (Mazri et al. 2018a).

In olive adventitious regeneration, GA_3 was used to promote the germination of somatic embryos of cv. Dahbia. GA_3 was used in combination with NAA at 0.1 mg L^{-1} each (Mazri et al. 2020).

3 Date Palm (*Phoenix Dactylifera* L.)

Date palm (*Phoenix dactylifera* L.) is a plant species of the Arecaceae family. It is originated from Mesopotamia (modern Iraq) and widely cultivated in the arid and semi-arid regions of the Middle-East and North-Africa (Johnson et al. 2013). For decades, date palm has been propagated conventionally by offshoots. Today, date palm is mainly propagated by tissue culture, either through somatic embryogenesis or organogenesis (Mazri and Meziani 2015). Indeed, these methods allow for the production of a large number of date palm plants in a small space and a short period of time. To achieve somatic embryogenesis, 2,4-dichlorophenoxyacetic acid (2,4-D) has been widely used by researchers and practitioners as it showed a potent ability to induce callogenesis and somatic embryos. Regarding organogenesis, a wide range of auxin-cytokinin combinations were recommended, depending on the cultivar.

3.1 Somatic Embryogenesis

In date palm, 2,4-D is the most commonly used auxin to induce embryogenic calli (Fig. 2). Indeed, since the 70th of the last century, many authors have used 2,4-D to achieve somatic embryogenesis from different date palm cultivars. Tisserat (1979) described a somatic embryogenesis process in which callogenesis was achieved on plant growth media containing 10–100 mg L^{-1} 2,4-D. Generally, a high level of 2,4-D (100 mg L^{-1}) was used to induce embryogenic callus (Eke et al. 2005; Eshraghi et al. 2005; Al-Khayri 2011). However, some authors were able to achieve this process in media containing lower 2,4-D concentrations ($\leq$10 mg L^{-1}) (El Hadrami et al. 1995; Fki et al. 2003; Zouine and El Hadrami 2007; Othmani et al. 2009; Abohatem et al. 2011; Mazri et al. 2017). 2,4-D is a synthetic auxin and the most frequently used in tissue culture to induce callogenesis (Machakova et al. 2008). It was discovered by Zimmerman and Hitchcock (1942) and has also been used as an herbicide to control annual and perennial weeds (Song 2014).

Other auxins were also used to induce embryogenic calli in date palm, but only to a limited extent. For example, picloram (Khierallah et al. 2015; Mazri et al. 2017,

Fig. 2 2,4-dichlorophenoxyacetic acid (2,4-D) induces embryogenic calli in date palm (*Phoenix dactylifera* L.)

2018b). A cytokinin, generally 2iP or BAP, was added to culture medium at a low concentration (Eshraghi et al. 2005; Al-Khayri 2011; Zouine and El Hadrami 2007; Abohatem et al. 2011; Al-Khayri and Al-Bahrany 2012; Mazri et al. 2017, 2018b).

After callus induction, somatic embryo formation was achieved either on a medium containing lower PGR concentrations; for example, 0.5 mg L^{-1} 2,4-D and 0.1 mg L^{-1} BAP (Zouine and El Hadrami 2007; Abohatem et al. 2011); or on a PGR-free medium (Mazri et al. 2018b).

In date palm somatic embryogenesis, GA_3 (0.5–1 mg L^{-1}) was used to promote somatic embryo germination and conversion (Mazri et al. 2019a). An auxin-cytokinin combination has also been suggested for somatic embryo germination. For example, 0.1 mg L^{-1} NAA, 0.1 mg L^{-1} IBA and 0.05 mg L^{-1} BAP (Zouine and El Hadrami 2007), 0.4–1 mg L^{-1} NAA and 0.5–1.2 mg L^{-1} BAP (Mazri et al. 2018b, 2019b).

3.2 *Adventitious Organogenesis*

Date palm propagation by organogenesis is the method used in Morocco for grove rehabilitation and the creation of new orchards. For adventitious bud induction, the combination of 3 mg L^{-1} 2-naphthoxyacetic acid (NOA), 1 mg L^{-1} NAA, 1 mg L^{-1} indole-3-acetic acid (IAA) and 0.1 mg L^{-1} 2iP showed good results in many Moroccan cultivars (e.g. cv. Najda, Mazri and Meziani (2013); cv. Boufeggous, Mazri (2015); cv. Al-Fayda, Mazri et al. (2019c)). The combination of 2.5 mg L^{-1} IAA, 2.5 mg L^{-1} NAA and 0.1 mg L^{-1} 2iP was also suggested for cv. Mejhoul (Meziani et al. 2016). For adventitious shoot bud multiplication, different auxin-cytokinin combinations were recommended, depending on the cultivar: 0.5 mg L^{-1} NOA and 0.5 mg L^{-1} kinetin for cvs. Najda and Al-Fayda; 0.6 mg L^{-1} IBA and 0.7 mg L^{-1} BAP for cv. Boufeggous; 0.2 mg L^{-1} NOA, 0.2 mg L^{-1} IAA, 0.4 mg L^{-1} kinetin and 0.4 mg L^{-1} 2iP for cv. Mejhoul (Mazri 2015; Mazri and Meziani 2013; Mazri et al. 2019c; Meziani et al. 2015). The use of different PGR combinations may reflect different hormonal requirements among date palm cultivars. On the other hand, our pervious experiments revealed the necessity to enrich the culture medium with an auxin-cytokinin combination, which supported better shoot bud multiplication than either PGR group used singly (Mazri 2015; Mazri and Meziani 2013).

In addition to the above studies carried out in Morocco, authors from other countries recommended different PGR combinations to achieve organogenesis in date palm. For example, for adventitious bud induction, the following PGRs were used: 1.6 mg L^{-1} IAA in cv. Khenezi (Al Kaabi et al. 2001); 1 mg L^{-1} NAA and 2 mg L^{-1} 2iP in cv. Alshakr (Al-Mayahi 2016), 4 mg L^{-1} IBA and 1 mg L^{-1} BAP in cvs. Asil, Hussaini, Zaidi (Hussain et al. 2001), 1 mg L^{-1} NAA, 1 mg L^{-1} NOA, 1 mg L^{-1} BAP, 2–4 mg L^{-1} 2iP in cvs. Barhee and Maktoom (Jazinizadeh et al. 2015; Khierallah and Bader 2007).

For shoot bud multiplication, the following PGRs were suggested: 0.4 mg L^{-1} IAA, 0.1 mg L^{-1} NAA, 0.1 mg L^{-1} kinetin, 1.5 mg L^{-1} 2iP for cv. Khenezi (Al Kaabi et al. 2001); 1 mg L^{-1} NAA, 0.5 mg L^{-1} BAP, 0.5 mg L^{-1} kinetin for cvs. Alshakr and Dhakki (Khan and Bi Bi 2012; Al-Mayahi 2016); 5 mg L^{-1} 2iP and 2 mg L^{-1} kinetin for cv. Zaghlool (Bekheet 2013); 1 mg L^{-1} NAA, 1.5 mg L^{-1} 2iP, 1 mg L^{-1} BAP for cv. Barhee (Jazinizadeh et al. 2015); and 1 mg L^{-1} NAA, 1 mg L^{-1} NOA, 4 mg L^{-1} 2iP and 2 mg L^{-1} BAP for cv. Maktoom (Khierallah and Bader 2007).

The use of TDZ for adventitious shoot bud induction and multiplication showed contradictory results. In fact, a study carried out on cv. Boufeggous revealed that TDZ provokes severe tissue browning, leading to explant death (Mazri 2015). However, Al-Mayahi (2014) reported that in cv. Hillawi, the combination of 1 mg L^{-1} BAP and 0.5 mg L^{-1} TDZ promoted adventitious bud induction and multiplication. A recent study by Taha et al. (2021) also reported that TDZ (either alone or in combination with BAP) induced direct organogenesis from immature inflorescence of date palm cvs. Selmi, Barhee and Medjool. These results confirm again the necessity to optimize the medium components, particularly in terms of PGRs, for each genotype.

For shoot rooting, the use of PGR-free medium was suggested for many cultivars (Mazri and Meziani 2013; Mazri 2015; Mazri et al. 2016; Meziani et al. 2019). However, some authors suggested the addition of 0.1–1.5 mg L^{-1} NAA to the culture medium (Al Kaabi et al. 2001; Khierallah and Bader 2007; Khan and Bi Bi 2012; Bekheet 2013; Jazinizadeh et al. 2015). NAA is a strong synthetic auxin involved in cell division and growth, and widely used for callogenesis, somatic embryogenesis and rhizogenesis (Machakova et al. 2008).

4 Cactus Pear (*Opuntia* spp.)

Cactus pear (*Opuntia* spp.) is a multipurpose tree native to the tropical and subtropical regions of America (Mazri 2021). Today, cactus pear is found in many regions of the world. Indeed, this plant is characterized by its remarkable adaptation to different climatic and environmental conditions (Mazri 2018; Vekiari and Ouzounidou 2018). In some Mediterranean countries such as Morocco, Tunisia and Italy, cactus pear plays important socio-economic and ecological roles. Italy is the main world exporter of cactus pear fruits (Albergamo et al. 2021).

4.1 Organogenesis

Cactus pear is generally propagated by cladodes. However, since the 1970s, many researchers have tried to develop in vitro culture systems for this plant to achieve rapid and large-scale propagation, combat desertification and as a tool for fundamental research (Mauseth and Halperin 1975; Mauseth 1976; Escobar et al. 1986; Khalafalla et al. 2007). In vitro propagation of cactus was mainly achieved through organogenesis. BAP has been used to stimulate areole activation and shoot growth. For example, in *O. lanigera* and *O. amyclaea* Tenore cv. Copena-5, 2.2–2.5 mg L^{-1} BAP was suggested (Escobar et al. 1986; Estrada-Luna et al. 2008). In *O. ellisiana*, Juarez and Passera (2002) recommended the combination of 2.2 mg L^{-1} BAP and 2 mg L^{-1} IBA. In *O. ficus indica*, BAP was used either alone (0.5 mg L^{-1}; Zoghlami et al. 2012) or in combination with GA_3 (0.5 mg L^{-1} BAP and 0.5 mg L^{-1} GA_3; Garcia-Saucedo et al. 2005), depending on the genotype.

For shoot proliferation, segments of axillary shoots established in vitro were used as explants. During this culture phase, BAP was added to the medium at different concentrations, depending on the species/cultivar used. For example, in *O. amyclaea* Tenore cv. Copena-5, Escobar et al. (1986) recommended 2.2 mg L^{-1} BAP. Estrada-Luna et al. (2008) observed that BAP at 5–7.5 mg L^{-1} promoted shoot proliferation in *O. lanigera* while in *O. ficus indica*, BAP was recommended at a concentration ranging from 0.1 to 0.5 mg L^{-1} (Garcia-Saucedo et al. 2005; Zoghlami et al. 2012).

The experiments carried out in our laboratory (data not published) showed that, in *O. ficus indica* explants, the use of BAP at a concentration ranging from 2.5 to 4 mg L^{-1} resulted in 100% shoot development (Fig. 3). Rham (2021) compared the effects of different cytokinins (TDZ, 2iP and BAP) on shoot formation from areoles and found that BAP was the most effective. Our experiments also indicated that BAP at 2–4 mg L^{-1} promoted shoot proliferation (Fig. 3).

As it could be seen from the above examples, BAP is necessary to induce and sustain organogenesis in cactus. BAP is a synthetic cytokinin used routinely by commercial and research laboratories due to its efficacy and affordability. It promotes cell division and has been established as a potent growth regulator of morphogenetic responses in many plant species (Van Staden et al. 2008).

Fig. 3 6-Benzylaminopurine (BAP) promotes areole activation and multiple shoot proliferation in cactus pear (*Opuntia ficus indica*)

For root induction from cactus shoots, IBA is the most commonly used PGR. This auxin was used alone at concentrations ranging from 1 to 10 mg L^{-1} depending on the species and genotype (Escobar et al. 1986; Juarez and Passera 2002; Garcia-Saucedo et al. 2005; Zoghlami et al. 2012).

4.2 Somatic Embryogenesis

Studies on somatic embryogenesis of cactus pear are scarce. Different PGRs were recommended depending on the explant. For example, in *O. ficus indica* cv. Gigante, direct somatic embryogenesis was achieved from shoot apices devoid of leaf primordia by supplementing picloram (1–4 mg L^{-1}) in the culture medium (Gomes et al. 2006). Somatic embryos were also obtained from callus cultures derived from immature anthers of *O. ficus indica* cvs. Moore and Gialla in media containing 2 mg L^{-1} 2,4-D and 2.5 mg L^{-1} TDZ (Bouamama et al. 2011). Jedidi et al. (2015) were able to induce callogenesis then somatic embryogenesis from ovules (10 days after anthesis) of *O. ficus indica* cultured on MS medium containing 1 mg L^{-1} GA_3.

These examples from the literature support the necessity of incorporating auxins, cytokinins and/or gibberellins into the culture medium to achieve somatic embryogenesis in cactus pear. On the other hand, GA_3 (0.1 mg L^{-1}) was also suggested for somatic embryo germination (Jedidi et al. 2015). GA_3 is a natural growth hormone that represses the inhibitory effect of endogenous ABA and promotes dormancy release (Skubacz and Daszkowska-Golec 2017; Li et al. 2021). Accordingly, GA_3 was used to promote somatic embryo germination of many plant species (Moshkov et al. 2008).

5 Fig (*Ficus Carica* L.)

Fig is a fruit tree probably native to north-eastern Asia or southern Arabia where the wild fig (i.e. caprifig) is still exist (Ikegami et al. 2009). Fig has been cultivated since ancient times and spread to several regions, especially those around the Mediterranean basin between 6000 and 4000 BC (Kislev et al. 2006). The common fig (*Ficus carica* L.) belongs to the Moraceae family. It is the only species cultivated for its edible fruits and that adapts well to the Mediterranean climate (Crisosto et al. 2020; Akin et al. 2021).

Fig can be propagated vegetatively by cuttings, air-layering or grafting. However, these methods are slow and not well suited for rapid and large-scale production of fig plants. The use of tissue culture techniques has many advantages for fig such as the mass propagation of elite genotypes (Kumar et al. 1998; Fraguas et al. 2004; Kim et al. 2007; Soliman et al. 2010), preservation of genetic material by long-term cryopreservation (Abd El-Wahab and Sayed 2019) sanitation/generation of pathogen-free plants (Comlekcioglu et al. 2007; Al-Shomali et al. 2017; Sahraroo et al. 2019), genetic transformation (Yancheva et al. 2005; Soliman et al. 2010), precise genome editing (Flaishman et al. 2020) and secondary metabolite production (Amani et al. 2020). The development and optimization of micropropagation systems for fig rely on many factors, among which PGRs, particularly auxins, cytokinins and gibberellins are the most important.

5.1 *Shoot Induction and Proliferation*

In vitro regeneration was successfully achieved from shoot tips of different fig cultivars. Generally, the cytokinin BAP was critical for shoot induction and proliferation (Sahraroo et al. 2019). Along this line, 0.5–3 mg L^{-1} BAP was recommended by Mustafa et al. (2013), Danial et al. (2014) and Ling et al. (2018).

The combination of BAP and GA_3 was also recommended for shoot formation in fig. Indeed, in cvs. Deanna and Conadria, Dhage et al. (2015) observed multiple shoot formation in a culture medium containing 2.5–3.5 mg L^{-1} BAP and 0.5 mg L^{-1} GA_3 while Darwesh et al. (2014) recommended the combination of 5.0 mg L^{-1} BAP and 1 mg L^{-1} GA_3. Other authors reported the beneficial effect of other cytokinins in stimulating shoot production from fig explants. For example, Abd El-Wahab and Sayed (2019) observed shoot proliferation from apical bud explants of cv. Sultani in all media supplemented with BAP or kinetin (0.5–1.0 mg L^{-1}). However, the highest number of shoots per explant was obtained in the medium containing 0.5 mg L^{-1} kinetin.

The optimal PGR combination for shoot induction and proliferation in fig may vary among cultivars. Shahcheraghi and Shekafandeh (2016) found that combining 0.5 mg L^{-1} BAP and 0.2 mg L^{-1} 2iP significantly improved shoot proliferation in cvs. Bargchenari and Runu. However, in cv. Dehdez, the combination of 6 mg L^{-1}

kinetin and 0.2 mg L^{-1} NAA gave the highest number of shoots per explant. This may reflect the different hormonal requirements for efficient shoot proliferation among fig cultivars.

5.2 *Adventitious Regeneration*

Many authors investigated the effects of PGRs on adventitious regeneration of fig (e.g. Kim et al. 2007; Soliman et al. 2010). The beneficial effect of TDZ (in combination with other PGRs) on callus induction and adventitious shoot formation was reported (Kim et al. 2007; Soliman et al. 2010; Mazri et al. 2018a). In cvs. Sabz and Torsh, the formation of morphogenetic calli from thin cell layer cultures was observed in media containing 2 mg L^{-1} TDZ and 2 mg L^{-1} IBA (Abdolinejad et al. 2020). Dhage et al. (2015) highlighted the beneficial effect of TDZ (7 mg L^{-1}) in combination with 0.25 mg L^{-1} NAA on shoot induction from callus of cv. Poona Fig. On the other hand, Abdolinejad et al. (2020) observed the best shoot regeneration response from callus on medium containing 4 mg L^{-1} BAP, 1 mg L^{-1} TDZ and 0.2 mg L^{-1} NAA.

After comparing the effects of BAP, kinetin and TDZ at different concentrations on multiple shoot induction from apical buds of cv. Black Jack, Parab et al. (2021) reported that TDZ and kinetin at 4.3–4.5 mg L^{-1} produced only a low average number of shoots per explant, not exceeding 5. On the other hand, BAP gave 37.8 shoots per explant, with no callus formation. These results may reflect different responses to exogenous cytokinins depending on the genotype and explant type. According to Soliman et al. (2010), in cv. Sultani, the combination of 2 mg L^{-1} TDZ and 4–6 mg L^{-1} 2iP promoted shoot production via direct embryogenesis.

5.3 *Rhizogenesis*

In fig, IBA can be considered as a potent growth regulator for root induction from in vitro shoots. In fact, in cvs. Sabz, Jaami-e-Kan and Sultani, Sahraroo et al. (2019) and Soliman et al. (2010) observed high root induction rates in media containing 1–2 mg L^{-1} IBA. Similarly, Dhage et al. (2015) reported that 1 mg L^{-1} IBA stimulated root induction in cvs. Conadria, Deanna, Brown Turkey and Poona Fig.

In addition to IBA, other PGRs were used either singly or in combination to induce rooting in fig. For example, 0.5 mg L^{-1} NAA (Abd El-Wahab and Sayed 2019), and the combination of 4.5 mg L^{-1} BAP and 1.4 mg L^{-1} IAA (Parab et al. 2021). This latter combination gave 100% rhizogenesis.

6 Pomegranate (*Punica granatum* L.)

Pomegranate (*Punica granatum* L.) is a plant native to the Himalayas, from northern India to Iran (Mazri et al. 2018a). It belongs to the Punicaceae family and is characterized by good adaptability to different soil and climate conditions (Ferrara et al. 2014). Pomegranate is cultivated on a commercial scale in a number of Mediterranean and Asian countries, with Iran being the world largest exporter (Gharaghani et al. 2017).

The global demand for pomegranate fruits has recently increased due to their high nutraceutical values (Jalikop 2010; Kotsampasi et al. 2021). However, pomegranate propagation is still a big challenge. Propagation by seeds is difficult to achieve since the seeds quickly lose their ability to germinate. Moreover, the plants produced by this method are characterized by high heterozygosity (Kanwar et al. 2010; Singh et al. 2013; Kahramanoğlu and Umar 2018). Vegetative propagation by cuttings enables the production of true-to-type plants at low cost. However, this method is hampered by high mortality and poor growth rates (Desai et al. 2018). *In vitro* propagation of pomegranate can be used for different purposes such as rapid and large-scale production of elite cultivars, genetic improvement and conservation of genetic resources. Along this line, several studies reported successful regeneration of pomegranate through organogenesis, somatic embryogenesis and micro-cuttings (e.g. Omura et al. 1987; Naik et al. 2000; Murkute et al. 2002; Shao et al. 2003; Terakami et al. 2007; Chauhan and Kanwar 2012; Desai et al. 2018; Kabir et al. 2021).

6.1 *Shoot Induction and Proliferation*

PGRs have played a major role in pomegranate micropropagation. Many authors tried to determine the optimal concentration of PGRs for shoot induction, proliferation and growth (e.g. El-Agamy et al. 2009; Naik and Chand 2011). ValizadehKaji et al. (2013) reported that the combination of 1–2 mg L^{-1} kinetin and 0.1 mg L^{-1} NAA promoted shoot induction in cvs. Malas Saveh and Yusef Khani. Indeed, the use of kinetin alone (auxin-free medium) decreased shoot elongation and proliferation by about 30%. However, in other pomegranate cultivars, exogenous auxins were not necessary for shoot proliferation, while the cytokinin effect varied among cultivars (Naik and Chand 2011).

The effects of cytokinins on shoot multiplication were compared by many authors. For example, Naik et al. (1999) evaluated the effects of zeatin riboside, BAP and TDZ on axillary shoot proliferation from nodal segments of an adult pomegranate tree, and found that zeatin riboside was the most effective cytokinin. Indeed, the use of 2 mg L^{-1} zeatin riboside gave the highest number of shoots per explant (5.2). On the other hand, the use of BAP at 1 mg L^{-1} resulted in higher bud break percentage (93%) than zeatin riboside (85%). However, BAP gave lower number of shoots per explant

(2.1). El-Agamy et al. (2009) compared the effects of BAP and kinetin on in vitro shoot proliferation from shoot tips of two pomegranate cultivars, Manfalouty and Nab El-Gamal, and highlighted the superior effect of BAP over kinetin. The beneficial effect of BAP on pomegranate shoot proliferation was also reported by Naik et al. (2000) and Desai et al. (2018).

6.2 *Rhizogenesis*

Many studies were carried out to evaluate the effects of PGRs on shoot rooting, and the use of auxins was found necessary (Naik et al. 1999; Deepika and Kanwar 2010; ValizadehKaji et al. 2013). Naik et al. (1999) reported that 1 mg L^{-1} IBA induced better rooting than IAA. Moreover, lower or higher IBA concentrations reduced rooting. According to Kanwar et al. (2010), no root induction was observed in auxin-free media. On the other hand, addition of NAA or IBA at 0.25 mg L^{-1} promoted rhizogenesis in cvs. Nab El Damal and Manfalouty (El-Agamy et al. 2009). Similar findings were reported by ValizadehKaji et al. (2013) who observed that adding 1 mg L^{-1} NAA or 1 mg L^{-1} IBA to culture medium stimulated root induction in cvs. Malas Saveh and Yusef Khani.

According to Mulaei et al. (2019), the use of 0.5 mg L^{-1} IBA promoted rhizogenesis in cvs. Malase Yazdi and Shirine Shahvar. Besides, IBA was found to have a better effect on root induction than NAA. In contrast, Naik et al. (2000) reported that in cv. Ganesh, a low NAA concentration (i.e. 0.1 mg L^{-1}) promoted rhizogenesis. These findings highlight the different auxin requirements for efficient rooting among pomegranate genotypes.

6.3 *Adventitious Regeneration*

In pomegranate, shoot and somatic embryo regeneration from callus induced in vitro was generally achieved by combining NAA and BAP. Kanwar et al. (2010) described indirect regeneration systems for pomegranate in which shoots and somatic embryos were obtained from zygotic embryo-derived calli. The highest induction frequency of organogenic calli was observed when combining 1.8 mg L^{-1} BAP, 1.1 mg L^{-1} NAA, and 2 mg L^{-1} GA_3. Regarding somatic embryogenesis, the highest mean number of globular and heart-shaped somatic embryos per callus was observed when combining 4 mg L^{-1} NAA and 2 mg L^{-1} BAP.

Murkute et al. (2002), Soukhak et al. (2011) and Deepika and Kanwar (2010) reported that the combination of 0.5–4 mg L^{-1} BAP and 0.4–2.4 mg L^{-1} NAA stimulated callus induction and shoot regeneration in cvs. Ganesh, MalasSaveh and Kandhari Kabuli. Direct organogenesis was also successfully achieved from hypocotyl and mature leaf explants of cv. Kandhari Kabuli in media containing the combination of 2–2.2 mg L^{-1} BAP and 0.4–1.5 mg L^{-1} NAA (Parmar et al. 2013; Verma et al. 2021).

All these findings highlight the beneficial effect of BAP and NAA on adventitious regeneration of pomegranate.

7 Argan (*Argania spinosa* L.)

Argan (*Argania spinosa* (L.) Skeels) is an endangered agroforestry species that belongs to the family Sapotaceae (Koufan et al. 2020a). It is endemic to Morocco where it covers an area of around 871,210 ha (Moukrim et al. 2018). Argan has a significant socio-economic impact in its area of cultivation, mainly through the production of edible and cosmetic oils. Indeed, argan oil is one of the most expensive oils in the world, with both nutritive and health promoting properties (El Kharrassi et al. 2018). Besides, the argan forest contributes to soil and biodiversity conservation and desertification prevention (Koufan et al. 2020b).

The sustainability of the argan ecosystem is threatened by several factors such as overexploitation and overgrazing, which causes a continuous degradation (De Waroux and Lambin 2012; Charrouf and Guillaume 2009). Propagation by tissue culture can be considered as a promising approach to rehabilitate and preserve the argan ecosystem. However, this species is highly recalcitrant to in vitro manipulations, and the published data on argan micropropagation are still insufficient. To date, only few in vitro regeneration protocols have been described, in which auxins, cytokinins and gibberellins were involved.

7.1 *Bud Break, Axillary Shoot Development and Rooting*

Argan propagation by in vitro shoot culture of adult origin was investigated by some authors. Generally, bud break and shoot growth were influenced by PGRs. Lamaoui et al. (2019) found that the combination of 1 mg L^{-1} IAA and 2.5 mg L^{-1} BAP was optimal for bud break, while BAP at 1–1.5 mg L^{-1} promoted shoot growth and multiplication. Bousselmame et al. (2001) used the combination of 1 mg L^{-1} IAA, 2 mg L^{-1} BAP and 0.5 mg L^{-1} kinetin for bud break and axillary shoot development, while Koufan et al. (2018, 2020b) used a PGR-free medium for bud break and 1 mg L^{-1} GA_3 for shoot growth (Fig. 4). These results highlight the importance of IAA, BAP and GA_3 in argan propagation by microcuttings. The different combinations and concentrations recommended by different authors may reflect different hormonal requirement depending on the genotype, mother tree provenance and age, and the explant type (i.e. hardwood, semi-hardwood and herbaceous). These factors are a constraint to the development of a standardized and reproducible micropropagation protocol for different argan genotypes.

The prominent role of auxins in shoot rooting was reported by different authors (e.g. Bousselmame et al. 2001; Koufan et al. 2018; Lamaoui et al. 2019). Indeed, the induction of roots from in vitro shoots is difficult to achieve. Thus, addition of

Fig. 4 Gibberellic acid (GA_3) promotes shoot elongation in argan (*Argania spinosa* L.)

auxins to the culture medium is necessary. Bousselmame et al. (2001) and Lamaoui et al. (2019) used high auxin concentrations to induce rooting. These authors recommended the combination of 5 mg L^{-1} IBA and 5 mg L^{-1} NAA (for 14 days), and that of 5 mg L^{-1} IBA and 1 mg L^{-1} NAA, respectively. On the other hand, Koufan et al. (2018) suggested the combination of 0.5 mg L^{-1} NAA and 0.5 mg L^{-1} BAP.

7.2 Adventitious Organogenesis

In vitro regeneration through organogenesis could be of great interest for rapid propagation and genetic improvement of argan. Recently, an in vitro regeneration system through adventitious organogenesis was described for argan (Amghar et al. 2021a, b). It was found that adventitious bud induction, shoot proliferation and elongation were highly influenced by PGR type and concentration. BAP and kinetin were compared at different concentrations and the superior effect of BAP at 2 mg L^{-1} on adventitious bud induction was demonstrated. On the other hand, the use of 2 mg L^{-1} GA_3, either alone or in combination with 1 mg L^{-1} BAP, promoted shoot bud multiplication (Amghar et al. 2021a). Root induction from adventitious shoots was directly linked to auxins and their concentrations. The highest rooting percentage was achieved by combining 0.5 mg L^{-1} NAA and 1.5 mg L^{-1} IBA. Besides, it was noticed that increasing NAA concentration decreased the rooting ability of argan adventitious shoots (Amghar et al. 2021b). These findings highlight the key role of auxins, cytokinins and gibberellins in the adventitious regeneration of argan, which is a highly recalcitrant species to in vitro culture.

8 Carob (*Ceratonia siliqua* L.)

Carob (*Ceratonia siliqua* L.) is an agroforestry species endemic to the Mediterranean region. It belongs to the family Fabaceae and is one of the most important species of the Mediterranean arid and semi-arid zones, due to its socio-economic, ecological, industrial and medicinal benefits (Aafi 1996; El Kahkahi et al. 2016). The main producing countries of carob fruits are Spain, Italy, Portugal and Morocco (Naggar and Lahssini 2015).

Carob is generally propagated by conventional methods. However, these methods fail to meet the growing demands of carob plants. The use of in vitro culture systems is a considerable challenge and a promising approach for rapid and large-scale propagation of elite genotypes (Nia et al. 2021). Although investigated by several researchers, carob micropropagation still presents many difficulties and is not currently a viable commercial practice. Thus, more investigations should be carried out to develop efficient micropropagation systems for carob.

Based on the data available in the literature, PGRs play a key role in carob micropropagation, particularly the cytokinin BAP, which has been widely used either alone or in combination with auxins and gibberellins. These PGRs have shown remarkable effects on seed germination and seedling development, shoot multiplication and rooting, and adventitious regeneration (Carbonaro 1999; Thomas and Mehta 1983; Naghmouchi et al. 2008; Lozzi et al. 2019; Zouari and El Mtili 2020, Nia et al. 2021).

8.1 *Seed Germination and Seedling Culture*

Carbonaro (1999) reported successful culture of carob seedlings established from immature seeds on MS medium supplemented with 0.5 mg L^{-1} BAP and 0.1 mg L^{-1} IAA. This medium also allowed for shoot multiplication and callus induction. Proliferation of hypocotyl-derived calli was observed following multiple subcultures on a medium containing 0.5–1.0 mg L^{-1} 2,4-D.

According to Hakim et al. (2010), the combination of BAP and GA_3, and that of BAP and IAA, had a remarkable effect on shoot development from nodal segments of seedlings. The combination of 1.5 mg L^{-1} BAP and 0.5 mg L^{-1} GA_3 was found to be the most effective for multiple shoot induction. Radi et al. (2013) found that the in vitro response of carob explants correlated positively with cytokinin concentrations. Indeed, high levels of cytokinins increased shoot formation from nodal segments of carob seedlings. On the other hand, zeatin and BAP increased shoot length, particularly at 2 mg L^{-1}.

8.2 Bud Break and Axillary Shoot Growth

Many authors demonstrated the beneficial effect of BAP on bud break and shoot growth in carob. According to Thomas and Mehta (1983), the combination of BAP (2 mg L^{-1}) and NAA (1 mg L^{-1}) gave the highest frequency of bud break after 4–5 weeks of culture. Higher BAP concentrations inhibited shoot growth, while the use of lower concentrations (with 1 mg L^{-1} IBA or NAA) promoted callogenesis. Naghmouchi et al. (2008) reported that the combination 0.5 mg L^{-1} BAP, 0.1 mg L^{-1} IBA and 0.5 mg L^{-1} GA_3 promoted bud break and shoot development, while 1–2 mg L^{-1} BAP stimulated shoot proliferation. Zouari and El Mtili (2020) indicated that the combination of 0.5 mg L^{-1} BAP and 0.2 mg L^{-1} IBA was the most effective for shoot formation. Besides, shoot proliferation was achieved on a medium containing 1.5 mg L^{-1} BAP. Addition of 0.2 mg L^{-1} GA_3 to the BAP-containing medium favored shoot elongation. Nia et al. (2021) also reported that BAP at 0.5 mg L^{-1} promoted bud break and development.

8.3 Rhizogenesis

Root induction from shoots obtained in vitro was achieved on media containing IBA (Naghmouchi et al. 2008; Hakim et al. 2010; Lozzi et al. 2019; Zouari and El Mtili 2020). According to Radi et al. (2013), the presence of 1–2 mg L^{-1} IBA in the culture medium promoted root formation and elongation. Naghmouchi et al. (2008) and Zouari and El Mtili (2020) also reported that 2 mg L^{-1} IBA promoted rhizogenesis. Lozzi et al. (2019) recommended shoot dipping in a solution of 4.8 mM IBA for 3 min to stimulate root initiation.

8.4 Adventitious Regeneration

BAP is the main PGR used to induce adventitious regeneration in carob. The combination of 1 mg L^{-1} BAP and 0.2 mg L^{-1} NAA promoted adventitious bud formation from immature embryo-derived cotyledons (El Bouzdoudi et al. 2017). Adventitious shoot multiplication was successfully achieved on MS medium supplemented with 0.5 mg L^{-1} BAP. Addition of 0.7 mg L^{-1} GA_3 promoted stem and leaf growth while 2 mg L^{-1} IBA promoted rhizogenesis. Saïdi et al. (2019) pointed out that organogenesis from seedling-derived explants was stimulated in the presence of BAP at 0.5 mg L^{-1}. However, bud growth was favored by the use of zeatin at 0.5 mg L^{-1}. The combination of 0.5 mg L^{-1} BAP and 0.1 mg L^{-1} IBA improved caulogenesis, while 0.5–1 mg L^{-1} BAP alone or in combination with 0.5 mg L^{-1} GA_3 promoted shoot proliferation. For rooting, 2 mg L^{-1} IBA was recommended.

Regarding somatic embryogenesis, this morphogenesis was achieved from cotyledonary explants derived from immature seeds when the combination of 1 mg L^{-1} BAP and 0.1 mg L^{-1} IBA was used (Canhoto et al. 2006). Carbonaro (1999) observed somatic embryo formation from hypocotyl-derived calli on MS medium containing 0.1 mg L^{-1} BAP.

9 Caper (*Capparis spinosa* L.)

Caper (*Capparis spinosa* L.) is a perennial shrub that belongs to the Capparidaceae family and *Capparis* genus, which contains about 250 species (Fici 2001). Caper is native to the Mediterranean countries where it spontaneously grows along roads, on slopes and rocky coasts, and is well adapted to the dry regions around the Mediterranean Sea (Chalak and Elbitar 2006; Chedraoui et al. 2017).

Caper is an aromatic and medicinal plant widely exploited for its flower buds and fruits (Saifi et al. 2014). It is one of the most socio-economically important species in arid and semi-arid regions of many countries of North and East Africa, Southern Europe, Southwest and Central Asia (Jiang et al. 2007). Caper is a rich source of bioactive compounds that can be used for pharmaceutical and culinary purposes (Ben Mansour et al. 2016; Chedraoui et al. 2017). Despite the medicinal virtues of caper, only a few studies were carried out on its micropropagation.

9.1 *Shoot Culture*

The effect of BAP on caper shoot multiplication was first discussed by Rodriguez et al. (1990), who described this PGR as a potent compound that promotes shoot induction, proliferation and growth. The beneficial effect of BAP on caper shoot formation and multiplication, either when used alone or in combination with other PGRs was confirmed by other researchers. Musallam et al. (2011) reported that BAP promoted shoot multiplication when used at 2 mg L^{-1}. According to Sottile et al. (2020), the combination of 1.3 mg L^{-1} BAP and 0.02 mg L^{-1} IBA gave optimal shoot proliferation and promoted their growth. Chalak et al. (2003) observed multiple shoot formation from nodal buds on a medium supplemented with 1.5 mg L^{-1} BAP, 0.05 mg L^{-1} IBA and 0.1 mg L^{-1} GA_3.

Meta-topolin is another plant growth regulator that was used for caper micropropagation (Kereša et al. 2019). This highly active cytokinin was reported to give better results than zeatin, 2iP and BAP in terms of in vitro shoot multiplication. Indeed, at low concentrations (0.2–0.4 mg L^{-1}), the effects of BAP and meta-topolin on shoot formation were almost identical. However, increasing their concentrations to 0.6 mg L^{-1} resulted in significant differences in their effects (Kereša et al. 2019).

9.2 Rooting

For root induction, different auxins were suggested (Chalak and Elbitar 2006; Musallam et al. 2011; Carra et al. 2012; El-Mekawy et al. 2013). Caglar et al. (2005) found that IBA pretreatment (5 mg L^{-1} for 10 min) improved the rooting ability of caper shoots. According to Attia et al. (2017), the use of 1.5 mg L^{-1} NAA promoted root induction while Carra et al. (2012) and Sottile et al. (2020) highlighted the beneficial effect of IBA (1 mg L^{-1}). The auxin IAA (5.2 mg L^{-1}) was also suggested (Rodriguez et al. 1990). On the other hand, Gianguzzi et al. (2020) recommended the combination of 0.75 mg L^{-1} NAA and 0.25 mg L^{-1} IBA. Although these findings are not conclusive regarding the optimal auxin type and concentration for root induction, they emphasize the necessity to use this PGR group to successfully achieve rooting in caper shoots.

9.3 Adventitious Organogenesis

Different PGR combinations were suggested for adventitious organogenesis in caper. Al-Safadi and Elias (2011) reported that the combination of 0.1 mg L^{-1} GA_3, 1 mg L^{-1} NAA and 2 mg L^{-1} zeatin riboside promoted adventitious shoot formation from stem cuttings. On the other hand, these authors observed callus induction from leaf and shoot segments in media containing 1 mg L^{-1} BAP and 0.1 mg L^{-1} NAA. Plant regeneration from callus was achieved on a medium supplemented with 1 mg L^{-1} kinetin and 0.1 mg L^{-1} IAA. Elmaghrabi et al. (2017) observed callogenesis in a medium containing 1.2 mg L^{-1} 2,4-D, while BAP (2 mg L^{-1}) was recommended for shoot and root regeneration. According to Movafeghi et al. (2008), the combination of 0.1 mg L^{-1} NAA and 0.5 mg L^{-1} BAP stimulated bud induction from hypocotyl explants. On the other hand, root induction was achieved in a medium containing only NAA at 0.5 mg L^{-1}.

10 Conclusions

This chapter highlighted the main roles played by auxins, cytokinins and gibberellins in the micropropagation of some economically important Mediterranean fruit species. These PGR groups significantly influence cell division, differentiation and growth. Hence, they were involved in different morphogenetic processes and played key roles in the development of efficient in vitro propagation protocols. The effects of auxins, cytokinins and gibberellins vary among species and genotypes, and also depends on the explant. Besides, within each PGR group; for example, auxins, the effect differs from one auxin to another. Auxins have been generally used to induce rhizogenesis while cytokinins have been used to stimulate axillary shoot growth. The combination of auxins and cytokinins was found to be essential for adventitious regeneration (i.e. somatic embryogenesis and organogenesis). Regarding GA_3, it was mainly involved in somatic embryo germination and shoot elongation. Concretely, the use of auxins, cytokinins and gibberellins has greatly helped in the rapid and large-scale propagation of some economically important Mediterranean fruit crops; for example, date palm. On the other hand, they were effective in inducing morphogenetic responses leading to plantlet regeneration in recalcitrant species. For example, argan and olive. Consequently, they allowed for expanded applications of tissue culture to rapid propagation, genetic transformation and cryopreservation.

References

Aafi A (1996) Note technique sur le caroubier (*Ceratonia siliqua* L.). Centre National de la Recherche Forestière, Rabat, Morocco, 10p

Abd El-Wahab EA, Sayed HA (2019) Micropropagation and cryopreservation of Sultani fig (*Ficus carica* L.) genotype. Middle East J Appl Sci 9:919–926

Abdolinejad R, Shekafandeh A, Jowkar A, Gharaghani A, Alemzadeh A (2020) Indirect regeneration of *Ficus carica* by the TCL technique and genetic fidelity evaluation of the regenerated plants using flow cytometry and ISSR. Plant Cell Tissue Organ Cult 143:131–144

Abohatem M, Zouine J, El Hadrami I (2011) Low concentrations of BAP and high rate of subcultures improve the establishment and multiplication of somatic embryos in date palm suspension cultures by limiting oxidative browning associated with high levels of total phenols and peroxidase activities. Sci Hortic 130:344–348

Akin M, Poljuha D, Eyduran SP, Ercisli S, Radunic M (2021) SSR based molecular characterization of local fig (Ficus carica L.) germplasm in northeastern Turkey. Erwerbs-Obstbau 63:387–392

Al Kaabi HH, Rhiss A, Hassan MA (2001) Effect of auxins and cytokinins on the *in vitro* production of date palm (*Phoenix dactylifera* L.) bud generative tissues and on the number of differentiated buds. In: Proceedings of the second international conference on date palm. Al Ain, UAE, pp 47–86

Al-Khayri JM (2011) Influence of yeast extract and casein hydrolysate on callus multiplication and somatic embryogenesis of date palm (*Phoenix dactylifera* L.). Sci Hortic 130:531–535

Al-Khayri JM, Al-Bahrany AM (2012) Effect of abscisic acid and polyethylene glycol on the synchronization of somatic embryo development in date palm (*Phoenix dactylifera* L.). Biotechnology 11:318–325

Al-Mayahi AMW (2014) Thidiazuron-induced *in vitro* bud organogenesis of the date palm (*Phoenix* dactylifera L.) cv Hillawi. Afr J Biotechnol 13:3581–3590

Al-Mayahi AMW (2016) Effect of red and blue light emitting diodes 'CRB-LED' on *in vitro* organogenesis of date palm (*Phoenix dactylifera* L.) cv. Alshakr. World J Microbiol Biotechnol 32:160

Al-Safadi B, Elias R (2011) Improvement of caper (*Capparis spinosa* L.) propagation using *in vitro* culture and gamma irradiation. Sci Hortic 127:290–297

Al-Shomali I, Sadder MT, Ateyyeha A (2017) Culture media comparative assessment of common fig (*Ficus carica* L.) and carryover effect. Jordan J Biol Sci 10:13–18

Albergamo A, Bartolomeo G, Messina L, Rando R, Di Bella G (2021) Traceability of *Opuntia* spp. In: *Opuntia* spp.: chemistry, bioactivity and industrial applications. Springer, Cham, pp 457–482

Ali A, Ahmad T, Abbasi AN, Hafiz IA (2009) Effect of different media and growth regulators on *in vitro* shoot proliferation of olive cultivar 'Moraiolo.' Pak J Bot 41:783–795

Amani S, Mohebodini M, Khademvatanb S, Jafaric M (2020) *Agrobacterium rhizogenes* mediated transformation of *Ficus carica* L. for the efficient production of secondary metabolites. J Sci Food Agric 100:2185–2197

Amghar I, Diria G, Boumlik I, Gaboun F, Iraqi D, Labhilili M, Mentag R, Meziani R, Mazri MA, Ibriz M, Abdelwahd R (2021a) An efficient regeneration pathway through adventitious organogenesis for the endangered *Argania spinosa* (L.) Skeels. Vegetos 34:355–367

Amghar I, Ibriz M, Ibrahimi M, Boudra A, Gaboun F, Meziani R, Iraqi D, Mazri MA, Diria G, Abdelwahd R (2021b) *In vitro* root induction from argan (*Argania spinosa* (L.) Skeels) adventitious shoots: influence of ammonium nitrate, auxins, silver nitrate and putrescine, and evaluation of plantlet acclimatization. Plants 10:1062

Attia AO, Dessoky EE, Al-Sodany YM, Ismail IA (2017) *Ex situ* preservation for some endemic and rare medicinal plants in Taif, KSA. Biotechnol Biotechnol Equip 31:912–920

Bajoub A, Bendini A, Fernández-Gutiérrez A, Carrasco-Pancorbo A (2018) Olive oil authentication: a comparative analysis of regulatory frameworks with especial emphasis on quality and authenticity indices, and recent analytical techniques developed for their assessment. A review. Crit Rev Food Sci Nutr 58:832–857

Bekheet S (2013) Direct organogenesis of date palm (*Phoenix dactylifera* L.) for propagation of true-to-type plants. Sci Agric 4:85–92

Ben Mansour R, Jilani IBH, Bouaziz M, Gargouri B, Elloumi N, Attia H, Ghrabi-Gammar Z, Lassoued S (2016) Phenolic contents and antioxidant activity of ethanolic extract of *Capparis spinosa*. Cytotechnology 68:135–142

Blakeslee JJ, Murphy AS (2016) Microscopic and biochemical visualization of auxins in plant tissues. In Environmental responses in plants. Springer, New York, pp 37–53

Bouamama B, Ben Salem A, Zoghlami N, Zemni H, Mliki A, Ghorbel A (2011) Somatic embryogenesis and plantlet regeneration from immature anthers of *Opuntia ficus-indica*. J Hortic Sci Biotechnol 86:313–318

Bousselmame F, Kenny L, Chlyah H (2001) Optimisation des conditions de cultures pour l'enracinement *in vitro* de l'arganier (*Argania spinosa* L.). Life Sci 324:995–1000

Caglar G, Caglar S, Ergin O, Yarim M (2005) The influence of growth regulators on shoot proliferation and rooting of *in vitro* propagated caper. J Environ Biol 26:479–485

Canhoto JM, Rama SC, Cruz GS (2006) Somatic embryogenesis and plant regeneration in carob (*Ceratonia siliqua* L.). In Vitro Cell Dev Biol Plant 42:514–519

Carbonaro M (1999) Studies on the micropropagation of carob: *Ceratonia siliqua* L. MSc dissertation, University of Malta, Malta

Carra A, Del Signore MB, Sottile F, Ricci A, Carimi F (2012) Potential use of new diphenylurea derivatives in micropropagation of *Capparis spinosa* L. Plant Growth Regul 66:229–237

Cerezo S, Mercado JA, Pliego-Alfaro F (2011) An efficient regeneration system via somatic embryogenesis in olive. Plant Cell Tissue Organ Cult 106:337–344

Chaari Rkhis A, Maalej M, Drira N, Standardi A (2011) Micropropagation of olive tree *Olea europaea* L. 'Oueslati.' Turk J Agric For 35:403–412

Chalak L, Elbitar A (2006) Micropropagation of *Capparis spinosa* L. subsp. *rupestris* Sibth. & Sm. by nodal cuttings. Indian J Biotechnol 5:555–558

Chalak L, Elbitar A, Cordahi N, Hage C, Chehade A (2003) *In vitro* propagation of *Capparis spinosa* L. Acta Hortic 616:335–338

Charrouf Z, Guillaume D (2009) Sustainable development in Northern Africa: the argan forest case. Sustainability 1:1012–1022

Chauhan RD, Kanwar K (2012) Biotechnological advances in pomegranate (*Punica granatum* L.). In Vitro Cell Dev Biol Plant 48:579–594

Chedraoui S, Abi-Rizk A, El-Beyrouthy M, Chalak L, Ouaini N, Rajjou L (2017) *Capparis spinosa* L. in a systematic review: a xerophilous species of multi values and promising potentialities for agrosystems under the threat of global warming. Front Plant Sci 8:1845

Comlekcioglu S, Kuden AB, Kacar YA, Kamberoglu MA (2007) Meristem culture of two fig cultivars in Turkey. Fruits 62:125–131

Crisosto CH, Corrales ML, Colelli G, Serradilla MJ (2020) Fresh fig. In: Manual on postharvest handling of mediterranean tree fruits and nuts. Cabi, London, pp 26–52

Danial G, Ibrahim D, Brkat S, Khalil BM (2014) Multiple shoots production from shoot tips of fig tree (*Ficus carica* L.) and callus induction from leaf segments. Int J Pure Appl Sci Technol 20:117–124

Darwesh HY, Bazaid SA, Abu Samra BN (2014) *In vitro* propagation method of *Ficus carica* at Taif governorate using tissue culture technique. Int J Adv Res 2:756–761

De Waroux YP, Lambin EF (2012) Monitoring degradation in arid and semi-arid forests and woodlands: the case of the argan woodlands (Morocco). Appl Geogr 32:777–786

Deepika R, Kanwar K (2010) *In vitro* regeneration of *Punica granatum* L. plants from different juvenile explants. J Fruit Ornam Plant Res 18:5–22

Desai P, Patil G, Dholiya B, Desai S, Patel F, Narayanan S (2018) Development of an efficient micropropagation protocol through axillary shoot proliferation for pomegranate variety 'Bhagwa.' Ann Agrar Sci 16:444–450

Dhage SS, Chimote VP, Pawar BD, Kale AA, Pawar SV, Jadhav AS (2015) Development of an efficient *in vitro* regeneration protocol for fig (*Ficus carica* L.). J Appl Hortic 17:160–164

Eke CR, Akomeah P, Asemota O (2005) Somatic embryogenesis in date palm (*Phoenix dactylifera* L.) from apical meristem tissues from 'zebia' and 'loko' landraces. Afr J Biotechnol 4:244–246

El Bouzdoudi B, Saïdi R, El Ansari ZN, El Kbiach ML, Martin P, Badoc A, Lamarti A (2017) Micropropagation of Carob (*Ceratonia siliqua* L.) through adventitious buds of immature embryonic cotyledons. Am J Plant Sci 8:78378

El Hadrami I, Cheikh R, Baaziz M (1995) Somatic embryogenesis and plant regeneration from shoot-tip explants in *Phoenix dactylifera* L. Biol Plant 37:205–211

El Kahkahi R, Moustaine M, Mouhajir A, Bachir S, Lemrhari A, Zouhair M, Ait Chitt M, Errakhi R (2016) Technical sheet on the culture carob tree (*Ceratonia Siliqua* L.) in Morocco. Hal Open Science, hal-01319583

El Kharrassi Y, Maata N, Mazri MA, El Kamouni S, Talbi M, El Kebbaj R, Moustaid K, Essamadi AK, Andreoletti P, El Mzouri EH, Cherkaoui-Malki M, Nasser B (2018) Chemical and phytochemical characterizations of argan oil (*Argania spinosa* L. skeels), olive oil (*Olea europaea* L. cv. Moroccan picholine), cactus pear (*Opuntia megacantha* salm-dyck) seed oil and cactus cladode essential oil. J Food Meas Charact 12:747–754

El-Agamy SZ, Mostafa RAA, Shaaban MM, El-Mahdy MT (2009) *In vitro* propagation of Manfalouty and Nab El-gamal pomegranate cultivars. Res J Agric Biol Sci 5:1169–1175

El-Mekawy MA, Ali MAM, Belal AH, Abdallah SAS (2013) Micropropgation of caper (Capparis spinosa, L.) from wild plants growing in North Sinai. Hortscience J Suez Canal Univ 1:15–20

Elmaghrabi AM, Abughnia E, Hamoud S (2017) *In vitro* propagation of the wild medicinal plant, caper (*Capparis spinosa* L.). Afr J Biotechnol

Escobar HAA, Villalobos AVM, Villegas MA (1986) *Opuntia* micropropagation by axillary proliferation. Plant Cell Tissue Organ Cult 7:269–277

Eshraghi P, Zarghami R, Mirabdulbaghi M (2005) Somatic embryogenesis in two Iranian date palm cultivars. Afr J Biotechnol 4:1309–1312

Estrada-Luna AA, Martínez-Hernández JDJ, Torres-Torres ME, Chablé-Moreno F (2008) *In vitro* micropropagation of the ornamental prickly pear cactus *Opuntia lanigera* Salm-Dyck and effects of sprayed GA_3 after transplantation to *ex vitro* conditions. Sci Hortic 117:378–385

Ferrara G, Giancaspro A, Mazzeo A, Giove SL, Matarrese AMS, Pacucci C, Punzi R, Trani A, Gambacorta G, Blanco A, Gadaleta A (2014) Characterization of pomegranate (*Punica granatum* L.) genotypes collected in Puglia region, Southeastern Italy. Sci Hortic 178:70–78

Fici S (2001) Intraspecific variation and evolutionary trends in *Capparis spinosa* L. (Capparaceae). Plant Syst Evol 228:123–141

Fki L, Masmoudi R, Drira N, Rival A (2003) An optimised protocol for plant regeneration from embryogenic suspension cultures of date palm, *Phoenix dactylifera* L., cv. Deglet Nour. Plant Cell Rep 21:517–524

Flaishman MA, Peer R, Raz A, Cohen O, Izhaki K, Bocobza S, Lama K, Pliner M, Levy A (2020) Advanced molecular tools for breeding in Mediterranean fruit trees: genome editing approach of *Ficus carica* L. Acta Hortic 1280:1–10

Fraguas CB, Pasqual M, Dutra LF, Cazetta JO (2004) Micropropagation of fig (*Ficus carica* L.) 'Roxo de Valinhos' plants. In Vitro Cell Dev Biol Plant 40:471–474

Garcia-Saucedo PA, Valdez-Morales M, Valverde ME, Cruz-Hernández A, Paredes-López O (2005) Plant regeneration of three *Opuntia* genotypes used as human food. Plant Cell Tissue Organ Cult 80:215–219

Gaspar T, Kevers C, Penel C, Greppin H, Reid DM, Thorpe TA (1996) Plant hormones and plant growth regulators in plant tissue culture. In Vitro Cell Dev Biol Plant 32:272–289

Gharaghani A, Soloklui AAG, Oraguzie N, Zare D (2017) Pollen source influences fruit quality, aril properties, and seed characteristics in pomegranate. Int J Fruit Sci 17:333–348

Gianguzzi V, Barone E, Sottile F (2020) *In vitro* rooting of *Capparis spinosa* L. as affected by genotype and by the proliferation method adopted during the multiplication phase. Plants 9:398

Gomes FLAF, Heredia FF, Silva PB, Facó O, Campos FAP (2006) Somatic embryogenesis and plant regeneration in *Opuntia ficus-indica* (L.) Mill. (Cactaceae). Sci Hortic 108:15–21

Hakim L, Islam MR, Mamun ANK, Ahmed G, Khan R (2010) Clonal propagation of carob (*Ceratonia siliqua* L., Fabaceae). Bangladesh J Bot 39:15–19

Hussain I, Rashid H, Muhammad A, Quraishi A (2001) *In vitro* multiplication of date palm. In: Proceedings of the second international conference on date palm. Al Ain, UAE, pp 432–438

Ikegami H, Nogata H, Hirashima K, Awamura M, Nakahara T (2009) Analysis of genetic diversity among European and Asian fig varieties (*Ficus carica* L.) using ISSR, RAPD, and SSR markers. Genet Resour Crop Evol 56:201–209

Jalikop SH (2010) Pomegranate breeding. Fruit Veg Cer Sci Biotechnol 4:26–34

Jazinizadeh E, Zarghami R, Majd A, Iranbakhsh A, Tajaddod G (2015) *In vitro* production of date palm (*Phoenix dactylifera* L.) cv. 'Barhee' plantlets through direct organogenesis. Biol Forum Int J 7:566–572

Jedidi E, Mahmoud KB, Kaaniche-Elloumi N, Jemmali A (2015) SEM and histological analysis of somatic embryogenesis performed on cactus pear (*Opuntia ficus-indica* (L.) Mill.) ovules explants. Acta Hortic 1067:231–237

Jiang HE, Li X, Ferguson DK, Wang YF, Liu CJ, Li CS (2007) The discovery of *Capparis spinosa* L. (Capparidaceae) in the Yanghai Tombs (2800 years B.P.), NW China, and its medicinal implications. J Ethnopharmacol 113:409–420

Johnson DV, Al-Khayri JM, Jain SM (2013) Seedling date palms (*Phoenix dactylifera* L.) as genetic resources. Emir J Food Agric 25:809–830

Juarez MC, Passera CB (2002) *In vitro* propagation of *Opuntia ellisiana* Griff. and acclimatization to field conditions. Biocell 26:319–324

Kabir MH, Das P, Mamun ANK, Islam MM, Islam MA (2021) *In vitro* micropropagation of pomegranate (*Punica granatum* L.) derived from cotyledon. Plant Tissue Cult Biotechnol 31:61–69

Kahramanoğlu I, Umar K (2018) Effects of cutting length, time and growing media on the sprouting of dormant semi-hardwood cuttings of pomegranate cv. Wonderful. Afr J Agric Res 13:2587–2590

Kanwar K, Joseph J, Deepika R (2010) Comparison of *in vitro* regeneration pathways in *Punica granatum* L. Plant Cell Tissue Organ Cult 100:199–207

Kereša S, Stanković D, Lodeta KB, Jerčić H, Bolarić S, Barić M, Mihovilovic AB (2019) Efficient protocol for the *in vitro* plantlet production of caper (*Capparis orientalis* Veill.) from the East Adriatic Coast. Agronomy 9:303

Khalafalla MM, Abdellatef E, Ahmed MMM, Osman MG (2007) Micropropagation of cactus (*Opuntia ficus-indica*) as strategic tool to combat desertification in arid and semi-arid regions. Int J Sustain Crop Prod 2:1–8

Khan S, Bi Bi T (2012) Direct shoot regeneration system for date palm (*Phoenix dactylifera* L.) cv. Dhakki as a means of micropropagation. Pak J Bot 44:1965–1971

Khierallah HSM, Bader SM (2007) Micropropagation of date palm (*Phoenix dactylifera* L.) var. Maktoom through direct organogenesis. Acta Hortic 736:213–224
Khierallah HSM, Al Hamdany MHS, Abdulkareem AA, Saleh FF (2015) Influence of sucrose and pacloburtazol on callus growth and somatic embryogenesis in date palm cv. Bream. Int J Curr Res Acad Rev 1:270–276
Kim KM, Kim MY, Yun PY, Chandrasekhar T, Lee HY, Song PS (2007) Production of multiple shoots and plant regeneration from leaf segments of fig tree (*Ficus carica* L.). J Plant Biol 50:440–446
Kislev ME, Hartmann A, Bar-Yosef O (2006) Early domesticated fig in the Jordan Valley. Science 312:1372–1374
Kotsampasi B, Christodoulou C, Mavrommatis A, Mitsiopoulou C, Bampidis VA, Christodoulou V, Chronopoulou EG, Labrou NE, Tsiplakou E (2021) Effects of dietary pomegranate seed cake supplementation on performance, carcass characteristics and meat quality of growing lambs. Anim Feed Sci Technol 273:114815
Koufan M, Belkoura I, Alaoui T (2018) The multiplication of the argane tree by microcutting (*Argania spinosa* L. Skeels). Eur J Biotechnol Biosci 6:47–52
Koufan M, Belkoura I, Mazri MA, Amarraque A, Essatte A, Elhorri H, Zaddoug F, Alaoui T (2020a) Determination of antioxidant activity, total phenolics and fatty acids in essential oils and other extracts from callus culture, seeds and leaves of *Argania spinosa* (L.) Skeels. Plant Cell Tissue Organ Cult 141:217–227
Koufan M, Mazri MA, Essatte A, Moussafir S, Belkoura I, El Rhaffari L, Toufik I (2020b) A novel regeneration system through micrografting for *Argania spinosa* (L.) Skeels, and confirmation of successful rootstock-scion union by histological analysis. Plant Cell Tissue Organ Cult 142:369–378
Kumar V, Radha A, Chita SK (1998) *In vitro* plant regeneration of fig (*Ficus carica* L. cv. Gular) using apical buds from mature trees. Plant Cell Rep 17:717–720
Lamaoui M, Chakhchar A, El Kharrassi Y, Wahbi S, Ferradous A, El Mousadik A, Ibnsouda-Koraichi S, Filali-Maltouf A, El Modafar C (2019) Selection and multiplication of argan (*Argania spinosa* L.) superior clones for conservation purposes. Acta Sci Agric 3:116–123
Lambardi M, Rugini E (2003) Micropropagation of olive (*Olea europaea* L.). In: Micropropagation of woody trees and fruits. Kluwer Academic Publishers, Dordrecht, pp 621–646
Letham DS (1963) Zeatin, a factor inducing cell division isolated from *Zea mays*. Life Sci 8:569–573
Li B, Zhang P, Wang F, Li R, Liu J, Wang Q, Liu W, Wang B, Hu G (2021) Integrated analysis of the transcriptome and metabolome revealed candidate genes involved in GA_3-induced dormancy release in *Leymus chinensis* seeds. Int J Mol Sci 22:4161
Libbenga KR, Mennes AM (1995) Hormone binding and signal transduction. In: Plant hormones: physiology, biochemistry and molecular biology. Kluwer Academic Publishers, Dordrecht, pp 272–297
Ling WT, Liew FC, Lim WY, Subramaniam S, Chew BL (2018) Shoot induction from axillary shoot tip explants of fig (*Ficus carica*) cv. Japanese BTM 6. Trop Life Sci Res 29:165–175
Lozzi A, Abdelwahd R, Mentag R, Abousalim A (2019) Development of a new culture medium and efficient protocol for *in vitro* micropropagation of *Ceratonia siliqua* L. In Vitro Cell Dev Biol Plant 55:615–624
Machakova I, Zazimalova E, George EF (2008) Plant growth regulators I. introduction; auxins, their analogues and inhibitors. In: Plant propagation by tissue culture. Springer, Dordrecht, pp 175–204
Mauseth JD (1976) Cytokinin- and gibberellic acid-induced effects on the structure and metabolism of shoot apical meristems in *Opuntia polyacantha* (Cactaceae). Am J Bot 63:1295–1301
Mauseth JD, Halperin W (1975) Hormonal control of organogenesis in *Opuntia polyacantha* (Cactaceae). Am J Bot 62:869–877
Mazri MA (2014) Effects of plant growth regulators and carbon source on shoot proliferation and regeneration in date palm (*Phoenix dactylifera* L.) '16-bis.' J Hortic Sci Biotechnol 89:415–422

Mazri MA (2015) Role of cytokinins and physical state of the culture medium to improve *in vitro* shoot multiplication, rooting and acclimatization of date palm (*Phoenix dactylifera* L.) cv. Boufeggous. J Plant Biochem Biotechnol 24:268–275
Mazri MA (2018) Cactus pear (*Opuntia* spp.) breeding. In: Advances in plant breeding strategies: fruits. Springer, Cham, pp 307–341
Mazri MA (2021) Cactus Pear (*Opuntia* spp.) species and cultivars. In: Opuntia spp.: chemistry, bioactivity and industrial applications. Springer, Cham, pp 83–107
Mazri MA, Meziani R (2013) An improved method for micropropagation and regeneration of date palm (*Phoenix dactylifera* L.). J Plant Biochem Biotechnol 22:176–184
Mazri MA, Meziani R (2015) Micropropagation of date palm: a review. Cell Dev Biol 4:160
Mazri MA, Elbakkali A, Belkoura M, Belkoura I (2011) Embryogenic competence of calli and embryos regeneration from various explants of Dahbia cv, a Moroccan olive tree (*Olea europaea* L.). Afr J Biotechnol 10:19089–19095
Mazri MA, Belkoura I, Pliego-Alfaro F, Belkoura M (2012) Embryogenic capacity of embryo-derived explants from different olive cultivars. Acta Hort 929:397–403
Mazri MA, Belkoura I, Pliego-Alfaro F, Belkoura M (2013) Somatic embryogenesis from leaf and petiole explants of the Moroccan olive cultivar Dahbia. Sci Hortic 159:88–95
Mazri MA, Meziani R, El Fadile J, Ezzinbi A (2016) Optimization of medium composition for *in vitro* shoot proliferation and growth of date palm cv. Mejhoul. 3 Biotech 6:111
Mazri MA, Belkoura I, Meziani R, Mokhless B, Nour S (2017) Somatic embryogenesis from bud and leaf explants of date palm (*Phoenix dactylifera* L.) cv. Najda. 3 Biotech 7:58.
Mazri MA, Belkoura I, Meziani R (2018a) Use of TDZ for micropropagation of some Mediterranean crop species. In: Thidiazuron: from urea derivative to plant growth regulator. Springer, Singapore
Mazri MA, Meziani R, Belkoura I, Mokhless B, Nour S (2018b) A combined pathway of organogenesis and somatic embryogenesis for an efficient large-scale propagation in date palm (*Phoenix dactylifera* L.) cv. Mejhoul. 3 Biotech 8:215
Mazri MA, Meziani R, Belkoura I, Elmaataoui S, Mokhless B, Nour S (2019a) Maturation and germination of date palm (*Phoenix dactylifera* L.) somatic embryos. Notulae Sci Biol 11:86–93
Mazri MA, Belkoura I, Meziani R, Es-Saoudy H, Rachad F, Elmaataoui S (2019b) Impact of osmotica and plant growth regulators on somatic embryogenesis of date palm. Curr Agric Res J 7:296–303
Mazri MA, Meziani R, Elmaataoui S, Alfeddy MN, Jaiti F (2019c) Assessment of genetic fidelity, biochemical and physiological characteristics of *in vitro* grown date palm cv. Al-Fayda. Vegetos 32:333–344
Mazri MA, Naciri R, Belkoura I (2020) Maturation and conversion of somatic embryos derived from seeds of olive (*Olea europaea* L.) cv. Dahbia: occurrence of secondary embryogenesis and adventitious bud formation. Plants 9:1489
Mencuccini M, Rugini E (1993) *In vitro* shoot regeneration from olive cultivar tissues. Plant Cell Tissue Organ Cult 32:283–288
Meziani R, Jaiti F, Mazri MA, Anjarne M, Ait Chitt M, El Fadile J, Alem C (2015) Effects of plant growth regulators and light intensity on the micropropagation of date palm (*Phoenix dactylifera* L.) cv. Mejhoul. J Crop Sci Biotechnol 18:325–331
Meziani R, Jaiti F, Mazri MA, Hassani A, Ben Salem S, Anjarne M, Ait Chitt M, Alem C (2016) Organogenesis of *Phoenix dactylifera* L. cv. Mejhoul: influences of natural and synthetic compounds on tissue browning, and analysis of protein concentrations and peroxidase activity in explants. Sci Hortic 204:145–152
Meziani R, Mazri MA, Arhazzal M, Belkoura I, Alem C, Jaiti F (2019) Evaluation of *in vitro* shoot elongation and rooting of date palm, and determination of physiological characteristics of regenerated plantlets. Notulae Sci Biol 11:77–85
Moshkov IE, Novikova GV, Hall MA, George EF (2008) Plant growth regulators III: gibberellins, ethylene, abscisic acid, their analogues and inhibitors; miscellaneous compounds. In: Plant propagation by tissue culture. Springer, Dordrecht, pp 227–281

Moukrim S, Lahssini S, Mharzi Alaoui H, Rifai N, Arahou M, Rhazi L (2018) Modélisation de la distribution spatiale des espèces endémiques pour leur conservation: cas de l'*Argania spinosa* (L.) Skeels. Revue d'ecologie (Terre Vie) 73:153–166

Movafeghi A, Habibi G, Ali AM (2008) Plant regeneration of *Capparis spinosa* L. using hypocotyl explants. Iran J Biol 21:289–297

Mulaei S, Jafari A, Shirmardi M, Kamali K (2019) Micropropagation of arid zone fruit tree, pomegranate, cvs. 'Malase Yazdi' and 'Shirine Shahvar.' Int J Fruit Sci 20:825–836

Murashige T, Skoog FA (1962) A revised medium for rapid growth and bioassays with tobacco tissue cultures. Physiol Plant 15:473–479

Murkute A, Patil S, Patil BN, Mayakumari MS (2002) Micropropagation in pomegranate, callus induction and differentiation. South Indian Hortic 50:49–55

Musallam I, Duwayri M, Shibli R (2011) Micropropagation of caper (*Capparis spinosa* L.) from wild plants. Funct Plant Sci Biotechnol 5:17–21

Mustafa NS, Taha RA, Hassan SAM, Zaied NSM (2013) Effect of medium strength and carbon source on *in vitro* shoot multiplication of two *Ficus carica* cultivars. J Appl Sci Res 9:3068–3074

Naggar M, Lahssini S (2015) La filière caroubier au service du développement socio-économique des territoires forestiers de montagne: cas de la province d'Azilal (Maroc). Hal Open Science, hal-01194408

Naghmouchi S, Khouja ML, Rejeb MN, Boussaid M (2008) Effect of growth regulators and explant origin on *in vitro* propagation of *Ceratonia siliqua* L. via cuttings. Biotechnol Agron Soc Environ 12:251–258

Naik SK, Chand PK (2011) Tissue culture-mediated biotechnological intervention in pomegranate: a review. Plant Cell Rep 30:707–721

Naik SK, Pattnaik S, Chand PK (1999) *In vitro* propagation of pomegranate (*Punica granatum* L. cv. Ganesh) through axillary shoot proliferation from nodal segments of mature tree. Sci Hortic 79:175–183

Naik SK, Pattnaik S, Chand PK (2000) High frequency axillary shoot proliferation and plant regeneration from cotyledonary nodes of pomegranate (*Punica granatum* L.). Sci Hortic 85:261–270

Narváez I, Martín C, Jiménez-Díaz RM, Mercado JA, Pliego-Alfaro F (2019) Plant regeneration via somatic embryogenesis in mature wild olive genotypes resistant to the defoliating pathotype of *Verticillium dahliae*. Front Plant Sci 10:1471

Nia S, Abid M, Belkoura I (2021) Carob tree micropropagation essays (*Ceratonia siliqua* L.). In: E3S web of conferences, vol 298, pp 03004

Omura M, Matsuta N, Moriguchi T, Kazaki I (1987) Adventitious shoot and plantlet formation from cultured pomegranate leaf explants. HortScience 22:133–134

Orinos T, Mitrakos K (1991) Rhizogenesis and somatic embryogenesis in calli from wild olive (*Olea europaea* var. sylvestris (Miller) Lehr) mature zygotic embryos. Plant Cell Tissue Organ Cult 27:183–187

Othmani A, Bayoudh C, Drira N, Marrakchi M, Trifi M (2009) Somatic embryogenesis and plant regeneration in date palm Phoenix dactylifera L., cv. Boufeggous is significantly improved by fine chopping and partial desiccation of embryogenic callus. Plant Cell Tissue Organ Cult 97:71–79

Oulbi S, Kohaich K, Baaziz M, Belkoura I, Loutfi K (2021) Peroxidase enzyme fractions as markers of somatic embryogenesis capacities in olive (*Olea europaea* L.). Plants 10:901

Parab AR, Chew BL, Yeow LC, Subramaniam S (2021) Organogenesis on apical buds in common fig (*Ficus carica*) var. Black Jack. Electron J Biotechnol 54:69–76

Parmar N, Kanwar K, Thakur AK (2013) Direct organogenesis in *Punica granatum* L. cv. Kandhari Kabuli from hypocotyl explants. Proc Natl Acad Sci India Sect B Biol Sci 83:569–574

Peixe A, Lourenço R, Raposo A, Jacob AP, Campos C, Macedo E, Cardoso H (2009) Simplifying procedures to increase competitiveness at *in vitro* propagation of the olive cultivar 'Galega Vulgar.' Acta Hort 812:277–282

Radi A, Echchgadda G, Ibijbijen J, Rochd M (2013) *In vitro* propagation of Moroccan carob (*Ceratonia siliqua* L.). J Food Agric Environ 11:1103–1107

Rham I (2021) Optimisation d'un milieu de culture pour la multiplication, l'élongation, l'enracinement et l'acclimatation des vitroplants du cactus (*Opuntia ficus indica*). MSc dissertation, University of Sciences of Kenitra, Morocco
Rodriguez R, Rey M, Cuozzo L, Ancora G (1990) *In vitro* propagation of caper (*Capparis spinosa* L.). In Vitro Cell Dev Biol Plant 26:531–536
Rostami AA, Shahsavar AR (2012) *In vitro* micropropagation of olive (*Olea europaea* L.) 'Mission' by nodal segments. J Biol Environ Sci 6:155–159
Roussos PA, Pontikis CA (2002) *In vitro* propagation of olive (*Olea europaea* L.) cv Koroneiki. Plant Growth Regul 37:295–304
Rugini E, Caricato G (1995) Somatic embryogenesis and plant recovery from mature tissues of olive cultivars (*Olea europaea* L.) 'Canino' and 'Moraiolo.' Plant Cell Rep 14:257–260
Sahraroo A, Zarei A, Babalar M (2019) *In vitro* regeneration of the isolated shoot apical meristem of two commercial fig cultivars 'Sabz' and 'Jaami-e-Kan.' Biocatal Agric Biotechnol 17:743–749
Saïdi R, Rahmouni S, El Ansari ZN, Maouni A, Badoc A, Lamarti A (2019) Effect of cytokinins on the micropropagation of carob (*Ceratonia siliqua* L.) through shoot tip culture. Am J Plant Sci 10:1469–1481
Saifi N, Echchgadda G, Nassiri L, Ibijbijen J (2014) Ability to root formation and germination of some Moroccan ecotypes of caper (*Capparis* spp.). ScienceLib Ed Mersenne 6:141201
Sghir S, Chatelet P, Ouazzani N, Dosba F, Belkoura I (2005) Micropropagation of eight Moroccan and French olive cultivars. HortScience 40:193–196
Shahcheraghi ST, Shekafandeh A (2016) Micropropagation of three endemic and endangered fig (*Ficus carica* L.) genotypes. Adv Hortic Sci 30:129–134
Shao TZ, Chen CL, Deng XX (2003) *In vitro* induction of tetraploid in pomegranate (*Punica granatum* L. var. Nana). Plant Cell Tissue Organ Cult 75:241–246
Shen Z, Zhang YH, Zhang L, Li Y, Sun YD, Li ZY (2020) Changes in the distribution of endogenous hormones in *Phyllostachys edulis* 'Pachyloen' during bamboo shooting. PLoS One 15:e0241806
Singh P, Patel RM, Kadam S (2013) *In vitro* mass multiplication of pomegranate from cotyledonary nodal explants cv. Ganesh. Afr J Biotechnol 12:2863–2868
Skubacz A, Daszkowska-Golec A (2017) Seed dormancy: the complex process regulated by abscisic acid, gibberellins, and other phytohormones that makes seed germination work. In: Phytohormones-signaling mechanisms and crosstalk in plant development and stress responses. InTech, New York, pp 77–100
Soliman H, Gabr M, Abdallah N (2010) Efficient transformation and regeneration of fig (*Ficus carica* L.) via somatic embryogenesis. GM Crops 1:42–53
Song Y (2014) Insight into the mode of action of 2,4-dichlorophenoxyacetic acid (2,4-D) as an herbicide. J Integr Plant Biol 56:106–113
Sottile F, Guiggioli NR, Marioni DT, Peano C, Del Signore MB (2020) Selection and micropropagation of valuable caper genotypes. Hortic Sci (Prague) 47:110–116
Soukhak F, Khalighi A, Ghaemmaghami SA (2011) Study of direct adventitious shoot regeneration in pomegranate (*Punica granatum* cv. MalasSaveh) through cotyledonary explants. Int J Agric Sci Res 2:19–26
Taha RA, Allam MA, Hassan SAM, Bakr BMM, Hassan MM (2021) Thidiazuron-induced direct organogenesis from immature inflorescence of three date palm cultivars. J Genet Eng Biotechnol 19:14
Terakami S, Matsuta N, Yamamoto T, Sugaya S, Gemma H, Soejima J (2007) *Agrobacterium*-mediated transformation of the dwarf pomegranate (*Punica granatum* L. var. nana). Plant Cell Rep 26:1243–1251
Thomas V, Mehta AR (1983) Effect of phloroglucinol on shoot growth and initiation of roots in carob tree cultures grown *in vitro*. In: Plant cell culture in crop improvement. Springer, Boston, pp 451–457
Tisserat B (1979) Propagation of date palm (*Phoenix dactylifera* L.) *in vitro*. J Exp Bot 30:1275–1283
Toufik I, Guenoun F, Belkoura I (2014) Embryogenesis expression from somatic explants of olive (*Olea europaea* L.) cv. Picual. Moroc J Biol 11:17–25

ValizadehKaji B, Ershadi A, Tohidfar M (2013) *In vitro* propagation of two Iranian commercial pomegranates (*Punica granatum* L.) cvs. 'Malas Saveh' and 'Yusef Khani.' Physiol Mol Biol Plants 19:597–603

Van Staden J, Zazimalova E, George EF (2008) Plant growth regulators II: cytokinins, their analogues and antagonists. In: Plant propagation by tissue culture. Springer, Dordrecht, pp 205–226

Vekiari S, Ouzounidou G (2018) An overview of the wild plants consumed in the island of Crete, Greece. In: Global perspectives on underutilized crops. Springer, Cham, pp 259–285

Verma V, Zinta G, Kanwar K (2021) Optimization of efficient direct organogenesis protocol for Punica granatum L. cv. Kandhari Kabuli from mature leaf explants. In Vitro Cell Dev Biol Plant 57:48–59

Wang Y, Zhou Y, Xu L, Han Z, Yin H, Ai S (2018) Photoelectrochemical apta-biosensor for zeatin detection based on graphene quantum dots improved photoactivity of graphite-like carbon nitride and streptavidin induced signal inhibition. Sens Actuators B Chem 257:237–244

Yancheva SD, Golubowicz S, Yablowicz Z, Perl A, Flaishman MA (2005) Efficient *Agrobacterium*-mediated transformation and recovery of transgenic fig (*Ficus carica* L.) plants. Plant Sci 168:1433–1441

Zimmerman PW, Hitchcock E (1942) Substituted phenoxy and benzoic acid growth substances and the relation of structure to physiological activity. Contrib Boyce Thompson Inst 12:321–343

Zoghlami N, Bouamama B, Khammassi M, Ghorbel A (2012) Genetic stability of long-term micropropagated *Opuntia ficus-indica* (L.) Mill. plantlets as assessed by molecular tools: perspectives for *in vitro* conservation. Ind Crops Prod 36:59–64

Zouari N, El Mtili N (2020) *In vitro* propagation of mature carob trees (*Ceratonia siliqua* L.) from the axillary buds. Am J Plant Sci 9:102917

Zouine J, El Hadrami I (2007) Effect of 2,4-D, glutamine and BAP on embryogenic suspension culture of date palm (*Phoenix dactylifera* L.). Sci Hortic 112:221–226

Integrative Approach of the Root Architecture by Interaction Between Auxin and Nutrients

Lucas Aparecido Gaion and Rogério Falleiros Carvalho

Abstract Throughout the plant development, the nutrient availability fluctuates along the soil profile. Thus, plants evolved a complex set of strategies to ensure nutrient uptake from the soil. Among these strategies, the modifications of the root system architecture (RSA) have been highlighted. These responses are under control of nutrient-specific signalling that generates systemic communications in the plant. Thus, nutritional deficiency is followed by modulation of root development in a nutrient-specific and dependent way. For example, the inorganic phosphate (Pi), potassium (K) and magnesium (Mg) deficiency causes the primary root (PR) elongation inhibition and induction of the development of lateral roots (LR). On the other hand, the PR and LR elongation is promoted by the deficiency of nitrogen (N) and iron (Fe). However, there is an intricate regulation of such responses because there are multiple signaling pathways which are triggered by nutritional condition. For example, auxin appears to be a key factor in the signaling pathways that lead to the optimization of the plants' ability to uptake nutrients. Thus, from the nutrient deficiency perception by plant's roots, a systemic signal is triggered, which will regulate genes of auxin biosynthesis (*YUC4*), signaling (*TIR1* and *ARF*) and transport (*PIN* and *AUX1*). Finally, differential auxin accumulation and/or perception in the root system's constituent portions, such as the PR, LR, and root hairs, will govern root development in response to nutrient availability. It is worth mentioning that plants can stimulate root growth (e.g. LR) in patches of nutrient-rich soil in response to local Fe, nitrate and ammonium supply. This demonstrates the plant's ability to fine-tune its root development to better explore the soil. Therefore, we aim in this review to provide an overview of the modulation of the auxin on RSA modification in response to soil nutrient availability.

L. A. Gaion
Agraian Sciences Center, Marilia University, Marília, Brazil

R. F. Carvalho (✉)
Department of Biology, São Paulo State University, Jaboticabal, Brazil
e-mail: rogerio.f.carvalho@unesp.br

T. Aftab (ed.), *Auxins, Cytokinins and Gibberellins Signaling in Plants*, Signaling and Communication in Plants,
https://doi.org/10.1007/978-3-031-05427-3_6

1 Introduction

Throughout the cycle life, plants must absorb elements {namely: macronutrients [nitrogen (N), potassium (K), phosphorus (P), calcium (Ca), magnesium (Mg) and sulfur (S)] and micronutrients [iron (Fe), zinc (Zn), manganese (Mn), boron (B), copper (Cu) and molybdenum (Mo)]} from the soil that are essential for their growth (Karthika et al. 2018; Uchida 2021). Although there is some variability in essentiality depending on the species, the elements necessary are rather conserved among cultivated species (Fageria 2001, 2009). It's worth noting that not all nutrients are required at similar levels; and if only one element is below the required amount, growth and production could be seriously limited. This principle has come to be known as the "Law of the Minimum" (Ploeg et al. 1999; Velayutham 2017). Furthermore, the availability of each nutrient varies greatly throughout the soil profile, both spatially and temporally (Malobane et al. 2020). For example, nitrate (NO_3^-), a form of N, and sulfate (SO_4^{2-}), a form of S, have a tendency to leach out and accumulate in the deeper soil layer, especially in tropical regions (Gaines and Gaines 2008; Wang et al. 2019c). On the other hand, inorganic phosphorus (Pi) tends to be fixed by other elements of soil, such as aluminum (Al), Ca and Fe, and so concentrate in the most superficial layers of soil (Arai & Livi 2013; Mbene et al. 2017). Therefore, the nutrient dynamic in the soil is clearly dependent on the element's properties as well as the local edaphoclimatic conditions.

Thus, in order to ensure access to essential nutrients from soil, plants have developed intricate pathways for modulating its development and metabolism in response to nutrient starvation and so stablish an efficient nutrient uptake system (Liang et al. 2013; Giehl & von Wirén 2014; Che et al. 2018; van der Bom et al. 2020). For instance, the activation of transporters, the release of organic acids and chelators by roots, and the remodeling of the root system architecture are all nutrient acquisition mechanisms (Pinto & Ferreira 2015; Che et al. 2018). Among these, root system architecture (RSA) remodeling is one of the main mechanisms that plants employ to improve nutrient acquisition efficiency. In fact, according to Lambers et al. (2007), the three primary parameters that most limit nutrient uptake by plants are: (i) the nutrient concentration close to the roots, (ii) the surface area of roots able to absorb the nutrient, and (iii) the distribution of roots along the soil; these parameters are directly or indirectly associate with root growth. Furthermore, the formation of a nutrient depletion zone in the rhizosphere is frequent as plants uptake nutrients (Syring & Claassem 1995). As a result, to maintain nutrient uptake, plants need to tightly control root growth, which allows plants to forage the soil for nutrient-rich zones.

More specifically, root growth has two main dimensions. Firstly, we can divide the root system longitudinally into three different portions: (i) the meristematic zone (with high cell division activity); (ii) the elongation zone (cell division gives rise to cell expansion); and (iii) maturation zone (cell differentiation region) (Petricka et al. 2012). Then, on a horizontal axis, we can observe the development of the lateral roots (LR) and root hairs (RH) that develop from the elongation and maturation zones, respectively (Petricka et al. 2012; Kazam 2013). Both the LR and the RH are responsible for most of the nutrient absorption activity in plants (Zhu & Lynch 2004; Leitner et al. 2010). Due to its importance for sustaining plant life, the control of root development at its various levels is under the intricate control of both external and internal signals, and the availability of nutrients has a special effect on root development (Gruber et al. 2013; Giehl & von Wirén 2014; Giehl et al. 2014) (Fig. 1). However, the nutrient does not appear to control root architecture directly.

Thus, some molecules must mediate the transition between the sense of nutrient deficiency and the adjustment of root growth (Kazan 2013; Saini et al. 2013; Hu et al. 2020). For example, hydrogen peroxide, nitric oxide, phytochromes, and plant hormones have been shown to act as root growth regulators in response to nutritional deficiency (Shin & Schachtman 2004; Buet et al. 2019; Romera et al. 2021;). Among these, auxin has deserved special attention due to its wide performance as a core player that integrates nutrient and other signals to regulate RSA in response to nutrient availability (Giehl et al. 2012; Kazan 2013; Saini et al. 2013). Indeed, from the nutrient deficiency perception by the plant's roots, a systemic signal is triggered, which will regulate genes of auxin biosynthesis (*YUC4*), signaling (*TIR1* and *ARF*) and transport (*PIN* and *AUX1*) (Sun et al. 2017a, b; Liu & von Wirén 2022). Finally, differential auxin accumulation and/or perception in the root system's regions, such as the primary root (PR), LR, and RH, will govern root development in response to nutrient availability (Gruber et al. 2013; Cavallari et al. 2021). Furthermore, the root growth adjustment should allow the plant to explore a larger volume of soil and, thus, stimulate root growth in the nutrient-rich soil patch. Given the importance of the interaction between auxin, root growth and nutrient acquisition, we aim in this review to provide an overview of the modulation of the auxin on RSA modification in response to soil nutrient availability.

2 Auxin Modulates Root System Architecture and Nutrient-Acquisition

2.1 Nitrogen (N)

N is the essential element required in greater amounts by plants and one that most limits the production of agricultural plants. In addition to being a component of several biomolecules, such as proteins, nucleic acids, hormones, and chlorophylls, N is an important regulator of physiological and biochemical processes, such as leaf expansion, gene expression, photosynthesis, and root development (Xuan et al. 2017; Liang et al. 2020). Plants uptake N preferentially in its inorganic form such as ammonium (NH_4^+) and NO_3^- (Masclaux Daubresse et al. 2010, Jia and Wirén 2020). However, factors like soil management and type, climate, fertilizer, and microbiological activity will influence the availability of N forms (Ukalska-Jaruga et al. 2020; Sorensen et al. 2021; Stoeckli et al. 2021). For example, NO_3^- tends to prevail

Fig. 1 An overview of schematic root growth under nutritional sufficiency conditions. Detail of the different dimensions of root growth divided into root tip, root hair, and lateral roots. In addition, the main modifications in the root system architecture in response to nutritional deficiency are presented. Nitrogen deficiency (−N): there is a higher elongation of the primary root accompanied by a decrease in the growth of the lateral roots. However, when the root system meets an NO_3^--rich soil patch, the multiplication and growth of lateral roots is stimulated in a localized way. Inorganic phosphate deficiency (−Pi): it is observed a reduction in the elongation of the primary root and an increase in lateral roots growth, allowing plants to explore a more superficial soil layer. Potassium deficiency (−K): it is verified a widespread decrease in root development. Calcium deficiency (−Ca): the elongation of the primary roots is inhibited, followed by a minor increase in the growth of the lateral roots. Magnesium deficiency (−Mg): there is a slight increase in the elongation of the primary root and a decrease in lateral roots growth. Sulphur deficiency (−S): there is a slight increase in the elongation of the primary root and a reduction in the growth of the lateral roots. Iron deficiency (−Fe): under severe Fe deficiency, a generalized reduction in root growth is observed

in well-aerated and non-acidic soils; on the other hand, NH_4^+ is the main form of N in flooded and/or acidic soils (Masclaux-Daubresse et al. 2010; Jia and Wirén 2020). Further, each N form tends to stratify differently along the soil profile, requiring the plant to successfully regulate its root growth to ensure N uptake.

Indeed, during their cycle life, plants must be flexible to modify their development and RSA as a function of external and internal N availability, which varies substantially in space (horizontally and vertically) and time (Jia and Wirén 2020). Thus, root architecture modulation helps plants optimize soil exploration, resulting in an increased capacity for N uptake from the soil. For example, plants with a deeper root system are more efficient in N prospecting (Heuermann et al. 2019). Actually, because NO_3^- is particularly sensitive to leaching, it accumulates in deeper soil layers. As a result, plants modulate the root growth angle to provide a deepening of the root system and so to explore subsurface soil layers with higher N availability (Ötvös et al. 2021). Furthermore, the multiplication and elongation of LR in N-rich layers contribute greatly to the supply of N to plants, particularly when N is scarce (Ma et al. 2013; Li et al. 2014; Wang et al. 2019b; Jia and von Wirén 2020).

In this way, the plants have evolved an intricate mechanism to regulate RSA, which is based on four basic characteristics: growth, branching, surface area and root angle insertion (Jia and von Wirén 2020). Nowadays, it is well established that plant hormonal homeostasis, particularly auxin, is linked to the regulation of RSA in N response (Krouk et al. 2010). For example, in Arabidopsis plants, NRT1.1/NPF6.3 is a dual affinity (i.e. low- and high-affinity) NO_3^- transporter that in NO_3^- absence transports auxin from the LR tip, which in turn leads to a decline in auxin levels and, consequently, a reduction in LR meristematic activity and elongation (Krouk et al. 2010; Maghiaoui et al. 2020; Ötvös et al. 2021; Vega et al. 2021; Wang et al. 2021) (Table 1). Furthermore, NO_3^- shortage may cause phosphorylation of PIN2, an auxin efflux transporter, resulting in increased auxin flow from the LR (Ötvös et al. 2021). Therefore, auxin content will be higher and lower in the PR and LR, respectively.

NRT1.1/NPF6.3 acts not only as a transporter but also as a sensor of NO_3^- availability in the soil, being able to trigger essential signaling pathways in response to N deficiency or scarcity. Indeed, NRT1.1/NPF6.3 modulates root development not only by auxin transport, but also by controlling auxin production and mechanical resistance to lateral root primordia (LRP) expansion under low NO_3^- circumstances (Ma et al. 2014; Zhang et al. 2019; Maghiaoui et al. 2020). For instance, when exposed to low N conditions, Arabidopsis exhibited a lower expression of the gene involved in auxin biosynthesis, *TRYPTOPHAN AMINOTRANSFERASE RELATED2* (*TAR2*), in the root stele (Maghiaoui et al. 2020). In other words, NRT1.1/NPF6.3 suppresses *TAR2*, reducing local auxin production and acropetal auxin supply to the LRP. Moreover, NRT1.1/NPF6.3 also acts as a negative regulator of *LAX3* (*Like AUX1*), an auxin influx carrier, involved in LR emergence by increasing auxin content in LRP. *LAX3* plays a crucial role in stimulating cell wall remodeling, allowing LR emergence (Swarup et al. 2008). These findings demonstrate that NRT1.1/NPF6.3 plays a multifaceted role in modulating plant RSA as a function of N availability in the soil.

Table 1 Summary of the modulation of the auxin on the root architecture in response to nutritional supply in Arabidopsis plants

Nutrient supply	Auxin response	Root system architecture modifications	References
Nitrogen deficiency	Auxin reduction in lateral root and allocation in primary root	Inhibition of lateral root elongation and increased primary root growth	Krouk et al. (2010)
Localized NO_3^-	Auxin transport to the lateral root	Increased lateral root proliferation in the NO_3^--rich soil zone	Mounier et al. (2014)
Phosphate deficiency	Reduced auxin content and activity in the primary root by redistribution to the lateral root	Reduced elongation of the primary root and promotion of lateral root and root hair emergence and elongation	Svistoonoff et al. (2007)
Localized Pi	Auxin redistribution to the lateral root	Increased lateral root proliferation in the Pi-rich soil zone	Wang et al. (2020)
Potassium deficiency	Less auxin delivery to the root system	Lower primary root and lateral root growth	Song et al. (2015) and Li et al. (2017)
Calcium deficiency	Putative interaction with auxin	Reduced primary root growth and a slight increase in lateral root growth	Giehl et al. (2014)
Sulphur deficiency	Increased auxin content in the root system	Increased primary root growth	López-Bucio et al. (2003)
Magnesium deficiency	Putative interaction with auxin	General reduction of root growth	Guo et al. (2015)
Iron deficiency	Decreasing auxin content in the lateral root	Inhibition of lateral root elongation	Long et al. (2020)

These modifications are followed by an increase in PR elongation, which is detrimental to LR initiation and development. As a result, deeper roots allow the plant to explore deeper soil horizons, increasing the efficiency of N uptake. In fact, Arabidopsis *nia1/nia2*, which is nitrate reductase deficient, proved insensitive to changes in PR length in response to N deficiency, although the double mutant accumulates NO_3^- in its tissues (Fu et al. 2020). This behavior is due to lower auxin biosynthesis in mutant plants when compared to wild type (Fu et al. 2020). Nevertheless, the *dgt* mutant tomato, which is insensitive to auxin, when exposed to N deficiency, displayed an enhanced number of LR and no change in other root growth traits (Santos et al. 2020a). Surprisingly, *dgt* grown under N deficiency accumulated more dry matter and used N more efficiently in their roots (Santos et al. 2020a, 2020b).

However, not only does low N availability change root architecture, but N excess also governs root growth. For instance, a large supply of NO_3^- promotes the formation and growth of LR in maize (*Zea mays*) plants by increasing auxin levels in the roots (Sattelmacher et al. 1993; Jia and von Wirén 2020). More recently, it was discovered that the favorable effect on LR development under high NO_3^- conditions is linked to greater auxin efflux in the phloem, which is mediated by ZmPIN-9, which favorably influences cell cycle activation (Yu et al. 2015b). Indeed, auxin was found to be differently distributed 12 h after NO_3^- stimulation, being carried from the root tip and cortex to the stele (Yu et al. 2015b).

Indeed, plants tends to allocate root multiplication in NO_3^--rich soil layers when NO_3^- is locally supplied to a previously N-deficient root system. In this condition, NO_3^- application stimulates LR initiation and extension in a species- and age-dependent way (Yu et al. 2015a; Jia and von Wirén 2020; Waidmann et al. 2020). Currently, several works have demonstrated that in NO_3^--rich soil zones, NO_3^- can acts as a signal, promoting LR growth (Sun et al. 2017a, b; Asim et al. 2020; Jia and von Wirén 2020). For instance, Arabidopsis plants growth in a split-root system confirmed that positive LR development responses to localized NO_3^- supply are influenced by local and systemic signals (Mounier et al. 2014; Asim et al. 2020; Jia and Wirén 2020). In another example, LR elongation and seminal root density of rice plants were stimulated by a localized supply of NO_3^- (Song et al. 2013). In short, when NO_3^- is unevenly supplied, the nitrate transceptor NRT1.1/NPF6.3 modifies auxin distribution throughout the root system, which, in turn, controls meristematic activity and so elongation of LR (Krouk et al. 2010; Mounier et al. 2014; Jia and Wirén 2020).

In addition, it has been demonstrated that NO_3^- supply can also modulate plant PR development. In Arabidopsis, high NO_3^- supply inhibits PR elongation by regulating the N-responsive *miRNA393/AFB3* module, an auxin signaling component, which is controlled by external and internal N availability (Vidal et al. 2010; Garrido-Vargas et al. 2020). These findings indicate that NO_3^- availability influences both the distribution and sensitivity to auxin in the root system. Consequently, this causes a differential development of the lateral (promoted) and primary (inhibited) roots, resulting in RSA change.

Nevertheless, NO_3^- is not the only N form uptaken by plants; NH_4^+ is the primary N source for countless cultivated plants (Gu et al. 2013), especially in flooded or acidic soils. Some works has evidenced that differential NH_4^+ availability in the soil can stimulate LR proliferation (Drew 1975; Hodge 2004). Even though increased LR growth in response to localized NH_4^+ supply is essential for improving the fertilizer uptake efficiency, the mechanism involved in this adaptive response is yet to be determined (Ma et al. 2013).

On the other hand, the elongation of PR and LR is impaired when NH_4^+ is used as the sole source of N for plants due to its toxicity effect. NH_4^+ is perceived in the root apex, according to evidence from providing NH_4^+ to distinct root regions (Li et al. 2010). NH_4^+ significantly inhibits cell proliferation and expansion at the cellular level (Liu et al. 2013). Auxin may be implicated in NH_4^+-mediated reduction of root elongation, according to a previous study, because mutants *aux1* (*axr1*, *axr2* and *dgt*), which impair auxin transport as well as disrupt auxin signaling, are more tolerant to root elongation inhibition by NH_4 (Cao et al. 1993; Santos et al. 2020b). However, Liu et al. (2013) used the *aux1* mutant to show that the inhibition of PR elongation by NH_4^+ could occur in an auxin-independent way.

Furthermore, the N source mix can also modulate RSA. For example, when rice plants were cultivated with partial nitrate (75/25 NH_4^+/NO_3^-), auxin accumulation was observed in the roots, particularly in the root tip, when compared to plants maintained only with NH_4^+ as a N source (Vega et al. 2019). Under mixed N sources, auxin accumulation seems to be related to enhanced auxin synthesis via the shikimic acid pathway by the higher levels of phosphoenol pyruvate and tryptophan as compared with forms isolates from NO_3^- and NH_4^+ (Wang et al. 2019a). Moreover, two tobacco (*Nicotiana tabacum*) cultivars (NC89 and Zhongyan 100), with different growth features, were exposed to N sources combinations (Lin et al. 2019). Then, exclusively in the NC89 cultivar, solo and low NO_3^- nutrition resulted in reduced N accumulation and inhibited elongation and development of first order LR (Lin et al. 2019). Less auxin was found in the roots of NC89 plants, which were shown to be sensitive to N sources. These modifications could be related to the regulation of auxin transporters by the PIN family (Lin et al. 2019).

In conclusion, N deficit promotes the elongation of PR while inhibiting the growth of LR. Plants can then explore deeper soil layers in search of N. The formation of LR that enhances N absorption is increased when the root system detects N-rich soil patches (Fig. 1). This demonstrates how root growth can be "intelligently" modulated to improve N uptake efficiency.

2.2 *Phosphorus (P)*

P is a macronutrient that is required for plant growth and development. It is found in a variety of biomolecules in plants, including nucleic acid, ATP, and NADPH, as well as bilayers of phospholipids that make up biomembranes (Epstein and Bloom 2005). P also has a role in several pivotal processes in plant metabolism, such as photosynthesis, respiration, N_2 fixation and protein phosphorylation, all of which are essential for plant survival (Ticconi and Abel 2004; Epstein and Bloom 2005). Due to its low availability in soils, P is the second most limiting element for crop production, closely after N (Cramer 2010; Richardson and Simpson 2011; Antonangelo et al. 2019).

P has a complex dynamic in soils, and it can be found in different forms, as inorganic and organic compounds. The predominance of one form is determined by a set of factors, including soil management and type, biochemical processes, fertilizer and limestone addition (Frossard et al. 2000). Furthermore, only a small portion of total soil P is bioavailable to plants, as P can be precipitated via interactions with Al, Ca, or Fe depending on soil pH (Penn and Camberato 2019). Thus, P concentrations in soil solutions are generally low, ranging from 0.1 to 10 μmol L^{-1} (Niu et al. 2013). In general, most of the P in the soil solution is in the form of $H_2PO_4^-$, also known as inorganic phosphorus (Pi); being the major form absorbed by plants at pHs below 6.0 (Schachtman et al. 1998; Shen et al. 2011; Penn and Camberato 2019).

Moreover, the plant's rapid uptake creates a Pi depletion zone in its rhizosphere due to the relatively low soil Pi diffusion coefficient (Hummel et al. 2021). As a result, after a few days of rapid uptake, the concentration of Pi in the rhizosphere can be drastically reduced, and the depletion zone can extend to about 2 mm from the root surface (Hummel et al. 2021). Furthermore, because Pi diffusion is affected by soil moisture, soils with low moisture content limit P migration to the root surface (McDowell et al. 2001; Hummel et al. 2021; Mardamootoo et al. 2021). These processes combine to produce significant fluctuation in Pi-available distribution in soil, which can lead to the heterogeneous Pi distribution in soil, mainly close to the plant root system (Werner et al. 2017). Thus, its low mobility and high fixation in the soil, makes the Pi especially concentrated in the upper layers of the soil and poorly available in soil solution. This suggests that plants should explore the topsoil layers to improve their Pi absorption efficiency. Indeed, in order to ensure Pi acquisition, plants have evolved a complex mechanism for controlling local and systemic responses to Pi shortage (Bhosale et al. 2018; Wang et al. 2020). Pi deficient response mechanisms in plants include molecular, biochemical, and physiological responses that promote morphological changes (Péret et al. 2011; Ham et al. 2018). Among these, remodeling root system growth to improve Pi uptake efficiency is the most significant change (Niu et al. 2013; Aslam et al. 2021; Lazali and Drevon 2021).

The ability of the plant to utilize a larger amount of soil is, in fact, intimately linked to Pi uptake. In general, under Pi deprivation, plants, such as Arabidopsis, maize, and rice, show reduced PR growth and an increased number and length of LR (Sato and Miura 2011; Niu et al. 2013; Jia et al. 2017; Wang et al. 2020) (Table 1). Then, RSA modulation allows the plants to explore more soil volume in the shallowest layer of soil and identify Pi-rich regions. Therefore, plants with more broad root systems may access more available Pi. These responses are being fine-tuned by the plant hormone auxin, which has its distribution and sensitivity altered in the root system due to Pi deprivation (Svistoonoff et al. 2007; Wang et al. 2020). Svistoonoff et al. (2007) demonstrated that when the root apex of Arabidopsis plants meets Pi-deficient soil, PR development is suppressed. Indeed, the suppression of PR growth is influenced by changes in auxin sensitivity and distribution at the root apex, and it involves a large number of genes whose expression is influenced by Pi availability.

For example, Arabidopsis defective for the *LPR1* gene (*LOW PHOSPHATE ROOT1*) and its paralog *LPR2* (*LOW PHOSPHATE ROOT2*) show a lower inhibition of PR growth under a condition of Pi deficiency (Svistoonoff et al. 2007). The authors additionally propose that the LPR1 protein found in the root cap regulates the activity and distribution of hormonal substances like auxin. Furthermore, the *PSI* gene (*PHOSPHORUS STARVATION INSENSITIVE*), an allele of *LPR1*, is essential for auxin sensitivity in the root as well as the occurrence of PR elongation reduction under Pi starvation conditions. When exposed to Pi deficiency stress, Arabidopsis *psi* mutants are less sensitive to auxin and have a greater ability to maintain PR growth than wild type (Wang et al. 2010). Therefore, loss of auxin sensitivity would impair the plant's responses to Pi deprivation, particularly due to the inability to remodel its root system, resulting in less exploration of the soil surface layer, where Pi is normally more plentiful.

The suppression of PR growth in plants exposed to Pi deficiency is associated with three key processes: (i) reduced cell elongation; (ii) lower cell division; and (iii) premature cell differentiation, which leads to root meristem exhaustion (Gutiérrez-Alaníz et al. 2018). According to Miura et al. (2011), there is an accumulation of auxin in the PR because of the root apex's perception of Pi insufficiency, which reaches toxic levels and limits PR elongation. The second step is auxin redistribution to the LR (Nacry et al. 2005; Miura et al. 2011). The redistribution of auxin to the LR, in turn, promotes their emergence and elongation; process that is under strong genetic control (Liu 2021). Thus, the increase in the formation of LR, in response to a Pi deficiency, is mediated by the auxin co-receptor TRANSPORT INHIBITOR RESPONSE 1/AUXIN SIGNALING F-BOX (TIR1/AFB) that act in nuclear auxin signaling (Narise et al. 2010; Shu et al. 2015). Furthermore, the TIR1/AFB requires the presence of transcription factors ARF7 and ARF19 (AUXIN RESPONSE FACTOR7 and 19), which increases auxin responses and induces to the formation of new roots (Narise et al. 2010). Thus, in the scarcity of Pi, an increase in the expression of the auxin receptor TIR1 raises the sensitivity of the pericycle's cells to auxin, resulting in the activation of transcription factors ARF, which promote the expression of genes responsible for the initiation and emergence of LR (Narise et al. 2010; Wu et al. 2020; Liu 2021). Indeed, the Pi uptake capability of the roots is reduced in the *arf7*, *arf19*, and *arf7 arf19* Arabidopsis mutants (Huang et al. 2018).

In order to investigate how plants respond to uneven Pi supply, Wang et al. (2020) carried out an elegant experiment in which they grew maize plants in a split-root system. This allowed them to evaluate the plants in two conditions: (i) a homogeneous Pi condition, in which both pots had the same Pi amount (sufficiency: 500 μM Pi, or deficiency: 0 μM Pi); and a heterogeneous Pi condition, in which one pot had 500 μM Pi and the other 0 μM Pi. They demonstrated that maize plants efficiently sense and signal the presence of Pi in the soil, promoting preferential root development in Pi-rich soil. Furthermore, plants with heterogeneous Pi supplies increased the number of LR threefold (Wang et al. 2020) (Table 1). Further, in this condition, when the auxin transport inhibitor *N*-1-naphthylphthalamic acid (NPA) was applied, there was inhibition of the proliferation of LR, indicating that auxin transport is an essential event in the modulation of RSA under environments with heterogeneous Pi distribution (Wang et al. 2020).

In addition to auxin redistribution, it has been found that increased LR growth is accompanied by increased sensitivity to auxin in the apical meristem and vascular tissue of the root tip. In fact, only four days after the start of cultivation under heterogeneous Pi there is an increased expression of *ZmPIN9* and *ZmARF2* in the roots of the heterogeneous high Pi (Wang et al. 2020). *ZmPIN9* is a monocot-specific PIN-FORMED9 that is only expressed in the endoderm and radicular pericycle that acts in the auxin redistribution. Whilst *ZmARF2*, a gene that encodes an auxin response factor, was strongly expressed in the LR of heterogeneous high Pi, when compared to heterogeneous low Pi or homogeneous Pi (Wang et al. 2020). Actually, ZmARF2 seems acts promoting auxin-mediated tissue growth rather by cell expansion than cell division, which could result in increased LR elongation (McSteen 2010). In rice plants, another monocot, the overexpression of the phosphate transporter *OsPHT1;8* (*OsPT8-Oe*) results in the accumulation of auxin in young LR, which is detected in greater numbers than in wild plants even in conditions of adequate Pi (Jia et al. 2017). The authors suggest that OsPT8 acts as a link between auxin and Pi starvation signaling. On the other hand, in rice plants, the auxin response factor OsARF12 operates as a negative regulator of Pi uptake and transport. Then, when the plant is supplied with Pi, the loss of function of the *OsARF12* gene enhances Pi absorption and transport, resulting in increased Pi content in the roots and leaves of rice plants (Wang et al. 2014). Thus, it is evidenced that the auxin redistribution under Pi deprivation is a conserved response among various species, such as Arabidopsis, maize, and rice. However, the role of ARF transcription factors seems to be species-dependent, exerting positive and negative effects on maize and rice plants, respectively, grown in Pi deficiency conditions. The negative effect of OsARF12 seems to involve more complex signaling pathways as it interacts with the cytokinin plant hormone to inhibit Pi uptake and transport by inhibition exactly of *OsPHT1* gene expression (Shen et al. 2014).

Furthermore, the formation and growth of LR by auxin redistribution in response to Pi deprivation is also followed by increased RH development (Jia et al. 2017; Bhosale et al. 2018). Thus, in order to increase auxin-dependent RH growth in response to Pi shortage, auxin must be redistributed from the PR tip to the zone of differentiation (Bhosale et al. 2018). For instance, in Arabidopsis, beginning with the biosynthesis of auxin, after Pi shortage perception, the auxin influx transporter Aux1 is activated, ensuring the auxin transport throughout the root system until the differentiation zone (Bhosale et al. 2018). This causes the formation and growth of RH in the differentiation zone of the root. Furthermore, low Pi stress promotes the expression of auxin-induced transcription factors such as *ARF9*, *RSL2* (*ROOT HAIR DEFECTIVE 6-LIKE 2*), and *RSL4* (*ROOT HAIR DEFECTIVE 6-LIKE 4*) (Bhosale et al. 2018), which generate the necessary stimulus for the elongation of the RH. Moreover, in the modulation of the LR, under Pi deprivation, the expression of TIR1, an important component of the auxin perception, is stimulated, elevating the auxin sensitivity in the roots. Finally, the combination of auxin redistribution and its enhanced sensitivity results in the initiation and elongation of RH under Pi deficiency.

These results demonstrate that the roots act as sensors of phosphorus availability along the soil profile, and then modulate RSA in response to Pi starvation through the regulation of auxin biosynthesis, transport and signaling. In general, Pi deprivation promotes the formation and growth of LR and RH while inhibiting PR elongation and decreasing basal root development angle (Miura et al. 2011; Péret et al. 2011; Huang et al. 2018); it is worth noting that these responses are extremely accurate. In fact, when the plant detects a Pi-rich soil patch, the plants are unable to stimulate the LR growth and so improve the Pi uptake. Overall, this results in a more superficial root system, which allows the plant to exploit a more superficial soil layer, which is often richer in Pi (Fig. 1).

2.3 Potassium (K)

K is a macronutrient and the most abundant cation in plants, and it is involved in the energy status, assimilate transport and storage, and plant osmotic regulation (Prajapati and Modi 2012). K is not found in any plant structure or chemical molecule, although it is involved in a variety of physiological functions (White and Karley 2010; Wang et al. 2013). For example, K is required for photosynthetic activity, maintains cell turgor, regulates of stomatal movements, promotes water uptake, regulates nutrient translocation in the plant, favors carbohydrate transport and storage, increases N uptake and protein synthesis, and participates in starch synthesis in leaves (Wang et al. 2013; Anschütz et al. 2014; Srivastava et al. 2020; Sardans and Peñuelas 2021).

Plants uptake K in the form of monovalent K^+, which is dissolved in soil solution (Yadav and Sidhu 2016). Furthermore, K^+ in the soil solution is in equilibrium with K^+ electrostatically attached to the negative charges of soil colloids, acting as a repository for the plants (Yadav and Sidhu 2016). K can be lost because of erosion of clay mineral particles that hold the nutrient in the soil, as well as leaching in soils with a poor cation exchange capacity (CEC) (Alfaro et al. 2004; Rosolem and Steiner 2017). For example, the latter can transport K to deeper layers of soil outside the root development zone of the plant (Alfaro et al. 2004; Ma et al. 2007; Zhang et al. 2013). Furthermore, K availability is particularly low in acidic soils (Dotaniya et al. 2016; Dhillon et al. 2019). In fact, low K stress is a common condition in agricultural soils (Kanai et al. 2011; Srivastava et al. 2020).

Thus, to ensure soil exploration and K acquisition efficiency, plants must evolve mechanisms that adjust the RSA in response to K availability; it has been shown that the majority of these mechanisms are controlled by auxin and K crosstalk (Sustr et al. 2019). Li et al. (2017) demonstrated that reduced K availability restricted PR formation in Arabidopsis plants. Actually, plants under low K stress accumulated less auxin at the tip root, which slows PR growth (Li et al. 2017). Further, the degradation of PIN1 proteins, which are important for auxin polar transport from the shoot to the root, appears to be the cause of the reduced auxin concentration in the tip root. This degradation of PIN1 is controlled by AKT1, a K transporter, which is required for K-dependent regulation of root growth; because *akt1* mutant plants showed no change in root growth in response to K supply (Li et al. 2017). A similar process was observed in tobacco plants, where plants subjected to low K exhibited reduced root system growth, particularly of the LR (Song et al. 2015). The response in this case was also due to a lower concentration of auxin in the root system of K-deficient plants (Song et al. 2015) (Table 1).

Furthermore, the K transporter TINY ROOT HAIR 1 (TRH1), which belongs to the KT/KUP/HAK (K^+ TRANSPORTER/ K^+ UPTAKE/HIGH AFFINITY-K^+

TRANSPORTER) family of K transporters, regulates RH growth in response to environmental cues (Vicente-Agullo et al. 2004; Dolan 2013; Daras et al. 2015). RH are extensions of the root epidermal cells that play a key role in microbial interactions and nutrient uptake. TRH1 activity, in turn, has been shown to be required for the polar localization of PIN proteins, and its absence impairs auxin transport (Dolan 2013). In fact, *trh1* Arabidopsis mutant plants have a tiny root hair phenotype and an impaired response to gravitropism (Dolan 2013; Rigas et al. 2013). Another member of the KT/KUP/HAK family of K transporters, OsHAK5, an H/K symporter, acts to modulate the polar auxin transport (PAT) (Yang et al. 2020). Thus, regardless of K supply, rice plants with *OsHAK5* loss of function demonstrated reduced auxin transport to the roots, as well as shorter LR and RH (Yang et al. 2020). Plants with lower *OsHAK5* expression exhibited reduced LR and RH elongation. Curiously, there was no detectable difference in K levels in the root system of wild-type and mutants with *OsHAK5* function loss (Yang et al. 2020).

Therefore, although some studies have shown increased root growth when plants are exposed to low K availability, what appears to predominate is root growth retardation when K is limited (Sustr et al. 2019). Under K deficiency, in addition to auxin balance disruption, the limitation of root growth appears also to be associated with decreased shoot-to-root transport of carbohydrates via phloem (Cakmak et al. 1994). In fact, K adequate nutrition is essential for the sugars shoot-to-root transport via phloem (Koch et al. 2019; Du et al. 2021). Moreover, the response to localized K, unlike N and P, does not appear to be related to the localized proliferation of lateral roots to explore K-rich soil patches (Fig. 1). On the other hand, when plants are subjected to K deficient conditions, their major approach is to improve the efficiency of K uptake, transport, and utilization by the plant rather than to promote root growth. Furthermore, these responses may be influenced by auxin, at least in part. For example, auxin can favorably regulate the K transporters AKT1 and AKT2 in Arabidopsis plants (Philippar et al. 2004; Fuchs et al. 2005; Shin 2017). However, the processes behind auxin's control of K metabolism need to be better understood.

2.4 Calcium, Sulfur and Magnesium

It is not surprising that other nutrients could be part of the modulation of the root architecture through auxin, even because plant nutrition is a complex event that integrates a multifaceted coordination over nutrient uptake and transport. Thus, we will discuss in this topic calcium (Ca), sulfur (S) and magnesium (Mg) which are macronutrients that play essential roles in the plant metabolism, signaling or structure.

Ca is a macronutrient with three major functions in plants: structural, enzyme activator and secondary messenger (White & Broadley 2003; Demidchik et al. 2018; Klimecka & Muszyńska 2007). For instance, it is required for the structural and functional integrity of membranes and cell walls, as well as the activation of enzymes such as ATPase, α-amylase, and phospholipase-D, hormone signaling and transport (White & Broadley 2003; Schapire et al. 2009; Thor 2019).

Although Ca levels in calcareous and arid soils can reach 250 g kg^{-1}, Ca contents in tropical soils are rather modest (Meriño-Gergichevich et al. 2010). Ca is typically provided to soil through the application of limestone (Scott et al. 1992; Castro et al. 2015). Then, the limited solubility of limestone causes Ca accumulation on the topsoil, particularly in conservationist systems with no soil tillage. Plants uptake Ca from the soil solution primarily in its ionic form (Ca^{2+}) (Yang & Jie 2005). Since the Ca present in the soil solution contacts roots mostly through mass flow and root growth (Ca intercept), soil moisture is a critical factor in the plant's Ca uptake (Kirkby 1979). In addition to soil moisture, external Ca concentration and the presence of other ions such as NH^{4+}, K^{+}, Mg^{2+}, and Al^{3+} reduce Ca absorption (Kirkby 1979; Mitra 2015).

Ca is a key player in various signaling pathways in plants, acting as a secondary messenger, including auxin signaling pathways (Tuteja & Mahajan 2007; Vanneste & Friml 2013; Choi et al. 2016). However, we will not be discussing Ca's role as a signaling molecule here, but rather how auxin coordinates root system changes in response to a Ca shortage in order to ensure Ca acquisition.

Primarily, Ca deprivation strongly inhibits PR elongation, while LR development and density are slightly increased (Giehl & von Wirén 2014; Giehl et al. 2014) (Table 1). However, the entire length of the root system tends to remain constant (Cao et al. 2013; Giehl et al. 2014). For instance, young plants of *Poncirus trifoliate* L., an orange rootstock, exhibited increased RH density and length when grown in a Ca-deficient condition (Cao et al. 2013). On the other hand, when Liu et al. (2019) cultivated this same species in Ca deficiency, they noticed an overall decrease in root development. However, these authors did not evaluate the various components of root architecture, such as primary and lateral roots, and root hair, individually. This could be explained by differences between these two experiments, such as deficiency intensity, culture mean and age of plants.

These changes would lead to the establishment of a horizontal root system that explores mostly the most superficial layers of the soil, which makes sense given that Ca tends to concentrate on the soil surface due to its low mobility (Gruber et al. 2013) (Fig. 1). Despite the fact that these responses are usually linked with auxin, no clear relationship has been established. Thus, the processes by which Ca deprivation influences root growth, particularly the role of auxin, remain unknown (Table 1).

Regarding S, this macronutrient is required at lower levels than other macronutrients by plants (Bender et al. 2015). Despite this, it plays crucial structural and metabolic functions in plants (Rennenberg et al. 2007; Moniuszko & Sirko 2008). Therefore, after its uptake, the S is reduced and integrated into amino acids, proteins, and coenzymes; where it participates in processes, such as photosynthesis, respiration, N metabolism, and biological N fixation (Kopriva et al. 2007; Bohrer & Takahashi 2016). Due to the poor availability of S in soils, plants must actively uptake S to meet their demands (Fuentes-Lara et al. 2019). Plants uptake S, preferably in the form of sulfate (SO_4^{2-}), which is then transported to the shoot of the plants via xylem and so integrated into plant metabolism (Takahashi et al. 1997; Rouached et al. 2009).

SO_4^{2-} meets the plant roots primarily through mass flow (Fuentes-Lara et al. 2019). This is due to its high mobility in the soil and low adsorption to the colloids of soil (Fuentes-Lara et al. 2019; Lucheta et al. 2021). Its high mobility may also be connected to the occurrence of a SO_4^{2-} shortage in soils from tropical and subtropical regions, with high annual precipitation rates (Karimian et al. 2018). Indeed, heavy rains could easily leach the SO_4^{2-} out of the plant's root system exploration zone (Dick et al. 2008). Moreover, the availability of SO_4^{2-} is affected by a variety of factors, including climate, soil management, soil type and correction, and nutrient addition (Dick et al. 2008; Scherer 2009; Carciochi et al. 2016). Liming, for example, can cause an enhancement in SO_4^{2-} leaching. In effect, plants have evolved mechanisms to modify root development in response to soil SO_4^{2-} deficiency via the action of auxin (Cui 2012; Giehl et al. 2014).

Therefore, SO_4^{2-} leaching and, consequently, its accumulation in deeper layers of soil, may explain why the SO_4^{2-} shortage leads to an increased elongation of PR and a reduction of LR near the root base, while stimulating greater proliferation of LR near the root tip (López-Bucio et al. 2003) (Table 1). In Arabidopsis plants, these changes seem to be directly associated with greater auxin levels in the root systems of SO_4^{2-}-starved plants because it was shown that SO_4^{2-} deficiency induced the expression of the *NITRILASE3* (*NIT3*) gene in the roots, which is responsible for converting indole-3-acetonitrile to indole-3-acetic acid (IAA) (Kutz et al. 2002). *Short hypocotyl 2* (*SHY2*) is another gene that regulates root growth under SO_4^{2-} shortage (Aarabi et al. 2020). This gene is involved in the apical meristem's cellular differentiation control. So, when the root grows in a SO_4^{2-}-deficient environment, *SHY2* is suppressed, producing a delay in cell differentiation, and allowing the PR to elongate for a longer period (Aarabi et al. 2020) (Fig. 1). Another key response to S deprivation is the induction of high-affinity SO_4^{2-} transporters (SULTRs) *SULTR1;1* and *SULTR1;2* in the RH and epidermis and cortex of roots, which increases the capacity of SO_4^{2-} uptake by plants (Giehl et al. 2014; Li et al. 2020). Nevertheless, the link between auxin and SULTRs activation has yet to be proven, necessitating additional research to determine if auxin biosynthesis in roots in response to SO_4^{2-} deficiency is responsible for the activation of genes encoding high-affinity SO_4^{2-} transporters.

Magnesium (Mg) is a macronutrient that is required for plant growth and development. Mg is particularly known as the core atom of the chlorophyll molecule, but it also plays an important role in activating enzymes involved in carbon and carbohydrate metabolism (Chen et al. 2018; Shaul 2002). Mg encounters plant roots primarily through the mass flow of ions solubilized in the soil solution, where it is absorbed by the plant as Mg^{2+} (Christenson et al. 1973; Gransee & Führs 2013; Ogura et al. 2018). So far, little is known regarding the involvement of auxin in the remodeling of RSA in the condition of Mg^{2+} insufficiency. Indeed, Mg^{2+} deficiency is more frequent in plants grown in tropical soils due to high temperatures, abundant rainfall, soil acidity, and competition with other elements such as K^+, Ca^{2+} and Al^{3+} (Mayland & Wilkinson 1989; Haby et al. 1990).

According to our current knowledge, Arabidopsis plants under mild Mg^{2+} deficiency conditions (0.05 Mg mM) tend to show reduced root growth (primary and lateral roots) accompanied by increased initiation, density, and elongation of RH (Niu et al. 2014a; Guo et al. 2015). On the other hand, when cultivated in the complete Mg^{2+} absence, Arabidopsis plants showed severe and complete reduction in root growth (Fig. 1), which appears to be due to impaired sucrose transport from the shoot to the root system (Guo et al. 2015). However, these results, particularly RH formation, appear to be influenced by factors such as culture media, genotype, and plant age. In Mg-deficient plants, radicular remodeling, mainly RH growth, was linked to redistribution of cytosolic Ca^{2+} and reactive oxygen species accumulation in roots (Niu et al. 2014b). Although the relationship between auxin biosynthesis or signaling and the root response to Mg^{2+} deprivation remains unknown, given its broad involvement in the control of root growth, it is extremely likely that these alterations in the root system in response to Mg deprivation involve the participation of auxin at some level (Table 1).

2.5 Iron (Fe)

Among all the micronutrients, Fe has received the greatest attention in terms of the consequences of its deficiency on root growth, in which it has been shown that auxin strongly acts (Giehl et al. 2014; Long et al. 2020). Furthermore, Fe is the micronutrient that higher plants require in larger quantities (Gupta et al. 2008). Fe is required in various electron transfer processes because of its capacity to interconvert between reduced and oxidized forms via redox reactions (Briat et al. 2007, 2015). The following are some of the functions of Fe in plants: (i) component of enzymes involved in oxi-reduction reactions; (ii) component of electron transfer mechanisms; (iii) catalyst for chlorophyll production; (iv) constituent of enzymes that engage in the reduction of nitrite and sulfite; and (v) required for the action of nitrogenize, which performs atmospheric N_2 fixation (Kerkeb & Connolly 2006; Briat et al. 2007, 2015).

In the soil, although total Fe is a reasonably plentiful element in cultivated soils (20–40 g kg^{-1}), its limited solubility makes it unavailable to plants (Colombo et al. 2014). The concentration of Fe in its ionic forms (Fe^{2+} and Fe^{3+}); preferentially uptaken by plants; in well-aerated soil solutions is around 10^{-10} M (Colombo et al. 2014; Mendoza et al. 2020; Qi et al. 2020). This concentration is significantly below what the plants require, namely 10^{-5} to 10^{-6} M (Konrad 1994; Colombo et al. 2014).

Therefore, plants have evolved a set of adaptive mechanisms to improve Fe uptake from the soil. These mechanisms are generally divided into two groups. Thus, strategy I, which is found in non-grass plants, involves acidification of the rhizosphere via H^+ release, which makes Fe more soluble (Flannery et al. 2013; Jeong et al. 2017). Another process implicated in strategy I is the exudation of organic acids, phenolic compounds, and flavin, which bind Fe and keep it soluble, allowing the Fe to move more easily to the roots (Tsai & Schmidt 2017; Harbort et al. 2020). Plants with strategy I for Fe acquisition can only uptake Fe in its reduced state (Fe^{2+}), so Fe^{3+} must be reduced to Fe^{2+} (Tsai & Schmidt 2017). Finally, the IRON-REGULATED

TRANSPORTER 1 (IRT1), a high-affinity Fe transporter, is involved in Fe^{2+} absorption (Tsai & Schmidt 2017). On the other hand, grass plants have the strategy II for Fe acquisition. These plants, unlike non-grass plants, can uptake Fe in its unreduced form (Fe^{3+}), and the key mechanism for increasing Fe absorption efficiency is Fe^{3+} chelation with phytosiderophores (Tsai & Schmidt 2017; Khan et al. 2018). This is due to mugineic acid exudation via the cell root epidermis (Kim & Guerinot 2007).

The central point is that all these processes, at least partially, are under auxin modulation because this hormone is also involved in the remodeling of root growth in response to Fe deficiency in order to better explore the Fe-rich soil patches (Giehl et al. 2012, 2014; Shen et al. 2015). In fact, the centerpiece of regulation of RSA under Fe deficiency is the redistribution of auxin throughout the root system. Thus, a common response to Fe deficiency is the reduction of root system growth by decreasing the concentration of auxin, mainly in the LR, which inhibits LR elongation (Giehl et al. 2014; Guo et al. 2020; Long et al. 2020). However, the localized supply of Fe increases AUX1 activity, an auxin efflux transporter; exclusively in the LR with access to Fe resulting in a larger concentration of auxin in those LR apices, which stimulates their elongation (Giehl et al. 2012). Therefore, based on their observations, the authors propose that, from its perception, the localized Fe supply acts as a systemic signal, stimulating the growth of LR specifically in Fe-rich soil patches (Giehl et al. 2014).

Similarly, auxin accumulation prevents cucumber (*Cucumis sativus*) plants from the detrimental consequences of Fe deprivation. Guo et al. (2020) revealed that gamma-aminobutyric acid (GABA) application acts as a Fe deficiency ameliorator by favorably regulating biosynthetic (*YUC4*) and auxin transport (*PIN1*) genes, causing auxin accumulation in both shoots and roots. In fact, GABA application induces increased root growth and Fe uptake under Fe deficiency conditions. However, the use of NPA, an auxin transport inhibitor, reversed the beneficial effects of GABA (Guo et al. 2020).

On the other hand, when rice plants (a Poaceae family species that uses Fe acquisition strategy II) were subjected to Fe deprivation, the initiation and elongation of RH and LRs were strongly reduced (Shen et al. 2015) (Fig. 1). Therefore, unlike Arabidopsis plants, where the reduction in LR elongation is attributed to a decrease in auxin accumulation (Table 1), rice plants exhibit the exact reverse. In other words, Fe shortage causes an excess of auxin accumulation in the root system, possibly reaching toxic levels for its development (Shen et al. 2015). Indeed, the responses to Fe shortage were largely recovered in plants of the *osarf16* mutant, which are insensitive to auxin; this included the formation and elongation of RH and LR (Shen et al. 2015).

3 Epilogue

So far, significant effort has been expended to better understand the role of auxin as a regulator of RSA modifications in response to nutrient depletion, particularly for the macronutrients N and Pi, as well as the micronutrient Fe (Fig. 1). It is worth noting that auxin does not act alone; rather, auxin interacts with other molecules, including other plant hormones, secondary messengers (hydrogen peroxide and nitric oxide) and phytochromes, to regulate root growth depending on nutrient availability (Chen et al. 2010; Celletti et al. 2020; Maciel et al. 2021; Raya-González et al. 2021; Soares et al. 2021). A classic example is the interaction between auxin and ethylene under Fe deficiency. In this case, the root remodeling is orchestrated by pathways involving ethylene biosynthesis and auxin transport inhibition from the shoot-to-root (Celletti et al. 2020). More specifically, Fe shortage causes an increase in ethylene biosynthesis, which is accompanied by a decrease in auxin delivery to the root system, resulting in a reduction in lateral root growth (Celletti et al. 2020; Giehl et al. 2014). However, in addition to root system remodeling, more research is needed to better clarify how root modifications is associate with nutrient availability in the soil, uptake, transport and accumulation. Moreover, it is important to consider the fact that roots interact with other factors, such as abiotic stress. For example, a common plant response to Pi deprivation is inhibition of PR growth, which results in a shorter or shallower root system, which could make the plants more susceptible to drought stress (Fig. 1). This is especially concerning in a climate change scenario, where water deficit events will be more frequent and intense. Certainly, an intricate hormonal control can be triggered.

References

Aarabi F, Naake T, Fernie AR, Hoefgen R (2020) Coordinating sulfur pools under sulfate deprivation. Trends Plant Sci 25(12). https://doi.org/10.1016/j.tplants.2020.07.007

Alfaro MA, Jarvis SC, Gregory PJ (2004) Factors affecting potassium leaching in different soils. Soil Use Manag 20:182–189

Anschütz U, Becker D, Shabala S (2014) Going beyond nutrition: regulation of potassium homoeostasis as a common denominator of plant adaptive responses to environment. J Plant Physiol 171:670–687

Antonangelo JA, Firmano RF, Alleoni LRF, Oliveira Junior A, Zhang H (2019) Soybean yield response to phosphorus fertilization in an oxisol under long-term no-till management. Soil Sci Soc Am J 83:173–180. https://doi.org/10.2136/sssaj2018.07.0251

Arai Y, Livi KJ (2013) Underassessed phosphorus fixation mechanisms in soil sand fraction. Geoderma 192:422–429

Asim M, Ullah Z, Xu F, An L, Aluko OO, Wang Q, Liu H (2020) Nitrate signaling, functions, and regulation of root system architecture: insights from *Arabidopsis thaliana*. Genes 11(6):633. https://doi.org/10.3390/genes11060633

Aslam MM, Karanja JK, Yuan W, Zhang Q, Zhang J, Xu W (2021) Phosphorus uptake is associated with the rhizosheath formation of mature cluster roots in white lupin under soil drying and phosphorus deficiency. Plant Physiol Biochem 166:531–539. https://doi.org/10.1016/j.plaphy.2021.06.022

Bender RR, Haegele JW, Below FE (2015) Nutrient uptake, partitioning, and remobilization in modern soybean varieties. Agron J 107(2):10. https://doi.org/10.2134/agronj14.0435

Bhosale R, Giri J, Pandey BK, Giehl RFH, Hartmann A, Traini R, Truskina J, Leftley N, Hanlon M, Swarup K, Rashed A, Voß U, Alonso J, Stepanova A, Yun J, Ljung K, Brown KM, Lynch JP, Dolan L, Vernoux T, Bishopp A, Wells D, von Wirén N, Bennett MJ, Swarup R (2018) A

mechanistic framework for auxin dependent Arabidopsis root hair elongation to low external phosphate. Nat Commun 9:1409. https://doi.org/10.1038/s41467-018-03851-3
Bohrer AS, Takahashi H (2016) Compartmentalization and regulation of sulfate assimilation pathways in plants. Int Rev Cell Mol Biol 326(17):1–31. https://doi.org/10.1016/bs.ircmb.2016.03.001
Briat JF, Curie C, Gaymard F (2007) Iron utilization and metabolism in plants. Curr Opin Plant Biol 10(3). https://doi.org/10.1016/j.pbi.2007.04.003
Briat JF, Dubos C, Gaymard F (2015) Iron nutrition, biomass production, and plant product quality. Trends Plant Sci 20(1). https://doi.org/10.1016/j.tplants.2014.07.005
Buet A, Galatro A, Ramos-Artuso F, Simontacchi M (2019) Nitric oxide and plant mineral nutrition: current knowledge. J Exp Bot 70(17):4461–4476. https://doi.org/10.1093/jxb/erz129
Cakmak I, Hengeler C, Marschner H (1994) Changes in phloem export of sucrose in leaves in response to phosphorus, potassium and magnesium deficiency in bean plants. J Exp Bot 45(278):1251–1257
Cao X, Chen C, Zhang D, Shu B, Xiao J, Xia R (2013) Influence of nutrient deficiency on root architecture and root hair morphology of trifoliate orange (*Poncirus trifoliata* L. Raf.) seedlings under sand culture. Sci Hortic 162(5)
Cao Y, Glass AD, Crawford NM (1993) Ammonium inhibition of Arabidopsis root growth can be reversed by potassium and by auxin resistance mutations *aux1*, *axr1*, and *axr2*. Plant Physiol 102:983–989
Carciochi WD, Wyngaard N, Divito GA, Calvo NIR, Cabrera ML, Echeverría HE (2016) Diagnosis of sulfur availability for corn based on soil analysis. Biol Fertil Soils 52(7):917–926. https://doi.org/10.1007/s00374-016-1130-8
Castro GSA, Crusciol CAC, Filho UEPJ, d. M. (2015) Effects of surface application of dolomitic limestone and calcium-magnesium silicate on soybean and maize in rotation with green manure in a tropical region. Bragantia 74:311–321. https://doi.org/10.1590/1678-4499.0346
Cavallari N, Artner C, Benkova E (2021) Auxin-regulated lateral root organogenesis. Cold Spring Harb Perscpectives Biol 1:a089941. https://doi.org/10.1101/cshperspect.a039941
Celletti S, Pii Y, Valentinuzzi F, Tiziani R, Fontanella MC, Beone GM, Mimmo T, Cesco S, Astolfi S (2020) Physiological responses to Fe deficiency in split-root tomato plants: possible roles of auxin and ethylene? Agronomy 10(7):1000. https://doi.org/10.3390/agronomy10071000
Che J, Yamaji N, Ma JF (2018) Efficient and flexible uptake system for mineral elements in plants. New Phytolog 219(2). https://doi.org/10.1111/nph.15140
Chen WW, Yang JL, Qin C, Jin CW, Mo JH, Ye T, Zheng SJ (2010) Nitric oxide acts downstream of auxin to trigger root ferric-chelate reductase activity in response to iron deficiency in Arabidopsis. Plant Physiol 154(2). https://doi.org/10.1104/pp.110.161109
Chen ZC, Peng WT, Li J, Liao H (2018) Functional dissection and transport mechanism of magnesium in plants. Semin Cell Dev Biol 74:142–152. https://doi.org/10.1016/j.semcdb.2017.08.005
Choi W-G, Hilleary R, Swanson SJ, Kim S-H, Gilroy S (2016) Rapid, long-distance electrical and calcium signaling in plants. Annu Rev Plant Biol 29(67):287–307
Christenson DR, White RP, Doll EC (1973) Yields and magnesium uptake by plants as affected by soil pH and calcium levels. Agron J 65(2):1. https://doi.org/10.2134/agronj1973.00021962006500020006x
Colombo C, Palumbo G, He J, Pinton R, Cesco S (2014) Review on iron availability in soil: interaction of Fe minerals, plants, and microbes. J Soils Sediments 14:538–548
Cramer MD (2010) Phosphate as a limiting resource: introduction. Plant Soil 334:1–10. https://doi.org/10.1007/s11104-010-0497-9
Cui X (2012) Nutrient sensing in plants. Mol Plant 5(6). https://doi.org/10.1093/mp/sss107
Daras G, Rigas S, Tsitsekian D, Iacovides TA, Hatzopoulos P (2015) Potassium transporter TRH1 subunits assemble regulating root-hair elongation autonomously from the cell fate determination pathway. Plant Sci 231:131–137. https://doi.org/10.1016/j.plantsci.2014.11.017

Demidchik V, Shabala S, Isayenkov S, Cuin T, Pottosin I (2018) Calcium transport across plant membranes: mechanisms and functions. New Phytol 220(1). https://doi.org/10.1111/nph.15266
Dhillon JS, Eickhoff EM, Mullen RW, Raun WR (2019) World potassium use efficiency in cereal crops. Agron J 111(2):889–896. https://doi.org/10.2134/agronj2018.07.0462
Dick WA, Kost D, Chen L (2008) Availability of sulfur to crops from soil and other sources. Crop Sci 50. https://doi.org/10.2134/agronmonogr50.c5
Dolan L (2013) Pointing PINs in the right directions: a potassium transporter is required for the polar localization of auxin efflux carriers. New Phytol 197:1027–1028
Dotaniya ML, Meena VD, Basak BB, Meena RS (2016) Potassium uptake by crops as well as microorganisms. In: Meena VS et al (eds) Potassium solubilizing microorganisms for sustainable agriculture. https://doi.org/10.1007/978-81-322-2776-2_19
Drew MC (1975) Comparison of effects of a localized supply of phosphate, nitrate, ammonium and potassium on growth of seminal root system, and shoot, in barley. New Phytol 75:479–490
Du F, Liu H, Yin X, Zhao Q, Shi C (2021) Potassium-mediated regulation of sucrose metabolism and storage root formation in sweet potato. Arch Agron Soil Sci 67(5):703–713. https://doi.org/10.1080/03650340.2020.1751824
Epstein E, Bloom AJ (2005) Mineral nutrition of plants: principles and perspectives, 2nd edn. Sinauer Associates, Sunderland
Fageria NK (2009) The use of nutrients in crop plants, 1 ed, vol 1. CRC Press
Fageria VD (2001) Nutrient interactions in crop plants. J Plant Nutr 24(8):21. https://doi.org/10.1081/PLN-100106981
Flannery AR, Renberg RL, Andrews NW (2013) Pathways of iron acquisition and utilization in Leishmania. Curr Opin Microbiol 16(6). https://doi.org/10.1016/j.mib.2013.07.018
Frossard E, Condron LM, Oberson A, Sinaj A, Fardeau JC (2000) Processes governing phosphorus availability in temperate soils. J Environ Qual 29:15–23
Fu Y, Zhang Z, Yang X, Wang C, Lan T, Tang X, Chen G, Zeng J, Yuan S (2020) Nitrate reductase is a key enzyme responsible for nitrogen-regulated auxin accumulation in Arabidopsis roots. Biochem Biophys Res Commun 532(4):633–639. https://doi.org/10.1016/j.bbrc.2020.08.057
Fuchs I, Stolzle S, Ivashikina N, Hedrich R (2005) Rice K^+ uptake channel OsAKT1 is sensitive to salt stress. Planta 221:212–221
Fuentes-Lara LO, Medrano-Macías J, Pérez-Labrada F, Rivas-Martínez EN, García-Enciso EL, González-Morales S, Juárez-Maldonado A, Rincón-Sánchez F, Benavides-Mendoza A (2019) From elemental sulfur to hydrogen sulfide in agricultural soils and plants. Molecules 24(12). https://doi.org/10.3390/molecules24122282
Gaines TP, Gaines ST (2008) Soil texture effect on nitrate leaching in soil percolates. Commun Soil Sci Plant Anal 25(13):9. https://doi.org/10.1080/00103629409369207
Garrido-Vargas F, Godoy T, Tejos R, O'Brien JA (2020) Overexpression of the auxin receptor AFB3 in arabidopsis results in salt stress resistance and the modulation of NAC4 and SZF1. Int J Mol Sci 21:9528. https://doi.org/10.3390/ijms21249528
Giehl RFH, Gruber BD, von Wirén N (2014) It's time to make changes: modulation of root system architecture by nutrient signals. J Exp Bot 65(3):769–778. https://doi.org/10.1093/jxb/ert421
Giehl RF, von Wirén N (2014) Root nutrient foraging. Plant Physiol 166(2). https://doi.org/10.1104/pp.114.245225
Giehl R, Lima JE, von Wirén N (2012) Localized iron supply triggers lateral root elongation in Arabidopsis by altering the AUX1-mediated auxin distribution. Plant Cell 24(1). https://doi.org/10.1105/tpc.111.092973
Gransee A, Führs H (2013) Magnesium mobility in soils as a challenge for soil and plant analysis, magnesium fertilization and root uptake under adverse growth conditions. Plant Soil 368(1):5–21. https://doi.org/10.1007/s11104-012-1567-y
Gruber BD, Giehl RF, Friedel S, von Wirén N (2013) Plasticity of the Arabidopsis root system under nutrient deficiencies. Plant Physiol 163(1). https://doi.org/10.1104/pp.113.218453
Gu R, Duan F, An X, Zhang F, von Wirén N, Yuan L (2013) Characterization of AMT-mediated high-affinity ammonium uptake in roots of maize (*Zea mays* L.). Plant Cell Physiol 54:1515–1524

Guo W, Chen S, Hussain N, Cong Y, Liang Z, Chen K (2015) Magnesium stress signaling in plant. just a beginning. Plant Signal Behav 10(3). https://doi.org/10.4161/15592324.2014.992287

Guo Z, Du N, Li Y, Zheng S, Shen S, Piao F (2020) Gamma-aminobutyric acid enhances tolerance to iron deficiency by stimulating auxin signaling in cucumber (*Cucumis sativus* L.). Ecotoxicol Environ Saf 192. https://doi.org/10.1016/j.ecoenv.2020.110285

Gupta UC, Wu K, Liang S (2008) Micronutrients in soils, crops, and livestock. Earth Sci Front 15(5):110–125

Gutiérrez-Alanís D, Ojeda-Rivera JO, Yong-Villalobos L, Cárdenas-Torres L, Herrera-Estrella L (2018) Adaptation to phosphate scarcity: tips from Arabidopsis roots. Trends Plant Sci 23(8). https://doi.org/10.1016/j.tplants.2018.04.006

Haby VA, Russelle MP, Skogley EO (1990) Testing soils for potassium, calcium, and magnesium. In: RL Westerman (ed) Soil testing and plant analysis, vol 3. https://doi.org/10.2136/sssabookser3.3ed.c8

Ham B-K, Chen J, Yan Y, Lucas WJ (2018) Insights into plant phosphate sensing and signaling. Curr Opin Biotechnol 49:1–9. https://doi.org/10.1016/j.copbio.2017.07.005

Harbort CJ, Hashimoto M, Inoue H, Niu Y, Guan R, Rombolà AD, Kropiva S, Voges MJEEE, Sattely ES, Garrido-Oter R, Schulze-Lefert P (2020) Root-secreted coumarins and the microbiota interact to improve iron nutrition in Arabidopsis. Cell Host Microbe 28(6). https://doi.org/10.1016/j.chom.2020.09.006

Heuermann D, Gentsch N, Boy J, Schweneker D, Feuerstein U, Groß J, Bauer B, Guggenberger G, von Wirén N (2019) Interspecifc competition among catch crops modifies vertical root biomass distribution and nitrate scavenging in soils. Sci Rep (9):11531. https://doi.org/10.1038/s41598-019-48060-0

Hodge A (2004) The plastic plant: root responses to heterogeneous supplies of nutrients. New Phytol 162:9–24

Hu S, Zhang M, Yang Y, Xuan W, Zou Z, Arkorful E, Chen Y, Ma Q, Jeyaraj A, Chen X, Li X (2020) A novel insight into nitrogen and auxin signaling in lateral root formation in tea plant [*Camellia sinensis* (L.) O. Kuntze]. BMC Plant Biol 20(1). https://doi.org/10.1186/s12870-020-02448-7

Huang K-L, Ma G-J, Zhang M-L, Xiong H, Wu H, Zhao C-Z, Liu C-S, Jia H-X, Chen L, Kjorven JO, Li X-B, Ren F (2018) The ARF7 and ARF19 transcription factors positively regulate phosphate starvation response1 in Arabidopsis roots. Plant Physiol 178:413–427

Hummel C, Boitt G, Santner J, Lehto NJ, Condron L, Wenzel WW (2021) Co-occurring increased phosphatase activity and labile P depletion in the rhizosphere of *Lupinus angustifolius* assessed with a novel, combined 2D-imaging approach. Soil Biol Biochem 153:107963. https://doi.org/10.1016/j.soilbio.2020.107963

Iron transport and metabolism in plants. Genet Eng 27. https://doi.org/10.1007/0-387-25856-6_8

Jeong J, Merkovich A, Clyne M, Connolly EL (2017) Directing iron transport in dicots: regulation of iron acquisition and translocation. Curr Opin Plant Biol 39https://doi.org/10.1016/j.pbi.2017.06.014

Jia Z, von Wirén N (2020) Signaling pathways underlying nitrogen-dependent changes in root system architecture: from model to crop species. J Exp Bot 71(15):4393–4404. https://doi.org/10.1093/jxb/eraa033

Jia H, Zhang S, Wang L, Yang Y, Zhang H, Cui H, Shao H, Xu G (2017) OsPht1;8, a phosphate transporter, is involved in auxin and phosphate starvation response in rice. J Exp Bot 68(18):5057–5068. https://doi.org/10.1093/jxb/erx317

Kanai S, Moghaieb RE, El-Shemy HA, Panigrahi R, Mohapatra PK, Ito J, Nguyen NT, Saneoka H, Fujita K (2011) Potassium deficiency affects water status and photosynthetic rate of the vegetative sink in green house tomato prior to its effects on source activity. Plant Sci 180:368–374

Karimian N, Johnston SG, Burton ED (2018) Iron and sulfur cycling in acid sulfate soil wetlands under dynamic redox conditions: a review. Chemosphere 197https://doi.org/10.1016/j.chemosphere.2018.01.096

Karthika KS, Rashmi I, Parvathi MS (2018) Biological functions, uptake and transport of essential nutrients in relation to plant growth. In: Hasanuzzaman M, Fujita M, Oku H, Nahar K, Hawrylak-Nowak B (eds) Plant nutrients and abiotic stress tolerance, p 49. SpringerLink. https://doi.org/10.1007/978-981-10-9044-8_1

Kazam K (2013) Auxin and the integration of environmental signals into plant root development. Ann Bot 112(9):1655–1665. https://doi.org/10.1093/aob/mct229

Khan A, Singh P, Srivastava A (2018) Synthesis, nature and utility of universal iron chelator—Siderophore: a review. Microbiol Res 212–213. https://doi.org/10.1016/j.micres.2017.10.012

Kim SA, Guerinot ML (2007) Mining iron: iron uptake and transport in plants. FEBS Lett 581(12). https://doi.org/10.1016/j.febslet.2007.04.043

Kirkby EA (1979) Maximizing calcium uptake by plants. Commun Soil Sci Plant Anal 10(1–2):89–113

Klimecka M, Muszyńska G (2007) Structure and functions of plant calcium-dependent protein kinases. Acta Biochim Pol 54(2):219–233

Koch M, Busse M, Naumann M, Jákli B, Smit I, Cakmak I, Hermans C, Pawelzik E (2019) Differential effects of varied potassium and magnesium nutrition on production and partitioning of photoassimilates in potato plants. Physiol Plant 166(4):921–935. https://doi.org/10.1111/ppl.12846

Konrad M (1994) Iron availability in plant tissues-iron chlorosis on calcareous soils. Plant Soil 165(2):275–283. https://doi.org/10.1007/BF00008070

Kopriva S, Wiedemann G, Reski R (2007) Sulfate assimilation in basal land plants—what does genomic sequencing tell us? Plant Biol 9(5). https://doi.org/10.1055/s-2007-965430

Krouk G, Lacombe B, Bielach A, Perrine-Walker F, Malinska K, Mounier E, Hoyerova K, Tillard P, Leon S, Ljung K, Zazimalova E, Benkova E, Nacry P, Gojon A (2010) Nitrate-regulated auxin transport by NRT1.1 defines a mechanism for nutrient sensing in plants. Dev Cell 18:927–937. https://doi.org/10.1016/j.devcel.2010.05.008

Kutz A, Müller A, Hennig P, Kaiser WM, Piotrowski M, Weiler EW (2002) A role for nitrilase 3 in the regulation of root morphology in sulphur-starving *Arabidopsis thaliana*. Plant J 30(1). https://doi.org/10.1046/j.1365-313x.2002.01271.x

Lambers H, Raven JA, Shaver GR, Smith SE (2007) Plant nutrient-acquisition strategies change with soil age. Trends Ecol Evol 23(2):95–103

Lazali M, Drevon JJ (2021) Mechanisms and adaptation strategies of tolerance to phosphorus deficiency in legumes. Commun Soil Sci Plant Anal. https://doi.org/10.1080/00103624.2021.1885693

Leitner D, Klepsch S, Ptashnyk M, MArchant A, Kirk GJD, Schnepf A, Roose T (2010) A dynamic model of nutrient uptake by root hairs. New Phytol 185(3):792–802.https://doi.org/10.1111/j.1469-8137.2009.03128.x

Li H, Ma G, Li H, Zhang F, Rengel Z, Shen J (2014) Root morphological responses to localized nutrient supply differ among crop species with contrasting root traits. Plant Soil 376:151–163. https://doi.org/10.1007/s11104-013-1965-9

Li J, Wu W, Wang Y (2017) Potassium channel AKT1 is involved in the auxin-mediated root growth inhibition in Arabidopsis response to low K^+ stress. J Integr Plant Biol 59(12):895–909. https://doi.org/10.1111/jipb.12575

Li Q, Gao Y, Yang A (2020) Sulfur homeostasis in plants. Int J Mol Sci 21(23). https://doi.org/10.3390/ijms21238926

Li Q, Li BH, Kronzucker HJ, Shi WM (2010) Root growth inhibition by NH4+ in Arabidopsis is mediated by the root tip and is linked to NH4+ efflux and GMPase activity. Plant Cell Environ 33:1529–1542

Liang C, Tian J, Liao H (2013) Proteomics dissection of plant responses to mineral nutrient deficiency. Proteomics Syst Biol 12. https://doi.org/10.1002/pmic.201200263

Liang X, Zhang T, Lu X, Ellsworth DS, BassiriRad H, You C, Wang D, He P, Deng Q, Liu H, Mo J, Ye Q (2020) Global response patterns of plant photosynthesis to nitrogen addition: a meta-analysis. Glob Change Biol 26:3585–3600. https://doi.org/10.1111/gcb.15071

Lin M, Jian-xin D, Shu-sheng W, Ke S, Ai-fen L, Jin-guang Y, Zhi-xin X, Wei L, Wen-jing S, Hong bo L (2019) Differential responses of root growth to nutrition with different ammonium/nitrate ratios involve auxin distribution in two tobacco cultivars. J Integr Agric 18(12):2703–2715
Liu Y, von Wirén N (2022) Integration of nutrient and water availabilities via auxin into the root developmental program. Curr Opin Plant Biol 65:102117
Liu D (2021) Root developmental responses to phosphorus nutrition. J Integr Plant Biol 63:1065–1090. https://doi.org/10.1111/jipb.13090
Liu Y, Lai N, Gao K, Chen F, Yuan L, Mi G (2013) Ammonium inhibits primary root growth by reducing the length of meristem and elongation zone and decreasing elemental expansion rate in the root apex in *Arabidopsis thaliana*. PLoS One 8:e61031
Liu Y, Riaz M, Yan L, Zeng Y, Cuncang J (2019) Boron and calcium deficiency disturbing the growth of trifoliate rootstock seedlings (*Poncirus trifoliate* L.) by changing root architecture and cell wall. Plant Physiol Biochem 144. https://doi.org/10.1016/j.plaphy.2019.10.007
Long W, Li Q, Wan N, Feng D, Kong F, Zhou Y, Yuan J (2020) Root morphological and physiological characteristics in maize seedlings adapted to low iron stress. PLoS One 15(9). https://doi.org/10.1371/journal.pone.0239075
López-Bucio J, Cruz-Ramírez A, Herrera-Estrella L (2003) The role of nutrient availability in regulating root architecture. Curr Opin Plant Biol 6(3). https://doi.org/10.1016/s1369-5266(03)00035-9
Lucheta AR, Lambais MR, Paulo UdS (2021) Sulfur in agriculture. Revista Brasileira de Ciência do Solo 36:1369–1379. https://doi.org/10.1590/S0100-06832012000500001
Ma QH, Zhang FS, Rengel Z, Shen JB (2013) Localized application of NH^{4+}-N plus P at the seedling and later growth stages enhances nutrient uptake and maize yield by inducing lateral root proliferation. Plant Soil 372:65–80. https://doi.org/10.1007/s11104-013-1735-8
Ma Q, Rengel Z, Bowden B (2007) Heterogeneous distribution of phosphorus and potassium in soil influences wheat growth and nutrient uptake. Plant Soil 291:301–309. https://doi.org/10.1007/s11104-007-9197-5
Ma W, Li J, Qu B, He X, Zhao X, Li B, Fu X, Tong Y (2014) Auxin biosynthetic gene TAR2 is involved in low nitrogen-mediated reprogramming of root architecture in Arabidopsis. Plant J 78:70–79. https://doi.org/10.1111/tpj.12448
Maciel GHP, Ferreira IEdP, Baron D (2021) Phytoregulators and mineral nutrition interactions to the establishment of *Passiflora alata* seedlings. J Plant Nutr 44(19):2825–2839. https://doi.org/10.1080/01904167.2021.1918715
Maghiaoui A, Bouguyon E, Cuesta C, Perrine-Walker F, Alcon C, Krouk G, Benková E, Nacry P, Gojon A, Bach L (2020) The Arabidopsis NRT1.1 transceptor coordinately controls auxin biosynthesis and transport to regulate root branching in response to nitrate. J Exp Bot 71(15):4480–4494. https://doi.org/10.1093/jxb/eraa242
Malobane ME, Nciizah AD, Mudau FN, Wakindiki IIC (2020) Tillage, crop rotation and crop residue management effects on nutrient availability in a sweet sorghum-based cropping system in marginal soils of South Africa. Agronomy 10(6):776. https://doi.org/10.3390/agronomy10060776
Mardamootoo T, Du Preez CC, Barnard JH (2021) Agricultural phosphorus management for environmental protection: a review. J Geosci Environ Prot 9:48–81. https://doi.org/10.4236/gep.2021.98004
Masclaux-Daubresse C, Daniel-Vedele F, Dechorgnat J, Chardon F, Gaufichon L, Suzuki A (2010) Nitrogen uptake, assimilation and remobilization in plants: challenges for sustainable and productive agriculture. Ann Bot 105(7):1141–1157. https://doi.org/10.1093/aob/mcq028
Mayland HF, Wilkinson SR (1989) Soil factors affecting magnesium availability in plant-animal systems: a review. J Anim Sci 67(12):3437–3444. https://doi.org/10.2527/jas1989.67123437x
Mbene K, Tening AS, Suh CE, Fomenky NN, Che VB (2017) Phosphorus fixation and its relationship with physicochemical properties of soils on the Eastern flank of Mount Cameroon. Afr J Agric Sci 12(36):11. https://doi.org/10.5897/AJAR2017.12530

McDowell RW, Sharpley AN, Condron LM, Haygarth PM, Brookes PC (2001) Processes controlling soil phosphorus release to runoff and implications for agricultural management. Nutr Cycl Agroecosystems 59:269–284

McSteen P (2010) Auxin and monocot development. Cold Spring Harb Perspect Biol 2(3):a001479. https://doi.org/10.1101/cshperspect.a001479

Mendoza B, Guananga N, Melendez JR, Lowy DA (2020) Differences in total iron content at various altitudes of Amazonian Andes soil in Ecuador. F10000 Res 9. https://doi.org/10.12688/f1000research.22411.1

Meriño-Gergichevich C, Alberdi M, Ivanov AG, Reyes-Diaz M (2010) Al $^{3+}$—Ca^{2+} interaction in plants growing in acid soils: Al-phytotoxicity response to calcareous amendments. Journal Soil Science and Plant Nutrition 10(3):217–243

Mitra GN (2015) Calcium (Ca) uptake. In: Regulation of nutrient uptake by plants. Springer. https://doi.org/10.1007/978-81-322-2334-4_5

Miura K, Lee J, Gong Q, Ma S, Jin JB, Yoo CY, Miura T, Sato A, Bohnert HJ, Hasegawa PM (2011) SIZ1 regulation of phosphate starvation-induced root architecture remodeling involves the control of auxin accumulation. Plant Physiol 155:1000–1012

Moniuszko G, Sirko A (2008) Sulfur metabolism and its regulation in plants. Postepy Biochem 54(4):402–411

Mounier E, Pervent M, Ljung K, Gojon A, Nacry P (2014) Auxin mediated nitrate signalling by NRT1.1 participates in the adaptive response of Arabidopsis root architecture to the spatial heterogeneity of nitrate availability. Plant Cell Environ 37:162–174

Nacry P, Canivenc G, Muller B, Azmi A, Onckelen HV, Rossignol M, Doumas P (2005) A role for auxin redistribution in the responses of the root system architecture to phosphate starvation in Arabidopsis. Plant Physiol 138:2061–2074

Narise T, Kobayashi K, Baba S, Shimojima M, Masuda S, Fukaki H, Ohta H (2010) Involvement of auxin signaling mediated by IAA14 and ARF7/19 in membrane lipid remodeling during phosphate starvation. Plant Mol Biol 72:533–544. https://doi.org/10.1007/s11103-009-9589-4

Niu YF, Chai RS, Jin GL, Wang H, Tang CX, Zhang YS (2013) Responses of root architecture development to low phosphorus availability: a review. Ann Bot 112:391–408. https://doi.org/10.1093/aob/mcs285

Niu Y, Chai R, Liu L, Jin G, Tang C, Zhang Y (2014a) Magnesium availability regulates the development of root hairs in *Arabidopsis thaliana* (L.) Heynh. Plant Cell Environ 37(12). https://doi.org/10.1111/pce.12362

Niu Y, Jin G, Zhang YS (2014b) Root development under control of magnesium availability. Plant Signal Behav 9(9). https://doi.org/10.4161/psb.29720

Ogura T, Kobayashi NI, Suzuki H, Iwata R, Nakanishi TM, Tanoi K (2018) Magnesium uptake characteristics in Arabidopsis revealed by 28 Mg tracer studies. Planta 248(3). https://doi.org/10.1007/s00425-018-2936-4

Penn CJ, Camberato JJ (2019) A critical review on soil chemical processes that control how soil ph affects phosphorus availability to plants. Agriculture 9:120. https://doi.org/10.3390/agriculture9060120

Péret B, Clément M, Nussaume L, Desnos T (2011) Root developmental adaptation to phosphate starvation: better safe than sorry. Trends Plant Sci 16(8):442–450. https://doi.org/10.1016/j.tplants.2011.05.006

Petricka JJ, Winter CM, Benfey PN (2012) Control of Arabidopsis root development. Annu Rev Plant Biol 63:563–590. https://doi.org/10.1146/annurev-arplant-042811-105501

Philippar K, Ivashikina N, Ache P, Christian M, Lüthen H, Palme K, Hedrich R (2004) Auxin activates KAT1 and KAT2, two K^+-channel genes expressed in seedlings of *Arabidopsis thaliana*. Plant J 37:815–827. https://doi.org/10.1111/j.1365-313x.2003.02006.x

Pinto E, Ferreira IM (2015) Cation transporters/channels in plants: tools for nutrient biofortification. J Plant Physiol 179. https://doi.org/10.1016/j.jplph.2015.02.010

Ploeg RR, Bohm W, Kirkham MB (1999) On the origin of the theory of mineral nutrition of plants and the law of the minimum. Soil Sci Soc Am J 63(5):1055–1062. https://doi.org/10.2136/sssaj1999.6351055x
Prajapati K, Modi HA (2012) The importance of potassium in plant growth—a review. Indian J Plant Sci 1(02–03):177–186
Qi Y, Cheng W, Nan X-Y, Yang F, Li J, Li D-C, Lundstrom CC, Yu H-M, Zhang G-L, Huang F (2020) Iron stable isotopes in bulk soil and sequential extracted fractions trace Fe redox cycling in paddy soils. J Agric Food Chem 68(31):8143–8150. https://doi.org/10.1021/acs.jafc.0c02515
Raya-González J, Ortiz-Castro R, Ruíz-Herrera LF, Kazan K, López-Bucio J (2021) Phytochrome and flowering time1/mediator25 regulates lateral root formation via auxin signaling in Arabidopsis. Plant Physiol 165(2):880–894. https://doi.org/10.1104/pp.114.239806
Rennenberg H, Herschbach C, Haberer K, Kopriva S (2007) Sulfur metabolism in plants: are trees different? Plant Biol 9:620–637. https://doi.org/10.1055/s-2007-965248
Richardson AA, Simpson RJ (2011) Soil microorganisms mediating phosphorus availability. Plant Physiol 156:989–996. https://doi.org/10.1104/pp.111.175448
Rigas S, Ditengou FA, Ljung K, Daras G, Tietz O, Palme K, Hatzopoulos P (2013) Root gravitropism and root hair development constitute coupled developmental responses regulated by auxin homeostasis in the Arabidopsis root apex. New Phytol 197(4):1130–1141. https://doi.org/10.1111/nph.12092
Romera FJ, Lucena C, García MJ, Alcántara E, Angulo M, Aparício MA, Pérez-Vicente R (2021) Plant hormones and nutrient deficiency responses. In: Gupta DK, Corpas FJ (eds) Hormones and plant response. Plant Challenging Environ 2. SpringerLink. https://doi.org/10.1007/978-3-030-77477-6_2
Rosolem CA, Steiner F (2017) Effects of soil texture and rates of K input on potassium balance in tropical soil. Eur J Soil Sci 68:658–666. https://doi.org/10.1111/ejss.12460
Rouached H, Secco D, Arpat AB (2009) Getting the most sulfate from soil: regulation of sulfate uptake transporters in Arabidopsis. J Plant Physiol 166(9). https://doi.org/10.1016/j.jplph.2009.02.016
Saini S, Sharma I, Kaur N, Pati PK (2013) Auxin: a master regulator in plant root development. Plant Cell Rep 32(6):741–757. https://doi.org/10.1007/s00299-013-1430-5
Santos LCN, Barreto RF, Prado RM, Silva GP, Santos JMR, Lasa B, Aparicio-Tejo PM, Michelena IA, Carvalho RF (2020a) The auxin-resistant *dgt* tomato mutant grows less than the wild type but is less sensitive to ammonium toxicity and nitrogen deficiency. J Plant Physiol 252:153243
Santos LCN, Gaion LA, Prado RM, Barreto RF, Carvalho RF (2020b) Low auxin sensitivity of diageotropica tomato mutant alters nitrogen deficiency response. Ann Braz Acad Sci 92(3):e20190254
Sardans J, Peñuelas J (2021) Potassium control of plant functions: ecological and agricultural implications. Plants 10:419. https://doi.org/10.3390/plants10020419
Sato A, Miura K (2011) Root architecture remodeling induced by phosphate starvation. Plant Signal Behav 6(8).1122–1126. https://doi.org/10.4161/psb.6.8.15752
Sattelmacher B, Gerendas J, Thoms K, Brück H, Bagdady NH (1993) Interaction between root-growth and mineral-nutrition. Environ Exp Bot 33:63–73. https://doi.org/10.1016/0098-8472(93)90056-L
Schachtman DP, Reid RJ, Ayling SM (1998) Phosphorus uptake by plants: from soil to cell. Plant Physiol 116:447–453
Schapire A, Valpuesta V, Botella M (2009) Plasma membrane repair in plants. Trends Plant Sci 14(12). https://doi.org/10.1016/j.tplants.2009.09.004
Scherer HW (2009) Sulfur in soils. J Plant Nutr Soil Sci 172(3):9. https://doi.org/10.1002/jpln.200900037
Scott B, Conyers M, Fisher R, Lill W (1992) Particle size determines the efficiency of calcitic limestone in amending acidic soil. Aust J Agric Res 43(5):1175–1185. https://doi.org/10.1071/ar9921175

Shaul O (2002) Magnesium transport and function in plants: the tip of the iceberg. Biometals 15(3). https://doi.org/10.1023/a:1016091118585

Shen C, Yue R, Sun T, Zhang L, Yang Y, Wang H (2015) OsARF16, a transcription factor regulating auxin redistribution, is required for iron deficiency response in rice (*Oryza sativa* L.). Plant Sci 231. https://doi.org/10.1016/j.plantsci.2014.12.003

Shen C, Yue R, Yang Y, Zhang L, Sun T, Tie S, Wang H (2014) OsARF12 is involved in cytokinin-mediated inhibition of phosphate transport and phosphate signaling in rice (*Oryza sativa* L.). PLoS One 9:e112906

Shen J, Yuan L, Zhang J, Li H, Bai Z, Chen X, Zhang W, Zhang F (2011) Phosphorus dynamics: from soil to plant. Plant Physiol 156:997–1005

Shin R (2017) Potassium sensing, signaling, and transport: toward improved potassium use efficiency in plants. In: Hossain MA, Kamiya T, Burritt DJ, Tran LP, Fujiwara T (eds) Plant macronutrient use efficiency. Academic Press, pp 149–163. https://doi.org/10.1016/B978-0-12-811308-0.00008-9

Shin R, Schachtman DP (2004) Hydrogen peroxide mediates plant root cell response to nutrient deprivation. PNAS 101:8827–8832. https://doi.org/10.1073/pnas.0401707101

Shu W, Liu Y, Guo Y, Zhou H, Zhang J, Zhao S, Lu M (2015) A Populus TIR1 gene family survey reveals differential expression patterns and responses to 1-naphthalene acetic acid and stress treatments. Front Plant Sci 6:719. https://doi.org/10.3389/fpls.2015.00719

Soares MB, Prado RM, Tenesaca LFL, Lúcio JBC, Carvalho RF (2021) Tomato phytochromes B1 and B2 are part of the responses to the nutritional stress induced by NPK deficiency. Physiol Plant 173(4):2238–2247

Song W, Liu S, Meng L, Xue R, Wang C, Liu G, Dong C, Wang S, Dong J, Zhang Y (2015) Potassium deficiency inhibits lateral root development in tobacco seedlings by changing auxin distribution. Plant Soil 396:163–173. https://doi.org/10.1007/s11104-015-2579-1

Song W, Sun H, Li J, Gong X, Huang S, Zhu X, Zhang Y, Xu G (2013) Auxin distribution is differentially affected by nitrate in roots of two rice cultivars differing in responsiveness to nitrogen. Ann Bot 112:1383–1393. https://doi.org/10.1093/aob/mct212

Sorensen PO, Beller HR, Bill M, Bouskill NJ, Hubbard SS, Karaoz U, Polussa A, Steltzer H, Wang S, Williams KH, Wu Y, Brodie E (2021) The snowmelt niche differentiates three microbial life strategies that influence soil nitrogen availability during and after winter. Front Microbiol 11:871. https://doi.org/10.3389/fmicb.2020.00871

Srivastava AK, Shankar A, Chandran AKN, Sharma M, Jung K-H, Suprasanna P, Pandey GK (2020) Emerging concepts of potassium homeostasis in plants. J Exp Bot 71(2):608–619. https://doi.org/10.1093/jxb/erz458

Stoeckli JL, Sharifi M, Hooker DC, Thomas BW, Khaefi F, Stewart G, MsDonald I, Deem B, Drury CF, Ma BL, Motaghian HR (2021) Predicting soil nitrogen availability to grain corn in Ontario, Canada. Can J Soil Sci 101(3). https://doi.org/10.1139/cjss-2020-0104

Sun C, Yu J, Hu D (2017a) Nitrate: a crucial signal during lateral roots development. Front Plant Sci 8:485. https://doi.org/10.3389/fpls.2017.00485

Sun H, Feng F, Liu J, Zhao Q (2017b) The interaction between auxin and nitric oxide regulates root growth in response to iron deficiency in rice. Front Plant Sci 8:2169. https://doi.org/10.3389/fpls.2017.02169

Sustr M, Soukup A, Tylova E (2019) Potassium in root growth and development. Plants 8:435. https://doi.org/10.3390/plants8100435

Svistoonoff S, Creff A, Reymond M, Sigoillot-Claude C, Ricaud L, Blanchet A, Nussaume L, Desnos T (2007) Root tip contact with low-phosphate media reprograms plant root architecture. Nat Genet 39(6):792–796

Swarup K, Benková E, Swarup R, Casimiro I, Péret B, Yang Y, Parry G, Nielsen E, de Smet I, Vanneste S, Levesque M, Carrier D, James N, Calvo V, Ljung K, Kramer E, Roberts R, Graham N, Marillonnet S, Patel K, Jones JDG, Taylor CG, Schachtman DP, May S, Sandberg G, Benfey P, Friml F, Kerr I, Beeckman T, Laplaze L, Bennett MJ (2008) The auxin influx carrier LAX3 promotes lateral root emergence. Nat Cell Biol 10(8):946–954

Syring KM, Claassem N (1995) Estimation of the influx and the radius of the depletion zone developing around a root during nutrient uptake. Plant Soil 175:115–123

Takahashi H, Yamazaki M, Sasakura N, Watanabe A, Leustek T, Engler JdA, Saito K (1997) Regulation of sulfur assimilation in higher plants: a sulfate transporter induced in sulfate-starved roots plays a central role in *Arabidopsis thaliana*. Proc Natl Acad Sci United State Am 94:11102–11107. https://doi.org/10.1073/pnas.94.20.11102

Thor K (2019) Calcium-nutrient and messenger. Front Plant Sci 10. https://doi.org/10.3389/fpls.2019.00440

Ticconi CA, Abel S (2004) Short on phosphate: plant surveillance and countermeasures. Trends Plant Sci 9(11):548–555. https://doi.org/10.1016/j.tplants.2004.09.003

Tsai HH, Schmidt W (2017) One way or another? Iron uptake in plants. New Phytol 214(2). https://doi.org/10.1111/nph.14477

Tuteja N, Mahajan S (2007) Calcium signaling network in plants. Plant Signal Behav 2(2):79–85. https://doi.org/10.4161/psb.2.2.4176

Uchida R (2021) Essential nutrients for plant growth: nutrient functions and deficiency symptoms. In: Silva JA, Uchida R (eds) Plant nutrient management in hawaii's soils, approaches for tropical and subtropical agriculture. College of tropical agriculture and human resources, University of Hawaii at Manoa

Ukalska-Jaruga A, Siebielec G, Siebielec S, Pecio M (2020) The impact of exogenous organic matter on wheat growth and mineral nitrogen availability in soil. Agronomy 10(9):1314. https://doi.org/10.3390/agronomy10091314

van der Bom FJT, Williams A, Bell M (2020) Root architecture for improved resource capture: trade-offs in complex environments. J Exp Bot 7(19):5752–5763. https://doi.org/10.1093/jxb/eraa324

Vanneste S, Friml J (2013) Calcium: the missing link in auxin action. Plants 2(4):650–675. https://doi.org/10.3390/plants2040650

Vega A, Fredes I, O'Brien J, Shen Z, Ötvös K, Abualia R, Benkova E, Briggs SP, Gutiérrez RA (2021) Nitrate triggered phosphoproteome changes and a PIN2 phosphosite modulating root system architecture. EMBO Rep e51813. https://doi.org/10.15252/embr.202051813

Vega A, O'Brien JA, Gutiérrez RA (2019) Nitrate and hormonal signaling crosstalk for plant growth and development. Curr Opin Plant Biol 52:155–163

Velayutham M (2017) The law of optimum—a unified concept in plant nutrition bridging the law of minimum and the law of the maximum. Madras Agric J 104(10–12):309–314. https://doi.org/10.29321/MAJ.2017.000067

Vicente-Agullo F, Rigas S, Desbrosses G, Dolan L, Hatzopoulos P, Grabov A (2004) Potassium carrier TRH1 is required for auxin transport in Arabidopsis roots. Plant J 40:523–535. https://doi.org/10.1111/j.1365-313X.2004.02230.x

Vidal EA, Araus V, Lu C, Parry G, Green PJ, Coruzzi GM, Gutiérrez RA (2010) Nitrate-responsive miR393/AFB3 regulatory module controls root system architecture in Arabidopsis thaliana. Proc Natl Acad Sci USA 107(9):4477–4482. https://doi.org/10.1073/pnas.0909571107

Waidmann S, Sarkel E, Kleine-Vehn J (2020) Same, but different: growth responses of primary and lateral roots. J Exp Bot 71(8):2397–2411. https://doi.org/10.1093/jxb/eraa027

Wang B, Zhu X, Guo X, Qi X, Feng F, Zhang Y, Zhao Q, Han D, Sun H (2021) Nitrate modulates lateral root formation by regulating the auxin response and transport in rice. Genes 12:850. https://doi.org/10.3390/genes12060850

Wang M, Zheng Q, Shen Q, Guo S (2013) The critical role of potassium in plant stress response. Int J Mol Sci 14:7370–7390. https://doi.org/10.3390/ijms14047370

Wang P, Wang Z, Pan Q, Sun X, Chen H, Chen F, Yuan L, Mi G (2019a) Increased biomass accumulation in maize grown in mixed nitrogen supply is mediated by auxin synthesis. J Exp Bot 70(6):1859–1873. https://doi.org/10.1093/jxb/erz047

Wang S, Zhang S, Sun C, Xu Y, Chen Y, Yu C, Qian Q, Jiang D, Qi Y (2014) Auxin response factor (OsARF12), a novel regulator for phosphate homeostasis in rice (*Oryza sativa*). New Phytol 201:91–103. https://doi.org/10.1111/nph.12499

Wang W, Mo Q, Han X, Hui D, Shen W (2019b) Fine root dynamics responses to nitrogen addition depend on root order, soil layer, and experimental duration in a subtropical forest. Biol Fertil Soils 55:723–736. https://doi.org/10.1007/s00374-019-01386-3

Wang X, Du G, Wang X, Meng Y, Li Y, Wu P, Yi K (2010) The function of LPR1 is controlled by an element in the promoter and is independent of SUMO E3 Ligase SIZ1 in response to low Pi stress in Arabidopsis thaliana. Plant Cell Physiol 51(3):380–394. https://doi.org/10.1093/pcp/pcq004

Wang X, Feng J, White PJ, Shen J, Cheng L (2020) Heterogeneous phosphate supply influences maize lateral root proliferation by regulating auxin redistribution. Ann Bot 125:119–130. https://doi.org/10.1093/aob/mcz154

Wang Y, Ying H, Yin Y, Zheng H, Cu Z (2019c) Estimating soil nitrate leaching of nitrogen fertilizer from global meta-analysis. Sci Total Environ 657:96–102

Werner F, Mueller CW, Thieme J, Gianoncelli A, Rivard C, Höschen C, Prietzel J (2017) Sci Rep 7:3203. https://doi.org/10.1038/s41598-017-03537-8

White PJ, Broadley MR (2003) Calcium in plants. Ann Bot 92(4). https://doi.org/10.1093/aob/mcg164

White PJ, Karley AJ (2010) Potassium. Cell Biol Metals Nutr Plant Cell Monogr 17:199–224. https://doi.org/10.1007/978-3-642-10613-2_9

Wu P, Li H, Zao H, Wang Y, Yang J, Tang L, Fan W (2020) Physiological and molecular mechanisms of plant co-evolution responses to phosphorous deficiency and aluminum toxicity. Biotechnol Bull 36(7):170–181

Xuan W, Beeckman T, Xu G (2017) Plant nitrogen nutrition: sensing and signaling. Curr Opin Plant Biol 39:57–65. https://doi.org/10.1016/j.pbi.2017.05.010

Yadav BK, Sidhu AS (2016) Dynamics of potassium and their bioavailability for plant nutrition. In: Meena VS et al (eds) Potassium solubilizing microorganisms for sustainable agriculture. https://doi.org/10.1007/978-81-322-2776-2_14

Yang H-Q, Jie Y-L (2005) Uptake and transport of calcium in plants. J Plant Physiol Mol Biol 31(3):227–234

Yang T, Feng H, Zhang S, Xiao H, Hu Q, Chen G, Xuan W, Moran N, Murphy A, Yu L, Xu G (2020) The potassium transporter OsHAK5 alters rice architecture via ATP-dependent transmembrane auxin fluxes. Plant Commun 1:100052

Yu P, Eggert K, von Wirén N, Li C, Hochholdinger F (2015a) Cell type-specific gene expression analyses by RNA sequencing reveal local high nitrate-triggered lateral root initiation in shoot-borne roots of maize by modulating auxin-related cell cycle regulation. Plant Physiol 169:690–704

Yu P, Hochholdinger F, Li C (2015b) Root-type-specific plasticity in response to localized high nitrate supply in maize (*Zea mays*). Ann Bot 116:751–762

Zhang S, Zhang X, Liu X, Liu W, Liu Z (2013) Spatial distribution of soil nutrient at depth in black soil of Northeast China: a case study of soil available potassium. Nutr Cycl Agroecosyst 95:319–331. https://doi.org/10.1007/s10705-013-9565-x

Zhang X, Cui Y, Yu M, Su B, Gong W, Baluška F, Komis G, Šamaj J, Shan X, Lin J (2019) Phosphorylation-mediated dynamics of nitrate transceptor NRT1.1 regulate auxin flux and nitrate signaling in lateral root growth. Plant Physiol 181:480–498. https://doi.org/10.1104/pp.19.00346

Zhu J, Lynch JP (2004) The contribution of lateral rooting to phosphorus acquisition efficiency in maize (*Zea mays*) seedlings. Funct Plant Biol 31(10):949–958. https://doi.org/10.1071/FP04046

Ötvös K, Marconi M, Vega A, O'Brien J, Johnson A, Abualia R, Antonielli L, Montesinos JC, Zhang Y, Tan S, Cuesta C, Artner C, Bouguyon E, Gojon A, Friml J, Gutiérrez RA, Wabnik K, Benková E (2021) Modulation of plant root growth by nitrogen source-defined regulation of polar auxin transport. EMBO J 40:e106862. https://doi.org/10.15252/embj.2020106862

Insights into Biosynthesis and Signaling of Cytokinins During Plant Growth, Development and Stress Tolerance

Ravinderjit Kaur, Nandni Sharma, Raman Tikoria, Mohd Ali, Sandeep Kour, Deepak Kumar, and Puja Ohri

Abstract Cytokinins are diversified signaling molecules present in all plant tissues, higher concentrations being found in continuous growth areas such as meristematic regions, roots, young leaves, developing fruits and seeds. These phytohormones mediate a wide array of physiological processes and environmental cues throughout the life of a plant by acting either locally and distantly via the vascular system. Innumerable studies conducted so far have elucidated their underlying functional roles in cell division, shoot induction, leaf senescence, apical dominance, source/sink relationships, absorption of nutrients and embryo development in plants. Additionally, it plays a crucial role during environmental strains by having both positive and negative impacts on stress tolerance endogenously or through exogenous application in a variety of plant taxa. Cytokinin signaling involves the coordinating roles of three major proteins including histidine kinase receptors (accepts the signal), histidine phosphotransfer proteins (transfers the signal) and response regulators that provides signal output through a modified bacterial two-component pathway which functions via a multi-step phosphorelay. Therefore, in the current book chapter an attempt has been made to get insights of biosynthesis and signaling of this phytohormone in the various aspects of plant growth and in combatting stress in plants.

1 Introduction

Cytokinin (Ck), a classic phytohormone first discovered in maize and has been found to control diverse facets of plant growth and development, physiology, metabolism as well as signal transduction at the tissue and organ levels (Werner and Schmülling 2009; Keshishian and Rashotte 2015; Pavlů et al. 2018; Hai et al. 2020). These phytohormones are present in all plant tissues being abundant in immature seeds, root

R. Kaur
Department of Zoology, S.R. Government College (W), Amritsar, Punjab, India

N. Sharma · R. Tikoria · M. Ali · S. Kour · D. Kumar · P. Ohri (✉)
Department of Zoology, Guru Nanak Dev University, Amritsar, Punjab, India
e-mail: ohri_puja_11@rediffmail.com

T. Aftab (ed.), *Auxins, Cytokinins and Gibberellins Signaling in Plants*,
Signaling and Communication in Plants,
https://doi.org/10.1007/978-3-031-05427-3_7

tip and shoot apex. Naturally occurring cytokinins are the derivatives of the purine adenine with isoprenoid or aromatic side chains. Of the two types, the isoprenoid is more prevalent in plants than the later (Kiba et al. 2013; Schmülling 2013; Liu et al. 2020). However, in different plant species, the type and activity of cytokinins varies remarkably because of their ability to conjugate with sugars such as riboside, sugar phosphates (ribonucleotide) and even with amino acids (for e.g. lupinic acid) by substituting at the C2 in the adenine ring (Radhika et al. 2015; Frébortová and Frébort 2021). The variety of activities performed by Ck includes embryogenesis, inhibited lateral root initiation, discerning phloem and metaxylem in roots, directing cell division, photomorphogenic cell differentiation in developing leaves and shoots, leaf senescence inhibition etc. (Bishopp et al. 2011; Bielach et al. 2012; Chiang et al. 2012; Efroni et al. 2013; Zwack and Rashotte 2015; Zürcher and Müller 2016; Akhtar et al. 2020). In turn, levels of Ck are maintained by various enzymes involved in its biosynthesis, metabolism, inter-conversion between its types and degradation (Frébort et al. 2011; Thu et al. 2017) while the signaling pathway is prompted by alterations in temperature, nutrition levels and osmotic conditions, but, the expression and regulation of the genes involved in plant adaptation is initiated by the downstream segments of Ck signaling (Thu et al. 2017; Pavlů et al. 2018).

In addition, Cks plays a twin role in abiotic/biotic stress tolerance as evidenced from the endogenous levels and the exogenous formulations of Cks in plants (Grosskinsky et al. 2011; Argueso et al. 2012; Zwack and Rashotte 2015). For instance, in transgenic plants recent developments suggested Ck as a potent mitigator in combatting environmental challenges and even show different responses in distinct stresses while for non-plant phytosphere members, it interacts differently with several invaders by inducing plant immunity (Spallek et al. 2018; Cortleven et al. 2019; Jameson 2019; Ngyuen et al. 2020; Singh and Roychoudhury 2021). Therefore, the current chapter highlights primarily the biosynthesis, metabolism, signaling and role of Cks in regulating the growth, development and stress tolerance in plants.

2 Biosynthesis

Cytokinin is produced by almost all organisms. Naturally occurring Cks are the derivative of 'adenine' with aromatic or isoprenoid chain attached specifically at its N^6 position (Kiba et al. 2013). On the basis of the side chain attached, they are categorized into two types, Aromatic Cks and Isoprenoid Cks. Isoprenoid Cks are usually present in abundance in plants (Sakakibara 2006). This category includes Cis-zeatin (cZ), Trans-zeatin (tZ), Dihydrozeatin (dhZ) and isopentenyladenine (iP). Among all these, the most commonly found types are tZ and iP (Sakakibara 2006; Kudo et al. 2010), whereas aromatic form encompasses Benzyladenin (BA), Mesotopolin (mT), Orthotopolin (oT) etc. (Sakakibara 2006; Kudo et al. 2010). In addition to natural Cks, synthetic ones are also available that can be applied exogenously to plant such as

Benzyladenine, Kinetin, Trans-zeatin riboside and 6-benzylaminopurine (Liu et al. 2020).

The process of biosynthesis of Cks rely on two major catalysts, *LONELYGUY* (LOG) and *ISOPENTENYL TRANSFERASE* (IPT). IPTs are broadly categorized into *t-RNA ISOPENTENYL TRANSFERASE* (tRNA-IPTs) and adenosine phosphate-isopentenyl transferase. Both types exhibit a conserved domain known as *ISOPENTENYLPYROPHOSPHATE TRANSFERASE* (IPPT-binding) domain. Furthermore, multiple IPTs encoding genes have been reported from different crops e.g., 7 FvI IPTs genes in strawberry, 9 IPTs in *Arabidopsis* and 10 in rice have been reported (Zürcher and Müller 2016; Mi et al. 2017; Sakamoto et al. 2006). The process of biosynthesis starts with IPT catalyzed addition of Dimethyl Allyl Diphosphate (DMAPP) based prenyl group to the N^6 position of AMP/ADP/ATP resulting in the generation of iP ribotides, which further undergo hydroxylation of isoprenoid side chain in the presence of Cytochrome P450 monoxygenase enzymes (CYP735As) to form trans-Zeatin Cks. This whole process resulted in shoot growth in case of *Arabidopsis* (Takei et al. 2004a, b; Sakakibara 2006). Contrary to this, the process of biosynthesis of cis-Zeatin is not well documented, but few studies report tRNA-IPT catalyzed DMAPP based prenylation at N^6 position of adenine present on tRNA, resulting in the generation of cis-Zeatin ribotides (cZ-ribotides) (Sakakibara 2006). Further, the inactive ribotide forms are converted into their active forms in a single step catalyzed by LOGs, which works by its cytokinin specific phosphohydrolase activity (Kudo et al. 2010). Further, level of Ck is maintained either by glycosylation (conjugation of Ck with sugar) or by Cytokinin oxidase dependent irreversible cleavage of Ck (Werner et al. 2006) (Fig. 1).

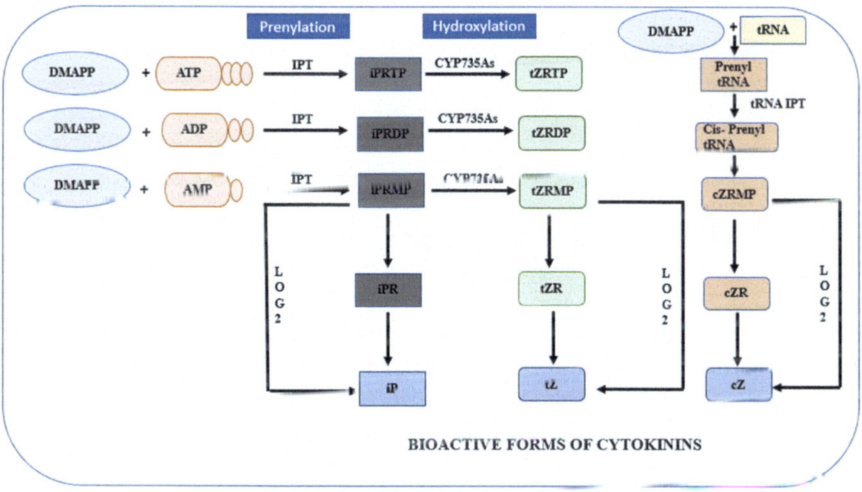

Fig. 1 Biosynthesis of Cytokinin in plants

3 Metabolism

Ethereal control of biosynthesis and cytokinin-metabolizing enzymes in addition to cytokinin synthetase are required for Ck maintenance in plants (Frébort et al. 2011; Liu et al. 2020). In a cell, level of active cytokinins can be decreased either through conjugation of CKs to glucose in a glycosyltransfer reaction or irreversible dehydrogenation (Zalabák et al. 2013). The binding of glucose to Cks occurs at hydroxyl group of side chain of zeatin or dihydrozeatin resulting in *O*-glycosylation that can be reversed by β-glucosidase or N^3, N^7 and N^9 positions of purine ring resulting in irreversible *N*-glycosylation (Sakakibara et al. 2006; Kudo et al. 2010). Glucosyl conjugates are dormant in bioassays and these bound Cks fail to bind to histidine kinase cytokinin receptors (Spíchal et al. 2004). In *Arabidopsis*, five cytokinin glucosyl transferase encoding genes (*UGT76C1, UGT76C2, UGT85A1, UGT73C5, UGT73C1*) have been identified. Two genes (*UGT76C1, UGT76C2*) encode proteins that cause *N*-glycosylation of most cytokinin species primarily at N^7 and N^9 position of adenine; three encodes protein that causes *O*-glycosylation of tZ and dhZ (Hou et al. 2004; Wang et al. 2011; Jin et al. 2013; Li et al. 2015). In *Arabidopsis*, overexpression of *UGT85A1* resulted in increased level of tZ O-glucosides and insensitivity to exogenous tZ, with no effect on growth and development and level of free, active cytokinin (Jin et al. 2013). Similarly, overexpression of *UGT76C2* resulted in an increase in cytokinin *N*-glycosides and insensitivity to exogenous cytokinin, while disruption has opposite effects (Wang et al. 2011). Surprisingly, growth and development of *Arabidopsis* was not affected by alteration in *UGT76C1* function because of the atoning changes in the expression of cytokinin signalling and metabolism to maintain consistent level of cytokinin function (Wang et al. 2013). Thereby, proposing that plants have a fairly high ability to maintain an appropriate level of cytokinin function concerning agitation through changes in cytokinin sensitivity and metabolism (Kieber and Schaller 2014).

CKXs (cytokinin dehydrogenases/oxidases) are the only known enzymes that catalyzes the irreversible dehydrogenation of cytokinin (Galuszka et al. 2007; Kudo et al. 2010). Earlier, it was believed that CKX have exclusive oxidase activity and was incorrectly classified as amine oxidase containing copper but later it was shown that flavin adenine dinucleotide (FAD) covalently bound to CKXs making the enzyme to act more efficiently as dehydrogenase under in vitro conditions (Bilyeu et al. 2001; Galuszka et al. 2001; Frébort et al. 2011). For normal plant development, homeostasis is maintained by natural catabolism of Cks by a small family of CKXs. Identification and characterization of CKXs have been done in *Arabidopsis* (Bilyeu et al. 2001; Werner et al. 2001), maize (Massonneau et al. 2004; Vyroubalová et al. 2009) and rice (Ashikari et al. 2005). In *Arabidopsis*, there are seven different CKXs (*AtCKX1-7*) (Bilyeu et al. 2001; Schmülling et al. 2003; Kowalska et al. 2010). Unsaturated N^6side chain of zeatin isoforms and isopentenyladenine (iP) are cleaved by CKXs, while dhZ and BA are resistant to CKXs action (Galuszka et al. 2007; Frébort et al. 2011). Substrate specificity was revealed in plants with *AtCKX* overexpression. iP and its ribosides are more susceptible to *AtCKX2* and *AtCKX4* than other isoforms

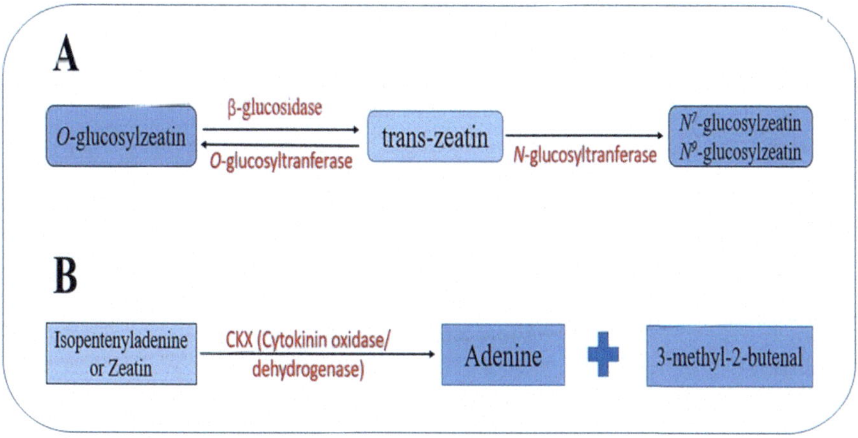

Fig. 2 Metabolism of cytokinins. **a** Glycosylation, **b** Dehydrogenation

(Galuszka et al. 2007). Cis-zeatin is digested preferably by AtCKX7, efficiently by AtCKX1 and almost resistant to deactivation by *AtCKX2, 3* and *4* (Gajdošová et al. 2011). Certain developmental stages of plants are also influenced by CKXs expression. Overexpression of CKXs causes decrease in endogenous cytokinin levels resulting in various growth and developmental defects (Wang et al. 2014) (Fig. 2).

4 Transport

Traditionally, it was assumed that synthesis of Cks occur in roots and then transported to shoots, but more recent studies made it clear that Cks are synthesized in various parts of plants, including aerial parts (Sakakibara 2006; Hirose et al. 2008; Kamada-Nobusada and Sakakibara 2009). Therefore, complex patterns of intercellular movement along with local (short-distance) transport and long-distance translocation of Cks is required to maintain Ck homeostasis and signal perception (Skalicky et al. 2018; Liu et al. 2019). Both passive diffusion and active transport mechanisms are used by different Cks for their translocation from the site of their biosynthesis to the site of their action. For example, tZ-type Cks which promote growth are synthesised in roots and transported to the shoots by apoplastic pathway (Beveridge et al. 1997; Hirose et al. 2008). Thus, tZ andtZR (tZ riboside) are the main transporters that transport Cks from root to shoot through xylem sap (Beveridge et al. 1997; Hirose et al. 2008; Kuroha et al. 2009; Osugi et al. 2017). On the contrary, cZ and iP-type Cks are integrated in shoots and transited to roots via phloem (Corbesier et al. 2003; Kudo et al. 2010).

Intercellular translocation and trans-membrane transport of Cks is allied by three types of transporters. Among these, Purine Permeases (PUP) (Bürkle et al. 2003; Zürcher et al. 2016) and Equilibrative Nucleoside Transporters (ENT) import apoplastic bioactive Cks (nucleobase) and Cks-nucleoside into the cytosol, respectively (Hirose et al. 2005; Sun et al. 2005; Durán-Medina et al. 2017). Studies in yeast cells suggested that PUPs transporter of *A. thaliana* and *Oryza sativa* participate in the uptake of tZ and iP in a proton coupled manner and ENTs transport iP-ribosides and tZ-riboside (Bürkle et al. 2003; Sun et al. 2005; Qi and Xiong 2013). PUPs and ENTs are not specific transporter of Cks as they are involved in the transit of other molecules as well (Gillissen et al. 2000; Girke et al. 2014; Durán-Medina et al. 2017). For instance, PUP14 determines the availability of Cks in apoplast where they are recognized by Arabidopsis Histidine Kinase (AHK) receptors and harmonizes the cytokinin signaling (Zürcher et al. 2016). AHK2-4 proteins cited in cell membrane and endoplasmic reticulum are involved in transcriptional exhilaration of numerous target genes (Yamada et al. 2001; Caesar et al. 2011; Wulfetange et al. 2011; Lomin et al. 2017; Pernisova et al. 2018). Recently, a specific Ck transporter viz. *AZA-GUANINE RESISTANT2* (AZG2) has been recognized which transport cytokinin nucleobase independent of energy source, and the direction of transport is regulated by the concentration gradient (Tessi et al. 2021). Additionally, other specific Cks transporters such as AtAZG1, AtABCG14, and AtPUP14 also play vital role in Ck signalization (Nedvěd et al. 2021; Romanov and Schmülling 2021).

Acropetal transport of Cks from root to shoot is affected by ATP-binding cassette (ABC) transporters, G subfamily, a third type of Cks transporter (Ko et al. 2014; Zhang et al. 2014). Notably, AtABCG14 and OsABCG18 proteins helps in xylem loading and are vital for distal transport of Cks synthesized in roots and affects shoot growth (Kang et al. 2017; Zhao et al. 2019). Conversely, shoot derived Cks are translocated by symplastic connections through phloem. A highway is formed between neighboring cells by plasmodesmata for the movement of endogenous Cks from the site of synthesis to phloem and finally to target cells (Romanov et al. 2018). Number of homologs in each family of PUP, ENT and ABCG-type transporter varies intensely between different plant species (Liu et al. 2019). In rice, 4 ENTs (Hirose et al. 2005) while in *Arabidopsis*, 8 ENTs, 21 PUPs and 28 half-size ABCG-type transporters have been identified (Li et al. 2003; Kang et al. 2011; Zürcher et al. 2016) (Fig. 3).

5 Signalling

Existed knowledge suggest a tightly regulated mechanism of cytokinin signaling pathway initiated by binding of cytokinin molecule with the receptor of histidine kinase (HK) and merging with the transcription of cytokinin responsive genes in the nucleus (Argueso et al. 2010). Signal transduction of cytokinin involve a photo transfer cascade model system similar to two component system of bacteria and fungi (To and Kieber 2008). Hybrid HK receptors—CRE1/WOL/AHK4, AHK3 and AHK2, binds to cytokinin and induces autophosphorylation on a histidine residue inside the kinase domain. This phosphate group is further relocated to a conserved aspartate residue within the receiver domain of AHK proteins (AHPs) and moves

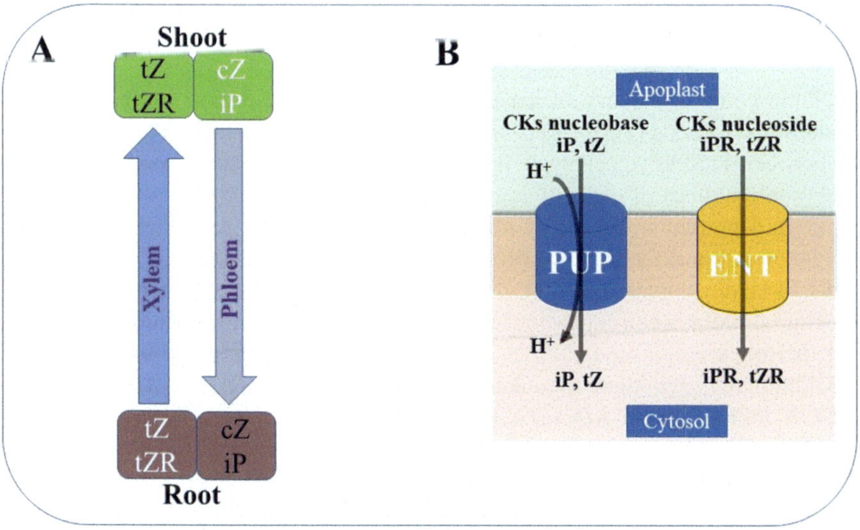

Fig. 3 Possible model of Cytokinin transport. **a** Long-distance translocation. Type of cytokinin shown in white are at site of their synthesis while cytokinin shown in black are at site of their action after transport. **b** Transport across plasma membrane

in and out of nucleus independent of phosphorylation and alteration by the exogenous cytokinins (Punwani et al. 2010). In the nucleus, phosphorylated AHP transfer one phosphate to regulators called Arabidopsis Response Regulators (ARRs). These ARRs are classified on the basis of their C-terminal domain. For example—Type A ARRs contains short C-terminal domains and are quickly transcriptionally activated by cytokinin administration while C-terminal of type-B ARRs possess DNA-binding and transactivating domains that control transcription of cytokinin-activated targets such as type-A ARRs. Type-C ARRs differ from type-A and type-B ARRs in that they lack DNA-binding domains and are not transcriptionally controlled by cytokinin. The conserved amino acids necessary for receiver domain phosphorylation are found in all ARRs (Ishidaet al. 2008).

6 Plant Growth and Development

Plant physiology includes a steady regulation of cell division, expansion and differentiation. Cytokinins influences and controls these varied processes thereby playing a multilayered role in all growth processes of plants ranging from cellular metabolism to their interactions with the environmental challenges (Kieber and Schaller 2018; Akhtar et al. 2020; Emery and Kisiala 2020).

7 Cell Division

Cytokinin controls the size of leaves by regulating cell division and proliferation in addition to auxins (Perrot-Rechenmann 2010). Also, cytokinin shorten the time gap between successive cell cycle thereby expanding the proliferation by providing a delay in the onset of cell differentiation (Zhang et al. 2005). This is brought by cytokinin degradation caused by upregulation of CKX3; a gene that slows down the rate of cell proliferation and prolongs the onset of cell expansion (Skalák et al. 2019). Once developmental process of leaf enters into cell expansion period, excess cytokinin stimulates cell expansion leading to an increase in the shoot biomass (Efroni et al. 2013).

Additionally, cytokinin modulates the expression of *CYCLIN D3* (CYCD3), *CYCLIN-DEPENDENT KINASES* (CDKs) and *AINTEGUMENTA* (ANT) in cell division phase. These three factors encode a protein that regulates cell cycle, serine threonine kinase and transcription factors respectively (Dewitte et al. 2007; Schaller et al. 2014; Randall et al. 2015). Cytokinin controls cell division by promoting the expression of CYCD3 and CDK which activate the transition through growth and synthesis phases (i.e., G1/S and G2/M transition) (Zhang et al. 2005). Overexpression of CYCD3 is enough to induce cytokinin dependent shoot formation in callus while its loss reduces the ability of exogenous cytokinin to promote cell division in shoot (Dewitte et al. 2007). Further, CYCD3 promotes mitotic cell division and restrict further cell division. Therefore, CYCD3 is considered the main factor through which cytokinin interact with cell cycle mechanism (Wu et al. 2021) (Fig. 4).

8 Leaf Senescence

Cytokinins have long been recognized to slow down the ageing of leaves in monocotyledons/dicotyledons. During the onset of senescence, there is a reduction in Ck levels thereby acting as a key signal during its initiation. However, transgenic expression of Ck biosynthesis genes or its exogenous application delays the process. This is achieved by preventing chlorophyll disassembly, degradation of photosynthetic proteins, lipids and RNA (Woo et al. 2013; Wu et al. 2021). In entire *Arabidopsis* plants, the ore12 mutation causes delayed leaf senescence caused by a recessive, gain-of-function missense mutation at AHK3's presumed extracellular domain. The recessive aspect of the gain-of-function ore12 allele is most likely due to dosage requirement for this mutation to have an effect on leaf senescence. AHK3 disruption resulted in early leaf senescence and genetic investigation of three cytokinin AHK receptors revealed that AHK3 controls leaf senescence the most. Phosphorylation and activation of ARR2 by AHK3 influence leaf senescence (Kim et al. 2006). AHK3 is the most essential cytokinin receptor in this activity, whereas, the specificity of type-B ARRs is yet to be studied. Indeed, a loss-of-function ARR2 mutant has no

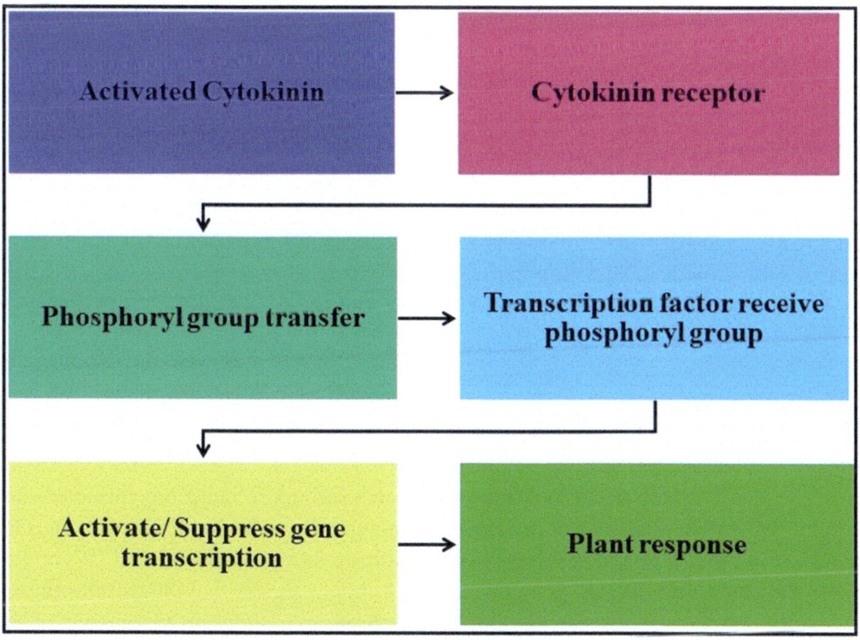

Fig. 4 Generalized mechanism of cytokinin signaling in plants

impact on leaf senescence, suggesting that other type-B ARRs are likely involved in the process control (Kieber and Schaller 2014).

Cytokinin Response Factors (CRFs), that are cytokinin-regulated AP2/ERF (*APETALA2/ETHYLENE RESPONSIVE FACTOR*) transcription factors, also plays significant role in leaf senescence regulation (Zwack et al. 2013). CRF6 disruption reduces leaf sensitivity to the inhibitory action of cytokinin in dark-induced leaf senescence and *crf6* mutants exhibit enhanced leaf senescence in intact plants (Zwack et al. 2013). Unexpectedly, overexpression of CRF6 resulted in an even faster onset of senescence (Kieber and Schaller 2014).

9 Apical Dominance

Actively growing primary roots of dicot plants may exhibit apical dominance and prevents the initiation of lateral roots. Phenomenon of root apical dominance can be compared to shoot apical dominance whereby growing primary leader prevented the lateral organ initiation and development (Lloret and Casero 2002). Cytokinin has been detected by the expression of ARR5 and GUS (a CK-activated promoter sequence of a response regulator fused to b-glucuronidase). Both reflect the transcriptional activation of cytokinin sensitive promoter which fuses to GUS reporter gene. Developing leader reacts to cytokinin in a concentration dependent manner and shows that the cap of primary root of *Arabidopsis* produces an increased level of free cytokinin higher than roots (Aloni et al. 2004, 2005). Additionally, application of root apical dominance enables growth of primary root which go deeper in soil in search of water during drought. It also governs the root architecture by regulating the promotion of auxins in secondary roots. This mechanism reduces the number of secondary roots, their development and requirement at the cost of primary root growth. However, these secondary roots grow by local stimulations like nitrate and phosphate availability and by down regulation of IPT expression. Such a system increases the growth and development of secondary and tertiary roots but may disrupt cytokinin dependent root apical dominance (Miyawaki et al. 2004).

10 Shoot Initiation

Earlier research stated that cytokinins together with auxin had a big influence on shoot development (Schaller et al. 2015). Plants grown under higherlevels of auxin and cytokinin multiply and dedifferentiate, resulting into callus formation (Wybouw and Rybel 2019). The capacity of cytokinins to begin shoots from undifferentiated callus cultures, as well as the beginning of ectopic meristems in transgenic plant breed to overexpress cytokinins, revealed that this hormone may have a role in Shoot Apical Meristem (SAM) formation. A number of subsequent studies have shed an insight on the function of cytokinin in SAM function as well as its interactions with other hormonal and developmental signalling pathways (Hwang and Sheen 2012). Cytokinin levels can be suppressed by overexpression of CKX gene, loss of various IPT genes, or interruption inside the cytokinin receptors, leading to smaller SAM, stating that cytokinin is a positive modulator of cell proliferation in the SAM. The *KNOTTED-LIKE* (KNOX) homeobox transcription factors are essential for the SAM to be established and maintained. The KNOX genes are triggered in SAM but not in developing leaf primordial (Kieber and Schaller 2014). Further, KNOX transcription factors govern the relative amounts of cytokinin and GA, the main way by which they regulate SAM function. It has been shown that *SHOOT-MERISTEMLESS* (STM) genes impedes cellular differentiation and endo-reduplication by operating through cytokinin and induced CYCD3 regulates cell cycle development. Furthermore, STM have been ascertained to regulate SAM resulting in the creation of ectopic meristems by a process independent of the rise in cytokinin levels, suggesting that not all of STM's actions in the SAM need the control of cytokinin levels (Scofield et al. 2013).

11 Source-Sink Relationships

Cytokinins play a role in modulating the source/sink interactions but the mechanism of cytokinin-mediated source and sink control remains unknown, as it occurs between different types of organs. Recent research, however, suggested that cytokinin may play a role in the stimulation of natural activity in both source and sink organs. This is achieved by improving the photosynthetic ratein the source (leaves), stomatal conductance, decreased starch creation and increased sucrose formation. As a result, the molar concentration of photo-assimilates rises, lowering water potential, promoting osmotic water inflow and increasing source pressure. Furthermore, cytokinins enhances the functional activity at the sink (tissue) via increasing cell enlargement, recycling of soluble substances, sucrose unloading, increasing water potential, decreasing pressure at the sink portion and in promoting assimilated influx to sink organs. All of these intricate processes are the consequence of decreasing water potential at the start and increasing water potential at the conclusion of the transport phase, which establishes a pressure gradient between source and sink. This procedure serves as the foundation for various aspects of a consistent and well-balanced system (Ronzhina 2007; Glanz-Idan et al. 2020; McIntyre et al. 2021).

Any modulation in the source-sink activity has been linked to the endogenous variables such as leaf age and cytokinin quantities (Lubovska et al. 2014). Also, endogenous levels of Ck were hypothesized to be substantially linked with Delayed Leaf Senescence (DLS) morphology and grain output as they impact both the leaf senescence and crop productivity (Jameson and Song 2016). The first Ck root-to-shoot translocation happens via its entranceinto the xylem in which these are first loaded into the xylem sap by ABCG proteins (ATP-binding Cassette G subfamily) forfurther transport through the plant (Liu et al. 2019). Furthermore, a recent study in *Arabidopsis* showed that following translocation of root-synthesized Cks in the xylem, it must primarily betransported to the phloem for proper source-to-sink dispersion. AtABCG14 mediated phloem discharge at the target shoot organs is the final stage. This highlights the need of phloem-directed Ck redistribution, which is ultimately regulated by AtABCG14, for successful, long-distance acropetal transport of root-synthesized Cks (Zhao et al. 2021). Cks play an important role in morphogenesis and plant development because of their presence throughout the vegetative and reproductive stages of plant life. However, the precise method by which Cks interact remains a mystery, including molecular players and hubs that could be engaged at the junction of the sucrose and Ck signalling pathways. As the Ck interaction can result in either antagonistic or agonistic effects, its regulation network is however, expected to be complex and multifactorial based on physiological and environmental inputs (Wang et al. 2021).

12 Nutrient Uptake

Cytokinin also regulates the ability of plants to absorb a variety of nutrients including nitrate, phosphate, sulfur and iron from soil. In return, nutrient status of plant too regulates the cytokinin function and growth of plant (Argueso et al. 2009). Level of cytokinin is determined by the nitrate availability in plants i.e., plants grown on low nitrogen levels tend to have low cytokinin while addition of exogenous nitrate elevates cytokinin activity by inducing the expression of *AtIPT3* and *AtIPT5* especially in the roots (Miyawaki et al. 2004; Takei et al. 2004a, b). Any disruption in IPT3 gene attenuates the induction of cytokinin in response to nitrate further affirming IPT3 as the primary target for nitrate-induced cytokinin biosynthesis (Takei et al. 2004a, b). However, IPT3 is also regulated by the level of phosphate, sulphate, and iron thus indicating it as the site for a variety of nutrient signals (Kiba et al. 2011). However, in *Arabidopsis*, addition of ammonium in nitrogen starved plants enhanced the level of *AtIPT5* instead of *AtIPT3* (Takei et al. 2004a, b). Expression of *CYP735A2* gene (that encodes for enzymes involved in the synthesis of cytokinin) is too managed by nitrate level (Wang et al. 2004). Correlation between nitrate and cytokinin level and their effect on expression of gene concluded Ck as a root to shoot signal to coordinate tissue specific nitrogen metabolism. The balance between different types of cytokinin anticipated by two component system indicate towards the availability of nitrate form, leading to the expression of cytokinin-responsive and nitrate-responsive genes to govern the metabolism of nitrate in the plants (Sakakibara 2006).

Sulfate responsive genes are too upregulated by Ck under both sulfur deficiency/availability condition and are slightly upregulated by other phytohormones (Ohkama et al. 2002). Expression of *APR1* that encodes for enzyme involved in sulfate assimilation is also induced by cytokinin as well as by sucrose and nitrate (Maruyama-Nakashita et al. 2004; Rouached et al. 2008). However, cytokinin concentration remains unaltered in sulfate-deficient plants (Ohkama et al. 2002), but *IPT3* was upregulated in response to sulfate (Hirose et al. 2008). Farther more, level of Ck do not alter with the change in the O-acetyl-L-serine level (a cysteine precursor that positively upregulates sulfate starvation-responsive genes). Thus, suggesting an indirect action of cytokinin towards regulation of sulfate uptake and their genes (Ohkama et al. 2002; Kieber and Schaller 2014).

Cytokinin levels repress phosphates (Pi) in Pi-starved plants while its exogenous application downregulates genes responsible for Pi starvation (Hou et al. 2004; Wang et al. 2006; Kieber and Schaller 2014). Two *Arabidopsis* mutants, *pho1* and *pho2* fail to accumulate and hyper-accumulate phosphates in shoots, respectively and displayed an altered sensitivity for Ck. Insufficient Pi causes complicated alterations in gene expression along with an early temporary alteration in the expression of genes encoding broad stress response components, followed by the activation of genes directly engaged in the response to Pi deficiency (Hou et al. 2004). Microarray investigation of rice plants subjected to Pi deficiency further supported these findings, even identified genes that were either up-regulated or remained unaltered after cytokinin addition, demonstrating a complicated influence of cytokinin on Pi-deficiency gene expression (Wang et al. 2006).

In case of iron, cytokinins withhold the expression of a subgroup of iron-responsive genes which are required for AHK3 and AHK4 receptors but is unaffected by iron status or FIT1 (*FER-LIKE IRON DEFICIENCY-INDUCED TRANSCRIPTION FACTOR*) that is a transcription factor with a fundamental helix-loop-helix (bHLH) structure that regulates a subset of iron responsive genes (Briat et al. 2007; Séguéla et al. 2008). A momentary upsurge in *IPT3* and type-A ARR gene expression was seen in iron-starved plants in response to iron replenishment, which is similar to the stimulation of the same genes in response to nitrogen resupply. Other substances like mannitol and NaCl that impede root development, were discovered to suppress iron-starvation responsive genes. It was also postulated that cytokinin inhibits iron-responsive gene expression via a growth-dependent mechanism which might explain cytokinin's influence on other nutrient absorption pathways (Kieber and Schaller 2014).

13 Phyllotaxis

Cytokinin play a crucial but independent role during leaf development and maintenance in addition to auxins. On one hand, where auxin have role in leaf development initiation and organogenesis, cytokinins play their role in meristem maintenance. However, both hormones act together in multiple cells, tissue and organs having both antagonistic and synergistic effects (El-Showk et al. 2013; Schaller et al. 2014). Primary function of cytokinin is to maintain the size, shape and structure of shoot apical meristem (Werner and Schmülling 2009). Any deduction in the concentration of cytokinin by mutation in *IPT* or overexpression of Cytokinin Oxidase gene (CKX) or by modulation in transporter signal may lead to decrease in the size and activity of SAM (Higuchi et al. 2004). In maize, phyllotaxis was altered in its mutant aberrant phyllotaxy1 (*abph1*) (Jackson and Hake 1999) and rice mutant decussate (*dec*) (Itoh et al. 2012). Both these mutants encode for protein that have function in signaling of cytokinin. Further, abph1 and dec mutants have enlarged shoot apical meristem. However, if any mutant has disruption in cytokinin signaling pathway, then it does not exhibit any phyllotactic shift (Zhao et al. 2010). In *abph1* mutants, *PINFORMED1* (PIN1) expression is greatly reduced along with auxin at leaf primordium because cytokinin promotes the expression of PIN1 (Lee et al. 2009).

14 Gametophyte and Embryo Development

During the development of female gametophyte, cytokinin plays a vital role in the cell fate specification. Connections between female gametophyte and phosphorelay were however, established after mutation in cytokinin insensitive (*cki*) gene was found to be lethal to plants (Zürcher and Müller 2016). Many genes such as *cki*, *arr7*, *arr15 double*, and *ahp2-2*, *ahp3*, *ahp5-2*, triple mutants are found lethal to female gametophyte (Yuan et al. 2016; Liu et al. 2017). CKI denotes a histidine kinase that triggers the cytokinin response in the absence of cytokinins (Hwang and Sheen 2012). Loss of function in mutant of *cki* leads to misspecification of cell fate resulting into egg cell fate adopting by antipodal and central cells. In these cells, *TCS* expression (TRICHOSANTHIN) was found to be reduced or absent due to *cki* mutant (Yuan et al. 2016). *AHP1*, *AHP2* and *AHP5* acts downstream of *cki*, thus standard cytokinin signaling pathway is often used for cell fate specification (Liu et al. 2017). Over expression of *cki* resulted into ectopic expression of *TCS* and specification of egg cell into central cell. After fertilization, sperm cell fuses with this mis specified egg cell and develops a diploid endosperm instead of embryo. Thereby suggesting that *cki* is very important in terms of providing antipodal and central cell fate whereas repression is required for specification of synergid and egg cell (Wybouw and Rybel 2019).

The formation of female gametophytes, however, is not retained in all plants. Gymnosperms, for instance, encode a *cki* ortholog but lack central cells and endosperm. The *cki* ortholog for example does not completely rescue in *A. thaliana cki* mutant but in *Ginkgo biloba*, it was unable to provide central cell specification even when cytokinin signaling was increased thereby implying that throughout angiosperm development, neo functionalization of CKI aided in the production of central cells and the establishment of endosperm (Yuan et al. 2018) (Fig. 5).

There is an asymmetrical distribution of Ck activity in female gametophyte being higher at chalazal end and further supported by an increased expression of *IPT1* and *AHK4* (Cheng et al, 2013). Ck also affects the ovule development by deregulation of auxin efflux carrier PIN1 because the addition of exogenous cytokinin altered the expression of *PIN1* and formation of altered ovule (Ceccato et al. 2013). While the functionality of male gametophyte depends upon atleast one functional receptor because in triple mutants, anther fails to dehisce and pollen does not reach to maturation properly (Kinoshita-Tsujimura and Kakimoto 2011). Also, a rise in ovule number has been erected in the *ckx5ckx6* mutant impaired in cytokinin breakdown (Bartrina

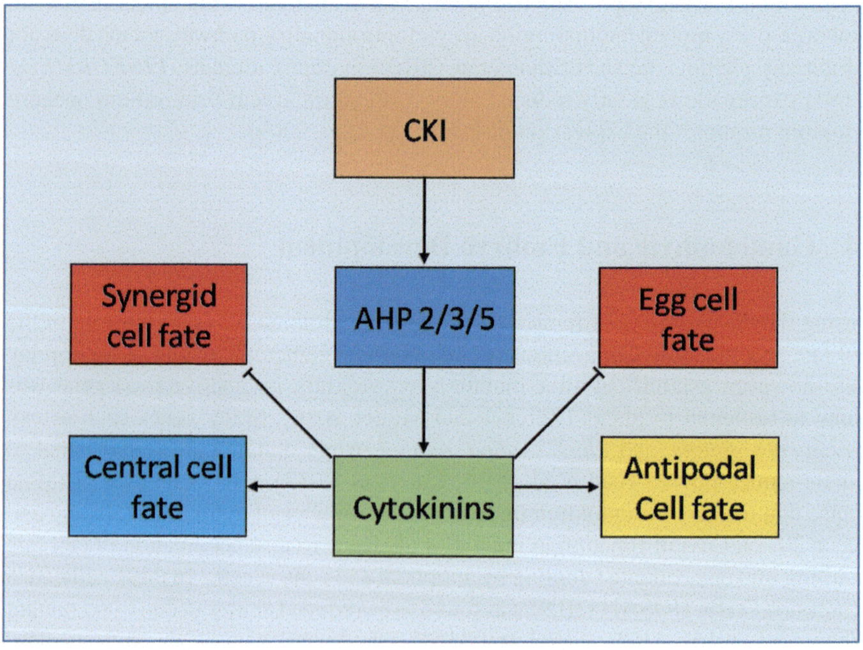

Fig. 5 Cytokinin signaling in female gametophyte development

et al. 2011). Besides, exogenous cytokinin causes abnormalities in the patterning of the gynoecium, which in extreme cases resulted to valveless siliques (Zúñiga-Mayo et al. 2014).

Cytokinins play a vital role in embryogenesis also. This has been established by tissue culture experiments where cytokinin promotes shoot fate, growth and inhibits root formation (Wybouw and De Rybel 2019). Recent research analyzes the expression profile of *LOG* genes and discovered that *LOG3* and *LOG4* are equally expressed in the provasculature, thereby explaining the profile of cytokinin activity (De Rybel et al. 2014). Surprisingly, *LOG3* and *LOG4* are regionally stimulated by auxin through the transcription factor dimer *TARGET OF MONOPTER OS5/ LONESOME HIGHWAY*, resulting in a close relationship between the two plant hormones (Ohashi-Ito et al. 2014). During the early embryonic developmental stages, cytokinin signalling occurs inside the hypophysis, the founder cell of the embryo root meristem. After an initial division, Ck response is retained in the apical, lens-shaped cell, while signalling is inhibited in the lower, basal cell. This inhibition is discovered to be dependent to auxin signaling, which directly stimulates the production of type-A *ARR7* and *ARR15*. While its ectopic signalling mediates an inducible *arr7arr15* double mutant leading to pattern defect (Muller and Sheen 2008). Surprisingly, a stable *arr7arr15* double mutant exhibited milder symptoms in comparison to inducible mutant indicating that embryos can particularly correct for the irreversible loss of *ARR7*and *ARR15* (Zhang et al. 2011).

15 During Abiotic Stress

For proper growth and development, plants require a number of elements which include both micronutrients (boron, iron silicon and selenium) and macronutrients (potassium, nitrogen, Sulphur and phosphorous). Cytokinin play a vital role in the uptake of both micronutrients and macronutrients by plants and their response to a variety of heavy metals such as cadmium, aluminum, arsenate etc. Salinity and drought stress are one of the most common abiotic stresses that reduces production of crop on global scale. Cytokinin is found to enhance *Arabidopsis* performance and promote transcriptional reprograming under salt and drought stress conditions. Heat stress decrease photochemical efficiency and chlorophyll content of leaves, while, the endogenous cytokinin can increase heat tolerance (Nam et al. 2012; Schaller et al. 2015). Hence, various studies have been reported to demonstrate the role of Cks in mitigation of abiotic stress caused by heavy metals, drought, heat etc. Some of the studies showing the effect of Cks under abiotic stress factors like metal stress, drought, salinity and temperature have been listed in Table 1.

Table 1 Effect of Cytokinins and the receptor molecules/ genes involved in abiotic stress factors in different plant species

Mode of application	Type of abiotic stress	Plant species	Receptor molecules/Genes involved	Reported effects	References
Endogenous	Flooding/Cold	Tomato	CRF	Enhanced stress tolerance	Shi et al. (2014)
Endogenous	Drought	*Arabidopsis*	CKX; IPT	Enhanced stress tolerance	Nishiyama et al. (2011)
Endogenous	Drought	Chickpea	CKX	Enhanced growth, stress tolerance and yield	Khandal et al. (2020)
Exogenous	Heat	*Agrostis stolonifera*	IPT	Enhanced heat tolerance, delayed leaf senescence	Xu et al. (2009)
Exogenous	Drought	*Gossypium hirsutum*	IPT	Enhanced drought tolerance, delayed leaf senescence	Kuppu et al. (2013)
Endogenous	Drought	Chickpea	CKX	Enhanced growth, stress tolerance and yield	Khandal et al. (2020)
Endogenous	Salt	*Nicotiana tabacum*	IPT	Enhanced salt tolerance, delayed leaf senescence	Qiu et al. (2012)
Endogenous	Salt	*Centaurium erythraea*Rafn	AtCKX	CK content in the root is reduced by AtCKX transgenic line, affects the growth of the root system, and enhances salt tolerance	Trifunović-Momčilov et al. (2020)
Endogenous	Salt	*Oryza sativa*	CYP71D8L-OE CYP71D8L	The application of CKs biosynthesis inhibitors can change the phenotypes of CYP71D8L-OE and CYP71D8L root dysplasia and show higher salt tolerance	Zhou et al. (2020)

(continued)

Table 1 (continued)

Mode of application	Type of abiotic stress	Plant species	Receptor molecules/Genes involved	Reported effects	References
Endogenous	Cold	*Arabidopsis*	ARR5	Enhanced cold tolerance	Jeon et al. (2010)
Endogenous	Drought	Barley	Beta-glucosidase (bGLU)	Enhanced growth, maintained higher water content and improved drought stress tolerance	Pospíšilová et al. (2016)
Endogenous	Drought	*O. sativa*	IPT	Improved drought tolerance, enhanced sink strength and improved drought tolerance	Peleg et al. (2011)
Endogenous	Drought	*Arachis stolonifera*	IPT	Maintain stomatal conductance, higher transpiration photosynthetic rate and improved drought tolerance	Qin et al. (2011)
Endogenous	Drought	*Gossypium hirsutum*	IPT	Enhanced shoot and root biomass, higher chlorophyll content, delayed leaf senescence and maintain photosynthetic rate under water deficit condition	Kuppu et al. (2013)
Exogenous Cytokinin	Nitrogen-deficiency	Tobacco plants	IPT	Delayed leaf senescence in transgenic plants	Rubio-Wilhelmi et al. (2011)
Cytokinin	Salt stress	Tobacco plants	IPT	Enhanced salt stress tolerance in transgenic tobacco plants	Qiu et al. (2012)

(continued)

Table 1 (continued)

Mode of application	Type of abiotic stress	Plant species	Receptor molecules/Genes involved	Reported effects	References
Exogenous Cytokinin	Drought, Salinity	*O. sativa* L	OsRR6, a type-A response regulator	reduced cytokinin sensitivity, adventitious root formation and enhanced anthocyanin accumulation in seeds, more tolerant to drought and salinity	Bhaskar et al. (2021)
Exogenous Cytokinin	Salinity	*Vicia faba* L		Enhanced fresh and dry biomass of root/shoot, escalated activities of Catalase, Superoxide dismutase, Peroxidase and Ascorbate peroxidase	Abdel Latef et al. (2021)
Cytokinin	Drought	*Arabidopsis*	CKX	Over expressed cytokinin oxidase in transgenic plants	Werner et al. (2010)
Cytokinin	Drought	Barley	AtCKX1	Showed better drought tolerance by transgenic plants via better dehydration avoidance	Pospíšilová et al. (2016)
Cytokinin	Drought	Rice	ERF-1(ETHYLENE RESPONSE FACTOR)	Increased drought tolerance	Zhang et al. (2010)
Exogenous Cytokinin	Cadmium	*Arabidopsis*	SHY2 (SHORT HYPOCOTYL2)	Negative regulation of PIN1,3,7 genes, increased level of SHY2, inhibited auxin transport to the root apex	Bruno et al. (2017)

(continued)

Table 1 (continued)

Mode of application	Type of abiotic stress	Plant species	Receptor molecules/Genes involved	Reported effects	References
Cytokinin	Arsenic	*Arabidopsis*, Tobacco		Enhanced stress tolerance due to sequestration of arsenic mediated by thiol compounds including glutathione and phytochelatins	Mohan et al. (2016)
Cytokinin	Zinc	Tobacco	IPT	Increased level of cytokinin IPT genes transformed tobacco plants, showed more transpiration and photosynthesis	Pavlíková et al. (2014)
Cytokinin	Cadmium	*Arabidopsis*	CKX IPT	Down regulation of CKX gene in shoot and upregulation in root regulated level of cytokinin Increased transcription level of IPT gene enhanced cytokinin content in shoot	Vitti et al. (2013)
Cytokinin	Cadmium	Wheat	Cytokinin oxidase	Decreased cytokinin level by enhancing the activity of cytokinin oxidase	Veselov et al. (2003)
Endogenous	Heat stress	*Arabidopsis*	tZR, DZR, and iPR	Reduce heat stress	Prerostova et al. (2020)
Endogenous	Stress	*Physcomitrella Patens*	PpCKX1	PpCKX1 increases dehydration resistance and salt tolerance, upregulation of stress-related genes, increased	Hyoung et al. (2020)

(continued)

Table 1 (continued)

Mode of application	Type of abiotic stress	Plant species	Receptor molecules/Genes involved	Reported effects	References
Endogenous	Salt	*Oryza sativa*	GlycosylTransferase-encoding gene	ROS scavenging activity, decreased ion leakage, more proline and soluble sugar accumulation,	Li et al. (2020)
Endogenous	Salt	*Oryza sativa*	CKX3	Increased CKs in roots but decreased in shoots	Yin et al. (2020)
Endogenous	Stress response	*Arabidopsis*	ERF115	Block adventitious root initiation downstream of jasmonate-induced ERF 115	Lakehal et al. (2020)
Endogenous	Arsenate	*Arabidopsis*	ASA1 and ASB1 (ANTHRANILATE SYNTHASE ALPHA SUBUNIT 1) and BETA SUBUNIT	Induced root growth inhibition Induced inhibition of primary root elongation	Tu et al. (2021)
Endogenous	NaCl	Pepper plants	CKs	Enhanced plant growth and development and stress responses,	Gálvez et al. (2021)
Endogenous	NaCl	*Solanum chilense*	DEGs (Differentially Expressed Genes)	ROS scavenging system, transporters, osmotic regulation, defence and stress response,	Kashyap et al. (2020)

16 During Biotic Stress

Cytokinins have been employed to regulate plant defense against microbial pathogens like bacteria, fungi, nematodes etc. during their growth and development by inducing plant immunity and pathogen resistance (Shigenaga et al. 2017; Checker et al. 2018; Gupta et al. 2020a). Their role in providing immunity in various plant-pathogen interactions have been highly acknowledged (Naseem et al. 2014; Siddique et al. 2015; Spallek et al. 2018; Cortleven et al. 2019). Apart from plants, plant associated microorganisms, microalgae and insects too produces Cks but the types and activities of these molecules vary with different plant species, tissues, stages of development and various environmental conditions (Akhtar et al. 2020). For instance, in *Arabidopsis,* the highly abundant and most bio-active Cks are tZ and iP whereas cZ, although less bioactive in *Arabidopsis*, is as active as tZ in rice (Miyawaki et al. 2006; Kudo et al. 2012). These variations in bioactivities correlate with different binding affinities to cytokinin receptors (Spallek et al. 2018). Activation of Ck receptors leads to a cascade of phosphorylation events that results in cellular reprogramming thus influencing a wide range of plant processes (Kieber and Schaller 2014). The effect of Cks against biotrophic pathogens depends on their concentration at the infection site. High Ck concentration reduces the growth of pathogens, while low Ck concentration results in increased pathogen growth. Initially, the role of Cks against biotic stress in plants was studied in *Arabidopsis*, where exogenous application of Ck attenuated the growth of pathogens *Hyaloperonospora, arabidopsidis* and *Pseudomonas syringae* (Shigenaga and Argueso 2016). Since then, various studies have reported the role of Cks in mitigating the biotic stress caused by various plant pathogens. Some examples of the effect of Cks against biotic stress factors like bacteria, fungi, nematodes, viruses and herbivores are listed in Table 2.

17 Conclusion

Cytokinins plays vital as well as pleiotropic role in plant taxa ranging from overall growth to signaling during environmental extremities along with the expression of genes which help in maintaining homeostasis in plants. However, further investigations are required to explore and decode the additional facets of Cks synthesis, circuitry, molecular regulation in plant growth and development and its interaction with other hormones to combat harsh conditions in plants. Also, the new knowledge gained through genomic tools will further help in understanding the hidden aspects of Ck signaling models and its interplaywith other plant growth regulating hormones will pave a way for the development of stress resistant plant varieties thereby elevating the agricultural productivity.

Table 2 Effect of Cks against varied biotic stress factors such as bacteria, fungi, nematodes, viruses and herbivores in plant species

Biotic stress factor		Mode of application	Plant species	Receptor molecules/Genes involved	Reported effects	References
Bacteria	*Xanthomonas campestris* and *Pseudomonas syringae*	Exogenous application	*Solanum lycopersicum*	Type-A Tomato response regulators (TRRs) and Cytokinin oxidase (CKX)	Resistance to bacterial pathogens was enhanced through a process that depends on Salicylic acid and Ethylene signaling	Gupta et al. (2020a)
	Ralstoniasolancearum	Exogenous application	*Arabidopsis thaliana*	Cytokinin oxidases, *Arabidopsis* response regulators (ARR)	Enhanced root immunity against pathogen	Alonso-Diaz et al. (2021)
	P. syringae pv *tabaci* (*Pst T*)	Pathogen-dependent upregulation of CK biosynthesis and Exogenous supply of kinetin (1–18 μM)	*Nicotiana tobacum*	Pathogenesis Related Gene 1 (PR1) and Enhanced Disease Susceptibility1 (EDS1)	Tobacco leaves treated with kinetin (10 μM) showed increased resistance against *Pst T* by 95%	Grosskinsky et al. (2011)
Nematodes	*Heteroderaschachtii*	Endogenous	*Arabidopsis*	*Arabidopsis* histidine kinase (AHKs), type-A and type-B *Arabidopsis* response regulators (ARR), histidine-containing phosphotransfer proteins (AHPs)	Cytokinin signaling was elevated that impeded nematode development at the site of infection	Shanks et al. (2016)
	Meloidogyne incognita and *H. schachtii*	Endogenous	*Arabidopsis*	Isopentenyl transferases (IPTs), CKX, AHKs	Reduced susceptibility to nematode infection	Dowd et al. (2017)

(continued)

Table 2 (continued)

Biotic stress factor		Mode of application	Plant species	Receptor molecules/Genes involved	Reported effects	References
Viruses	*Potato Virus Y^{NTN}* (PVY^{NTN})	Endogenous	*N. tabacum*	NADP-malic enzyme, Phosphoenolpyruvate carboxylase (PEPC), Pyruvate orthophosph-ate dikinase (PPDK)	Enhanced tolerance to PVY^{NTN}, phenolic and lignin content was increased, oxidative stress tolerance was improved due to higher activities of antioxidant enzymes	Spoustova et al. (2015)
	Chilli veinal mottle virus (ChiVMV)	*Agrobacterium*-mediated transformation of Cytokinin receptor 1 (CRE_1)	*N. tobacum*	Cytokinin receptor 1 (CRE_1)	CRE_1 defective mutants showed more severe symptoms of viral infection than wild type) (WT) plants; virus concentration was higher in CRE_1 deficient mutants as compared to WT plants	Zou et al. (2020)
Fungi	*Magnaportheoryzae*	Exogenous treatment	*O. sativa*	WRKY45 transcription factors, Diterpenoid phytoalexin (DP) biosynthetic genes	Cytokinin induced upregulation of DP biosynthetic gene increased immunity to fungal pathogen	Akagi et al. (2014)

(continued)

Table 2 (continued)

Biotic stress factor		Mode of application	Plant species	Receptor molecules/Genes involved	Reported effects	References
	Plasmodiophorabrassicae	Endogenous treatment	*Brassica oleracea*	Cytokinin dehydrogenase/oxidase (CKX) genes	Increased resistance to club root disease	Zhu et al. (2021)
	Botrytis cinerea and *Oidium neolycopersici*	Exogenous treatment	*S. lycopersicum*	Pattern recognition receptor (PRR) LeEIX2	CK induced systemic immunity in tomato and enhanced resistance to pathogens in SA- and ET-dependent mechanisms	Gupta et al. (2020b)
	Plasmodiophorabrassicae	Exogenous treatment	*A. thaliana*	SYNERGISTIC ON AUXIN AND CYTOKININ 1 (*syac1-3 and syac 1–5*)	Auxin-cytokinin crosstalk component (SYAC1) enhanced tolerance to pathogenic protist	Hurny et al. (2020)
	B. cnerea	Exogenous treatment	*S. lycoersicum*	Pattern Recognition Receptor (PRR), *LeEIX1, SIFLS2*	Disease was reduced by ~ 50% in soil grown and ~ 20% in sterile grown plants	Gupta et al. (2021)
Insects	*Manduca sexta*	Foliar spray with 1 μM of bioactive CK *tZ* (trans-zeatin)	*N. attenuate*	Chase-domain containing His Kinase 2 (NaCHK2) and NaCHK3	Increased CK concentrations amplified JA-pathway signaling against herbivore	Schafer et al. (2015)

(continued)

Table 2 (continued)

Biotic stress factor		Mode of application	Plant species	Receptor molecules/Genes involved	Reported effects	References
	Lymantria dispar	Exogenous application of 100 μM of benzylaminopurine (BAP)	*Populas*	Allene oxide synthase (AOS), *win3* trypsin inhibitor (TI), and *win8* chitinase (CHI)	Reduced larval weight gain, deterrence of herbivory, wound-inducible accumulation of JA and Linoleic acid	Dervinis et al. (2010)

Conflicts of Interest The authors declare no conflict of interest.

References

Abdel Latef AAA, Hasanuzzaman M, Tahjib-Ul-Arif M (2021) Mitigation of salinity stress by exogenous application of cytokinin in faba bean (*Vicia faba* L.). NotulaeBotanicae Horti Agrobotanici Cluj-Napoca 49(1):12192

Akagi A, Fukushima S, Okada K, Jiang CJ, Yoshida R, Nakayama A, Shimono M, Sugano S, Yamane H, Takatsuji H (2014) WRKY45-dependent priming of diterpenoid phytoalexin biosynthesis in rice and the role of cytokinin in triggering the reaction. Plant Mol Biol 86:171–183

Akhtar SS, Mekureyaw MF, Pandey C, Roitsch T (2020) Role of cytokinins for interactions of plants with microbial pathogens and pest insects. Front Plant Sci 10:1777

Aloni R, Langhans M, Aloni E, Dreieicher E, Ullrich CI (2005) Root-synthesized cytokinin in *Arabidopsis* is distributed in the shoot by the transpiration stream. J Exp Bot 56(416):1535–1544

Aloni R, Langhans M, Aloni E, Ullrich CI (2004) Role of cytokinin in the regulation of root gravitropism. Planta 220(1):177–182

Alonso-Diaz A, Satbhai SB, de Pedro-Jove R, Berry HM, Goschl C, Argueso CT, Novak O, Busch W, Valls M, Coll NS (2021) A genome-wide association study reveals cytokinin as a major component in the root defense responses against *Ralstonia solanacearum*. J Exp Bot 72(7):2727–2740

Argueso CT, Ferreira FJ, Hutchison CE, To JPC, Epple P, Mathews DE, Schaller GE, Dangl JL, Kieber JJ (2012) Two-component elements mediate interactions between cytokinin and salicylic acid in plant immunity. PLoS Genet 8:e1002448

Argueso CT, Ferreira FJ, Kieber JJ (2009) Environmental perception avenues: the interaction of cytokinin and environmental response pathways. Plant Cell Environ 32(9):1147–1160

Argueso CT, Raines T, Kieber JJ (2010) Cytokinin signaling and transcriptional networks. Curr Opin Plant Biol 13(5):533–539

Ashikari M, Sakakibara H, Lin S, Yamamoto T, Takashi T, Nishimura A, Angeles ER, Qian Q, Kitano H, Matsuoka M (2005) Cytokinin oxidase regulates rice grain production. Science 309(5735):741–745

Bartrina I, Otto E, Strnad M, Werner T, Schmülling T (2011) Cytokinin regulates the activity of reproductive meristems, flower organize, ovule formation, and thus seed yield in *Arabidopsis thaliana*. Plant Cell 23:69–80

Beveridge CA, Murfet IC, Kerhoas L, Sotta B, Miginiac E, Rameau C (1997) The shoot controls zeatin riboside export from pea roots. Evidence from the branching mutant rms4. Plant J 11(2):339–345

Bielach A, Podlešáková K, Marhavý P, Duclercq J, Cuesta C, Müller B, Benková E (2012) Spatiotemporal regulation of lateral root organogenesis in *Arabidopsis* by cytokinin. Plant Cell 24(10):3967–3981

Bilyeu KD, Cole JL, Laskey JG, Riekhof WR, Esparza TJ, Kramer MD, Morris RO (2001) Molecular and biochemical characterization of a cytokinin oxidase from maize. Plant Physiol 125(1):378–386

Bhaskar A, Paul L, Sharma E, Jha S, Jain M, Khurana P (2021) OsRR6, a type-A response regulator in rice, mediates cytokinin, light and stress responses when over-expressed in *Arabidopsis*. Plant Physiol Biochem 161:98–112

Bishopp A, Help H, El-Showk S, Weijers D, Scheres B, Frim J, Helariutta Y (2011) A mutually inhibitory interaction between auxin and cytokinin specifies vascular pattern in roots. Curr Biol 21:917–926

Briat JF, Curie C, Gaymard F (2007) Iron utilization and metabolism in plants. Curr Opin Plant Biol 10:276–282

Bruno L, Pacenza M, Forgione I, Lamerton LR, Greco M, Chiappetta A, Bitonti MB (2017) In *Arabidopsis* thaliana cadmium impact on the growth of primary root by altering SCR expression and auxin-cytokinin cross-talk. Front Plant Sci 8:1323

Bürkle L, Cedzich A, Döpke C, Stransky H, Okumoto S, Gillissen B, Kühn C, Frommer WB (2003) Transport of cytokinins mediated by purine transporters of the PUP family expressed in phloem, hydathodes, and pollen of *Arabidopsis*. Plant J 34(1):13–26

Caesar K, Thamm AM, Witthöft J, Elgass K, Huppenberger P, Grefen C, Horak J, Harter K (2011) Evidence for the localization of the *Arabidopsis* cytokinin receptors AHK3 and AHK4 in the endoplasmic reticulum. J Exp Bot 62(15):5571–5580

Ceccato L, Masiero S, Sinha Roy D, Bencivenga S, Roig-Villanova I, Ditengou FA, Palme K, Simon R, Colombo L (2013) Maternal control of PIN1 is required for female gametophyte development in *Arabidopsis*. PLoS ONE 8(6):e66148

Checker VG, Kushwaha HR, Kumari P, Yadav S (2018) Role of phytohormones in plant defense: signaling and cross talk. In: Singh A, Singh I (eds) Molecular aspects of plant-pathogen interaction. Springer, Singapore, pp 159–184

Cheng CY, Mathews DE, Eric Schaller G, Kieber JJ (2013) Cytokinin-dependent specification of the functional megaspore in the *Arabidopsis* female gametophyte. Plant J 73(6):929–940

Chiang YH, Zubo YO, Tapken W, Kim HJ, Lavanway AM, Howard L, Schaller GE (2012) Functional characterization of the GATA transcription factors GNC and CGA1 reveals their key role in chloroplast development, growth, and division in *Arabidopsis*. Plant Physiol 160:332–348

Corbesier L, Prinsen E, Jacqmard A, Lejeune P, Van Onckelen H, Périlleux C, Bernier G (2003) Cytokinin levels in leaves, leaf exudate and shoot apical meristem of *Arabidopsis thaliana* during floral transition. J Exp Bot 54(392):2511–2517

Cortleven A, Leuendorf JE, Frank M, Pezzetta D, Bolt S, Schmülling T (2019) Cytokinin action in response to abiotic and biotic stresses in plants. Plant Cell Environ 42:998–1018

De Rybel B, Adibi M, Breda AS, Wendrich JR, Smit ME, Novák O, Yamaguchi N, Yoshida S, Van Isterdael G, Palovaara J, Nijsse B, Boekschoten MV, Hooiveld G, Beeckman T, Wagner D, Ljung K, Fleck C, Weijers D (2014) Integration of growth and patterning during vascular tissue formation in *Arabidopsis*. Science (new York) 345(6197):1255215

Dervinis C, Frost CJ, Lawrence SD, Novak NG, Davis JM (2010) Cytokinin primes plant responses to wounding and reduces insect performance. J Plant Growth Regul 29:289–296

Dewitte W, Scofield S, Alcasabas AA, Maughan SC, Menges M, Braun N, Collins C, Nieuwland J, Prinsen E, Sundaresan V, Murray JA (2007) *Arabidopsis* CYCD3 D-type cyclins link cell proliferation and endocycles and are rate-limiting for cytokinin responses. Proc Natl Acad Sci 104(36):14537–14542

Dowd CD, Chronis D, Radakovic ZS, Siddique S, Schmülling T, Werner T, Kakimoto T, Grundler FMW, Mitchum MG (2017) Divergent expression of cytokinin biosynthesis, signaling and catabolism genes underlying differences in feeding sites induced by cyst and root-knot nematodes. Plant J 92(2):211–228

Durán-Medina Y, Díaz-Ramírez D, Marsch-Martínez N (2017) Cytokinins on the move. Front Plant Sci 8:146

Efroni I, Han SK, Kim HJ, Wu MF, Steiner E, Birnbaum KD, Wagner D (2013) Regulation of leaf maturation by chromatin-mediated modulation of cytokinin responses. Dev Cell 24:438–445

El-Showk S, Ruonala R, Helariutta Y (2013) Crossing paths: cytokinin signalling and crosstalk. Development 140(7):1373–1383

Emery RJN, Kisiala A (2020) The roles of cytokinins in plants and their response to environmental stimuli. Plants 9(9):1158

Frébort I, Kowalska M, Hluska T, Frébortová J, Galuszka P (2011) Evolution of cytokinin biosynthesis and degradation. J Exp Bot 62(8):2431–2452

Frébortová J, Frébort I (2021) Biochemical and structural aspects of cytokinin biosynthesis and degradation in bacteria. Microorganisms 9:1314

Gajdošová S, Spíchal L, Kamínek M, Hoyerová K, Novák O, Dobrev PI, Galuszka P, Klíma P, Gaudinová A, Žižková E, Hanuš J (2011) Distribution, biological activities, metabolism, and the conceivable function of cis-zeatin-type cytokinins in plants. J Exp Bot 62(8):2827–2840

Galuszka P, Frébort I, Šebela M, Sauer P, Jacobsen S, Peč P (2001) Cytokinin oxidase or dehydrogenase? Mechanism of cytokinin degradation in cereals. Eur J Biochem 268(2):450–461

Galuszka P, Popelková H, Werner T, Frébortová J, Pospíšilová H, Mik V, Köllmer I, Schmülling T, Frébort I (2007) Biochemical characterization of cytokinin oxidases/dehydrogenases from *Arabidopsis thaliana* expressed in *Nicotiana tabacum* L. J Plant Growth Regul 26(3):255–267

Gálvez A, Albacete A, Martínez-Andújar C, Del Amor FM, López-Marín J (2021) Contrasting rootstock-mediated growth and yield responses in salinized pepper plants (*Capsicum annuum*L.) are associated with changes in the hormonal balance. Int J. Mol Sci 22(7):3297

Gillissen B, Bürkle L, André B, Kühn C, Rentsch D, Brandl B, Frommer WB (2000) A new family of high-affinity transporters for adenine, cytosine, and purine derivatives in *Arabidopsis*. Plant Cell 12(2):291–300

Girke C, Daumann M, Niopek-Witz S, Möhlmann T (2014) Nucleobase and nucleoside transport and integration into plant metabolism. Front Plant Sci 5:443

Glanz-Idan N, Tarkowski P, Turečková V, Wolf S (2020) Root-shoot communication in tomato plants: cytokinin as a signal molecule modulating leaf photosynthetic activity. J Exp Bot 71(1):247–257

Grosskinsky DK, Naseem M, Abdelmohsen UR, Plickert N, Engelke T, Griebel T, Zeier J, Novak O, Strnad M, Pfeifhofer, van der Graaff E, Simon U, Roitsch T (2011) Cytokinins mediate resistance against *Pseudomonas syringae* in tobacco through increased antimicrobial phytoalexin synthesis independent of salicylic acid signaling. Plant Physiol 157:815–830

Gupta R, Elkabetz D, Leibman-Markus M, Sayas T, Schneider A, Jami E, Kleiman M, Bar M (2021) Cytokinin drives assembly of the phyllosphere microbiome and promotes disease resistance through structural and chemical cues. ISME J. https://doi.org/10.1038/s41396-021-01060-3

Gupta R, Leibman-Markus M, Pizarro L, Bar M (2020a) Cytokinin induces bacterial pathogen resistance in tomato. Plant Pathol 70(2):318–325

Gupta R, Pizarro L, Leibman-Markus M, Marash I, Bar M (2020b) Cytokinin response induces immunity and fungal pathogen resistance, and modulates trafficking of the PRR LeEIX2 in tomato. Mol Plant Pathol 21(10):1287–1306

Hai NN, Chuong NN, Tu N, Kisiala A, Hoang X, Thao NP (2020) Role and regulation of cytokinins in plant response to drought stress. Plants (basel, Switzerland) 9(4):422

Higuchi M, Pischke MS, Mähönen AP, Miyawaki K, Hashimoto Y, Seki M, Kobayashi M, Shinozaki K, Kato T, Tabata S, Helariutta Y (2004) In planta functions of the *Arabidopsis* cytokinin receptor family. Proc Natl Acad Sci 101(23):8821–8826

Hirose N, Makita N, Yamaya T, Sakakibara H (2005) Functional characterization and expression analysis of a gene, OsENT2, encoding an equilibrative nucleoside transporter in rice suggest a function in cytokinin transport. Plant Physiol 138(1):196–206

Hirose N, Takei K, Kuroha T, Kamada-Nobusada T, Hayashi H, Sakakibara H (2008) Regulation of cytokinin biosynthesis, compartmentalization and translocation. J Exp Bot 59(1):75–83

Hou B, Lim EK, Higgins GS, Bowles DJ (2004) *N*-glucosylation of cytokinins by glycosyltransferases of *Arabidopsis thaliana*. J Biol Chem 46:47822–47832

Hurny A, Cuesta C, Cavallari N, Otvos K, Duclercq J, Dokladal L, Montesinos C, Gallemí M, Semerádová H, Rauter T, Stenzel I, Persiau G, Benade F, Bhalearo R, Sýkorová E, Gorzsás A, Sechet J, Mouille G, Heilmann I, De Jaeger G, Ludwig-Müller J, Benkova E (2020) SYNERGISTIC ON AUXIN AND CYTOKININ 1 positively regulates growth and attenuates soil pathogen resistance. Nat Commun 11(1):2170

Hwang I, Sheen J (2012) Cytokinin signaling networks. Annu Rev Plant Biol 63:353–380

Hyoung S, Cho SH, Chung JH, So WM, Cui MH, Shin JS (2020) Cytokinin oxidase PpCKX1 plays regulatory roles in development and enhances dehydration and salt tolerance in *Physcomitrella patens*. Plant Cell Rep 39(3):419–430

Ishida K, Yamashino T, Yokoyama A, Mizuno T (2008) Three type-B response regulators, ARR1, ARR10 and ARR12, play essential but redundant roles in cytokinin signal transduction throughout the life cycle of *Arabidopsis thaliana*. Plant Cell Physiol 49(1):47–57

Itoh JI, Hibara KI, Kojima M, Sakakibara H, Nagato Y (2012) Rice DECUSSATE controls phyllotaxy by affecting the cytokinin signaling pathway. Plant J 72(6):869–881

Jackson D, Hake S (1999) Control of phyllotaxy in maize by the abphyl1 gene. Development 126(2):315–323

Jameson PE (2019) Virulent *Rhodococcusfasciars* produce unique methylated cytokinins. Plants 8(12):582

Jameson PE, Song J (2016) Cytokinin: a key driver of seed yield. J Exp Bot 67:593–606

Jeon J, Kim NY, Kim S, Kang NY, Novák O, Ku SJ, Cho C, Lee DJ, Lee EJ, Strnad M, Kim JA (2010) Subset of cytokinin two-component signaling system plays a role in cold temperature stress response in *Arabidopsis*. J Biol Chem 285(30):23371–23386

Jin SH, Ma XM, Kojima M, Sakakibara H, Wang YW, Hou BK (2013) Overexpression of glucosyltransferase UGT85A1 influences trans-zeatin homeostasis and trans-zeatin responses likely through *O*-glucosylation. Planta 237(4):991–999

Kamada-Nobusada T, Sakakibara H (2009) Molecular basis for cytokinin biosynthesis Phytochemistry 70(4):444–449

Kang J, Lee Y, Sakakibara H, Martinoia E (2017) Cytokinin transporters: GO and STOP in signaling. Trends Plant Sci 22(6):455–461

Kang J, Park J, Choi H, Burla B, Kretzschmar T, Lee Y, Martinoia E (2011) Plant ABC transporters. Arabidopsis Book 9:e0153

Kashyap SP, Prasanna HC, Kumari N, Mishra P, Singh B (2020) Understanding salt tolerance mechanism using transcriptome profiling and de novo assembly of wild tomato *Solanum chilense*. Sci Rep 28 10(1):1–20

Keshishian EA, Rashotte AM (2015) Plant cytokinin signalling. Essays Biochem 58:13–27

Khandal H, Gupta SK, Dwivedi V, Mandal D, Sharma NK, Vishwakarma NK, Pal L, Choudhary M, Francis A, Malakar P, Singh NP (2020) Root-specific expression of chickpea cytokinin oxidase/dehydrogenase 6 leads to enhanced root growth, drought tolerance and yield without compromising nodulation. Plant Biotechnol J 18(11):2225–2240

Kiba T, Kudo T, Kojima M, Sakakibara H (2011) Hormonal control of nitrogen acquisition: roles of auxin, abscisic acid, and cytokinin. J Exp Bot 62(4):1399–1409

Kiba T, Takei K, Kojima M, Sakakibara H (2013) Side-chain modification of cytokinins controls shoot growth in *Arabidopsis*. Dev Cell 27:452–461

Kieber JJ, Schaller GE (2014) Cytokinins. The Arabidopsis Book 12:e0168

Kieber JJ, Schaller GE (2018) Cytokinin signaling in plant development. Development 145:dev149344

Kim HJ, Ryu H, Hong SH, Woo HR, Lim PO, Lee IC, Sheen J, Nam HG, Hwang I (2006) Cytokinin-mediated control of leaf longevity by AHK3 through phosphorylation of ARR2 in *Arabidopsis*. Proc Natl Acad Sci 103(3):814–819

Kinoshita-Tsujimura K, Kakimoto T (2011) Cytokinin receptors in sporophytes are essential for male and female functions in *Arabidopsis thaliana*. Plant Sig Behav 6(1):66–71

Ko D, Kang J, Kiba T, Park J, Kojima M, Do J, Kim KY, Kwon M, Endler A, Song WY, Martinoia E (2014) *Arabidopsis* ABCG14 is essential for the root-to-shoot translocation of cytokinin. Proc Natl Acad Sci (USA) 111(19):7150–7155

Kowalska M, Galuszka P, Frébortová J, Šebela M, Béres T, Hluska T, Šmehilová M Bilyeu KD, Frébort I (2010) Vacuolar and cytosolic cytokinin dehydrogenases of *Arabidopsis thaliana*: heterologous expression, purification and properties. Phytochemistry 71(17–18):1970–1978

Kudo T, Kiba T, Sakakibara H (2010) Metabolism and long-distance translocation of cytokinins. J Integr Plant Biol 52(1):53–60

Kudo T, Makita N, Kojima M, Tokunaga H, Sakakibara H (2012) Cytokinin activity of cis-zeatin and phenotypic alterations induced by overexpression of putative cis-zeatin-O-glucosyltransferase in rice. Plant Physiol 160(1):319–331

Kuppu S, Mishra N, Hu R, Sun L, Zhu X, Shen G, Blumwald E, Payton P, Zhang H (2013) Water-deficit inducible expression of a cytokinin biosynthetic gene IPT improves drought tolerance in cotton. PLoS ONE 8(5):e64190

Kuroha T, Tokunaga H, Kojima M, Ueda N, Ishida T, Nagawa S, Fukuda H, Sugimoto K, Sakakibara H (2009) Functional analyses of LONELY GUY cytokinin-activating enzymes reveal the importance of the direct activation pathway in *Arabidopsis*. Plant Cell 10:3152–3169

Lakehal A, Dob A, Rahneshan Z, Novák O, Escamez S, Alallaq S, Strnad M, Tuominen H, Bellini C (2020) ETHYLENE RESPONSE FACTOR 115 integrates jasmonate and cytokinin signaling machineries to repress adventitious rooting in *Arabidopsis*. BioRxiv. 20:42

Lee BH, Johnston R, Yang Y, Gallavotti A, Kojima M, Travençolo BA, Costa LD, Sakakibara H, Jackson D (2009) Studies of aberrant phyllotaxy1 mutants of maize indicate complex interactions between auxin and cytokinin signaling in the shoot apical meristem. Plant Physiol 150(1):205–216

Li G, Liu K, Baldwin SA, Wang D (2003) Equilibrative nucleoside transporters of *Arabidopsis thaliana* cDNA cloning, expression pattern, and analysis of transport activities. J Biol Chem 278(37):35732–35742

Li Y, Liu F, Li P, Wang T, Zheng C, Hou B (2020) An *Arabidopsis* cytokinin-modifying glycosyltransferase UGT76C2 improves drought and salt tolerance in rice. Front Plant Sci 11:560696

Li YJ, Wang B, Dong RR, Hou BK (2015) AtUGT76C2, an *Arabidopsis* cytokinin glycosyltransferase is involved in drought stress adaptation. Plant Sci 236:157–167

Liu CJ, Zhao Y, Zhang K (2019) Cytokinin transporters: multisite players in cytokinin homeostasis and signal distribution. Front Plant Sci 10:693

Liu Y, Zhang M, Meng Z, Wang B, Chen M (2020) Research progress on the roles of cytokinin in plant response to stress. Int J Mol Sci 21(18):6574

Liu Z, Yuan L, Song X, Yu X, Sundaresan V (2017) AHP2, AHP3, and AHP5 act downstream of CKI1 in *Arabidopsis* female gametophyte development. J Exp Bot 68(13):3365–3373

Lloret PG, Casero PJ (2002) Lateral root initiation. In: Waisel Y, Eshel A, Kafkafi U (eds) Plant roots: the hidden half. Marcel Dekker Pub, New York, pp 127–156

Lomin SN, Myakushina YA, Arkhipov DV, Leonova OG, Popenko VI, Schmülling T, Romanov GA (2017) Studies of cytokinin receptor-phosphotransmitter interaction provide evidences for the initiation of cytokinin signalling in the endoplasmic reticulum. Funct Plant Biol 45(2):192–202

Lubovska Z, Dobra J, Storchova H, Wilhelmova N Vankova R (2014) Cytokinin oxidase/dehydrogenase overexpression modifies antioxidant defense against heat, drought and their combination in Nicotiana tabacum plants. J Plant Physiol 171:1625–1633

Marcel Dekker Pub, New York, pp 127–156

Maruyama-Nakashita A, Nakamura Y, Watanabe-Takahashi A, Yamaya T, Takahashi H (2004) Induction of SULTR1; 1 sulfate transporter in *Arabidopsis* roots involves protein phosphorylation/dephosphorylation circuit for transcriptional regulation. Plant Cell Physiol 45(3):340–345

Massonneau A, Houba-Hérin N, Pethe C, Madzak C, Falque M, Mercy M, Kopecny D, Majira A, Rogowsky P, Laloue M (2004) Maize cytokinin oxidase genes: differential expression and cloning of two new cDNAs. J Exp Bot 55(408):2549–2557

McIntyre KE, Bush DR, Argueso CT (2021) Cytokinin regulation of source-sink relationships in plant-pathogen interactions. Front Plant Sci 12:677585

Mi X, Wang X, Wu H, Gan L, Ding J, Li Y (2017) Characterization and expression analysis of cytokinin biosynthesis genes in *Fragaria vesca*. Plant Growth Regul 82:139–149

Miyawaki K, Matsumoto-Kitano M, Kakimoto T (2004) Expression of cytokinin biosynthetic isopentenyl transferase genes in *Arabidopsis*: tissue specificity and regulation by auxin, cytokinin, and nitrate. Plant J 37(1):128–138

Miyawaki K, Tarkowski P, Matsumoto-Kitano M, Kato T, Sato S, Tarkowska D, Tabata S, Sandberg G, Kakimoto T (2006) Roles of *Arabidopsis* ATP/ADP isopentenyl transferases and tRNA isopentenyl transferases in cytokinin biosynthesis. Proc Natl Acad Sci USA 103(44):16598–16603

Mohan TC, Castrillo G, Navarro C, Zarco-Fernández S, Ramireddy E, Mateo C, Zamarreño AM, Paz- Ares J, Muñoz R, García-Mina JM, Hernández LE (2016) Cytokinin determines thiol-mediated arsenic tolerance and accumulation. Plant Physiol 171(2):1418–1426

Muller B, Sheen J (2008) Cytokinin and auxin interaction in root stem-cell specification during early embryogenesis. Nature 453:1094–1097

Nam Y-J, Tran LSP, Kojima M, Sakakibara H, Nishiyama R, Shin R (2012) Regulatory roles of cytokinins and cytokinin signaling in response to potassium deficiency in Arabidopsis. PLoS ONE 7(10):e47797

Naseem M, Wolfling M, Dandekar T (2014) Cytokinins for immunity beyond growth, galls and green islands. Trends Plant Sci 19:481–484

Nedvěd D, Hošek P, Klíma P, Hoyerová K (2021) Differential subcellular distribution of cytokinins: how does membrane transport fit into the Big Picture? Int J Mol Sci 22(7):3428

Ngyuen HN, Chuong NN, Tu NHC, Kisiala A, Hoang XLT, Thao NP (2020) Role and regulation of cytokinins in plant response to drought stress. Plants 9:422

Nishiyama R, Watanabe Y, Fujita Y, Le DT, Kojima M, Werner T, Vankova R, Yamaguchi-Shinozaki K, Shinozaki K, Kakimoto T, Sakakibara H (2011) Analysis of cytokinin mutants and regulation of cytokinin metabolic genes reveals important regulatory roles of cytokinins in drought, salt and abscisic acid responses, and abscisic acid biosynthesis. Plant Cell 23(6):2169–2183

Ohashi-Ito K, Saegusa M, Iwamoto K, Oda Y, Katayama H, Kojima M, Sakakibara H, Fukuda H (2014) A bHLH complex activates vascular cell division via cytokinin action in root apical meristem. Current Biol 24(17):2053–2058

Ohkama N, Takei K, Sakakibara H, Hayashi H, Yoneyama T, Fujiwara T (2002) Regulation of sulfur-responsive gene expression by exogenously applied cytokinins in *Arabidopsis thaliana*. Plant Cell Physiol 43(12):1493–1501

Osugi A, Kojima M, Takebayashi Y, Ueda N, Kiba T, Sakakibara H (2017) Systemic transport of trans-zeatin and its precursor have differing roles in *Arabidopsis* shoots. Nat Plants 3(8):1–6

Pavlíková D, Pavlík M, Procházková D, Zemanová V, Hnilička F, Wilhelmová N (2014) Nitrogen metabolism and gas exchange parameters associated with zinc stress in tobacco expressing an ipt gene for cytokinin synthesis. J Plant Physiol 171(7):559–564

Pavlů J, Novák J, Koukalová V, Luklová M, Brzobohatý B, Černý M (2018) Cytokinin at the crossroads of abiotic stress signalling pathways. Int J Mol Sci 19:2450

Peleg Z, Reguera M, Tumimbang E, Walia H, Blumwald E (2011) Cytokinin-mediated source/sink modifications improve drought tolerance and increase grain yield in rice under water-stress. Plant Biotechnol J 9(7):747–758

Pernisova M, Grochova M, Konecny T, Plackova L, Harustiakova D, Kakimoto T, Heisler MG, Novak O, Hejatko J (2018) Cytokinin signalling regulates organ identity via the AHK4 receptor in *Arabidopsis*. Development (Cambridge, England) 14:dev163907

Perrot-Rechenmann C (2010) Cellular responses to auxin: division versus expansion. Cold Spring Harb Perspect Biol 2(5):a001446

Pospíšilová H, Jiskrova E, Vojta P, Mrizova K, Kokáš F, Čudejková MM, Bergougnoux V, Plíhal O, Klimešová J, Novák O, Dzurova L (2016) Transgenic barley overexpressing a cytokinin dehydrogenase gene shows greater tolerance to drought stress. New Biotechnol 33(5):692–705

Prerostova S, Dobrev PI, Kramna B, Gaudinova A, Knirsch V, Spichal L, Zatloukal M, Vankova R (2020) Heat acclimation and inhibition of cytokinin degradation positively affect heat stress tolerance of *Arabidopsis*. Front Plant Sci 11:87

Punwani JA, Hutchison CE, Schaller GE, Kieber JJ (2010) The subcellular distribution of the Arabidopsis histidine phosphotransfer proteins is independent of cytokinin signaling. Plant J 62(3):473–482

Qi Z, Xiong L (2013) Characterization of a Purine Permease family gene OsPUP7 involved in growth and development control in rice. J Int Plant Biol 55(11):1119–1135

Qin H, Gu Q, Zhang J, Sun L, Kuppu S, Zhang Y, Burow M, Payton P, Blumwald E, Zhang H (2011) Regulated expression of an isopentenyl transferase gene (IPT) in peanut significantly improves drought tolerance and increases yield under field conditions. Plant Cell Physiol 52(11):1904–1914

Qiu W, Liu M, Qiao G, Jiang J, Xie L, Zhuo R (2012) An isopentyl transferase gene driven by the stress-inducible rd29A promoter improves salinity stress tolerance in transgenic tobacco. Plant Mol Biol Rep 30(3):519–528

Radhika V, Ueda N, Tsuboi Y, Kojima M, Kikuchi J, Kudo T, Sakakibara H (2015) Methylated cytokinins from the phytopathogen*Rhodococcusfascians* mimic plant hormone activity. Plant Physiol 169:1118–1126

Randall RS, Miyashima S, Blomster T, Zhang J, Elo A, Karlberg A, Immanen J, Nieminen K, Lee JY, Kakimoto T, Blajecka K (2015) AINTEGUMENTA and the D-type cyclin CYCD3; 1 regulate root secondary growth and respond to cytokinins. Biol Open 4(10):1229–1236

Romanov GA, Lomin SN, Schmülling T (2018) Cytokinin signaling: from the ER or from the PM? That is the question! New Phytol 218(1):41–53

Romanov GA, Schmülling T (2021) opening doors for cytokinin trafficking at the ER membrane. Trends Plant Sci 26(4):305–308

Ronzhina ES (2007) Source and sink effects of cytokinins in potato (*Solanum tuberosum* L.) plants. In: International symposium on source-sink relationships in plants, vol 835, pp 39–48

Rouached H, Wirtz M, Alary R, Hell R, Arpat AB, Davidian JC, Fourcroy P, Berthomieu P (2008) Differential regulation of the expression of two high-affinity sulfate transporters, SULTR1. 1 and SULTR1. 2, in *Arabidopsis*. Plant Physiol 147(2):897–911

Rubio-Wilhelmi MM, Sanchez-Rodriguez E, Rosales MA, Begona B, Rios JJ, Romero L, Blumwald E, Ruiz JM (2011) Effect of cytokinins on oxidative stress in tobacco plants under nitrogen deficiency. Environ Exp Bot 72(2):167–173

Sakakibara H (2006) Cytokinins: Activity, biosynthesis, and translocation. Annu Rev Plant Biol 57:431–449

Sakakibara H, Takei K, Hirose N (2006) Interactions between nitrogen and cytokinin in the regulation of metabolism and development. Trends Plant Sci 11(9):440–448

Sakamoto T, Sakakibara H, Kojima M, Yamamoto Y, Nagasaki H, Inukai Y, Sato Y, Matsuoka M (2006) Ectopic expression of KNOTTED1-like homeobox protein induces expression of cytokinin biosynthesis genes in rice. Plant Physiol 142:54–62

Schafer M, Meza-Canales ID, Brutting C, Baldwin IT, Meldau S (2015) Cytokinin concentrations and CHASE-DOMAIN CONTAINING HIS KINASE 2 (NaCHK2)- and NaCHK3-mediated perception modulate herbivory-induced defense signaling and defenses in *Nicotianaattenuata*. New Phytol 207(3):645–658

Schaller GE, Bishopp A, Kieber JJ (2015) The yin-yang of hormones: cytokinin and auxin interactions in plant development. Plant Cell 27(1):44–63

Schaller GE, Street IH, Kieber JJ (2014) Cytokinin and the cell cycle. CurrOpin Plant Biol 21:7–15

Schmülling T (2013) Cytokinins. In: Lennarz, WJ, Lane, MD (eds) Encyclopedia of biological chemistry. Academic Press, pp 627–631

Schmülling T, Werner T, Riefler M, Krupková E, Manns IB (2003) Structure and function of cytokinin oxidase/dehydrogenase genes of maize, rice, *Arabidopsis* and other species. J Plant Res 116(3):241–252

Scofield S, Dewitte W, Nieuwland J, Murray JA (2013) The *Arabidopsis* homeobox gene SHOOT MERISTEMLESS has cellular and meristem-organisational roles with differential requirements for cytokinin and CYCD3 activity. Plant J 75:53–66

Séguéla M, Briat JF, Vert G, Curie C (2008) Cytokinins negatively regulate the root iron uptake machinery in *Arabidopsis* through a growth-dependent pathway. Plant J 55:289–300

Shanks CM, Rice JH, Zubo Y, Schaller GE, Hewezi T, Kieber JJ (2016) The role of cytokinin during infection of *Arabidopsis thaliana* by the cyst nematode *Heteroderaschachtii*. Mol Plant Microbe Interact 29(1):57–68

Shi X, Gupta S, Rashotte AM (2014) Characterization of two tomato AP2/ERF genes, SlCRF1 and SlCRF2 in hormone and stress responses. Plant Cell Rep 33(1):35–45

Shigenaga AM, Argueso CT (2016) No hormone to rule them all: interactions of plant hormones during the responses of plants to pathogen. Semin Cell Dev Biol 56:174–189

Shigenaga AM, Berens ML, Tsuda K, Argueso CT (2017) Towards engineering of hormonal crosstalk in plant immunity. CurrOpin Plant Biol 38:164–172

Siddique S, Radakovic ZS, Carola M, Chronis D, Novak O, Ramireddy E, Holbein J, Matera C, Hutten M, Gutbrod P, Anjam MS, Rozanska E, Habash S, Elashry A, Sobczak M, Kakimoto T, Strnad M, Schmulling T, Mitchum MG, Grundler FMW (2015) A parasitic nematode releases cytokinin that controls cell division and orchestrates feeding site formation in host plants. Proc Nat Acad Sci 112:12669–12674

Singh A, Roychoudhury A (2021) Cytokinin-mediated signalling during environmental stress in plants. In: Aftab T, Hakeem KR (eds) Plant growth regulators—signalling under stress conditions. Springer Nature, Switzerland, pp133–152

Skalák J, Vercruyssen L, Claeys H, Hradilová J, Černý M, Novák O, Plačková L, Saiz-Fernández I, Skaláková P, Coppens F, Dhondt S (2019) Multifaceted activity of cytokinin in leaf development shapes its size and structure in *Arabidopsis*. Plant J 97(5):805–824

Skalicky V, Kubes M, Napier R, Novák O (2018) Auxins and cytokinins-the role of subcellular organization on homeostasis. Int J Mol Sci 19(10):3115

Spallek T, Gan P, Kadota Y, Shirasu K (2018) Same tune, different song-cytokinins as virulence factors in plant-pathogen interactions? CurrOpin Plant Biol 44:82–87

Spíchal L, Rakova NY, Riefler M, Mizuno T, Romanov GA, Strnad M, Schmülling T (2004) Two cytokinin receptors of *Arabidopsis thaliana*, CRE1/AHK4 and AHK3, differ in their ligand specificity in a bacterial assay. Plant Cell Physiol 45(9):1299–1305

Spoustova P, Hyskova V, Muller K, Schnablova R, Ryslavs H, Cerovska N, Malbeck J, Cvikrova M, Synkova H (2015) Tobacco susceptibility to Potato virus YNTN infection is affected by grafting and endogenous cytokinin content. Plant Sci 235:25–36

Sun J, Hirose N, Wang X, Wen P, Xue L, Sakakibara H, Zuo J (2005) *Arabidopsis*SCI33/AtENT8 gene encodes a putative equilibrative nucleoside transporter that is involved in cytokinin transport in planta. J Integr Plant Biol 47(5):588–603

Takei K, Ueda N, Aoki K, Kuromori T, Hirayama T, Shinozaki K, Yamaya T, Sakakibara H (2004a) AtIPT3 is a key determinant of nitrate-dependent cytokinin biosynthesis in *Arabidopsis*. Plant Cell Physiol 45(8):1053–1062

Takei K, Yamaya T, Sakakibara H (2004b) *Arabidopsis* CYP735A1 and CYP735A2 encode cytokinin hydroxylases that catalyze the biosynthesis of trans-zeatin. J Biol Chem 279:41866–41872

Tessi TM, Brumm S, Winklbauer E, Schumacher B, Pettinari G, Lescano I, González CA, Wanke D, Maurino VG, Harter K, Desimone M (2021) *Arabidopsis* AZG2 transports cytokinins in vivo and regulates lateral root emergence. New Phytol 229(2):979–993

Thu NBA, Hoang XLT, Truc MT, Sulieman S, Thao NP, Tran LSP (2017) Cytokinin signaling in plant response to abiotic stresses. In: Pandey GK (ed) Mechanism of plant hormone signaling under stress. Wiley, Hoboken, NJ, USA, pp 71–100

To JP, Kieber JJ (2008) Cytokinin signaling: two-components and more. Trends Plant Sci 13(2):85–92

Trifunović-Momčilov M, Paunović D, Milošević S, Marković M, Jevremović S, Dragićević IČ, Subotić A (2020) Salinity stress response of non-transformed and AtCKX transgenic centaury (*Centaurium erythraea* Rafn.) shoots and roots grown in vitro. Ann Appl Biol 177(1):74–89

Tu T, Zheng S, Ren P, Meng X, Zhao J, Chen Q, Li C (2021) Coordinated cytokinin signaling and auxin biosynthesis mediates arsenate-induced root growth inhibition. Plant Physiol 185(3):1166–1181

Veselov D, Kudoyarova G, Symonyan M, Veselov S (2003) Effect of cadmium onion uptake, transpiration and cytokinin content in wheat seedlings. Bulg J Plant Physiol 29(3–4):353–359

Vitti A, Nuzzaci M, Scopa A, Tataranni G, Remans T, Vangronsveld J, Sofo A (2013) Auxin and cytokinin metabolism and root morphological modifications in *Arabidopsis thaliana* seedlings infected with *Cucumber mosaic virus* (CMV) or exposed to cadmium. Int J Mol Sci 14(4):6889–6902

Vyroubalová S, Vaclavikova K, Turecková V, Novák O, Šmehilová M, Hluska T, Ohnoutková L, Frébort I, Galuszka P (2009) Characterization of new maize genes putatively involved in cytokinin metabolism and their expression during osmotic stress in relation to cytokinin levels. Plant Physiol 151(1):433–447

Wang J, Ma XM, Kojima M, Sakakibara H, Hou BK (2011) *N*-glucosyltransferase UGT76C2 is involved in cytokinin homeostasis and cytokinin response in *Arabidopsis thaliana*. Plant Cell Physiol 52(12):2200–2213

Wang J, Ma XM, Kojima M, Sakakibara H, Hou BK (2013) Glucosyltransferase UGT76C1 finely modulates cytokinin responses via cytokinin *N*-glucosylation in *Arabidopsis thaliana*. Plant PhysiolBiochem 65:9–16

Wang M, Le Gourrierec J, Jiao F, Demotes-Mainard S, Perez-Garcia MD, Ogé L, Hamama L, Crespel L, Bertheloot J, Chen J, Grappin P (2021) Convergence and divergence of sugar and cytokinin signaling in plant development. Int J Mol Sci 22(3):1282

Wang R, Tischner R, Gutiérrez RA, Hoffman M, Xing X, Chen M, Coruzzi G, Crawford NM (2004) Genomic analysis of the nitrate response using a nitrate reductase-null mutant of *Arabidopsis*. Plant Physiol 136(1):2512–2522

Wang XM, Yi KK, Tao Y, Wang F, Wu ZC, Jiang D, Chen X, Zhu LH, Wu P (2006) Cytokinin represses phosphate-starvationresponse through increasing of intracellular phosphate level. PlantCell Environ 29:1924–1935

Wang Y, Liu H, Xin Q (2014) Genome-wide analysis and identification of cytokinin oxidase/dehydrogenase (CKX) gene family in foxtail millet (*Setariaitalica*). Crop J 2(4):244–254

Werner T, Kollmer I, Bartrina I, Holst K, Schmulling T (2006) New insights into the biology of cytokinin degradation. Plant Biol 8:371–381

Werner T, Motyka V, Strnad M, Schmülling T (2001) Regulation of plant growth by cytokinin. Proc Natl Acad Sci (USA) 98:10487–10492

Werner T, Nehnevajova E, Köllmer I, Novák O, Strnad M, Krämer U, Schmülling T (2010) Root-specific reduction of cytokinin causes enhanced root growth, drought tolerance, and leaf mineral enrichment in *Arabidopsis* and tobacco. Plant Cell 22(12):3905–3920

Werner T, Schmülling T (2009) Cytokinin action in plant development. Curr Opin Plant Biol 12(5):527–538

Woo HR, Kim HJ, Nam HG, Lim PO (2013) Plant leaf senescence and death—regulation by multiple layers of control and implications for aging in general. J Cell Sci 126:4823–4833

Wu W, Du K, Kang X, Wei H (2021) The diverse roles of cytokinins in regulating leaf development. Hortic Res 8(1):1–3

Wulfetange K, Lomin SN, Romanov GA, Stolz A, Heyl A, Schmülling T (2011) The cytokinin receptors of *Arabidopsis* are located mainly to the endoplasmic reticulum. Plant Physiol 156(4):1808–1818

Wybouw B, De Rybel B (2019) Cytokinin-a developing story. Trends Plant Sci 24(2):177–185

Xu Y, Tian J, Gianfagna T, Huang B (2009) Effects of SAG12-ipt expression on cytokinin production, growth and senescence of creeping bentgrass (*Agrostis stolonifera* L.) under heat stress. Plant Growth Regul 57(3):281–291

Y, Eshel A, Kafkafi U (eds) Plant roots: the hidden half.

Yamada H, Suzuki T, Terada K, Takei K, Ishikawa K, Miwa K, Yamashino T, Mizuno T (2001) The *Arabidopsis* AHK4 histidine kinase is a cytokinin-binding receptor that transduces cytokinin signals across the membrane. Plant Cell Physiol 42(9):1017–1023

Yin W, Xiao Y, Niu M, Meng W, Li L, Zhang X, Liu D, Zhang G, Qian Y, Sun Z, Huang R (2020) ARGONAUTE2 enhances grain length and salt tolerance by activating BIG GRAIN3 to modulate cytokinin distribution in rice. Plant Cell 32(7):2292–2306

Yuan L, Liu Z, Song X, Jernstedt J, Sundaresan V (2018) The gymnosperm ortholog of the angiosperm central cell-specification gene CKI1 provides an essential clue to endosperm origin. New Phytol 218(4):1685–1696

Yuan L, Liu Z, Song X, Johnson C, Yu X, Sundaresan V (2016) The CKI1 histidine kinase specifies the female gametic precursor of the endosperm. Dev Cell 37(1):34–46

Zalabák D, Pospíšilová H, Šmehilová M, Mrízová K, Frébort I, Galuszka P (2013) Genetic engineering of cytokinin metabolism: prospective way to improve agricultural traits of crop plants. Biotechnol Adv 31(1):97–117

Zhang K, Diederich L, John PC (2005) The cytokinin requirement for cell division in cultured *Nicotiana plumbaginifolia* cells can be satisfied by yeast Cdc25 protein tyrosine phosphatase. Implications for mechanisms of cytokinin response and plant development. Plant Physiol 137(1):308–316

Zhang K, Novak O, Wei Z, Gou M, Zhang X, Yu Y, Yang H, Cai Y, Strnad M, Liu CJ (2014) *Arabidopsis* ABCG14 protein controls the acropetal translocation of root-synthesized cytokinins. Nat Commun 5(1):1–2

Zhang W, To JP, Cheng CY, Eric Schaller G, Kieber JJ (2011) Type-A response regulators are required for proper root apical meristem function through post-transcriptional regulation of PIN auxin efflux carriers. Plant J 68(1):1–10

Zhang Z, Li F, Li D, Zhang H, Huang R (2010) Expression of ethylene response factor JERF1 in rice improves tolerance to drought. Planta 232(3):765–774

Zhao J, Ding B, Zhu E, Deng X, Zhang M, Zhang P, Wang L, Dai Y, Xiao S, Zhang C, Liu CJ (2021) Phloem unloading via the apoplastic pathway is essential for shoot distribution of root-synthesized cytokinins. Plant Physiol 186(4):2111–2123

Zhao J, Yu N, Ju M, Fan B, Zhang Y, Zhu E, Zhang M, Zhang K (2019) ABC transporter OsABCG18 controls the shootward transport of cytokinins and grain yield in rice. J Exp Bot 70(21):6277–6291

Zhao Z, Andersen SU, Ljung K, Dolezal K, Miotk A, Schultheiss SJ, Lohmann JU (2010) Hormonal control of the shoot stem-cell niche. Nature 465(7301):1089–1092

Zhou Y, Zhang Y, Wang X, Han X, An Y, Lin S, Shen C, Wen J, Liu C, Yin W, Xia X (2020) Root-specific NF-Y family transcription factor, PdNF-YB21, positively regulates root growth and drought resistance by abscisic acid-mediated indoylacetic acid transport in Populus. NewPhytol. 227(2):407–426

Zhu M, Wang Y, Lu S, Yang L, Zhuang M, Zhang Y, Lv H, Fang Z, Hou X (2021) Genome—wide identification and analysis of cytokinin dehydrogenase/oxidase (CKX) family genes in *Brassica oleracea* L. reveals their involvement in response to *Plasmodiophorabrassicae* infections. Hortic Plant J. https://doi.org/10.1016/j.hpj.2021.05.003

Zou W, Chen L, Zou J, Han H, Fei C, Lin H, Xi D (2020) Cytokinin receptor CRE1 is required for the defense response of *Nicotiana tabacum* to *Chilliveinal mottle virus*. Plant Growth Regul 90(3):545–555

Zúñiga-Mayo VM, Reyes-Olalde JI, Marsch-Martínez N, deFolter S (2014) Cytokinin treatments affect the apical-basal patterning of the *Arabidopsis gynoecium* and resemble the effects of polar auxin transport inhibition. Front Plant Sci 5:191

Zürcher E, Liu J, di Donato M, Geisler M, Müller B (2016) Plant development regulated by cytokinin sinks. Science 353(6303):1027–1030

Zürcher E, Müller B (2016) Cytokinin synthesis, signaling, and function—advances and New Insights. Int Rev Cell Mol Biol 324:1–38

Zwack PJ, Rashotte AM (2015) Interactions between cytokinin signalling and abiotic stress responses. J Exp Bot 66:4863–4871

Zwack PJ, Robinson BR, Risley MG, Rashotte AM (2013) Cytokinin response factor 6 negatively regulates leaf senescence and is induced in response to cytokinin and numerous abiotic stresses. Plant Cell Physiol 54(6):971–981

Cytokinin Signaling in Plants Under Salt Stress

Kazem Ghassemi-Golezani and Samira Samea-Andabjadid

Abstract Salt stress negatively affects plant growth by impairing biochemical and physiological processes. Appropriate modulation of cytokinin (CK) metabolism and signaling can improve salt tolerance in plants. Protection of the photosynthetic apparatus, promotion of antioxidant systems, improvement of plant growth and differentiation, and crosstalk with stress-related phytohormones are important mechanisms that may contribute to cytokinin-mediated enhancement of salt tolerance. CKs mainly trigger plant environmental stress responses through the regulation of gene expression. A two-component system is employed to transduce the cytokinin signal to the target genes. CKs are perceived by membrane-localized histidine kinase receptors. The signal is transduced through a His-Asp phosphorelay (Histidine-aspartate phosphorelays) to activate a family of transcription factors in the nucleus. CKs cause organ specific responses in plants. This hormone is a negative regulator of root growth. Root-specific overexpression of *CKX* (cytokinin oxidase/dehydrogenase) gene can enhance root growth, nutrient uptake and salt tolerance. In contrast, increasing cytokinin level (by overexpression of *IPT* genes) promotes shoot growth of salt stressed plants, by inducing the expression of genes that are involved in photosynthesis, chlorophyll levels, photochemical quenching, photochemical efficiency, electron transport rates and CO_2 assimilation. This chapter focuses on the cytokinin metabolism, transport and signaling, and discusses the role of this phytohormone in regulating changes in gene expression and physiological processes to mediate salt tolerance in plants.

1 Introduction

Salt stress is an important factor affecting plant growth and yield by influencing major biological processes, such as photosynthesis (Feng et al. 2014), energy metabolism (Song et al. 2016) and protein synthesis (Sui et al. 2018). Plant responses to salinity

K. Ghassemi-Golezani (✉) · S. Samea-Andabjadid
Department of Plant Eco-Physiology, Faculty of Agriculture, University of Tabriz, Tabriz, Iran
e-mail: golezani@gmail.com

T. Aftab (ed.), *Auxins, Cytokinins and Gibberellins Signaling in Plants*,
Signaling and Communication in Plants,
https://doi.org/10.1007/978-3-031-05427-3_8

have been divided into osmotic stress and ion toxicity, which can lead to oxidative stress, that limit plant growth and development (Liang et al. 2017). Thus, there is a notable overlap between osmotic and salt stresses in early and downstream signaling of plants.

Plants have developed diverse adaptive mechanisms such as hormonal regulation, redox change and epigenetic regulation of stress relevant genes to overcome environmental stresses. Many genes induced by salinity are differentially expressed after plants perceive the external signals of salt stress (Zhu 2001). Expressions of these genes can then regulate physiological and biochemical reactions of plants (Cui et al. 2018). So far, many stress-induced genes have been found to improve plants resistance against stress, which can be classified into four types: genes related to synthesis of osmotic regulators, such as proline biosynthesis genes of *OsP5CS1* and *OsP5CR* (Hu et al. 1992; Sripinyowanich et al. 2013); genes related to ion transportation, such as *SOS1* (Shi et al. 2000); antioxidant-linked genes such as ascorbate peroxidase, catalase and superoxide dismutase (Mhamdi et al. 2010; Verma et al. 2019); and genes regulating signaling cascades such as transcription factors (Xiong et al. 2014; Chen et al. 2015) and protein kinases (Kang et al. 2017).

Cytokinins (CKs) promote cell division and differentiation, delay leaf senescence, limit root growth, and enhance branching and nodulation (Werner and Schmulling 2009; Kieber and Schaller 2018). Variations in endogenous cytokinin in different plants are not similar under saline conditions. Thus, salt tolerance is improved either with upregulation or downregulation of CKs, depending on plant species, and the degree and duration of salt stress (Liu et al. 2020). The essential enzymes for CK metabolism are adenosine phosphate isopentenyltransferases (*IPTs*) and CK oxidases/dehydrogenases (*CKXs*) (Hirose et al. 2008; Werner and Schmulling 2009). High level of CK in tomato (*Solanum lycopersicum*) plants increased salt tolerance through overexpression of *SlIPT3*, that preserves photosynthetic pigments and maintains high K^+/Na^+ ratio (Ghanem et al. 2011). In addition, the ectopic expression of the *IPT* gene improved dehydration tolerance in transgenic maize (Leta et al. 2016), creeping bentgrass (Xu et al. 2016) and eggplant (Xiao et al. 2017) via enhancing endogenous CK level. Suppression of *CKX2* in rice resulted in a higher water content, chlorophyll content, photosynthesis rate, plant height, yield and lower oxidative damage under salt stress (Joshi et al. 2018).

Some studies have shown a negative effect of high CK on stress tolerance. It was reported that plants with low level of CK, due to the reduced synthesis or enhanced degradation, increases salinity resistance (Avalbaev et al. 2016; Ghanem et al. 2008). Overexpression of *AtIPT8* in *Arabidopsis*, with high CK content was led to a substantial reduction in the survival rate of plants under salt stress by downregulating the expression of stress-related genes, inhibiting the antioxidant defense and decreasing chlorophyll content (Wang et al. 2015). In addition, enhancing *CKX* gene expression has been shown to augment dehydration and salt tolerances of transgenic plants (Macková et al. 2013; Pospisilova et al. 2016). Overexpression of *MsCKX* improved salt tolerance of transgenic alfalfa (*Medicago sativa*) plants by keeping a high K^+/Na^+ ratio and boosting the antioxidant enzymes activity to scavenge ROS (Reactive oxygen species) (Li et al. 2019). Declined active CK level resulted from overexpression of *AtUGT76C2* (a cytokinin glycosyltransferase) has led to salt tolerance in rice plants through enhancing root growth, elevating proline and soluble sugar accumulation and ROS scavenging activity, reducing ion leakage, limiting stomatal opening and upregulating stress-responsive genes (*OsSOS1*, *OsPIP2.1*, *OsDREB2A*, *OsCOIN*, *OsABF2*, *OsRAB16*, *OsP5CR*, and *OsP5CS1*) (Li et al. 2020).

The CK signaling components display important roles in plant salt tolerance. All three receptors (AHK2, AHK3, and AHK4) have been shown to be negative regulators in dehydration and salt resistance. The *ahk2*, *ahk3* mutants, dwarfed plants with stronger root growth, exhibited high salt tolerance by overexpression of stress-related genes (Tran et al. 2007; Kang et al. 2012). The histidine phospho-transfer proteins (AHPs) also are negative regulators of dehydration stress in *Arabidopsis* (Nishiyama et al. 2013).

In addition to CK content, its distribution is an important factor affecting salinity resistance of plants. CK biosynthesis and degradation enzymes participate in stress responses, depending on their spatial and temporal expression patterns. For example, high CK content in tomato (*Solanum lycopersicum*) leaves by inducing *SlIPT3* expression, significantly enhanced plant performance under salinity by maintaining photosynthesis. However, its transcription is strongly suppressed in tomato roots (Žižková et al. 2015). These data show that cytokinin enhancement in shoots before stress occurrence, as a pre-adapted factor, stimulates the necessary morphological changes to prevent the negative effects of stress on plant physiology (Bielach et al. 2017). Due to the importance of cytokinins in salt stress, this chapter focuses on cytokinin metabolism and its role in regulating changes in gene expression and physiological processes that mediate salt tolerance in plants.

2 Cytokinin Biosynthesis and Metabolism

Natural CKs are adenine derivatives with isoprenoid side chains attached to the N^6 position of the adenine ring. Zeatin, as the most prevailing CK in plants, includes both *trans* and *cis* configurations. The *trans*-zeatin (*t*Z) is an active CK in all plant species (Gajdošová et al. 2011). At the beginning of cytokinin biosynthesis in *Arabidopsis*, a prenyl group derived from dimethylallyl diphosphate (DMAPP) is added to the N^6 position of AMP, ADP or ATP (with preferential use of ADP or ATP) that is catalyzed by an isopentenyltransferase (*IPT*) (Sakakibara 2006). The resulting product, isopentenyl adenosine 5′-phosphates (iP nucleotides), are then changed to *t*Z derivatives by *trans*-hydroxylases, the cytochrome P450 enzymes (CYP735A1 and CYP735A2) (Takei et al. 2004). Finally, LONELY GUY (LOG), a phosphoribohydrolase converts the iP-nucleotide 5′-monophosphate (iPRMP) and *t*Z- nucleotide 5′-mononophosphate (*t*ZRMP) to their active forms, iP and *t*Z, respectively (Kurakawa et al. 2007; Kuroha et al. 2009). These processes are presented in Fig. 1.

The CK content of plant tissues can also be altered via conjugation to a sugar, usually glucose, or through irreparable cleavage by cytokinin oxidases (*CKXs*) (Werner et al. 2006). Conjugated CKs are inactive and also unable to bind to CK

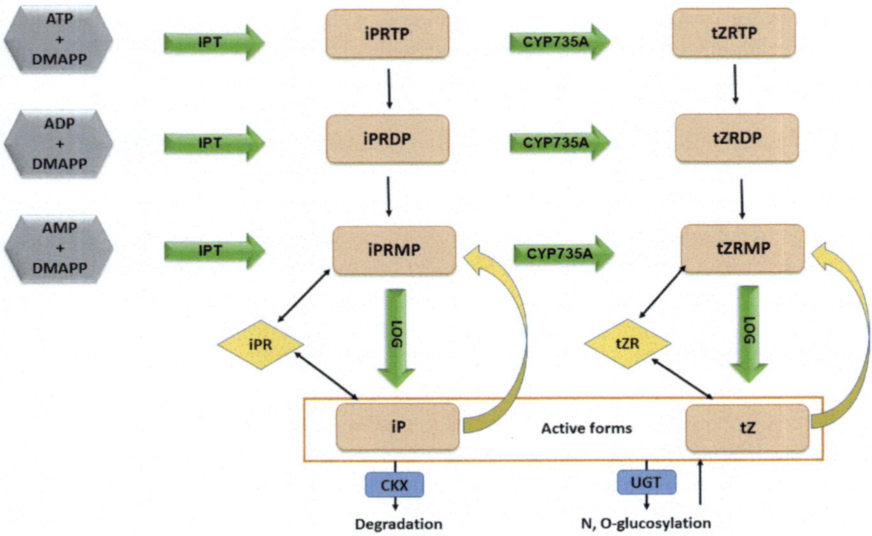

Fig. 1 Current model of iP and tZ biosynthesis and metabolic pathways in *Arabidopsis*. In *Arabidopsis*, *IPT* preferentially utilizes ATP and ADP, and *CYP735A* preferentially utilizes iPRMP and iPRDP, as substrates. LOG exclusively reacts with their monophosphate forms. Active cytokinins are degraded by *CKX*, glucosylated by UGT, or reverted to their precursors by the purine salvage pathway

receptors (Spichal et al. 2004). Both iP and *t*Z are cleaved by *CKXs*, but dihydrozeatin and the synthetic CKs such as 6-benzylaminopurine and kinetin are resistant to breakage by *CKXs* (Galuszka et al. 2007; Zalabák et al. 2014). The synthesis and degradation of CKs are regulated by endogenous developmental cues and also by biotic and abiotic factors (Werner et al. 2006).

3 Cytokinin Transport

Cytokinins, as chemical signals, are produced in both roots and shoots and are transported either in short distance among neighboring cells or in long distance between roots and shoots (Sakakibara 2006). The *t*Z-riboside and the active free-base *t*Z are mainly synthesized in roots and transported apoplastically to shoots, which increase the growth of the above-ground parts of the plants (Beveridge et al. 1997; Hirose et al. 2008). Plants modulate the ratio of *t*Z/*t*Z-riboside translocated from the root in order to regulate shoot growth in response to varying environmental conditions (Osugi et al. 2017). In contrast, the iP- and *c*Z type cytokinins are the major forms found in phloem and are translocated rootward to transmit messages from shoots to roots (Corbesier et al. 2003; Hirose et al. 2008). The shoot-produced iP-type cytokinins have been indicated to act as a signal of nitrogen satiety, regulating root architecture, suppressing nitrogen uptake in the root and modulating nodulation (Sasaki et al. 2014).

Three types of membrane transporters have been recognized for CK translocation. The purine permeases (PUPs) and equilibrative nucleoside transporters (ENTs) serve as influx transporters and carry CK nucleobases and nucleosides, respectively. The ATP-binding cassette transporter subfamily G14 (ABCG14) in *Arabidopsis*, as an efflux pump, facilitates long-distance translocation of the root-born CKs (Liu et al. 2019). ABCG14 is expressed mainly in roots and it is important for loading CK into the xylem sap for transport to the shoot. Disturbance of ABCG14 lead to a 90% reduction in CK contents in the xylem and a delay in shoot growth (Ko et al. 2014; Zhang et al. 2014).

4 Cytokinin Signaling Pathway

Cytokinins can regulate physiological responses through the regulation of gene expression. Transmission of the CK signal to the target genes takes place through a two-component system (TCS). Three groups of proteins are involved in CK signaling pathway in *Arabidopsis*: histidine kinases (AHKs), histidine-containing phosphotransfer proteins (AHPs), and type-B response regulators (type-B ARRs). In response to CKs, AHKs are auto-phosphorylated. AHPs transport the phosphoryl group to the type-B ARRs. Then, phosphorylated type-B ARRs bind to target DNA and encourage the expression of genes that respond to CKs. In addition to type-B ARRs, there are type-A ARR genes in the *Arabidopsis* genome (Kieber and Schaller 2014). Type-A ARRs, like type-B ARRs, have a receiver domain to get a phosphoryl group from AHPs, but they have no DNA-binding domain (GARP domain). Thus, type-A ARRs disrupt the CK signaling pathway by competing with type-B ARRs for phosphoryl group (Kiba et al. 2003). Existence of feedback loops in a cell or an organ specific manner is mainly important for balancing of CK signaling flux. In this regard, the expression of AHP6 in specific cells negatively affects CK response through competition with canonical AHPs, which may be inhibited by CK (Mähönen et al. 2006) (Fig. 2). Moreover, the CK receptor gene, AHK4/WOL1/CRE1, is induced by this hormone that might lead to enhanced sensitivity to CKs (Kiba et al. 2004).

5 Role of Cytokinin in Salt Stress Responses

5.1 Photosynthesis and Leaf Senescence

High NaCl destroys the structure of chloroplasts and decreases the chlorophyll content (Ma et al. 2012). In addition, salinity-induced water insufficiency and nutrient

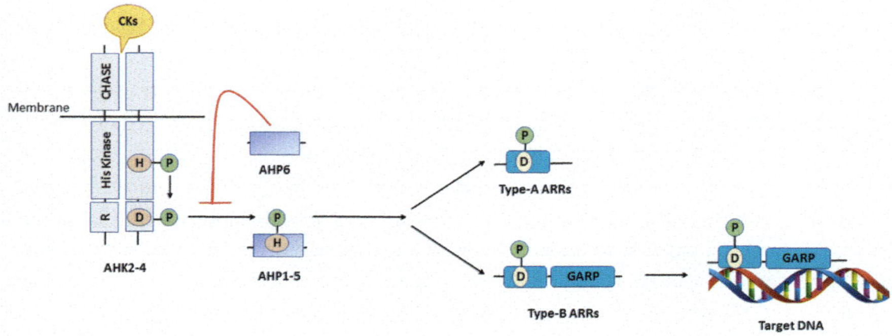

Fig. 2 Diagram of the cytokinin two component system (TCS). AHKs (AHK2, AHK3 and AHK4/WOL1/CRE1) are autophosphorylated in response to cytokinins. The phosphoryl group is transferred to type-B ARRs through AHPs. Phosphorylated type-B ARRs bind to target DNA and induce the expression of a set of genes involved in cytokinin primary response. Type-A ARRs repress cytokinin TCS signaling. Expression of AHP6, which inhibits phosphotransfer between AHKs and canonical AHPs, is repressed by cytokinin. Red T-end line indicates negative regulation

deficiency within plants (such as K^+, Mg^{2+}, Ca^{2+}, Fe^{2+} and Zn^{2+}) disrupt photosynthesis (Rahman et al. 2019). Salt stress significantly reduces leaf area, stomatal conductance, intercellular CO_2 concentration, PSII photochemical efficiency and electron transport rate, thereby impairing photosynthesis, reducing ATP and NADPH and inducing oxidative stress in salt-sensitive species (Sui et al. 2015). Additionally, salinity rapidly affects the activity and expression levels of enzymes involved in biosynthesis of chlorophyll and photosynthesis rate (Yang et al. 2020). Under salt stress, downregulation of many genes related to the photosynthesis pathway were documented, including genes involved in the photosystem complexes and electron transport chain components (Sui et al. 2015). Thus, in salt-tolerant species, increasing the expression of particular genes causes a protection by formation of photosynthetic assemblies (Yang et al. 2020).

Reduction of chlorophyll content and photosynthesis, macromolecules degradation, nutrients remobilization, cellular components destruction and finally cell death due to salinity result in leaf senescence (Krupinska et al. 2012). Leaves and flowers senescence occur by the coordinated function of numerous senescence-associated genes (SAGs) with cysteine proteases as key components (Díaz-Mendoza et al. 2014).

CKs play an important role in delaying senescence and protecting plant growth under stresses. They also activate plants post-stress restoration. Adequate cytokinin content triggers the conversion of etioplasts to chloroplasts accompanied by the greater presence of the metabolites engaged in the chlorophyll tetrapyrrole biosynthesis pathway (Yaronskaya et al. 2006) and synthesis of other components of electron transport system (Veselova et al. 2006). It has been reported that exogenous application of BAP (6-benzylaminopurine) on wheat significantly increases membrane stability index, photosynthetic pigments, chlorophyll stability index and other growth parameters under dehydration conditions (Kumari et al. 2018).

Increasing CK levels can improve photosynthesis under salt stress in many plant species by inducing overexpression of genes that are involved in synthesis of chlorophyll and electron transport chain proteins, photochemical efficiency, photochemical quenching, electron transport rates and CO_2 assimilation (Ghanem et al. 2011). Therefore, activation of RNA and protein production by CKs support plant stability under adverse environmental conditions (Chernyad'ev 2009).

It appears that CKs selectively affect the expression of certain genes, which is of special importance for stress protection. The results of a study on *pSAG12::IPT* transgenic creeping bentgrass under water deficit revealed higher content of proteins related to energy generation during photosynthesis and respiration (ribulose 1,5-bisphosphate carboxylase (RuBisCO) and glyceraldehyde phosphate dehydrogenase (GAPDH)) and proteins involved in chloroplastic elongation factor (EF-Tu), protein disulphide isomerases (PDIs) and antioxidant enzymes (Merewitz et al. 2011). Assessment of transgenic tobacco plants with high (*pSSU::IPT*) and low (*p35S:CKX1*) endogenous CK showed significant quantitative differences in stroma proteins with no qualitative change in chloroplast proteome (Cortleven et al. 2011).

CK signaling can affect photosynthesis and salinity tolerance through effects on sodium content. Given that type-A response regulators (type-A RRs) negatively regulate CK signaling, the *osrr9/osrr10* double mutants of rice are more tolerant to salinity than wild type seedlings. In the *osrr9/osrr10* mutants, high-affinity potassium transporter genes such as *OsHKT1;1*, *OsHKT1;3* and *OsHKT2;1* were overexpressed in response to salt stress, which play an important role in sodium and potassium homeostasis. Disruption of the genes *Osrr9* and *Osrr10* also influence the expression of multiple genes related to photosynthesis that delay chlorophyll degradation, and enhance electron transport rates and photon yield (Wang et al. 2019). Similarly, the *arr1* and *arr12* mutants decrease the sodium contents in the aerial parts and increase the salt stress tolerance by overexpression of *Arabidopsis* high-affinity K^+ transporter 1;1 (*AtHKT1;1*) in the roots (Mason et al. 2010). Exogenous application of benzyladenine on salt-stressed faba bean plants can induce leaf freshness via enhancing of K^+, Mg^{2+} and Ca^{2+} accumulation and reducing Na^+ content (Abdel Latef et al. 2021).

CKs were reported to trigger the synthesis of carotenoids. Carotenoids play an important role in the protection of photosynthesis against photo-oxidative damage. The higher carotenoids in transgenic *pSSU::IPT* tobacco plants with overproduction of CK indicated the CK-mediated activation of the xanthophyll cycle and subsequently its protective action against photo-oxidative damage (Cortleven and Valcke 2012).

Additional data demonstrated that reduced leaf senescence, changes in metabolism that promoted photorespiration and maintained photosynthetic function as well as sustained nitrogen assimilation are connected with the greater dehydration resistance of plants with stress-inducible *IPT* expression (Reguera et al. 2013; Rivero et al. 2010). Rivero et al. (2009) reported that CK-induced elevation in photorespiration of tobacco (*Nicotiana tabacum* cv SR1) transgenic plants expressing *PSARK::IPT* is another mechanism for protection of photosynthetic processes during water limitations. Under optimal conditions, photorespiration is considered as a negative process due to a reduction in photosynthesis and CO_2 assimilation, but it can provide RuBP to the Calvin-Benson cycle under salinity (Wingler et al. 2000). The metabolites derived from photorespiration may be used by other biosynthetic pathways (Noctor et al. 2002). For example, serine and glycine can be used for the synthesis of glutathione, that protects plants from oxidative damage (Foyer and Noctor 2000).

5.2 *Antioxidant Capacity*

Salt stress causes the accumulation of ROS, which could greatly damage to plants by attacking membrane structure (Wakeel et al. 2020). Enzymatic (superoxide dismutase: SOD, ascorbate peroxidase: APX, catalase: CAT and glutathione peroxidase: GPX) and non- enzymatic (ascorbate: AsA and glutathione: GSH) antioxidants are employed in plants to prevent ROS damages (Sui et al. 2015). Gene expression level or activity of these antioxidant components typically increases during short-term stress periods, whereas longer durations of salt stress may decrease antioxidant effectiveness or efficacy (Arghavani et al. 2012).

The ROS homeostasis is differently affected by CKs, depending on types of plant and stress. In transgenic *pSSU::IPT* tobacco (*Nicotiana tabacum*) plants overproduction of cytokinin promoted peroxidases and other antioxidant enzymes such as superoxide dismutase, catalase and glutathione-S-transferase, which play an important role in the detoxification of radicals (Cortleven and Valcke 2012). In another research, spraying INCYDE (cytokinin degradation inhibitor) on tomato under salt stress improved plant salt tolerance by increasing the activity of antioxidant enzymes (Aremu et al. 2014). Samea-Andabjadid et al. (2018) found that increasing CAT, APX and SOD activities due to benzyl-aminopurine application under salinity can scavenge ROS and protect cell membrane in faba bean (*Vicia faba*) plants. In *Solanum lycopersicum*, application of kinetin elevated the activities of antioxidant enzymes and enhanced the contents of AsA and glutathione (GSH and GSSG), thereby reducing ROS and enhancing membrane integrity. Declining lipid peroxidation in kinetin-supplemented plants results in the maintenance of cellular functioning, greater photoprotection and mineral uptake, which protect photosynthetic pigments. On the other hand, the greater PSII activity in kinetin-treated plants may prevent the formation of singlet oxygen, that protects the chloroplast structure from the oxidative damage. Regulation of $NADP^+$/NADPH ratio by kinetin application prohibits the flow of electrons to molecular oxygen, that eventually restricts the generation of superoxide radical and protects the photosynthetic electron transport chain of eggplant (*Solanum melongena* L.) seedlings (Ahanger et al. 2018). Cortleven et al. (2014) have demonstrated that *Arabidopsis* mutants with low CK production display greater intensity of stress-induced photodamage by exhibiting a reduction in D1 protein repair, which is associated with reduced synthesis of ascorbate and glutathione. In kinetin-treated rice plants, higher concentrations of GSH (glutathione) have largely contributed to the maintenance of the glyoxylase system for methylglyoxal (a highly reactive dicarbonyl compound) scavenging. Actually, enhanced activity of glyoxylase I and II (important enzymatic components of glyoxylase system) protects the electron transport system by inhibiting injury to chloroplast and mitochondrial ultrastructures (Gupta et al. 2017).

Induction of phenols and flavonoids synthesis in the CK treated plants is improved the antioxidant system. Accumulation of phenols due to external application of kinetin may contribute to reinforcing of cell wall structures and inhibition of oxidative damage to membrane lipids and proteins by modifying their peroxidation kinetics (Ahanger et al. 2018). It has also been reported that more phenolics production adjusts plant developmental processes such as lignin and pigment biosynthesis, thus providing structural integrity and protecting plants (Bhattacharya et al. 2010).

A number of other substances such as proline and dehydrin proteins may act as antioxidants under adverse conditions (Zhang et al. 2018; Sharma et al. 2019). Increased proline levels have been shown to play an important role in increasing antioxidant responses (Bhagyawant et al. 2019). The ROS levels and lipid peroxidation are also reduced in transgenic plants engineered for excessive accumulation of proline by overexpression of the proline biosynthesis gene (Guan et al. 2018). In addition to direct ROS scavenging, proline can protect and stabilize antioxidant enzymes (Szabados and Savouré 2010). The dehydrins are also known to be effective in ROS detoxification. They are also able to constrain lipid peroxidation (Hanin et al. 2011; Zhang et al. 2018). In transgenic rice plants, overexpression of dehydrins displays low ROS accumulation and malondialdehyde (MDA) content under salt and drought stresses (Kumar et al. 2014). CK has been shown to be involved in slight accumulation of ROS in wheat seedlings, causing an acclimation response through the activation of antioxidant enzymes (superoxide dismutase and peroxidase) and accumulation of defense compounds such as proline and dehydrin proteins (Avalbaev et al. 2020).

In contrary, some reports have shown that CK overproduction could enhance salt sensitivity in *Arabidopsis*. The CK-deficient mutants by low synthesis (*ipt1,3,5,7* and *35 s:CKXs*) and signaling (*AHKs* and *ARRs*) in *Arabidopsis* showed stronger salt-resistant phenotypes, as confirmed by lower electrolyte leakage and higher leaf relative water content and survival rates (Mason et al. 2010; Nishiyama et al. 2011; Tran et al. 2007). In another study, overproduction of endogenous CK by overexpression of *AtIPT8* led to a decrement in antioxidants activities and an increment in ROS contents, thereby enhancing salt sensitivity (Wang et al. 2015). According to Nishiyama et al. (2012) in CK-deficient *Arabidopsis* mutant, the genes involving in ROS depletion are significantly affected. Some studies suggest that CK signaling in abiotic stresses result in dysfunction of photosynthesis and ROS production by affecting the expression of PSII subunits genes (Yi et al. 2008; Kobayashi et al. 2012). Expression of the PSII subunits including PSAN, PSAK, PSBP and PSBQ with important role in oxygen evolution, are down-regulated by CK overproduction (Wang et al. 2015). Altering the function of chlorophyll-binding proteins affects the ABA and dehydration sensitivity in plants (Xu et al. 2012). Similarly, overexpression of *MsCKX* increases the salt tolerance of transgenic alfalfa plants by maintaining a high K^+/Na^+ ratio and increasing the activity of antioxidant enzymes (Li et al. 2019).

5.3 Organ-Specific Responses

Under saline conditions, plants adjust the distribution pattern of internal CK to increase adaptation to stress (Yin et al. 2020). CK is one of the imperative factors involved in regulating the architecture of the root system. Increasing CK can considerably reduce root growth and root/shoot ratio (Ghassemi-Golezani and Samea-Andabjadid, 2022), while decreasing CK levels or signaling could generate enlarged root system (Heyl et al. 2008). Overproduction of CK-degradation enzymes (*CKXs*) has been reported to reduce CK content in roots, which in turn increases root biomass and modifies root morphology and improves salt and drought resistance in plants (Werner et al. 2001; Mackova´ et al. 2013). The CK reduction in different compartments affects differently the root and aerial parts. The *CKX* isoenzymes differ in location within subcellular partitions (Schmulling et al., 2003; Werner et al., 2003) and in their time pattern of expression (Mrızova et al., 2013). *Arabidopsis* plants with overproduced vacuolar *AtCKX1*, apoplastic *AtCKX2* and cytosolic *AtCKX7* have diverse phenotypes, but with similar CK content. High expression of cytosolic *AtCKX7* in *Arabidopsis* has an adverse impact on the primary root development, and the root system is shaped only by a proliferation of adventitious roots (Kollmer et al. 2014). Conversely, *AtCKX* overexpression with other than cytosolic localization has enhanced root elongation and lateral branching (Werner et al. 2003). In another case, research on transgenic barley (*Hordeum vulgare*) plants with overexpression of *AtCKX1* gene in different subcellular compartments of roots under the control of the weak root-specific β-glucosidase promoter from maize (*Zea mays*) revealed that, even though cytosolic and vacuolar *AtCKX1* had a little impression on shoot growth, exudation of the *AtCKX1* protein into the apoplast showed a negative effect on the development of the aerial part and yield. In *Arabidopsis* plants with an overproduction of *AtCKX3* in the roots showed up to 40% increase in root mass without any adverse effect on inflorescence, fertility and seed formation. Furthermore, these plants demonstrated greater tolerance to stress (Werner et al. 2010). On the contrary, increased expression of *OsCKX4* under the ubiquitous promoter control in rice (*Oryza sativa*) led to the creation of a strong root system with large number of crown roots and low plant height and yield (Gao et al. 2014). Roots with a higher number of longer lateral roots are useful in two types of soil, a type in which root penetration is difficult to reach available water at depth and another type water is not available at depth and only a shallow layer of soil is subjected to seasonal wetting (Blum 2010).

5.4 Water Balance Regulation

Evidently, the plants with low CK levels or poor CK signaling usually have a higher relative water content (RWC) under stress than plants with higher CK content or stronger signaling (Nishiyama et al. 2011; Vojta et al. 2016). This could be due to the damaging effect of CK on root growth and lateral root formation (Ramireddy et al. 2018). In addition, reducing stomatal opening and consequently reducing transpiration rates in plants with low CK could protect them via reducing water losses under stressful conditions (Liao et al. 2017).

Accumulation of osmolytes such as soluble sugars, ammonium compounds and amino acids provides an adaptation under adverse environmental conditions such as salt stress (Samea-Andabjadid et al. 2018). For instance, increasing soluble sugars in faba bean roots and leaves under salinity and 6-benzylaminopurine and salicylic acid treatments was an important mechanism for osmotic adjustment and salt tolerance (Fig. 3).

Supporting the structure of membranes as a substitute for water (Mundree et al. 2002) and also maintaining water homeostasis among different parts of the cell

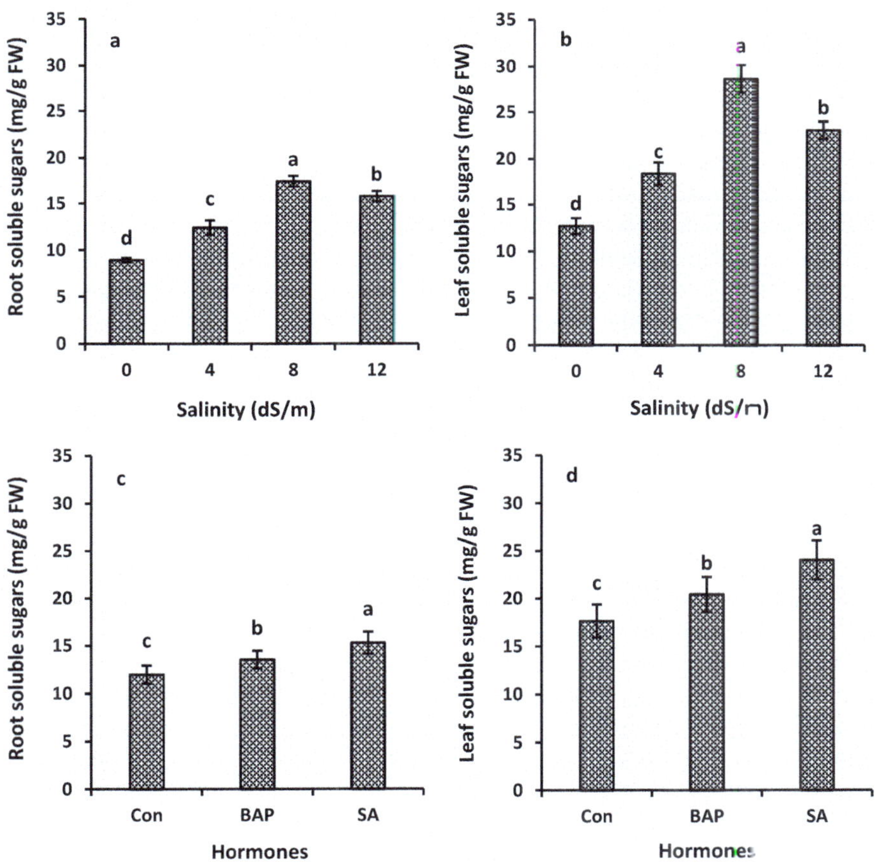

Fig. 3 Changes in faba bean root and leaf soluble sugars affected by salinity (**a** and **b**) and hormonal treatments (**c** and **d**). Con: control, BAP: 6-benzylaminopurine, SA: salicylic acid. Different letters indicate significant difference at $P \leq 0.05$

are the key functions of soluble sugars (Sidari et al. 2008). High soluble sugars in roots of plants enhance the ability of roots for water absorption from a water-stressed bed by increasing osmotic pressure (Hajiboland et al. 2014). Harinasut et al. (2000) have reported a large increase of proline in salt-stressed mulberry (*Morus alba*), which was probably related to up-regulation of proline-synthesizing enzymes, down-regulation of the catabolizing enzymes or reduction of its incorporation into proteins (Iqbal et al. 2015). In addition, salt-treated tomato plants promoted the accumulation of proline and glycine betaine and further increment occurred due to exogenous application of kinetin, which helped plants to prevent the salinity-induced decline in tissue RWC (Ahanger et al. 2018). Actually, CK increases the synthesis of proline and glycine betaine under both normal and saline conditions, indicating its protective role through osmoregulation. Proline and glycine betaine maintain plant water balance, thus minimize the deleterious effects of stress on metabolism, especially by protecting protein turnover, enzyme activities and the expression of protective proteins in plants (Ahanger et al. 2017). As a result, increasing the tissue water content by CK leads to maintaining cell wall extensibility and cell division, which increases growth and biomass production (Ahanger et al. 2018). CK affects the relative water content by regulating ionic balance. Exogenous application of kinetin not only averts Na^+ toxicity by decreasing its uptake, but it also causes a substantial increase in K^+ uptake, which is reflected in high ratio of K^+/Na^+ (Ahanger et al. 2018). According to Munns and Tester (2008), the K^+ ion is the main osmoticum in leaves and an adequate content of K^+ can facilitate osmotic adjustment that plays an important role in preserving cell turgor under saline conditions. It has been showed that, exogenous application of BAP improves the osmotic adjustment by significant accumulation of K^+ and enhancement of K^+/Na^+ ratio in leaves of faba bean plants under saline conditions (Samea-Andabjadid et al. 2018).

5.5 Cambial Activity

Vascular tissues, including the xylem and phloem, are responsible for transporting water and nutrients all over the plant. These tissues are constantly generated from stable population of stem cells, specifically the procambium during primary growth and the cambium during secondary growth (Campbell and Turner 2017). Cambial activity responds to the developmental signs of primary meristem and the ever-changing environment (Wang 2019). Salt stress as an adverse environmental factor affect the cambial activity. Hence, anatomical and chemical changes were detected in the xylem originating from salinity affected meristem (Lautner 2013). Moreover, the salt-exposed poplars exhibited a low number of cell layers in the cambial zone and a reduced growth compared to control plants (Escalante-Pérez et al. 2009). It has been shown that under salt stress the vessel lumina of poplars are reduced (Junghans et al. 2006) and the vessel frequency is increased (Janz et al. 2012). The extent of these anatomical changes depends on the sensitivity of the species to salinity. For example, the salt-sensitive poplar species *P. x canescens* responded extremely to moderate salinity, while the salt-tolerant *P. euphratica* showed reduced cambial activity and vessel lumina after long-term exposure to higher salinity levels (150 mM NaCl) (Junghans et al. 2006; Janz et al. 2012). In addition, nutrient deficiency, especially lower contents of calcium and potassium decreased xylem radial growth under saline conditions (Escalante-Pérez et al. 2009). In wheat plants, salt stress showed negative effects on conductive tissues of flag leaf, where reduction in the phloem area led to low translocation of photo-assimilates to the developing grains (Aldesuquy and Mickky 2014).

CKs play an important role in regulating cambium activity. In the primary root, disturbance of CK signaling or reduction of CK content by overexpressing *CKX* genes restrain periclinal cell division of pro-cambium cells and decrease the vasculature size (Nieminen et al. 2008; Werner et al. 2003). In addition, by interacting with auxins, CKs promote vascular differentiation and increase the phloem/xylem ratio (Aloni 1993). The positive role of CK on phloem regeneration was established in trees with high expression of CK. According to Chen et al. (2019) auxin alone stimulates both phloem and cambium formation, while CK only promotes phloem formation and hinders cambium regeneration, possibly by preventing auxin redistribution and signaling. On the other hand, auxin may induce phloem repatterning by affecting the CK signaling pathway through inducing *Populus response regulator 7* (PtRR7). Improving CK signaling or biosynthesis by increasing *AtCKI1* or *AtIPT7* expression led to the regeneration of phloem without any hormone treatment. In contrast, when the CK level was decreased in *AtCKX2*-overexpressed trees, a higher content of CK was essential for phloem recovery.

CK is an important negative regulator inhibiting xylem development in root vascular tissues. Jang and Choi (2018) found that exogenous application of CK suppresses xylem formation. Interestingly, it has been reported that CK induces the expression of auxin efflux carriers PIN3 and PIN7, which transfer auxin laterally into the xylem area. Likewise, auxin in the protoxylem positions induces *AHP6* as a negative regulator of CK signaling (Bishopp et al. 2011), that reduces CK response and limits periclinal cell divisions in the xylem axis. In the central xylem axis, auxin is involved in the promotion of HD-ZIP III transcription (Ursache et al. 2014), which probably contributes to the repression of CK signaling through the prevention of B-type response regulators (Sebastian et al. 2015). Seed priming with kinetin enhances phloem thickness in both leaf and peduncle of the main shoot and subsequently induces a fast rate of translocation of photo-assimilates from flag leaf to developing seeds in spikes and consequently increases the productivity of wheat plants irrigated by seawater (Aldesuquy and Mickky 2014).

5.6 Cytokinin Crosstalk with Stress-Related Phytohormones

5.6.1 Cytokinin-Abscisic Acid Crosstalk

The abscisic acid (ABA) accumulates rapidly in plants in response to osmotic stress and plays an important role in plant tolerance by regulating various processes such as stomatal closure, root growth and protective metabolites production (Tuteja 2007; Gomez-Cadenas et al. 2015). CK and ABA have also been shown to exert antagonistic activities during growth and physiological processes, including plant adaptation to stressful conditions (Huang et al. 2017a). Unlike CK, which delays stomatal closure as well as leaf senescence, ABA accumulation under stress helps plants to avoid stress by promoting stomatal closure to minimize water loss, accelerating leaf senescence, reducing plant growth and inducing protective substances biosyntheses (Pospisilova et al. 2005). *Arabidopsis* mutants with CK deficiency (*CKX* overexpressing or *ipt1,3,5,7* quadruple knockout) showed high sensitivity to ABA in seed germination and ABA-related gene expression (Nishiyama et al. 2011). High expression of the isopentenyl transferase gene under the control of a heat-shock-inducible promoter (*HSP70::ipt*) in tomato plants almost doubled the concentration of bioactive CKs in the xylem sap and reduced ABA by 30% under 100 mM NaCl stress (Ghanem et al. 2011). The *MsCKX* expression in leaves and particularly in roots of alfalfa (*Medicago sativa*) was significantly induced under salt stress and ABA treatment, indicating that *MsCKX* may function as a positive regulator in response to salinity and participated in ABA signaling pathway in alfalfa (Li et al. 2019). Furthermore, the CK receptor AHK2, AHK3 and AHK4 are negatively involved in ABA and osmotic stress signaling (Tran et al. 2010). However, AHK1 has been shown to act as a positive regulator in abscisic acid (ABA) and osmotic stress signaling (Wohlbach et al. 2008). CK signaling also reverses the inhibition of cotyledon greening induced by ABA through elevating the degradation of ABI5, a transcription factor that regulates ABA-induced genes (Guan et al. 2014). Interestingly, Huang et al. (2018) found that type-B ARRs, as positive regulators of CK signaling, act as inhibitors of ABA-induced SnRK2 kinases activity, which are central and positive regulators of the ABA signaling pathway. Conversely, the type-A protein ARR5, a negative regulator of CK signaling is phosphorylated by SnRK2s to stimulate ABA-responsive genes. Thus, the SnRK2-ARR regulatory unit as a signaling center balances growth and defense in response to environmental cues.

5.6.2 Cytokinin-Ethylene Crosstalk

Numerous evidences have shown the role of ethylene as an important regulator of salinity tolerance in plants. This gaseous plant hormone controls many major cellular processes from seed germination to photosynthesis for sustaining the plants growth and yield under salt stress. Ethylene maintains homeostasis of Na^+/K^+, nutrients, and ROS through inducing antioxidant defense, thereby modulating salt stress responses

(Riyazuddin et al. 2020). CK and ethylene often show antagonistic effects on the shoot, so that CK is associated with greening and cell proliferation, and ethylene is associated with ripening, senescence and the inhibition of cell proliferation (Hwang et al. 2012; Rai et al. 2015). On the other hand, these hormones function cooperatively in the regulation of root growth, where both hormones attend to inhibit root growth through impacts on cell proliferation and elongation (Qin et al. 2019; Zdarska et al. 2019). CK manages hormone crosstalk through action of the type-B ARRs. The binding sites of type-B ARRs are associated with genes involved in ethylene biosynthesis and signal transduction. Actually, ACS2 (1-aminocyclopropane-1-carboxylic acid synthases) expression is upregulated by CK in a type-B ARR-dependent manner (Zdarska et al. 2015). Ethylene production facilitates the ability of CK to prevent hypocotyl elongation in dark-grown seedlings as well as to prevent root growth (Hansen et al. 2009).

5.6.3 Cytokinin-Jasmonic Acid Crosstalk

Jasmonic acid (JA) is an important regulator of plant growth in response to stress (Huang et al. 2017b). Some studies have suggested that JA interacts antagonistically with CK in various aspects of plant development. For instance, JA inhibits CK-induced callus growth in soybean (*Glycine max*) plants (Ueda and Kato 1982). In addition, JA and CK differently adjust the expression of genes involved in chlorophyll development (Mukherjee et al. 2002; Liu et al. 2016). Moreover, an antagonistic effect has been revealed between CK and JA in xylem development of *Arabidopsis* roots. In fact, JA suppresses the procambium-specific CK response and the effect of JA on extra xylem formation is neutralized by CK (Jang et al. 2017). For example, *MYC2* mutant did not form extra xylem in response to exogenous JA. The *MYC2* negatively regulates CK response by high expression of *AHP6*, a CK signaling inhibitor (Jang et al. 2020). There are several reports confirming the negative effects of CK and the positive effects of JA on xylem differentiation. Exogenous application of CK prevented xylem development, and the *wooden leg* mutants with defects in CK signaling powerfully indicated an all-xylem phenotype and lack of procambial cells in their roots. In addition, mutants with no type-B *ARRs* transcription such as *ARR1*, *ARR10*, and *ARR12*, and also transgenic plants with overexpression of *AHP6*, a negative regulator of CK signaling, created further xylem (Yokoyama et al. 2007). Unlike JA signaling mutants, treatment of JA-deficient *OPDA reductase 3* (*opr3*) mutants by exogenous JA resulted in an extra xylem phenotype (Jang et al. 2017). It is likely that an antagonistic interaction between JA and CK is also involved in the regulation of JA-dependent stress responses. Given that the CK levels is affected by stress, regulation of JA and CK metabolism may also be involved in the JA-CK interaction (Pavlu et al. 2018).

6 Conclusions and Future Perspectives

The climate change, water deficit and human activities increased soil salinity and decreased cultivable lands. Salt stress is a serious threat to the growth and productivity of plants. This stress inhibits plants from achieving their genetic potential, that decreases yields and endangers food security. Generally, cytokinins metabolism and signaling play important roles in salt stress tolerance. The manipulation of these processes in crops can be useful for sustainable plant production. A review of the roles of cytokinin receptors and signaling proteins will help to understand the mechanisms involved in cytokinin induction of salt tolerance. Recent studies have mostly focused on transcriptomic, proteomic and metabolomic variations in various plant species with regulated cytokinin levels. However, additional detailed analysis is required to approve the significance of identified candidate genes/proteins and verify their roles in salt stress tolerance.

References

Abdel Latef AAH, Akter A, Tahjib-Ul-Arif M (2021) Foliar application of auxin or cytokinin can confer salinity stress tolerance in (*Vicia faba* L.). Agronomy 11, 790

Ahanger MA, Alyemeni MN, Wijaya L, Alamri SA, Alam P, Ashraf M et al (2018) Potential of exogenously sourced kinetin in protecting *Solanum lycopersicum* from NaCl-induced oxidative stress through up-regulation of the antioxidant system, ascorbate-glutathione cycle and glyoxalase system. PLoS ONE 13(9):e0202175

Ahanger MA, Tomar NS, Tittal M, Argal S, Agarwal RM (2017) Plant growth under water/salt stress: ROS production; antioxidants and significance of added potassium under such conditions. Physiol Mol Biol Plants 23:731–744

Aldesuquy HS, Mickky BM (2014) Interactive effects of kinetin and spermine on anatomical adaptations and productivity to seawater salinity in wheat. Int J Bioassays 3(11):3499–3508

Aloni R (1993) The role of cytokinin in organized differentiation of vascular tissues. Funct Plant Biol 20:601–608

Aremu AO, Masondo NA, Sunmonu TO, Kulkarni MG, Zatloukal M, Spichal L (2014) A novel inhibitor of cytokinin degradation (INCYDE) influences the biochemical parameters and photosynthetic apparatus in NaCl-stressed tomato plants. Planta 240:877–889

Arghavani M, Kafi M, Babalar M, Naderi R, Hoque MA, Murata Y (2012) Improvement of salt tolerance in kentucky bluegrass by trinexapac-ethyl. HortScience 47:1163–1170

Avalbaev A, Allagulova C, Maslennikova D, Fedorova K, Shakirova F (2020) Methyl jasmonate and cytokinin mitigate the salinity-induced oxidative injury in wheat seedlings. J Plant Growth Regul 40(4):1741–1752

Avalbaev A, Yuldashev RA, Fedorova KA, Somov K, Vysotskaya LB, Allagulova C et al (2016) Exogenous methyl jasmonate regulates cytokinin content by modulating cytokinin oxidase activity in wheat seedlings under salinity. J Plant Physiol 191:101–110

Beveridge CA, Murfet IC, Kerhoas L, Sotta B, Miginiac E, Rameau C (1997) The shoot controls zeatin riboside export from pea roots. Evidence from the branching mutant rms4. Plant J 11:339–345

Bhagyawant SS, Narvekar DT, Gupta N, Bhadkaria A, Koul KK, Srivastava N (2019) Variations in the antioxidant and free radical scavenging under induced heavy metal stress expressed as proline content in chickpea. Physiol Mol Biol Plants 25:683–696

Bhattacharya A, Sood P, Citovsky V (2010) The roles of plant phenolics in defense and communication during *Agrobacterium* and *Rhizobium* infection. Mol Plant Pathol 11:705–719
Bielach A, Hrtyan M, Tognetti VB (2017) Plants under stress: involvement of auxin and cytokinin. Int J Mol Sci 18:1427
Bishopp A, Help H, El- S, Weijers D, Scheres B, Friml J et al (2011) A mutually inhibitory interaction between auxin and cytokinin specifies vascular pattern in roots. Curr Biol 21:917–926
Blum A (2010) Drought resistance and its improvement. In: Blum A (eds) Plant breeding for water-limited environments. Springer, New York, USA, pp 53–152
Campbell L, Turner S (2017) Regulation of vascular cell division. J Exp Bot 68(1):27–43
Chen JJ, Wang LY, Immanen J, Nieminen K, Spicer R, Helariutta Y et al (2019) New Phytol 224:188–201
Chen M, Zhao Y, Zhuo C, Lu S, Guo Z (2015) Overexpression of a NFYC transcription factor from bermudagrass confers tolerance to drought and salinity in transgenic rice. Plant Biotechnol J 13:482–491
Chernyad'ev (2009) The protective action of cytokinins on the photosynthetic machinery and productivity of plants under stress. Appl Biochem Microbiol 45(4):351–362
Corbesier L, Prinsen E, Jacqmard A, Lejeune P, Van Onckelen H, Perilleux C et al (2003) Cytokinin levels in leaves, leaf exudate and shoot apical meristem of *Arabidopsis thaliana* during floral transition. J Exp Bot 54:2511–2517
Cortleven A, Valcke R (2012) Evaluation of the photosynthetic activity in transgenic tobacco plants with altered endogenous cytokinin content: lessons from cytokinin. Physiol Plant 144:394–408
Cortleven A, Nitschke S, Klaumunzer M, AbdElgawad H, Asard H, Grimm B et al (2014) A novel protective function for cytokinin in the light stress response is mediated by the ARABIDOPSIS HISTIDINE KINASE2 and ARABIDOPSIS HISTIDINE KINASE3 receptors. Plant Physiol 164:1470–1483
Cortleven A, Noben JP, Valcke R (2011) Analysis of the photosynthetic apparatus in transgenic tobacco plants with altered endogenous cytokinin content: a proteomic study. Proteome Sci 9:1
Cui F, Sui N, Duan G, Liu Y, Han Y, Liu S et al (2018) Identification of metabolites and transcripts involved in salt stress and recovery in peanut. Front Plant Sci 9:217
Díaz-Mendoza M, Velasco-Arroyo B, González P, Martínez M, Díaz I (2014) C1A cysteine protease–cystatin interactions in leaf senescence. J Exp Bot 65:3825–3833
Escalante-Pérez M, Lautner S, Nehls U, Selle A, Teuber M, Schnitzler JP et al (2009) Salt stress affects xylem differentiation of grey poplar (*Populus* x *canescens*). Planta 229:299–309
Feng ZT, Deng YQ, Fan H, Sun QJ, Sui N, Wang BS (2014) Effects of NaCl stress on the growth and photosynthetic characteristics of *Ulmus pumila* L. seedlings in sand culture. Photosynthetica 52:313–320
Foyer CH, Noctor G (2000) Oxygen processing in photosynthesis: regulation and signalling. New Phytol 146:359–388
Gajdošová S, Spıchal L, Kamınek M, Hoyerova K, Novák O, Dobrev PI et al (2011) Distribution, biological activities, metabolism, and the conceivable function of cis-zeatin-type cytokinins in plants. J Exp Bot 62:2827–2840
Galuszka P, Popelkova H, Werner T, Frebortova J, Pospisilova H, Mik V et al (2007) Biochemical characterization of cytokinin oxidases/dehydrogenases from *Arabidopsis thaliana* expressed in *Nicotiana tabacum* L. J Plant Growth Regul 26:255–267
Gao S, Fang J, Xu F, Wang W, Sun X, Chu J et al (2014) A cytokinin oxidase/dehydrogenase gene *OsCKX4* integrates cytokinin and auxin signaling to control rice crown root formation. Plant Physiol 165:1035–1046
Ghanem ME, Albacete A, Martinezandujar C, Acosta M, Romeroaranda MR, Dodd IC et al (2008) Hormonal changes during salinity-induced leaf senescence in tomato (*Solanum lycopersicum* L.). J Exp Bot 59:3039–3050
Ghanem ME, Albacete A, Smigocki AC, Frebort I, Pospısilova H, Martınez C et al (2011) Root-synthesized cytokinins improve shoot growth and fruit yield in salinized tomato (*Solanum lycopersicum* L.). J Exp Bot 62:125–140

Ghassemi-Golezani K, Samea-Andabjadid S (2022) Exogenous cytokinin and salicylic acid improve amino acid content and composition of faba bean seeds under salt stress. Gesunde Pflanzen. https://doi.org/10.1007/s10343-022-00673-8

Gomez-Cadenas A, Vives V, Zandalinas SI, Manzi M, SanchezPerez AM, Perez- et al (2015) Abscisic acid: a versatile phytohormone in plant signaling and beyond. Curr Protein Pept Sci 16:413–434

Guan C, Huang YH, Cui X, Liu SJ, Zhou YZ, Zhang YW (2018) Overexpression of gene encoding the key enzyme involved in proline biosynthesis (PuP5CS) to improve salt tolerance in switchgrass (*Panicum virgatum* L.). Plant Cell Rep 37:1187–1199

Guan C, Wang X, Feng J, Hong S, Liang Y, Ren B et al (2014) Cytokinin antagonizes abscisic acid-mediated inhibition of cotyledon greening by promoting the degradation of abscisic acid insensitive 5 protein in *Arabidopsis*. Plant Physiol 164:1515–1526

Gupta BK, Sahoo KK, Ghosh A, Tripathi AK, Anwar K, Das P et al (2017) Manipulation of glyoxalase pathway confers tolerance to multiple stresses in rice. Plant, Cell Environ 41:1186–1200

Hajiboland R, Radpour E, Pasbani B (2014) Effect of phosphorus deficiency on drought stress tolerance in two tomato (*Solanum lycopersum* L.) cultivars. Iran J Biol 27:788–803

Hanin M, Brini F, Ebel C, Toda Y, Takeda S, Masmoudi K (2011) Plant dehydrins and stress tolerance: versatile proteins for complex mechanisms. Plant Signal Behav 6:1503–1509

Hansen M, Chae HS, Kieber JJ (2009) Regulation of ACS protein stability by cytokinin and brassinosteroid. Plant J 57:606–614

Harinasut P, Srisunak S, Pitukchaisopol S, Charoensataporn R (2000) Mechanisms of adaptation to increasing salinity of mulberry: Proline content and ascorbate peroxidase activity in leaves of multiple shoots. Sci Asia 26:207–211

Heyl A, Ramireddy E, Brenner WG, Riefler M, Allemeersch J, Schmülling T (2008) The transcriptional silencer ARR1-SRDX suppresses pleiotropic cytokinin activities in *Arabidopsis*. Plant Physiol 147:1380–1395

Hirose N, Takei K, Kuroha T, Kamada T, Hayashi H, Sakakibara H (2008) Regulation of cytokinin biosynthesis, compartmentalization and translocation. J Exp Bot 59:75–83

Hu CA, Delauney AJ, Verma DP (1992) A bifunctional enzyme (delta 1-pyrroline-5-carboxylate synthetase) catalyzes the first two steps in proline biosynthesis in plants. Proc Natl Acad Sci USA 89:9354–9358

Huang H, Liu B, Liu L, Song S (2017a) Jasmonate action in plant growth and development. J Exp Bot 68:1349–1359

Huang X, Hou L, Meng J, You H, Li Z, Gong Z et al (2018) The antagonistic action of abscisic acid and cytokinin signaling mediates drought stress response in *Arabidopsis*. Mol Plant 11:970–982

Huang Y, Sun MM, Ye Q, Wu XQ, Wu WH, Chen YF (2017b) Abscisic acid modulates seed germination via ABA INSENSITIVE5-Mediated PHOSPHATE1. Plant Physiol 175:1661–1668

Hwang I, Sheen J, Müller B (2012) Cytokinin signaling networks. Annu Rev Plant Biol 63:353–380

Iqbal N, Umar S, Khan NA (2015) Nitrogen availability regulates proline and ethylene production and alleviates salinity stress in mustard (*Brassica juncea*). J Plant Physiol 178:84–91

Jang G, Choi YD (2018) Drought stress promotes xylem differentiation by modulating the interaction between cytokinin and jasmonic acid. Plant Signal Behav 13(3):e1451707

Jang G, Chang SH, Um TY, Lee S, Kim JK, Choi Y (2017) Antagonistic interaction between jasmonic acid and cytokinin in xylem development. Sci Rep 7:10212

Jang G, Yoon Y, Choi YD (2020) Crosstalk with jasmonic acid integrates multiple responses in plant development. Int J Mol Sci 21:305

Janz D, Lautner S, Wildhagen H, Behnke K, Schnitzler JP, Rennenberg H et al (2012) Salt stress induces the formation of a novel type of 'pressure wood' in two *Populus* species. New Phytol 194:129–141

Joshi R, Sahoo KK, Tripathi AK, Kumar R, Gupta BK, Pareek A et al (2018) Knockdown of an inflorescence meristem-specific cytokinin oxidase—*OsCKX2* in rice reduces yield penalty under salinity stress condition. Plant, Cell Environ 41:936–946

Junghans U, Polle A, Düchting P, Weiler E, Kuhlman B, Gruber F et al (2006) Adaptation to high salinity in poplar involves changes in xylem anatomy and auxin physiology. Plant Cell Environ 29:1519–1531

Kang J, Li J, Gao S, Tian C, Zha X (2017) Overexpression of the leucine rich receptor-like kinase gene LRK2 increases drought tolerance and tiller number in rice. Plant Biotechnol J 15:1175–1185

Kang NY, Cho C, Kim NY, Kim J (2012) Cytokinin receptor-dependent and receptor-independent pathways in the dehydration response of *Arabidopsis thaliana*. J Plant Physiol 169:1382–1391

Kiba T, Aoki K, Sakakibara H, Mizuno T (2004) Arabidopsis response regulator, ARR22, ectopic expression of which results in phenotypes similar to the wol cytokinin-receptor mutant. Plant Cell Physiol 45:1063–1077

Kiba T, Yamada H, Sato S, Kato T, Tabata S, Yamashino T et al (2003) The type-A response regulator, ARR15, acts as a negative regulator in the cytokinin-mediated signal transduction in *Arabidopsis thaliana*. Plant Cell Physiol 44:868–874

Kieber JJ, Schaller GE (2014) Cytokinins. In: The arabidopsis book. American Society of Plant Biologists, p 12

Kieber JJ, Schaller GE (2018) Cytokinin signaling in plant development. Development 145

Klein A, Keyster M, Ludidi N (2015) Response of soybean nodules to exogenously applied caffeic acid during NaCl-induced salinity. S Afr J Bot 96:13–18

Ko D, Kang J, Kiba T, Park J, Kojima M, Do J et al (2014) Arabidopsis ABCG14 is essential for the root-to-shoot translocation of cytokinin. Proc Natl Acad Sci USA 111:7150–7155

Kobayashi K, Baba S, Obayashi T, Sato M, Toyooka K, Keränen M et al (2012) Regulation of root greening by light and auxin/cytokinin signaling in *Arabidopsis*. Plant Cell 24:1081–1095

Kollmer I, Novak O, Strnad M, Schmulling T, Werner T (2014) Overexpression of the cytosolic cytokinin oxidase/dehydrogenase (CKX7) from *Arabidopsis* causes specific changes in root growth and xylem differentiation. Plant J 78:359–371

Krupinska K, Mulisch M, Hollmann J, Tokarz K, Zschiesche W, Kage H et al (2012) An alternative strategy of dismantling of the chloroplasts during leaf senescence observed in a high-yield variety of barley. Physiol Plant 144:189–200

Kumar M, Lee SC, Kim JY, Kim SJ, Aye SS, Kim SR (2014) Over-expression of dehydrin gene, OsDhn1, improves drought and salt stress tolerance through scavenging of reactive oxygen species in rice (*Oryza sativa* L.). J Plant Biol 57:383–393

Kumari S, Kumar S, Prakash P (2018) Exogenous application of cytokinin (6-BAP) ameliorates the adverse effect of combined drought and high temperature stress in wheat seedling. J Pharmaco Phytochem 7:1176–1180

Kurakawa T, Ueda N, Maekawa M, Kobayashi K, Kojima M, Nagato Y et al (2007) Direct control of shoot meristem activity by a cytokinin-activating enzyme. Nature 445:652–655

Kuroha T, Tokunaga H, Kojima M, Ueda N, Ishida T, Nagawa S et al (2009) Functional analyses of LONELY GUY cytokinin-activating enzymes reveal the importance of the direct activation pathway in Arabidopsis. Plant Cell 21:3152–3169

Lautner S (2013) Wood formation under drought stress and salinity. In: Fromm J (ed.) Cellular aspects of wood formation, vol 20, pp 187–202. Springer, Berlin, Germany

Leta TB, Miccah SS, Steven MR, Wondyifraw T, Francis W (2016) Drought tolerant tropical maize (*Zea mays* L.) developed through genetic transformation with isopentenyltransferase gene. Afr J Biotech 15:2447–2464

Li S, An Y, Hailati S, Zhang J, Cao Y, Liu Y et al (2019) Overexpression of the cytokinin oxidase/dehydrogenase (CKX) from *Medicago sativa* enhanced salt stress tolerance of Arabidopsis. J Plant Biol 62:374–386

Li Y, Liu F, Li P, Wang T, Zheng C, Hou B (2020) An Arabidopsis cytokinin-modifying glycosyltransferase UGT76C2 improves drought and salt tolerance in rice. Front Plant Sci 11:560696

Liang W, Ma X, Wan P, Liu L (2017) Plant salt-tolerance mechanism: a review. Biochem Biophys Res Commun 495:286–291

Liao X, Guo X, Wang Q, Wang Y, Zhao D, Yao L et al (2017) Overexpression of MsDREB6.2 results in cytokinin-deficient developmental phenotypes and enhances drought tolerance in transgenic apple plants. Plant J 89:510–526
Liu CJ, Zhao Y, Zhang K (2019) Cytokinin transporters: multisite players in cytokinin homeostasis and signal distribution. Front Plant Sci 10:693
Liu L, Li H, Zeng H, Cai Q, Zhou X, Yin C (2016) Exogenous jasmonic acid and cytokinin antagonistically regulate rice flag leaf senescence by mediating chlorophyll degradation, membrane deterioration, and senescence-associated genes expression. J Plant Growth Regul 35:366–376
Liu Y, Zhang M, Meng Z, Wang B, Chen M (2020) Research progress on the roles of cytokinin in plant response to stress. Int J Mol Sci 21:6574
Ma Q, Yue LJ, Zhang JL, Wu GQ, Bao AK, Wang SM (2012) Sodium chloride improves photosynthesis and water status in the succulent xerophyte Zygophyllum xanthoxylum. Tree Physiol 32:4–13
Macková H, Hronková M, Dobrá J, Turecková V, Novák O, Lubovská Z et al (2013) Enhanced drought and heat stress tolerance of tobacco plants with ectopically enhanced cytokinin oxidase/dehydrogenase gene expression. J Exp Bot 64:2805–2815
Mähönen AP, Bishopp A, Higuchi M, Nieminen KM, Kinoshita K, Törmäkangas K et al (2006) Cytokinin signaling and its inhibitor AHP6 regulate cell fate during vascular development. Science 311:94–98
Mason MG, Jha D, Salt DE, Tester M, Hill K, Kieber JJ et al (2010) Type-B response regulators ARR1 and ARR12 regulate expression of AtHKT1;1 and accumulation of sodium in Arabidopsis shoots. Plant J 64:753–763
Merewitz EB, Gianfagna T, Huang B (2011) Protein accumulation in leaves and roots associated with improved drought tolerance in creeping bentgrass expressing an ipt gene for cytokinin synthesis. J Exp Bot 62:5311–5333
Mhamdi A, Queval G, Chaouch S, Vanderauwera S, Van Breusegem F, Noctor G (2010) Catalase function in plants: a focus on *Arabidopsis* mutants as stress-mimic models. J Exp Bot 61:4197–4220
Mrızova K, Jiskrova E, Vyroubalova S, Novak O, Ohnoutkova L, Pospısilova H et al (2013) Overexpression of cytokinin dehydrogenase genes in barley (*Hordeum vulgare* cv. Golden Promise) fundamentally affects morphology and fertility. PLoS One 8, e79029
Mukherjee I, Reid D, Naik G (2002) Influence of cytokinins on the methyl jasmonate-promoted senescence in *Helianthus annuus* cotyledons. Plant Growth Regul 38:61–68
Mundree SG, Baker B, Sh, Mowla, Sh, Peters, Marais S, Willigen CV et al (2002) Physiological and molecular insights into drought tolerance. Afr J Biotech 1:28–38
Munns R, Tester M (2008) Mechanisms of salinity tolerance. Annu Rev Plant Biol 59:651–681
Nieminen K, Immanen J, Laxell M, Kauppinen L, Tarkowski P, Dolezal K et al (2008) Cytokinin signaling regulates cambial development in *poplar*. Proc Natl Acad Sci USA 105:20032–20037
Nishiyama R, Le DT, Watanabe Y, Matsui A, Tanaka M, Seki M et al (2012) Transcriptome analyses of a salt-tolerant cytokinin-deficient mutant reveal differential regulation of salt stress response by cytokinin deficiency. PLoS ONE 7:e32124
Nishiyama R, Watanabe Y, Fujita Y, Le DT, Kojima M, Werner T et al (2011) Analysis of cytokinin mutants and regulation of cytokinin metabolic genes reveals important regulatory roles of cytokinins in drought, salt and abscisic acid responses, and abscisic acid biosynthesis. Plant Cell 23:2169–2183
Nishiyama R, Watanabe Y, Leyva-Gonzalez MA, Ha CV, Fujita Y, Tanaka M et al (2013) Arabidopsis AHP2, AHP3, and AHP5 histidine phosphotransfer proteins function as redundant negative regulators of drought stress response. Proc Natl Acad Sci USA 110:4840–4845
Noctor G, Veljovic-Jovanovic S, Driscoll S, Novitskaya L, Foyer CH (2002) Drought and oxidative load in the leaves of C-3 plants: a predominant role for photorespiration? Annals Botany (London) 89:841–850
Osugi A, Sakakibara H (2015) Q&A: how do plants respond to cytokinins and what is their importance? BMC Biol 13:102

Osugi A, Kojima M, Takebayashi Y, Ueda N, Kiba T, Sakakibara H (2017) Systemic transport of trans-zeatin and its precursor have differing roles in Arabidopsis shoots. Nat Plants 3:17112

Oteiza PI, Erlejman AG, Verstraeten SV, Keen CL, Fraga CG (2005) Flavonoid membrane Interactions: a protective role of flavonoids at the membrane surface? Clin Dev Immunol 12:19–25

Pavlu J, Novak J, Koukalová V, Luklova M, Brzobohatý B, Cerný M (2018) Cytokinin at the crossroads of abiotic stress signaling pathways. Int J Mol Sci 19:2450

Pospisilova H, Jiskrova E, Vojta P, Mrizova K, Kokas F, Cudejkova MM et al (2016) Transgenic barley overexpressing a cytokinin dehydrogenase gene shows greater tolerance to drought stress. New Biotechnol 33:692–705

Pospisilova J, Vagner M, Malbeck J, Travniakova A, Batkova P (2005) Interactions between abscisic acid and cytokinins during water stress and subsequent rehydration. Biol Plant 49:533–540

Qin H, He L, Huang R (2019) The coordination of ethylene and other hormones in primary root development. Front Plant Sci 10:874

Rahman MM, Mostofa MG, Rahman MA, Islam MR, Keya SS, Das AK et al (2019) Acetic acid: a cost-effective agent for mitigation of seawater-induced salt toxicity in Mung Bean. Sci Rep 9:15186

Rai MI, Wang X, Thibault DM, Kim HJ, Bombyk MM, Binder BM et al (2015) The ARGOS gene family functions in a negative feedback loop to desensitize plants to ethylene. BMC Plant Biol 15:157

Ramireddy E, Hosseini SA, Eggert K, Gillandt S, Gnad H, von Wirén N et al (2018) Root Engineering in Barley: increasing cytokinin degradation produces a larger root system, mineral enrichment in the shoot and improved drought tolerance. Plant Physiol 177:1078–1095

Reguera M, Peleg Z, Abdel-Tawab YM, Tumimbang E, Delatorre CA, Blumwald E (2013) Stress-induced CK synthesis increases drought tolerance through the coordinated regulation of carbon and nitrogen assimilation in rice. Plant Physiol 163:1609–1622

Rivero RM, Gimeno J, Van Deynze A, Walia H, Blumwald E (2010) Enhanced cytokinin synthesis in tobacco plants expressing *p(SARK)::IPT* prevents the degradation of photosynthetic protein complexes during drought. Plant Cell Physiol 51:1929–1941

Rivero RM, Shulaev V, Blumwald E (2009) Cytokinin-dependent photorespiration and the protection of photosynthesis during water deficit. Plant Physiol 150:1530–1540

Riyazuddin R, Verma R, Singh K, Nisha N, Keisham M, Bhati KK et al (2020) Ethylene: a master regulator of salinity stress tolerance in plants. Biomolecules 10:959

Sakakibara H (2006) Cytokinins: activity, biosynthesis, and translocation. Annu Rev Plant Biol 57:431–449

Samea S, Ghassemi-Golezani K, Nasrollahzadeh S, Najafi N (2018) Exogenous salicylic acid and cytokinin alter sugar accumulation, antioxidants and membrane stability of faba bean. Acta Biol Hung 69:86–96

Sasaki T, Suzaki T, Soyano T, Kojima M, Sakakibara H, Kawaguchi M (2014) Shoot-derived cytokinins systemically regulate root nodulation. Nat Commun 5:4983

Schmulling T, Werner T, Riefler M, Krupkova E, Bartrina (2003) Structure and function of cytokinin oxidase/dehydrogenase genes of maize, rice, Arabidopsis and other species. J Plant Res 116:241–252

Sebastian J, Ryu KH, Zhou J, Tarkowská D, Tarkowski P, Cho YH et al (2015) PHABULOSA controls the quiescent center-independent root meristem activities in *Arabidopsis thaliana*. PLoS Genet 11:e1004973

Sharma A, Shahzad B, Kumar V, Kohli SK, Sidhu GPS, Bali AS et al (2019) Phytohormones regulate accumulation of osmolytes under abiotic stress. Biomolecules 9:285

Shi H, Ishitani M, Kim C, Zhu JK (2000) The *Arabidopsis thaliana* salt tolerance gene SOS1 encodes a putative Na^+/H^+ antiporter. Proc Natl Acad Sci USA 97:6896–6901

Sidari M, Santonoceto C, Anastasi U, Preiti G, Muscolo A (2008) Variations in four genotypes of lentil under NaCl-salinity stress. Am J Agric Biol Sci 3:410–416

Siddiqui MW, Singh JP, Nayyer MA, Barman K, Ahmad MS, Kumar V (2015) 6-Benzylaminopurine affects lipid peroxidation and membrane permeability and thereby preserves curd quality and antioxidants during storage of cauliflower. Acta Physiol Plant 37:96

Song J, Zhou J, Zhao W, Xu H, Wang F, Xu Y et al (2016) Effects of salinity and nitrate on production and germination of dimorphic seeds applied both through the mother plant and exogenously during germination in Suaeda salsa. Plant Species Biol 31:19–28

Spichal L, Rakova NY, Riefler M, Mizuno T, Romanov GA, Strnad M et al (2004) Two cytokinin receptors of *Arabidopsis thaliana*, CRE1/AHK4 and AHK3, differ in their ligand specificity in a bacterial assay. Plant Cell Physiol 45:1299–1305

Sripinyowanich S, Klomsakul P, Boonburapong B, Bangyeekhun T, Asami T, Gu H et al (2013) Exogenous ABA induces salt tolerance in indica rice (*Oryza sativa* L.): the role of OsP5CS1 and OsP5CR gene expression during salt stress. Environ Exp Bot 86:94–105

Sui N, Wang Y, Liu S, Yang Z, Wang F, Wan S (2018) Transcriptomic and physiological evidence for the relationship between unsaturated fatty acid and salt stress in peanut. Front Plant Sci 9:7

Sui N, Yang Z, Liu M, Wang B (2015) Identification and transcriptomic profiling of genes involved in increasing sugar content during salt stress in sweet sorghum leaves. BMC Genomics 16:534

Szabados L, Savouré A (2010) Proline: a multifunctional amino acid. Trends Plant Sci 15:89–97

Takei K, Yamaya T, Sakakibara H (2004) Arabidopsis *CYP735A1* and *CYP735A2* encode cytokinin hydroxylases that catalyze the biosynthesis of *trans*-zeatin. J Biol Chem 279:41866–41872

Tran LS, Shinozaki K, Yamaguchi-Shinozaki K (2010) Role of cytokinin responsive two-component system in ABA and osmotic stress signaling. Plant Signal Behav 5(2):148–150

Tran LS, Urao T, Qin F, Maruyama K, Kakimoto T, Shinozaki K et al (2007) Functional analysis of AHK1/ATHK1 and cytokinin receptor histidine kinases in response to abscisic acid, drought, and salt stress in *Arabidopsis*. Proc Natl Acad Sci USA 104:20623–20628

Tsai YC, Weir NR, Hill K, Zhang W, Kim HJ, Shiu SH et al (2012) Characterization of genes involved in cytokinin signaling and metabolism from rice. Plant Physiol 158:1666–1684

Tuteja N (2007) Abscisic acid and abiotic stress signaling. Plant Signal Behav 2:135–138

Ueda J, Kato J (1982) Inhibition of cytokinin-induced plant growth by jasmonic acid and its methyl ester. Physiol Plant 54:249–252

Ursache R, Miyashima S, Chen Q, Vatén A, Nakajima K, Carlsbecker A et al (2014) Tryptophan-dependent auxin biosynthesis is required for HD-ZIP III-mediated xylem patterning. Development 141:1250–1259

Verma D, Lakhanpal N, Singh K (2019) Genome-wide identification and characterization of abiotic-stress responsive SOD (superoxide dismutase) gene family in *Brassica juncea* and *B. rapa*. BMC Genomics 20, 227

Veselova SV, Farkhutdinov RG, Veselov DS, Kudoyarova GR (2006) Role of cytokinins in the regulation of stomatal conductance of wheat seedlings under conditions of rapidly changing local temperature. Russ J Plant Physiol 53:756–761

Vojta P, Kokáš F, Husicková A, Grúz J, Bergougnoux V, Marchetti CF et al (2016) Whole transcriptome analysis of transgenic barley with altered cytokinin homeostasis and increased tolerance to drought stress. New Biotechnol 33:676–691

Wakeel A, Xu M, Gan Y (2020) Chromium-induced reactive oxygen species accumulation by altering the enzymatic antioxidant system and associated cytotoxic, genotoxic, ultrastructural, and photosynthetic changes in plants. Int J Mol Sci 21:728

Wang H (2019) Regulation of vascular cambium activity. Plant Sci 291:110322

Wang WC, Lin TC, Kieber J, Tsai YC (2019) Response regulators 9 and 10 negatively regulate salinity tolerance in Rice. Plant Cell Physiol 60(11):2549–2563

Wang Y, Shen W, Chan Z, Wu Y (2015) Endogenous cytokinin overproduction modulates ROS homeostasis and decreases salt stress resistance in *Arabidopsis Thaliana*. Front Plant Sci 6:1004

Wang Z, Pote J, Huang B (2003) Responses of cytokinins, antioxidant enzymes, and lipid peroxidation in shoots of creeping bentgrass to high root-zone temperatures. J Am Soc Hortic Sci 128:648–655

Werner T, Schmulling T (2009) Cytokinin action in plant development. Curr Opin Plant Biol 12:527–538
Werner T, Köllmer I, Bartrina I, Holst K, Schmülling T (2006) New insights into the biology of cytokinin degradation. Plant Biol 8:371–381
Werner T, Motyka V, Laucou V, Smets R, Van Onckelen H, Schmulling T (2003) Cytokinin-deficient transgenic Arabidopsis plants show multiple developmental alterations indicating opposite functions of cytokinins in the regulation of shoot and root meristem activity. Plant Cell 15:2532–2550
Werner T, Motyka V, Strnad M, Schmulling T (2001) Regulation of plant growth by cytokinin. Proc Natl Acad Sci USA 98:10487–10492
Werner T, Nehnevajova E, Kollmer I, Novak O, Strnad M, Kramer U et al (2010) Root-specific reduction of cytokinin causes enhanced root growth, drought tolerance and leaf mineral enrichment in Arabidopsis and tobacco. Plant Cell 22:3905–3920
Wingler A, Lea PJ, Quick WP, Leegood RC (2000) Photorespiration: metabolic pathways and their role in stress protection. Philos Trans R Soc Lond B Biol Sci 355:1517–1529
Wohlbach DJ, Quirino BF, Sussman MR (2008) Analysis of the Arabidopsis histidine kinase ATHK1 reveals a connection between vegetative osmotic stress sensing and seed maturation. Plant Cell 20:1101–1117
Wu X, Zhu Z, Li X, Zha D (2012) Effects of cytokinin on photosynthetic gas exchange, chlorophyll fluorescence parameters and antioxidative system in seedlings of eggplant (*Solanum melongena* L.) under salinity stress. Acta Physiol Plant 34:2105–2114
Xiao XO, Zeng YM, Cao BH, Lei JJ, Cheng YI (2017) PSAG12-IPT overexpression in eggplant delays leaf senescence and induces abiotic stress tolerance. J Hortic Sci Biotechnol 92:1–9
Xiong H, Li J, Liu P, Duan J, Zhao Y, Guo X et al (2014) Overexpression of OsMYB48-1, a novel MYB-related transcription factor, enhances drought and salinity tolerance in rice. PLoS ONE 9:e92913
Xu YH, Liu R, Yan L, Liu ZQ, Jiang SC, Shen YY et al (2012) Light-harvesting chlorophyll a/b-binding proteins are required for stomatal response to abscisic acid in Arabidopsis. J Exp Bot 63:1095–1106
Xu Y, Burgess P, Zhang X, Huang B (2016) Enhancing cytokinin synthesis by overexpressing ipt alleviated drought inhibition of root growth through activating ROS-scavenging systems in *Agrostis stolonifera*. J Exp Bot 67:1979–1992
Yang Z, Li JL, Liu LN, Xie Q, Sui N (2020) Photosynthetic regulation under salt stress and salt-tolerance mechanism of sweet sorghum. Front Plant Sci 10:1722
Yaronskaya E, Vershilovskaya I, Poers Y, Alawady AE, Averina N, Grimm B (2006) Cytokinin effects on tetrapyrrole biosynthesis and photosynthetic activity in barley seedlings. Planta 224:700–709
Yi X, Hargett SR, Frankel LK, Bricker TM (2008) The effects of simultaneous RNAi suppression of PsbO and PsbP protein expression in photosystem II of Arabidopsis. Photosynth Res 98:439–448
Yin W, Xiao Y, Niu M, Meng W, Li L, Zhang X et al (2020) ARGONAUTE2 enhances grain length and salt tolerance by activating *BIG GRAIN3* to modulate cytokinin distribution in Rice. Plant Cell 32:2292–2306
Yokoyama A, Yamashino T, Amano YI, Tajima Y, Imamura A, Sakakibara H et al (2007) Type-B ARR transcription factors, ARR10 and ARR12, are implicated in cytokinin-mediated regulation of protoxylem differentiation in roots of *Arabidopsis thaliana*. Plant Cell Physiol 48:84–96
Zalabák D, Galuszka P, Mrızova K, Podlešáková K, Gu R, Frébortová J (2014) Biochemical characterization of the maize cytokinin dehydrogenase family and cytokinin profiling in developing maize plantlets in relation to the expression of cytokinin dehydrogenase genes. Plant Physiol Biochem 74:283–293
Zdarska M, Cuyacot AR, Tarr PT, Yamoune A, Szmitkowska A, Hrdinova V et al (2019) ETR1 integrates response to ethylene and cytokinins into a single multistep phosphorelay pathway to control root growth. Mol Plant 12:1338–1352

Zdarska M, Dobisova T, Gelova Z, Pernisova M, Dabravolski S, Hejatko J (2015) Illuminating light, cytokinin, and ethylene signaling crosstalk in plant development. J Exp Bot 66:4913–4931

Zhang H, Zheng J, Su H, Xia K, Jian S, Zhang M (2018) Molecular cloning and functional characterization of the dehydrin (IpDHN) gene from *Ipomoea pes-caprae*. Front Plant Sci 9:1454

Zhang K, Novak O, Wei Z, Gou M, Zhang X, Yu Y et al (2014) Arabidopsis ABCG14 protein controls the acropetal translocation of root-synthesized cytokinins. Nat Commun 5:3274

Zhu JK (2001) Plant salt tolerance. Trends Plant Sci 6(2):66–71

Žižková E, Dobrev PI, Muhovski Y, Hošek P, Hoyerová K, Haisel D et al (2015) Tomato (*Solanum lycopersicum* L.) SlIPT3 and SlIPT4 isopentenyltransferases mediate salt stress response in tomato. BMC Plant Biol 15(1), 1–20

Auxin and Cytokinin Signaling in Plant Stress Response

Ankita Mallick, Subhajoy Dey, Soustav Datta, Mainak Barman, Suman Samui, and Gopal Dutta

Abstract Under stress, plant growth is restricted to optimize response to stress and prevent death. Plants must promptly disable defensive stress responses and promote growth recovery processes once environmental stress is no longer present. Through their dynamical and complementary effects, the pathways of auxin and cytokinin control a number of developmental functions as well as their potential to crosstalk enables them excellent contenders to regulate stress-acclimatization retaliations. Another significant signalling molecule that contributes to the remarkable flexibility studied about plant morphological characters as well as abiotic stress is reactive oxygen species (ROS). The appropriate temporal and geographical dissemination of ROS and hormonal gradients is essential for survival of plants in adverse conditions. The combination of ROS and phytohormone networks serves like a consolidator of environmental and developmental markers into comprehensive reactions that allow plants to adjust to their environments in this way. The signalling mechanisms of auxin and cytokinin have been thoroughly explored. Despite this, we do not yet know how the profound crosstalk between the two hormones influences plant stress tolerance. The combined effect of these two hormones is crucial for controlling root meristem length and ensuring root growth. Meristems, which are a source of indivisible cells that evolve into mature vegetative organs, aid in post-embryonic plant development. Phytohormones cytokinin and auxin, which govern meristem activities, are known to

A. Mallick · S. Dey · G. Dutta
Department of Agricultural Meteorology and Physics, Bidhan Chandra Krishi Viswavidyalaya, Mohanpur, Nadia, West Bengal, India

S. Datta (✉)
Department of Fruit Science, Bidhan Chandra Krishi Viswavidyalaya, Mohanpur, Nadia, West Bengal, India
e-mail: sodpurarunachal@gmail.com

M. Barman
Department of Genetics and Plant Breeding, Bidhan Chandra Krishi Viswavidyalaya, Mohanpur, Nadia, West Bengal, India

S. Samui
Department of Agronomy, Bidhan Chandra Krishi Viswavidyalaya, Mohanpur, Nadia, West Bengal, India

T. Aftab (ed.), *Auxins, Cytokinins and Gibberellins Signaling in Plants*, Signaling and Communication in Plants,
https://doi.org/10.1007/978-3-031-05427-3_9

be hostile. In order to ascertain their plausible sites of crosstalk, we analyze current information about the involvement of auxin and cytokinin in steering growth driven by abiotic stress.

1 Introduction

Plant growth and development are significantly impacted by several abiotic factors. Plants have had to evolve sturdy adaptation systems to withstand unpredictable changes in their habitats. In nature, different inconveniences such as bacteria, fungi, viruses, oomycetes, nematodes and insect predators continuously endanger plants. To keep these potential invaders out, plants have a variety of structural defence and preformed antimicrobial properties. Plant vitality and growth are jeopardised by abiotic conditions including high temperatures, antagonistic water table and toxic metal concentrations. Phytohormones have significant contribution in plant development. Auxin (indole-3-acetic acid (IAA)), the first growth hormone derived from plants, possesses an exceptional position among all hormones. Among plant hormones, auxin and cytokinins are distinctive in view of the fact that they are both necessary to survive (Pandey et al. 2019).

Since the discovery of the Phytohormones half a century ago auxin and cytokinin have been considered to be important for cell divisions and proliferation in culturing plant tissue (Skoog and Miller 1957). These two phytohormones have been considered to be the primary moderators of plant growth (Schaller et al. 2015). Auxin is involved in plant growth in a numerous ways, including phototropism, cell elongation, apical dominance, parthenocarpy, fruit development, abscission and adventitious root formation (Taiz and Zeiger 2002); however, its function of this in various physiological processes remains elusive. In the last decade, auxin has been observed to play a function in a number of abiotic challenges. According to various studies, auxin signalling and transportation contributes significantly in plant abiotic stress tolerance (Krishnamurthy and Rathinasabapathi 2013). CKs, including auxin, are plant hormones that were recognised about 60 years back. Cytokinin was first found to aid in cellular division (the term "cytokinin" comes after "cytokinesis") and shoot meristem diversification (Miller et al. 1955, 1956; Su et al. 2011). According to recent studies, CKs govern apical dominance, lateral bud growth, shoot meristem formation and maintenance, root growth suppression, nitrogen (N) signaling, phyllotaxis, leaf expansion and abscission, among other important plant development cycles (Takei et al. 2002; Miyawaki et al. 2004; Frébort et al. 2011). In addition to their extensive use in plant tissue culture as a "balanced phytohormone" with auxin, CKs have been implicated in boosting plant abiotic stress resistance via altering transcription of genes encoding in the CK metabolic pathway (Rivero et al. 2007; O'Brien and Benková 2013).

Auxin and cytokinin crosstalk has already been widely investigated in several aspects (biosynthesis, detection, and transportation), and we will be starting to comprehend how such systems interrelate to govern several plant activities (El-Showk et al. 2013; Chandler and Werr 2015; Schaller et al. 2015) employing plant resilience against biotic (Großkinsky et al. 2011) and abiotic stresses (O'Brien and Benková 2013). To protect themselves from invading diseases, plants have evolved defense signalling systems. Plant hormones like ethylene, jasmonates, salicylic acid behave as indicators for eliciting and regulating wide range of defense responses. Additional hormones involved in pathogen defense signalling include auxin, abscisic acid, cytokinins, gibberellic acids, and brassinosteroids, those have already been linked to developmental and abiotic stress responses. The hormone signaling mechanisms interact in synergistic or antagonistic ways, providing plants a huge regulative aptitude for adapting quickly to their biotic abode and making optimal utilization of their finite resources for development and sustainability. In contrast, microorganisms have found tactics to affect the signaling system and enhance their pathogenicity (Takatsuji and Jiang 2014). In the biosynthetic pathways of many stress hormones like ethylene, auxin has both antagonistic and synergistic role (Stepanova and Alonso 2005; Ruzicka et al. 2007) and their interactions play a vital role in aiding auxin-conciliated stress reactions.

Being a sessile entity, plants are constantly exhibited to an extensive amount of external stimulus. External stimulus might comprise a huge spectrum of extreme weather that can arise unexpectedly, for example. Plants employ reactive oxygen species (ROS) as a particle that sends out signals since they are aerobic organisms. However, if the formation of ROS occurs under less-than-ideal development conditions, such as abiotic stressors, the consequences of oxidative stress upon plant tissues may become catastrophic. Plants are able to evolve mechanisms that allow for swift recognition, differentiation, and response during a critical event by keeping ROS levels under control via a diverse array of antioxidant defence system on either one hand, or ROS interfacing with plant hormones routes on the other (Mittler et al. 2011). In order to persist and adjust to new surroundings, plants use structural, morphological, and biochemical responses to reduce stress prominence, minimise damages, and promote the restoration processes. Plants have evolved a stress-resistance or survival strategy by altering some of their morphological and physiological characteristics. The simultaneous occurrence of these phytohormones auxin and cytokinin, as well as stress-impelled ROS impulses, are linked to plant growth and responses to environmental changes (Kazan 2013; O'Brien and Benková 2013; Zwack and Rashotte 2015; Verma et al. 2016). The consequent crosstalk between auxin, cytokinin and ROS enables plants in order to respond to adverse environmental inputs by adjusting their development and growth. A number of transcriptomic assays have revealed that auxin and cytokinin affect each others signaling pathways and/or metabolic activity (Rashotte et al. 2003; Goda et al. 2004). Despite this, scientists are only now starting to recognize the molecular pathways through which these hormones collaborate for creating a certain physiological outcome. Crosstalk between these phytohormones is evidently also mediated by common signalling components and co-regulated genes. Furthermore, crosstalk is regionally and temporally managed, allowing for response flexibility and exquisite. The biology of auxin and CKs, including its structure, metabolism, and signalling pathways, is discussed in this particular chapter. The role of auxin and CKs in crop stress response stresses, and also the crosstalk mechanism between the two hormones under stress, have been reviewed.

2 Auxin Signaling Pathway in Plants

Light and gravity responses, organ patterning, root and shoot architecture, and vascular development are all controlled by auxin signaling. Its effects differ depending on the cellular and developmental milieu in which it is received; auxin triggers responses such as cell division and expansion. Due to transcriptional and post-transcriptional regulation, the key components of auxin signalling have distinct expression patterns. Auxin, along with cytokinins, is unique among phytohormones in that it is essential for plant survival. Alterations in transcription are the most common way for the level of auxin to be turned into cellular responses. In response to the presence of exogenous auxin, many genes change their expression (Paponov et al. 2008). Auxin controls transcription by a simple and well-studied signal transduction pathway (Fig. 1) Leyser (2018) (Chapman and Estelle 2009; Salehin et al. 2015).

Auxin signalling includes the detection of auxin by receptors such as ABP1 (Auxin-Binding Protein), AFB proteins (Auxin F-Box) andTIR1 (Transport Inhibitor Response 1) (Taiz and Zeiger 2002) (Dharmasiri et al. 2005; Kazan 2013). The 26S proteasome degrades Aux/IAAs (Auxin/IAA) repressor when Auxin binds to its receptor. Auxin Response Factors (ARFs) are Transcription Factors (TFs) that tie up with auxin responsive elements (AREs) in the promoter site of auxin-induced genes, are repressed by Aux/IAA proteins (Kim et al. 1997). ARFs are released from repression when Aux/IAAs are degraded, and they attach to the ARE for controlling the expression of auxin-dependent genes (Kazan 2013).

In a nutshell, auxin binds F-box proteins and plays as molecular glue. It binds F-box proteins from the Transport Inhibitor Response 1/Auxin Signalling F-Box (TIR1/AFB) family to transcriptional repressors from the Aux/IAA family (Tan et al. 2007). Substrate selection subunits of SCF-type ubiquitin protein ligase complexes include Skp1, Cullin and an F-box protein (Smalle and Vierstra 2004). F-box proteins have a region at their N-terminus that allows them to link with Skp1, which in turn reacts with RBX1and a Cullin dimer. This dimer conjugates active ubiquitin to target proteins after receiving it from a ubiquitin activating enzyme. The interaction of the target protein with the F-box protein's C-terminal domain brings it to the SCF. In the case of TIR1/AFBs, this is made up of Leu-rich repetitions with a pocket for auxin bindings. The Aux/IAA protein docking throughout the pocket mouth, governed by a minor protein motif in the Aux/IAA usually known as domain II, considerably stabilizes auxin binding in this pocket (Tan et al. 2007). As a result, TIR1/AFB-Aux/IAA couples can be called auxin coreceptors. The binding of Aux/IAAs to TIR1/AFBs by auxin delivers them to the SCF, where they are ubiquitinated and eventually degraded (Gray et al. 2001). In this method, alterations in auxin levels are translated into variations in Aux/IAA levels.

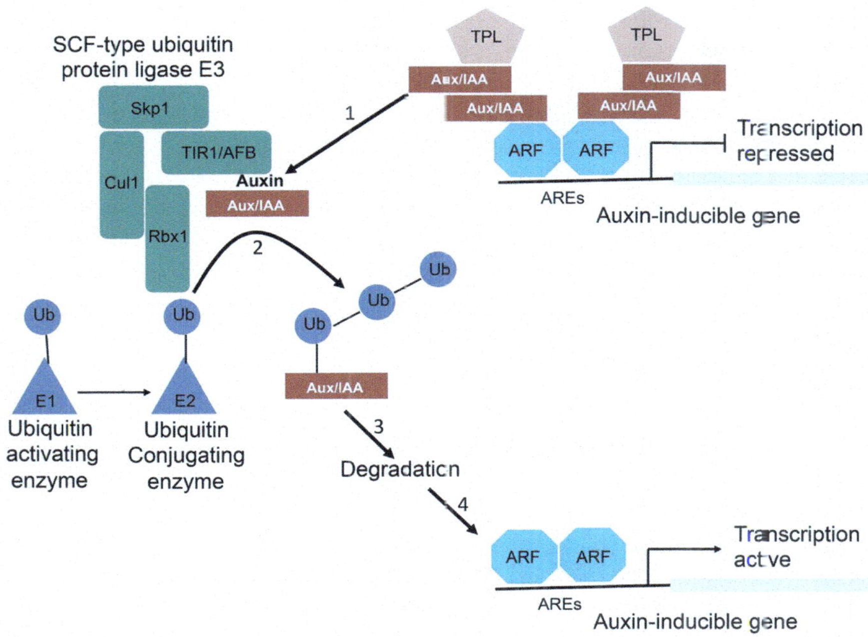

Fig. 1 **The principal pathway for regulating auxin transcription**. AREs in the promoters of auxin-inducible genes are bound by dimers of the ARF protein family. The appointment of Aux/IAA transcriptional repressors to these promoters through their interaction with the ARFs prevents gene expression. TPL family corepressors are recruited by Aux/IAAs, who then recruit chromatin modifying enzymes to stabilise the suppressed state. The numbered arrows represent the steps in the auxin response pathway. 1. Auxin works as molecular glue, binding Aux/IAAs and TIR1/AFB family F-box proteins together. 2. These F-box proteins are part of an E3 ubiquitin protein ligase complex of the SCF type, which transfers active ubiquitin (Ub) from an E1/E2 enzyme system. 3. The Aux/IAAs are degenerated as a result of polyubiquitination. 4. Repression at ARE-containing promoters is released

The transcriptional control of auxin sensitive genes is primarily regulated by the two big TF families, ARFs and Aux/IAAs. These genes' expression is influenced by environmental factors (Guilfoyle and Hagen 2007). Many ARF groups have been discovered in several crop species, including *Sorghum bicolor* (Wang et al. 2010), *Oryza sativa* (Jain and Khurana 2009; Song et al. 2009; Shen et al. 2010), *Zea mays* (Xing et al. 2011; Wang et al. 2012), *Solanum lycopersicum* (Wu et al. 2011). Similarly, the establishment of auxin-signalling mechanisms that govern plant response to natural conditions is dependent on transcriptional regulation of Aux/IAA genes. The DREB/CBF family (which is involved in stress tolerance) of transcription factors regulates the Aux/IAA genes and in response to abiotic stress, direct transcription of these genes is induced (Shani et al. 2017).

The auxin biosynthesis pathway including the YUCCA gene (which codes for a flavinmono oxygenase and is part of the tryptophan-dependent auxin biosynthetic mechanism) might be utilised to change plant responds to the environment, according to new study (Lee et al. 2012; Kim et al. 2013; Park et al. 2013). In Arabidopsis thaliana and *Hordeum vulgare*, the expression of the YUC2 and YUC6 genes has been identified to influence plant responses to high temperatures (Cheng et al. 2006; Sakata et al. 2010). Furthermore, Arabidopsis plants that over express YUC6 or transgenic poplars that express Arabidopsis YUC6 under the control of the SWPA2 promoter (stress-inducible) showed better resistance to drought and oxidative stress (Ke et al. 2015; Cha et al. 2015). Auxin boosted antioxidant enzyme activities and changed the expression of abiotic stress-related genes (RD22, RD29A, RD29B, RAB18, DREB2A and DREB2B) in a beneficial way; as a result, there is a greater tolerance for change in the environment (Shi et al. 2014a, b).

3 Cytokinin Signaling Pathway in Plants

It was discovered that the first gene implicated in cytokinin response is a histidine kinase (HK) after it was cloned, which is very comparable to the two-component prokaryotic signalling HK receptors (Kakimoto 1996). Later, bacterial homologues of response regulators were revealed to respond to cytokinin inclusion (Brandstatter and Kieber 1998), and mounting data led to the development of a two-component paradigm for cytokinin signalling. In general, a HK receptor and a response regulator (RR) protein are the two constituents of the bacterial two-component pathway which detects signal and modulates the signal's response respectively (Rowland and Deeds 2014; Stock et al. 2000). The HK receptor is a homodimer of an integral membrane-spanning protein. An input domain on one side of the membrane detects an environmental signal that causes an ATP-dependent process on another side of the membrane to be activated. One of the HK molecules catalyses the phosphorylation of a conserved histidine remainder on the oppositional HK molecule. This phosphate is subsequently transferred to an aspartic acid residue in the receiver domain of a RR protein (Stock et al. 2000). The active RR protein then reacts in this signalling pathway as a transcription factor. As a transcription factor, it activates or represses genes.

The Pathway of Cytokinin Signalling in plants works similarly to a modified bacterial two-component system, but with a few key variations, including new and altered components like signal relay to the nucleus. CHASE-domain included with hybrid histidine sensor kinases (CHKs) with the cytokinin receptors like a HK and a receiver domain. The HK domain is activated and auto-phosphorylated when cytokinin binds to the receptor and in the receiver province of the molecule the phosphate is shifted from a conserved histidine to a conserved aspartic acid residue. Then Phosphate is moved to a new component in the system, a histidine phosphotransfer protein (HPt), which goes into the RRs containing nucleus and the phosphorelay continues. The HPt delivers phosphate to an RR's receiver domain, which regulates cytokinin signalling output, once it is in the nucleus (Dortay et al. 2006) (Fig. 2) Keshishian and Rashotte (2015).

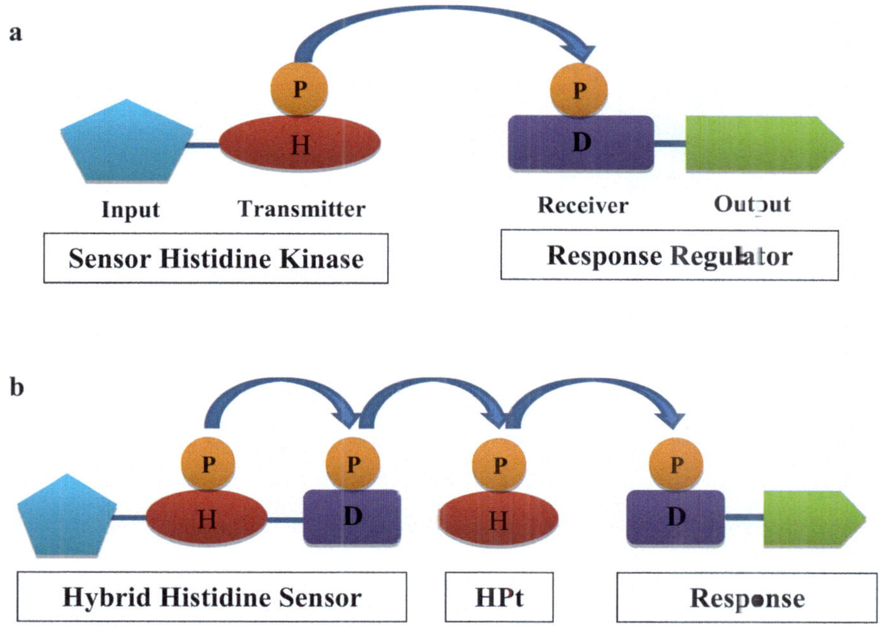

Fig. 2 Combination of two-component phosphorelay mechanism. a A sensor histidine kinase perceives input in a generic bacterial two-component system. A phosphorylated histidine residue (H) in the transmitter region of a response regulatory protein transmits the phosphate to an aspartic acid remainder (D) in the receiver site through phosphorelay. The output domain from this response regulator protein responds in the original discerned stimulus. **b** The multistep phosphorelay scheme of cytokinin. A hybrid histidine sensor kinase which has both a transmitter and receiver region with phosphorylated histidine and aspartic acid remainder detects the cytokinin input. The phosphate is subsequently transmitted to a histidine phospho transfer protein (HPt), which also has an output domain, and lastly to a response regulator

Membrane-bound CHK receptors respond to cytokinin. AHKs 2, 3 and 4 are the three primary CHK receptors in Arabidopsis, most of the other diploid angiosperm species appear to have the same traits (Pils and Heyl 2009). The CHASE domain of CHK receptors can detect cytokinin through a 200–230 amino-acid conserved region which may detect low-molecular-mass ligands like cytokinin derivatives (Anantharaman and Aravind 2001). It's thought that when the receptor dimer binds to cytokinin, it undergoes a conformational change and then auto-phosphorylation occurs and phosphate group from a particular histidine remainder in the CHK domains transferred to an aspartic acid remainder in the receiver region, the canonical phosphorelay begins with this step. Endomembranes are involved with cytokinin binding, which was very recently found. The CHK receptors were discovered to be linked with the endoplasmic reticulum after being tagged, gradient centrifuged, and immunoblotted (ER). At the ER, bimolecular fluorescence complementation revealed significant fluorescence. As a result, it's now considered that the CHASE region is

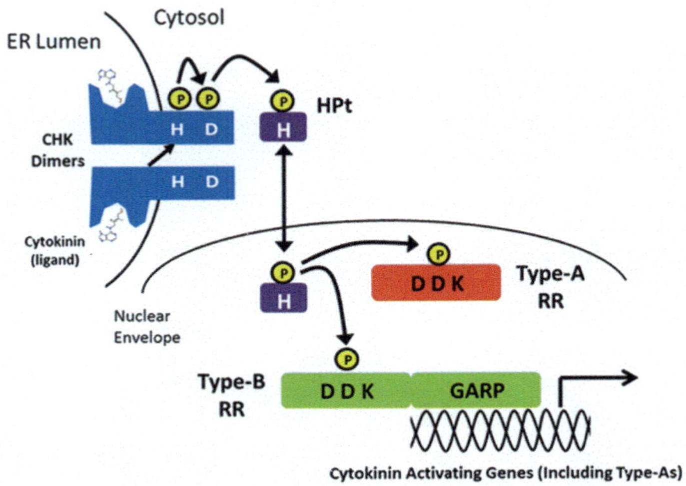

Fig. 3 Two-component cytokinin phosphorelay. Within the cell, the two-component cytokinin signalling (TCS) phosphorelay mechanism is depicted. Signalling pathway is initiated when cytokinin binds to the dimerized receptor (CHK) at the ER membrane in lumen of the ER. The CHK is subsequently auto-phosphorylated (P) at a preserved histidine residue (H) and relayed to a preserved aspartic acid remainder (D). The stimulus then passes via the cytosol to a histidine phosphotransfer protein (HPt) that then proceeds towards the nucleus, where the phosphate is transmitted to one of two types of response regulators (RRs). Both RRs contain conserved DDK amino acid sequences, and the type-B RR has a GARP transcription factor (outcome) region which connects to and stimulates cytokinin-regulated genes

detected in the ER lumen, while CHK receptors cross the ER membrane (Wulfetange et al. 2011) (Fig. 3) Keshishian and Rashotte (2015).

4 Role of Auxin and Cytokinin in Plant's Response to Abiotic Stress

Individual hormonal routes have been examined in diverse developmental situations through physiological and genetic research and current discoveries have revealed that hormones behave by generating a cascade of interrelated responses rather than by following distinct linear courses (Vanstraelen and Benková 2012). Abiotic stress has an effect on auxin and cytokinin pathway regulatory points that could illustrate some of changes in plant design and growth trends due to stress. The effects of abiotic stress on auxin and cytokinin concentrations, transportation as well as responses are discussed. This chapter summarises the genes involved in the cytokinin and auxin pathways that react to abiotic stress and their functions in stress tolerance.

4.1 Stress-Induced Modulation of Auxin and Cytokinin Biosynthetic Targets

Recent research suggests that a usable auxin production pathway could be used to manipulate Plant reactions to their surroundings (Park et al. 2013). High temperatures, for instance, block the manifestation of the YUC6 and YUC2 genes in Arabidopsis thaliana and barley (*Hordeum vulgare*), resulting in a local drop in the amounts of endogenous auxin in growing anthers and male sterility (Sakata et al. 2010). In both species, exogenic administration of indole-3-acetic acid (IAA) effectively restores male sterility (Sakata et al. 2010). The predominant response is tissue-specific reduction in auxin due to changes in the environment, like an increase in temperature, which causes pollen production will be stopped and as a result, return to be reduced. Furthermore, over-expression of the Arabidopsis YUC6 gene in potatoes (*Solanum tuberosum*) was found to be constitutive (Kim et al. 2013) or over-expression of the TRYPTOPHAN-2-MONOOXYGENASE called as iaaM in Arabidopsis improves drought resistance (Shi et al. 2014a, b). The biomass root: shoot proportion is well documented to be linked to drought resistance (Werner et al. 2010). Drought activates the YUC7 gene predominantly mostly in roots, with higher free auxin levels in the Arabidopsis activation-tagged mutant yuc7-1D promotes root development as well as architecture. As a result, yuc7-1D plants are drought tolerant and have drought-receptive genes over-expressed (Lee et al. 2012). A mutation in the CONSTITUTIVELY WILTED1 (COW1) gene of rice (*Oryza sativa*), which introduces a unique member of the YUC protein complex, results in a lower root:shoot biomass proportion and thus, improper water balance (Woo et al. 2007).

Recent research into cytokinin production and signalling has improved our knowledge about how environmental factors interplay with all these constituents to affect plant development, growth and physiology. A number of abiotic stressors have been connected to cytokinin function (Argueso et al. 2009). Endogenous cytokinin levels have been shown to decrease in response to stress over a long time (Shashidhar et al. 1996). In xylem sap, drought stress lowers zeatin riboside, trans-zeatin, isopentenyl adenosine, isopentenyl adenine and levels (Alvarez et al. 2008). Furthermore, trans-zeatin riboside transfer is considerably suppressed (Davies et al. 2005). The cytokinin 6-benzylaminopurine, on the other hand, is increased by the similar treatment, indicating that benzylaminopurine may contribute in respond to stress by retarding stress-impelled leaf senescence (McDavid et al. 1973). Furthermore, the elevated concentration of benzylaminopurine causes the osmolyte proline to accumulate (Thomas et al. 1992). A transcriptome investigation of barley plants with CKX1 in their roots revealed decreased cytokinin responsiveness via the HvHK3 cytokinin receptors, as well as activation of two stress-related transcription factors (Pospíšilová et al. 2016).

4.2 Regulation of Auxin and Cytokinin Metabolic Pathway by Stress

Extracellularly released plants peroxidases, that catalyse a peroxide breakdown, catalyse the catabolic oxidation of IAA. As seen in UV-treated duckweed and tobacco (*Lemna gibba* and *Spirodela punctate*) plants (Jansen et al. 2001), auxin oxidase activities connected with oxidases have an impact on auxin consistency and concentration (Kawano 2003). IAA peroxidases are also necessary in the regulation of IAA levels throughout root formation and growth (Vatulescu et al. 2004) and hypocotyls elongation (Cosio et al. 2009). Abiotic stress has an impact on auxin metabolism gene expression (Tognetti et al. 2012). TLD1, a GH3.13 gene that is normally repressed in terrestrial tissues but is significantly activated by water scarcity, amends for changes in rice plant structure and tissue patterns by increasing drought resistance (Zhang et al. 2009). Drought and/or salinity stress upregulate CaGH3-1 and -7 in chickpea (*Cicer arietinum*) and MtGH3-7, -8, and -9 in *Medicago truncatula*, upregulating the contributions in abiotic stress responses (Singh et al. 2015). The ability of ZmGH3 genes to respond to several abiotic stimuli and stress-associated hormones in maize demonstrates that they are auxin-stress crosstalk prudential sites.

During stress responses, glycosylation of cytokinin is also significant. In plants, cytokinins are mostly found as conjugations. The stable storage forms of cytokinins are created through reversible glycosylation (O-glycosylation and O-acetylation). Localized upregulation of intrinsic auxin and cytokinin is known to modulate stress-adaptation responses via influencing ROS homeostasis (Wang et al. 2015). Tissue-related ROS magnitude may act as a reconciler of reactions brought out via both routes, and they may be important for stress-elicited growth sculpting. Simultaneous cytokinin and auxin oxidation through sudden reactivity along with enhanced regional ROS deposition (Peer et al. 2013) could be a secondary ROS energy dissipation pathway, but its minor compared to enzyme-induced breakdown.

4.3 Stress-Induced Regulation of Auxin and Cytokinin Transport

Accumulation of flavonoids, that appear toward being unfavorable moderators of polar auxin transfer, is a characteristic of stressed plants (Peer et al. 2013) and causes auxin-dependent stress responses to be activated (Agati et al. 2013). In flavonoid-deficient mutants, changes in flavonoid accumulation alter lateral root initiation and root morphology (Buer et al. 2013). By affecting auxin transportation and distribution, flavonoids influence plant response to challenges and also the formation of stress-employed morphological reactions that modify development of plants in prior to lessen stress-induced injury to tissues or organs (Potters et al. 2009). Flavonols operate as positional signals in the root meristem, integrating auxin, cytokinin, and ROS signalling to drive root light aversion and root development, according to a recent theory (Silva-Navas et al. 2016). The discovery how flavonol glycosides affect auxin physiological productivity adds to the intricacy of flavonol control of auxin dispersal (Kuhn et al. 2011). Given that they are accountable for auxin's asymmetric distribution, it is not unexpected that constituents of the polar intercellular auxin transfer apparatus are key receivers of environmental cues. For example, the artificial polar auxin shipping inhibitors 2,3,5-triiodobenzoic acid and 1-N-naphthylphthalamic acid cause morphological alterations identical to those caused by strain, such as reduced root length, enhanced density of root hair, reduced leaf size, suppression of mesophyll cell growth as well as chlorophyll content instability (Tognetti et al. 2010).

Apart from auxin, we have limited information about cytokinin cell-to-cell transfer, so as a result, its participation in abiotic stress response in plants is uncertain. The existence of cytokinin in xylem and phloem sap suggests that cytokinin can travel extended distances both acropetally and basipetally (Bishopp et al. 2011). The xylem transports cytokinins because tZ-ribosides from root to shoot by acropetal transfer (Bürkle et al. 2003), whereas the phloem transports cytokinins as iP-type from shoot to root using basipetal transportation (Gillissen et al. 2000). Furthermore, basipetal cytokinin transport is facilitated by symplastic linkages inside the phloem, which helps to maintain the root vascular patterns (Bishopp et al. 2011) PUP1 and PUP2, two purine permeases engaged in cytokinin cellular distribution in Arabidopsis, have been identified. The existence of PUP2 in the phloem indicates that it is involved in cytokinin long way transmission (Bürkle et al. 2003). The access of zeatin-type cytokinins further into xylem delivery channel is similarly controlled by ABCG14. This envoy has been proposed for functioning as an efflux pump which is required for root-to-shoot transportation of cytokinins synthesised at the roots. It is found predominantly inside this plasma membranes of pericycle and stellar cells of root systems, overlying also with coding sequences of IPT3 and CYP73A2 (Ko et al. 2014).

Understanding how stress-adaptation responses link cytokinin and auxin transport and dispersal is a unique path with enormous potential for comprehending morphological changes and reduced rate of growth of plants subjected to surrounding challenges.

4.4 *Stress-Induced Auxin and Cytokinin Signaling Circuits*

Abiotic stresses alter the expression of various genes associated in nuclear auxin signaling, particularly auxin response factor (ARF) transcription factors and early auxin-responsive genes (Aux/IAA, SAUR and LBD). Furthermore, stress-induced auxin signalling appears to be conserved across plant species (Blomster et al. 2011). According to a thorough transcriptome analysis of auxin-linked genes in *Sorghum bicolor*, three genes (SbIAA1, SbGH3-13 and SbLBD32) were extremely elevated during salinity and drought treatment. Salinity up-regulates several ARF genes in leaves but down-regulates them through roots in sorghum. The SbARF16, SbARF10 and SbARF21 genes are persuaded by salinity in roots (Wang et al. 2010). Drought and salt stress, on the other hand, adversely regulate most Arabidopsis ARF genes (Matsui et al. 2008). OsARF11 and OsARF15 in rice, as well as GmARF33 and GmARF50 in soybean (*Glycine max*), are drought-responsive gene targets (Jain and Khurana 2009).

Arabidopsis has a multi-step phosphorelay for cytokinin signalling that involves histidine protein kinase (AHK), histidine phosphotransfer proteins (AHPs) and response regulators (ARRs) (Zürcher and Müller 2016). The two forms of ARRs are type-A partly negative regulators and type-B positive regulators (Argyros et al. 2008). Whereas phosphorylation of type-A ARRs stagnates them, it regulates transcription of cytokinin-activated sites, comprising type-A ARRs that are highly and quickly increased in relation to cytokinin (Sakai et al. 2001). Different stimuli have different effects on the genes of the cytokinin signalling component (Argueso et al. 2009). The expression of the Arabidopsis cytokinin receptors AHK2, AHK3 and AHK4 is quickly increased in respond to dehydration stress (Tran et al. 2007); increased cytokinin perception may have contribution in stress response, according to the findings. Cold, droughts and salt stress induce type-A ARR7 transcripts, while salinity and dehydration stress induce ARR5, ARR6, and ARR15 transcripts (Jiang and Deyholos 2006; Kang et al. 2013). Activation of the AHK2, AHK3 and AHK4 genes and also the type-A ARR8 and ARR9 and type-B ARR10 and ARR12 react to regulator genes, is reduced in leaves when exposed to heat (Skalák et al. 2016). Three Arabidopsis AHPs (AHP2, AHP3 and AHP5) have redundant and detrimental effects on drought stress responses. The inactivation of abovementioned three AHP genes caused in a strong, drought-resistant phenotype, which was connected to activation of defence mechanisms like improved cell membrane integrity (Nishiyama et al. 2013). Negative mediators of the osmotic stress reaction in rice have been identified as OsAHP1 and OsAHP2. The OsAHP RNAi rice plants demonstrated high osmotic resistance as their root fresh weight increased (Sun et al. 2014).

Apart from the fundamental constituents, several secondary targets of the cytokinin signalling mechanism have been connected to the abiotic stress response. APETALA2 (AP2) family members CYTOKININ RESPONSE FACTORS (CRFs) are transcriptionally upregulated via cytokinin and govern transcription of a wide range of cytokinin-response genes; several are simultaneously influenced differentially by type-B ARRs (Rashotte et al. 2006). During abiotic stress, analysis of tomato SlCRF1 and SlCRF2 transcripts indicated that the two genes have unique regulatory patterns, both within and between roots and shoot tissues. SlCRF1 expression was strongly stimulated in leaves and roots by cold, whereas it was inhibited in roots by heat. Oxidative stress, on the other hand, stimulated SlCRF2 expression in roots (Shi et al. 2014a, b). CRF6 has been suggested as a constituent of the ROS-cytokinin crosstalk regulation pathway in Arabidopsis, with the goal of attenuating cytokinin signalling as part of an adaptive stress response. CRF4 expression levels and stress tolerance were found to be positively correlated when exposed to freezing temperatures (Zwack et al. 2016). Finally, because T1R1 and AHPs are substrates of NO S-nitrosylation, ROS- and redox-responsive protein alteration during stress adjustment processes might intensify hormone signalling (Feng et al. 2013).

5 Role of Auxin and Cytokinin in Plant's Response to Biotic Stress

The main function of auxin is to control plant growth and development. Modern research, on the other hand, has emphasised the role of auxin homeostasis in plant-pathogen interconnections. A collection of auxin-inducible GH3 (Gretchen Hagen 3) family genes which encrypt auxin-conjugating enzymes regulate endogenous auxin levels in part by negative feedback (Staswick et al. 2005). Auxin-mediated disease susceptibility is often linked to a mutually antagonistic relationship between auxin and SA pathways (Pieterse et al. 2012). By stabilising Aux/IAA repressor proteins, that are part of the SA-mediated disease-tolerance system, salicylic acid inhibits auxin responses (Wang et al. 2007). Auxin signalling is critical for plant tolerance to necrotrophic fungus, in contrast to auxin-mediated vulnerability to biotrophs. The auxin signaling mutants axr (Arabidopsis auxin-resistance) 1, axr2 and axr6 all demonstrated enhanced vulnerability to the necrotrophic fungus *Plectosphaerella cucumerina* and *B. cinerea* (Llorente et al. 2008). During their infection processes, several bacterial as well as fungal microorganisms can generate auxin or alter auxin signalling in the host (Kazan and Manners 2009). IAA is produced and secreted by the rice pathogens Xoo, Xoc, and *M. oryzae* (Jiang et al. 2013). These viruses may employ IAA as a toxic factor to make rice tissues easier to infect.

Cytokinins are well-established plant growth hormones, but new research has linked them to a variety of plant-pathogen interconnections. Their impacts are frequently manifested as CK disorders, which are morphological abnormalities (Grant and Jones 2009). During infection of Arabidopsis, the fungal pathogen *Plasmodiophora brassicae*, which causes Brassicaceae clubroot disease, downregulates the CK degradation mechanism. Clubroot formation was inhibited by transgenic over-expression of CK oxidase/dehydrogenase, demonstrating the relevance of CKs in *P. Brassicaensis* pathogenicity (Siemens et al. 2006). CKs, on the contrary, have been found to have a crucial part in pathogen defence responds (Choi et al. 2011). Transgenic tobacco plants with a pathogen-inducible promoter and a bacterial ipt gene showed improved tolerace to virulent *P. syringae pv. tabaci* (Großkinsky et al. 2011). Cytokinins have also been linked to necrotrophic pathogen resistance. After infection with *Botrytis cinerea*, transgenic tomato plants with higher cytokinin levels displayed retarded leaf senescence as well as reduced disease symptoms (Swartzberg et al. 2008) and boosted CK levels in transgenic Arabidopsis increased tolerance to *Alternaria brassicicola* KACC40036 (Choi et al. 2010).

6 Auxin-Cytokinin Crosstalk

Auxin and cytokinin have physiologic effects that are mostly reliant on their concentrations; therefore, mechanisms that control their production and degradation are critical for various developmental stages. Furthermore, auxin and cytokinin interactions are mostly mediated through reciprocal affects on each other's metabolism (Jones and Ljung 2011). Co-regulated genes and similar signalling components are clearly involved in crosstalk between these phytohormones. Furthermore, crosstalk is geographically and temporally managed, allowing for response flexibility and fine-tuning. In current years, crosstalk between the two hormones has been extensively investigated at all levels: synthesis, perception and transportation (Chandler and Werr 2015). The current scenario on auxin-cytokinin crosstalk mechanisms is summarised here.

6.1 Signaling-Associated Auxin-Cytokinin Crosstalk Components

The interaction of auxin and cytokinin signalling mechanisms is important in controlling shoot apical meristem activity and embryonic root specialization. In the embryonic root and shoot-stem cell niche, the type-A ARR7 and ARR15 negative regulators of the cytokinin signalling pathway have been discovered to combine cytokinin and auxin signals. The functioning of the auxin-controlled ARR7and ARR15 genes is required for normal embryo development and Arabidopsis embryos lacking either gene show severe patterning problems (Müller and Sheen 2008). Cytokinin stimulates ARR7 and ARR15 expression in shoot meristem, while auxin has the opposite impact. This is regulated in part by the transcription factor AUXIN RESPONSE FACTOR5/MONOPTEROS (MP). These regulatory processes support auxin-cytokinin antagonistism in the root meristem while proposing a synergistic relationship between the two hormones in the shoot apical meristem, as evidenced by classic shoot regeneration experiments (Zhao et al. 2010). Aux/IAASHORT HYPOCOTYL2 (SHY2), in root apical meristems, a repressor of auxin signalling was discovered to mediate interrelation between auxin and cytokinin signalling mechanisms (Ioio et al. 2008). As a result, ARR1 and ARR12 increase SHY2 transcription in the vascular tissue of the root meristem's transition zone that inhibits the expression of PIN1, PIN3 and PIN7. A change in auxin levels causes cell differentiation and a reduction in the size of the root apical meristem (Moubayidin et al. 2010). In addition, auxin-dependent SHY2 degradation is necessary for ITP5 expression induction in the transition domain, as IPT5 activity is abolished in the shy2-2 mutant (Ioio et al. 2008). As a result, cytokinin has been demonstrated to inhibit auxin outflow from cultivated tobacco cells and to suppress the activation of most PIN genes in roots (Pernisová et al. 2009). PIN7 transcription, which is triggered by cytokinin in Arabidopsis roots, is an exception. Aside from transcriptional regulation, cytokinin has been discovered to regulate the auxin outflow carrier PIN1's endocytic recycling by diverting it to lytic destruction in vacuoles. In cytokinin-mediated developmental processes, stimulation of lytic PIN1 degeneration is a specialised method for quickly modifying distribution of auxin, rather than a default outcome of protein downregulation from plasma membranes (Marhavý et al. 2011). During lateral root primordia patterning, for example, The suppression of cytokinin signalling by AHP6 allows for correct PIN1 localization and as a result, the creation of the auxin gradient (Moreira et al. 2013). Some hormonal interactions are still poorly known, making it difficult to discern between primary, secondary and tertiary regulation. Crosstalk between auxin and cytokinin is geographically and completely appreciate developmental procedures and responds to the environment, interactions with other hormones must be taken into consideration.

6.2 *Transcriptional Crosstalk Networks of Auxin-Cytokinin in Response to Abiotic Stress*

Understanding the nature of plant growth which is tuned to a mix of environmental stimuli at the cellular and molecular levels, much as with crops in the field is crucial for developing specially customised biotechnology tools. Transcriptome investigation of the response of the entire set of genes involved in cytokinin signalling and metabolism to various environmental stresses previously shown that IPT3, IPT5, CYP735A2, LOG5, CKX4, ARR10 and CRF6 are the most responsive genes (Ramireddy et al. 2014). The effect of various abiotic stress conditions, as well as cytokinin and auxin treatments, on the whole collection of auxin and cytokinin pathway genes was investigated; They utilised gene expression data from Arabidopsis ecotypes Col-0, which are publically available in the Genevestigator database, to do the analysis (Hruz et al. 2008). Abiotic stress included oxidative stress caused by strong light (HL), heat, H_2O_2 or the ROS propagator methyl viologen (MV); osmotic stress generated by exogenous addition of mannitol, polyethylene glycol or NaCl; and dehydration and drought. Cytokinin and Auxin-related genes that are influenced by stressors and hormones at the same time were chosen. Depending on the array experiments, it is also feasible to contrast their expression in other tissues such as seedlings, leaves and roots. These target genes could be essential cytokinin and auxin interaction hubs that regulate the dynamic behaviour of cellular processes associated with stress-induced reorientation of growth, making them useful genetic tools for farmers looking for climate-resilient crops with higher yields and few stress-related morphological characters.

7 Conclusion

Different biotic and abiotic environmental conditions have a significant impact on crop productivity and development. Genetic engineering has aided modern agriculture greatly during the previous decade. Stress-suppressing proteins have greatly increased crop resilience to a variety of diseases and environmental conditions. Auxin is a phytohormone that controls a number of physiological functions in plants. Plants vary their quantity in reaction to external stimuli, allowing them to gain flexibility. It is a crucial signaling phytohormone that regulates plant development and growth in exposure to abiotic conditions such as heavy metals, nutritional shortage, drought, salinity, and temperature changes. Many auxin-signaling components and strategies, however, have still to be discovered in order to improve auxin-mediated resistance in plants. The production of stable auxin-engineered crops that function mostly under stressful and relaxed situations is a huge challenge that still has to be overcome. To obtain this, research should be concentrated on the production of transgenic crop plants with increased stress endurance while having negligible influence on produce in non-stressed conditions. The involvement of cytokinins in plant abiotic stress conditions is mostly because of their functions in cell division stimulation, meristematic cell authenticity maintenance and enhanced cellular redox opportunities throughout dry spell and nutrient disposal management (Gupta and Rashotte 2012). CKs have a significant and diverse role in growth and development of plants, particularly during direct exposure to abiotic stress, as evidenced by the literature. The utilization of cytokinin related alleles in genetic manipulation has shown considerable promise in terms of boosting crop output and stress tolerance, paving the way for more sustainable agriculture. We are grateful for the enhanced work of scientific community; a comparatively distinct vision of variability in cytokinin structure, homeostasis, signaling and cytokinin-based modulation has been established. Although, further research is required to answer the remaining queries.

Growth and development of plants are regulated by auxin and cytokinin, which have long been recognised as crucial signalling substances. In 1957, it had been found that the cytokinin:auxin ratio influences shoot and root advancement in tobacco pith tissue cultures, and also that cell discrimination could be controlled through adjustment of the respective intensities of those same two growth variables in the growth media: high amounts of cytokinin favoured shoot emergence and elevated levels of auxin favoured rooting, whereas tissue grown at equal concentrations of cytokinin and auxin grew in a poorly organized manner (Skoog and Miller 1957). In past few decades, the research community has compiled an incredible degree of expertise about the molecular and genetic processes that underlie the defence systems that plants use to sustain disruptions in their environments, as well as the physiological and developmental processes that phytohormone regulates. However, the necessity of combining the two pathways has only recently been apparent, given that stress and phytohormone modules have common elements in addition to interacting with one another. These components are part of complex signalling systems that regulate plant growth and development, preventing or attenuating cellular damage in response to stress. As a consequence, plants change their morphology to fit their new environment. In order to obtain a comprehensive knowledge regarding plant growth and development concerning environmental stimuli, future research combining genome-scale numerical simulations in cell biology, laboratory and field advancement trials, and large-scale mutational analysis will be critical. As a result, the intracellular and intercellular temporal and spatial distribution of ROS and plant hormones throughout plant organs and tissues might be a promising modification basis for enhancing agricultural yield.

References

Agati G, Brunetti C, Di Ferdinando M, Ferrini F, Pollastri S, Tattini M (2013) Functional roles of flavonoids in photoprotection: new evidence, lessons from the past. Plant Physiol Biochem 72:35–45

Alvarez S, Marsh EL, Schroeder SG, Schachtman DP (2008) Metabolomic and proteomic changes in the xylem sap of maize under drought. Plant Cell Environ 31(3):325–340

Anantharaman V, Aravind L (2001) The CHASE domain: a predicted ligand-binding module in plant cytokinin receptors and other eukaryotic and bacterial receptors. Trends Biochem Sci 26(10):579–582

Argyros RD, Mathews DE, Chiang YH, Palmer CM, Thibault DM, Etheridge N, Schaller GE et al (2008) Type B response regulators of Arabidopsis play key roles in cytokinin signaling and plant development. Plant Cell 20(8):2102–2116

Argueso CT, Ferreira FJ, Kieber JJ (2009) Environmental perception avenues: the interaction of cytokinin and environmental response pathways. Plant Cell Environ 32(9):1147–1160

Bishopp A, Lehesranta S, Vatén A, Help H, El- S, Scheres B, Helariutta Y et al (2011) Phloem-transported cytokinin regulates polar auxin transport and maintains vascular pattern in the root meristem. Curr Biol 21(11):927–932

Blomster T, Salojärvi J, Sipari N, Brosché M, Ahlfors R, Keinänen M, Kangasjärvi J et al (2011) Apoplastic reactive oxygen species transiently decrease auxin signaling and cause stress-induced morphogenic response in Arabidopsis. Plant Physiol 157(4):1866–1883

Brandstatter I, Kieber JJ (1998) Two genes with similarity to bacterial response regulators are rapidly and specifically induced by cytokinin in Arabidopsis. Plant Cell 10(6):1009–1019

Buer CS, Kordbacheh F, Truong TT, Hocart CH, Djordjevic MA (2013) Alteration of flavonoid accumulation patterns in transparent testa mutants disturbs auxin transport, gravity responses, and imparts long-term effects on root and shoot architecture. Planta 238(1):171–189

Bürkle L, Cedzich A, Döpke C, Stransky H, Okumoto S, Gillissen B, Frommer WB et al (2003) Transport of cytokinins mediated by purine transporters of the PUP family expressed in phloem, hydathodes, and pollen of Arabidopsis. Plant J 34(1):13–26

Cha JY, Kim WY, Kang SB, Im J, Baek D, Jung IJ, Yun DJ et al (2015) A novel thiol-reductase activity of Arabidopsis YUC6 confers drought tolerance independently of auxin biosynthesis. Nat Commun 6(1):1–13

Chandler JW, Werr W (2015) Cytokinin–auxin crosstalk in cell type specification. Trends Plant Sci 20(5):291–300

Chapman EJ, Estelle M (2009) Mechanism of auxin-regulated gene expression in plants. Annu Rev Genet 43:265–285

Cheng Y, Dai X, Zhao Y (2006) Auxin biosynthesis by the YUCCA flavin monooxygenases controls the formation of floral organs and vascular tissues in Arabidopsis. Genes Dev 20(13):1790–1799

Choi J, Huh SU, Kojima M, Sakakibara H, Paek KH, Hwang I (2010) The cytokinin-activated transcription factor ARR2 promotes plant immunity via TGA3/NPR1-dependent salicylic acid signaling in Arabidopsis. Dev Cell 19(2):284–295
Choi J, Choi D, Lee S, Ryu CM, Hwang I (2011) Cytokinins and plant immunity: old foes or new friends? Trends Plant Sci 16(7):388–394
Cosio C, Vuillemin L, De Meyer M, Kevers C, Penel C, Dunand C (2009) An anionic class III peroxidase from zucchini may regulate hypocotyl elongation through its auxin oxidase activity. Planta 229(4):823–836
Davies WJ, Kudoyarova G, Hartung W (2005) Long-distance ABA signaling and its relation to other signaling pathways in the detection of soil drying and the mediation of the plant's response to drought. J Plant Growth Regul 24(4):285–295
Dharmasiri N, Dharmasiri S, Estelle M (2005) The F-box protein TIR1 is an auxin receptor. Nature 435(7041):441–445
Dortay H, Mehnert N, Bürkle L, Schmülling T, Heyl A (2006) Analysis of protein interactions within the cytokinin-signaling pathway of Arabidopsis thaliana. FEBS J 273(20):4631–4644
El-Showk S, Ruonala R, Helariutta Y (2013) Crossing paths: cytokinin signalling and crosstalk. Development 140(7):1373–1383
Feng J, Wang C, Chen Q, Chen H, Ren B, Li X, Zuo J (2013) S-nitrosylation of phosphotransfer proteins represses cytokinin signaling. Nat Commun 4(1):1–9
Frébort I, Kowalska M, Hluska T, Frébortová J, Galuszka P (2011) Evolution of cytokinin biosynthesis and degradation. J Exp Bot 62(8):2431–2452
Gillissen B, Bürkle L, André B, Kühn C, Rentsch D, Brandl B, Frommer WB (2000) A new family of high-affinity transporters for adenine, cytosine, and purine derivatives in Arabidopsis. Plant Cell 12(2):291–300
Goda H, Sawa S, Asami T, Fujioka S, Shimada Y, Yoshida S (2004) Comprehensive comparison of auxin-regulated and brassinosteroid-regulated genes in Arabidopsis. Plant Physiol 134(4):1555–1573
Grant MR, Jones JD (2009) Hormone (dis) harmony moulds plant health and disease. Science 324(5928):750–752
Gray WM, Kepinski S, Rouse D, Leyser O, Estelle M (2001) Auxin regulates SCF TIR1-dependent degradation of AUX/IAA proteins. Nature 414(6861):271–276
Großkinsky DK, Naseem M, Abdelmohsen UR, Plickert N, Engelke T, Griebel T, Roitsch T et al (2011) Cytokinins mediate resistance against Pseudomonas syringae in tobacco through increased antimicrobial phytoalexin synthesis independent of salicylic acid signaling. Plant Physiol 157(2):815–830
Guilfoyle TJ, Hagen G (2007) Auxin response factors. Curr Opin Plant Biol 10(5):453–460
Gupta S, Rashotte AM (2012) Down-stream components of cytokinin signaling and the role of cytokinin throughout the plant. Plant Cell Rep 31(5):801–812
Hruz T, Laule O, Szabo G, Wessendorp F, Bleuler S, Oertle L, Zimmermann P (2008) Genevestigator v3: a reference expression database for the meta-analysis of transcriptomes. Adv Bioinform
Ioio RD, Nakamura K, Moubayidin L, Perilli S, Taniguchi M, Morita MT, Sabatini S et al (2008) A genetic framework for the control of cell division and differentiation in the root meristem. Science 322(5906):1380–1384
Jain M, Khurana JP (2009) Transcript profiling reveals diverse roles of auxin-responsive genes during reproductive development and abiotic stress in rice. FEBS J 276(11):3148–3162
Jansen MA, van den Noort RE, Tan MA, Prinsen E, Lagrimini LM, Thorneley RN (2001) Phenol-oxidizing peroxidases contribute to the protection of plants from ultraviolet radiation stress. Plant Physiol 126(3):1012–1023
Jiang Y, Deyholos MK (2006) Comprehensive transcriptional profiling of NaCl-stressed Arabidopsis roots reveals novel classes of responsive genes. BMC Plant Biol 6(1):1–20
Jiang CJ, Shimono M, Sugano S, Kojima M, Liu X, Inoue H, Takatsuji H et al (2013) Cytokinins act synergistically with salicylic acid to activate defense gene expression in rice. Mol Plant Microbe Interact 26(3):287–296

Jones B, Ljung K (2011) Auxin and cytokinin regulate each other's levels via a metabolic feedback loop. Plant Signal Behav 6(6):901–904

Kakimoto T (1996) CKI1, a histidine kinase homolog implicated in cytokinin signal transduction. Science 274(5289):982–985

Kang NY, Cho C, Kim J (2013) Inducible expression of Arabidopsis response regulator 22 (ARR22), a type-C ARR, in transgenic Arabidopsis enhances drought and freezing tolerance. PLoS ONE 8(11):e79248

Kawano T (2003) Roles of the reactive oxygen species-generating peroxidase reactions in plant defense and growth induction. Plant Cell Rep 21(9):829–837

Kazan K (2013) Auxin and the integration of environmental signals into plant root development. Ann Bot 112(9):1655–1665

Kazan K, Manners JM (2009) Linking development to defense: auxin in plant–pathogen interactions. Trends Plant Sci 14(7):373–382

Ke Q, Wang Z, Ji CY, Jeong JC, Lee HS, Li H, Kwak SS et al (2015) Transgenic poplar expressing Arabidopsis YUCCA6 exhibits auxin-overproduction phenotypes and increased tolerance to abiotic stress. Plant Physiol Biochem 94:19–27

Keshishian EA, Rashotte AM (2015) Plant cytokinin signalling. Essays Biochem 58:13–27

Kim JI, Baek D, Park HC, Chun HJ, Oh DH, Lee MK, Yun DJ et al (2013) Overexpression of Arabidopsis YUCCA6 in potato results in high-auxin developmental phenotypes and enhanced resistance to water deficit. Mol Plant 6(2) 337–349

Kim J, Harter K, Theologis A (1997) Protein–protein interactions among the Aux/IAA proteins. Proc Natl Acad Sci 94(22):11786–11791

Ko D, Kang J, Kiba T, Park J, Kojima M, Do J, Lee Y et al (2014) Arabidopsis ABCG14 is essential for the root-to-shoot translocation of cytokinin. Proc Natl Acad Sci 111(19):7150–7155

Krishnamurthy A, Rathinasabapathi B (2013) Auxin and its transport play a role in plant tolerance to arsenite-induced oxidative stress in *Arabidopsis thaliana*. Plant Cell Environ 36(10):1838–1849

Kuhn BM, Geisler M, Bigler L, Ringli C (2011) Flavonols accumulate asymmetrically and affect auxin transport in Arabidopsis. Plant Physiol 156(2):585–595

Lee M, Jung JH, Han DY, Seo PJ, Park WJ, Park CM (2012) Activation of a flavin monooxygenase gene YUCCA7 enhances drought resistance in Arabidopsis. Planta 235(5):923–938

Leyser O (2018) Auxin signaling. Plant Physiol 176(1):465–479

Llorente F, Muskett P, Sánchez-Vallet A, López G, Ramos B, Sánchez-Rodríguez C, Molina A et al (2008) Repression of the auxin response pathway increases Arabidopsis susceptibility to necrotrophic fungi. Mol Plant 1(3):496–509

Marhavý P, Bielach A, Abas L, Abuzeineh A, Duclercq J, Tanaka H, Benková E et al (2011) Cytokinin modulates endocytic trafficking of PIN1 auxin efflux carrier to control plant organogenesis. Dev Cell 21(4):796–804

Matsui A, Ishida J, Morosawa T, Mochizuki Y, Kaminuma E, Endo TA, Seki M et al (2008) Arabidopsis transcriptome analysis under drought, cold, high-salinity and ABA treatment conditions using a tiling array. Plant Cell Physiol 49(8):1135–1149

McDavid CR, Sagar GR, Marshall C (1973) The effect of root pruning and 6-benzyl-aminopurine on the chlorophyll content, 14co2 fixation and the shoot/root ratio in seedlings of *Pisvm sativum* L. New Phytol 72(3):465–470

Miller CO, Skoog F, Von Saltza MH, Strong FM (1955) Kinetin, a cell division factor from deoxyribonucleic acid1. J Am Chem Soc 77(5):1392–1392

Miller CO, Skoog F, Okumura FS, Von Saltza MH, Strong FM (1956) Isolation, structure and synthesis of kinetin, a substance promoting cell division 1, 2. J Am Chem Soc 78(7):1375–1380

Mittler R, Vanderauwera S, Suzuki N, Miller G, Tognetti VB, Vandepoele K, Gollery M, Shulaev V, Van Breusegem F (2011) ROS signaling: the new wave? Trends Plant Sci 16:300–309

Miyawaki K, Matsumoto-Kitano M, Kakimoto T (2004) Expression of cytokinin biosynthetic isopentenyltransferase genes in Arabidopsis: tissue specificity and regulation by auxin, cytokinin, and nitrate. Plant J 37(1):128–138

Moreira S, Bishopp A, Carvalho H, Campilho A (2013) AHP6 inhibits cytokinin signaling to regulate the orientation of pericycle cell division during lateral root initiation. PLoS ONE 8(2):e56370
Moubayidin L, Perilli S, Ioio RD, Di Mambro R, Costantino P, Sabatini S (2010) The rate of cell differentiation controls the Arabidopsis root meristem growth phase. Curr Biol 20(12):1138–1143
Müller B, Sheen J (2008) Cytokinin and auxin interaction in root stem-cell specification during early embryogenesis. Nature 453(7198):1094–1097
Nishiyama R, Watanabe Y, Leyva-Gonzalez MA, Van Ha C, Fujita Y, Tanaka M, Tran LSP et al (2013) Arabidopsis AHP2, AHP3, and AHP5 histidine phosphotransfer proteins function as redundant negative regulators of drought stress response. Proc Natl Acad Sci 110(12):4840–4845
O'Brien JA, Benková E (2013) Cytokinin cross-talking during biotic and abiotic stress responses. Front Plant Sci 4:451
Pandey V, Bhatt ID, Nandi SK (2019) Role and regulation of auxin signaling in abiotic stress tolerance. In: Plant signaling molecules. Woodhead Publishing, pp 319–331
Paponov IA, Paponov M, Teale W, Menges M, Chakrabortee S, Murray JA, Palme K (2008) Comprehensive transcriptome analysis of auxin responses in Arabidopsis. Mol Plant 1(2):321–337
Park HC, Cha JY, Yun DJ (2013) Roles of YUCCAs in auxin biosynthesis and drought stress responses in plants. Plant Signal Behav 8(6):337–349
Peer WA, Cheng Y, Murphy AS (2013) Evidence of oxidative attenuation of auxin signalling. J Exp Bot 64(9):2629–2639
Pernisová M, Klíma P, Horák J, Válková M, Malbeck J, Souček P, Hejátko J et al (2009) Cytokinins modulate auxin-induced organogenesis in plants via regulation of the auxin efflux. Proc Natl Acad Sci 106(9):3609–3614
Pieterse CM, Van der Does D, Zamioudis C, Leon-Reyes A, Van Wees SC (2012) Hormonal modulation of plant immunity. Annu Rev Cell Dev Biol 28:489–521
Pils B, Heyl A (2009) Unraveling the evolution of cytokinin signaling. Plant Physiol 151(2):782–791
Pospíšilová H, Jiskrova E, Vojta P, Mrizova K, Kokáš F, Čudejková MM, Galuszka P et al (2016) Transgenic barley overexpressing a cytokinin dehydrogenase gene shows greater tolerance to drought stress. New Biotechnol 33(5):692–705
Potters G, Pasternak TP, Guisez Y, Jansen MA (2009) Different stresses, similar morphogenic responses: integrating a plethora of pathways. Plant Cell Environ 32(2):158–169
Ramireddy E, Chang L, Schmülling T (2014) Cytokinin as a mediator for regulating root system architecture in response to environmental cues. Plant Signal Behav 9(1):5021–5032
Rashotte AM, Carson SD, To JP, Kieber JJ (2003) Expression profiling of cytokinin action in Arabidopsis. Plant Physiol 132(4):1998–2011
Rashotte AM, Mason MG, Hutchison CE, Ferreira FJ, Schaller GE, Kieber JJ (2006) A subset of Arabidopsis AP2 transcription factors mediates cytokinin responses in concert with a two-component pathway. Proc Natl Acad Sci 103(29):11081–11085
Rivero RM, Kojima M, Gepstein A, Sakakibara H, Mittler R, Gepstein S, Blumwald E (2007) Delayed leaf senescence induces extreme drought tolerance in a flowering plant. Proc Natl Acad Sci. USA 104:19631–19636
Rowland MA, Deeds EJ (2014) Crosstalk and the evolution of specificity in two-component signaling. Proc Natl Acad Sci 111(15):5550–5555
Ruzicka K, Ljung K, Vanneste S, Podhorská R, Beeckman T, Friml J, Benková E (2007) Ethylene regulates root growth through effects on auxin biosynthesis and transport-dependent auxin distribution. Plant Cell 19(7):2197–2212
Sakai H, Honma T, Aoyama T, Sato S, Kato T, Tabata S, Oka A (2001) ARR1, a transcription factor for genes immediately responsive to cytokinins. Science 294(5546):1519–1521
Sakata T, Oshino T, Miura S, Tomabechi M, Tsunaga Y, Higashitani N, Higashitani A et al (2010) Auxins reverse plant male sterility caused by high temperatures. Proc Natl Acad Sci 107(19):8569–8574
Salehin M, Bagchi R, Estelle M (2015) SCFTIR1/AFB-based auxin perception: mechanism and role in plant growth and development. Plant Cell 27(1):9–19

Schaller GE, Bishopp A, Kieber JJ (2015) The yin-yang of hormones: cytokinin and auxin interactions in plant development. Plant Cell 27(1):44–63
Shani E, Salehin M, Zhang Y, Sanchez SE, Doherty C, Wang R, Estelle M et al (2017) Plant stress tolerance requires auxin-sensitive Aux/IAA transcriptional repressors. Curr Biol 27(3):437–444
Shashidhar VR, Prasad TG, Sudharshan L (1996) Hormone signals from roots to shoots of sunflower (*Helianthus annuus* L.). Moderate soil drying increases delivery of abscisic acid and depresses delivery of cytokinins in xylem sap. Ann Botany 78(2):151–155
Shen C, Wang S, Bai Y, Wu Y, Zhang S, Chen M, Qi Y et al (2010) Functional analysis of the structural domain of ARF proteins in rice (*Oryza sativa* L.). J Exp Botany 61(14):3971–3981
Shi H, Chen L, Ye T, Liu X, Ding K, Chan Z (2014a) Modulation of auxin content in Arabidopsis confers improved drought stress resistance. Plant Physiol Biochem 82:209–217
Shi X, Gupta S, Rashotte AM (2014b) Characterization of two tomato AP2/ERF genes, SlCRF1 and SlCRF2 in hormone and stress responses. Plant Cell Rep 33(1):35–45
Siemens J, Keller I, Sarx J, Kunz S, Schuller A, Nagel W, Ludwig-Müller J et al (2006) Transcriptome analysis of Arabidopsis clubroots indicate a key role for cytokinins in disease development. Mol Plant Microbe Interact 19(5):480–494
Silva-Navas J, Moreno-Risueno MA, Manzano C, Téllez-Robledo B, Navarro-Neila S, Carrasco V, Del Pozo JC et al (2016) Flavonols mediate root phototropism and growth through regulation of proliferation-to-differentiation transition. Plant Cell 28(6):1372–1387
Singh VK, Jain M, Garg R (2015) Genome-wide analysis and expression profiling suggest diverse roles of GH3 genes during development and abiotic stress responses in legumes. Front Plant Sci 5:789
Skalák J, Černý M, Jedelský P, Dobrá J, Ge E, Novák J, Brzobohatý B et al (2016) Stimulation of ipt overexpression as a tool to elucidate the role of cytokinins in high temperature responses of *Arabidopsis thaliana*. J Exp Bot 67(9):2861–2873
Skoog F, Miller C (1957) Chemical regulation of growth and organ formation in plant tissues cultured. In Vitro Symp Soc Exp Biol 54:118–130
Smalle J, Vierstra RD (2004) The ubiquitin 26S proteasome proteolytic pathway. Annu Rev Plant Biol 55:555–590
Song Y, Wang L, Xiong L (2009) Comprehensive expression profiling analysis of OsIAA gene family in developmental processes and in response to phytohormone and stress treatments. Planta 229(3):577–591
Staswick PE, Serban B, Rowe M, Tiryaki I, Maldonado MT, Maldonado MC, Suza W (2005) Characterization of an Arabidopsis enzyme family that conjugates amino acids to indole-3-acetic acid. Plant Cell 17(2):616–627
Stepanova AN, Alonso JM (2005) Ethylene signalling and response pathway: a unique signalling cascade with a multitude of inputs and outputs. Physiol Plant 123(2):195–206
Stock AM, Robinson VL, Goudreau PN (2000) Two-component signal transduction. Annu Rev Biochem 69(1):183–215
Su YH, Liu YB, Zhang XS (2011) Auxin–cytokinin interaction regulates meristem development. Mol Plant 4(4):616–625
Sun L, Zhang Q, Wu J, Zhang L, Jiao X, Zhang S, Sun Y et al (2014) Two rice authentic histidine phosphotransfer proteins, OsAHP1 and OsAHP2, mediate cytokinin signaling and stress responses in rice. Plant Physiol 165(1):335–345
Swartzberg D, Kirshner B, Rav-David D, Elad Y, Granot D (2008) Botrytis cinerea induces senescence and is inhibited by autoregulated expression of the IPT gene. Eur J Plant Pathol 120(3):289–297
Taiz L, Zeiger E (2002) Plant physiology, 3rd edn Sinauer Associates, Inc., Sunderland, pp 424–460
Takatsuji H, Jiang CJ (2014) Plant hormone crosstalks under biotic stresses. In: Phytohormones: a window to metabolism, signaling and biotechnological applications, pp 323–350
Takei K, Takahashi T, Sugiyama T, Yamaya T, Sakakibara H (2002) Multiple routes communicating nitrogen availability from roots to shoots: a signal transduction pathway mediated by cytokinin. J Exp Bot 53(370):971–977

Tan X, Calderon-Villalobos LIA, Sharon M, Zheng C, Robinson CV, Estelle M, Zheng N (2007) Mechanism of auxin perception by the TIR1 ubiquitin ligase. Nature 446(7136):640–645

Thomas JC, McElwain EF, Bohnert HJ (1992) Convergent induction of osmotic stress-responses: abscisic acid, cytokinin, and the effects of NaCl. Plant Physiol 100(1):416–423

Tognetti VB, Van Aken O, Morreel K, Vandenbroucke K, Van De Cotte B, De Clercq I, Van Breusegem F et al (2010) Perturbation of indole-3-butyric acid homeostasis by the UDP-glucosyltransferase UGT74E2 modulates Arabidopsis architecture and water stress tolerance. Plant Cell 22(8):2660–2679

Tognetti VB, Mühlenbock PER, Van Breusegem F (2012) Stress homeostasis–the redox and auxin perspective. Plant Cell Environ 35(2):321–333

Tran LSP, Urao T, Qin F, Maruyama K, Kakimoto T, Shinozaki K, Yamaguchi-Shinozaki K (2007) Functional analysis of AHK1/ATHK1 and cytokinin receptor histidine kinases in response to abscisic acid, drought, and salt stress in Arabidopsis. Proc Natl Acad Sci 104(51):20623–20628

Vanstraelen M, Benková E (2012) Hormonal interactions in the regulation of plant development. Annu Rev Cell Dev Biol 28:463–487

Vatulescu AD, Fortunato AS, Sá MC, Amâncio S, Ricardo CP, Jackson PA (2004) Cloning and characterisation of a basic IAA oxidase associated with root induction in *Vitis vinifera*. Plant Physiol Biochem 42(7–8):609–615

Verma V, Ravindran P, Kumar PP (2016) Plant hormone-mediated regulation of stress responses. BMC Plant Biol 16(1):1–10

Wang D, Pajerowska-Mukhtar K, Culler AH, Dong X (2007) Salicylic acid inhibits pathogen growth in plants through repression of the auxin signaling pathway. Curr Biol 17(20):1784–1790

Wang S, Bai Y, Shen C, Wu Y, Zhang S, Jiang D, Qi Y et al (2010) Auxin-related gene families in abiotic stress response in Sorghum bicolor. Funct Integr Genomics 10(4):533–546

Wang Y, Deng D, Shi Y, Miao N, Bian Y, Yin Z (2012) Diversification, phylogeny and evolution of auxin response factor (ARF) family: insights gained from analyzing maize ARF genes. Mol Biol Rep 39(3):2401–2415

Wang Y, Shen W, Chan Z, Wu Y (2015) Endogenous cytokinin overproduction modulates ROS homeostasis and decreases salt stress resistance in *Arabidopsis thaliana*. Front Plant Sci 6:1004

Woo YM, Park HJ, Su'udi M, Yang JI, Park JJ, Back K, An G et al (2007) Constitutively wilted 1, a member of the rice YUCCA gene family, is required for maintaining water homeostasis and an appropriate root to shoot ratio. Plant Mol Biol 65(1):125–136

Wu J, Wang F, Cheng L, Kong F, Peng Z, Liu S, Lu G et al (2011) Identification, isolation and expression analysis of auxin response factor (ARF) genes in *Solanum lycopersicum*. Plant Cell Rep 30(11):2059–2073

Wulfetange K, Lomin SN, Romanov GA, Stolz A, Heyl A, Schmülling T (2011) The cytokinin receptors of Arabidopsis are located mainly to the endoplasmic reticulum. Plant Physiol 156(4):1808–1818

Xing H, Pudake RN, Guo G, Xing G, Hu Z, Zhang Y, Ni Z et al (2011) Genome-wide identification and expression profiling of auxin response factor (ARF) gene family in maize. BMC Genomics 12(1):1–13

Zhang SW, Li CH, Cao J, Zhang YC, Zhang SQ, Xia YF, Sun Y et al (2009) Altered architecture and enhanced drought tolerance in rice via the down-regulation of indole-3-acetic acid by TLD1/OsGH3. 13 activation. Plant Physiol 151(4):1889–1901

Zhao Z, Andersen SU, Ljung K, Dolezal K, Miotk A, Schultheiss SJ, Lohmann JU (2010) Hormonal control of the shoot stem-cell niche. Nature 465(7301):1089–1092

Zürcher E, Müller B (2016) Cytokinin synthesis, signaling, and function—advances and new insights. Int Rev Cell Mol Biol 324:1–38

Zwack PJ, Rashotte AM (2015) Interactions between cytokinin signalling and abiotic stress responses. J Exp Bot 66(16):4863–4871

Zwack PJ, Compton MA, Adams CI, Rashotte AM (2016) Cytokinin response factor 4 (CRF4) is induced by cold and involved in freezing tolerance. Plant Cell Rep 35(3):573–584

Gibberellins' Cross Talk and Signal Transduction in Plant Stress Response

Sicon Mitra, Mimosa Ghorai, Samapika Nandy, Rupa Sanyal, Abdel Rahman Al-Tawaha, Niraj Kumar Jha, Vineet Kumar, Potshangbam Nongdam, Mahipal S. Shekhawat, Arabinda Ghosh, Padmanabh Dwivedi, Devendra Kumar Pandey, and Abhijit Dey

Abstract Phytohormones play significant roles in plants for its growth and development. These phytohormones have been identified to regulate stress tolerance in plants. Gibberellic acid is one such phytohormones which carries out several important roles,

---

S. Mitra · N. K. Jha
Department of Biotechnology, School of Engineering & Technology (SET), Sharda University, Greater Noida 201310, India

M. Ghorai · S. Nandy · A. Dey (✉)
Department of Life Sciences, Presidency University, Kolkata, West Bengal, India
e-mail: abhijit.dbs@presiuniv.ac.in

R. Sanyal
Department of Botany, Bhairab Ganguly College, University of Calcutta, Kolkata, India

A. R. Al-Tawaha
Department of Biological Sciences, Al-Hussein Bin Talal University, Maan, Jordan

V. Kumar
Department of Botany, School of Life Science, Guru Ghasidas Vishwavidyalya, Bilaspur, Chhattisgarh, India

P. Nongdam
Department of Biotechnology, Manipur University, Canchipur, Imphal, Manipur, India

M. S. Shekhawat
Biotechnology Unit, Department of Botany, Kanchi Mamunivar Government Institute for Postgraduate Studies and Research, Pondicherry, India

A. Ghosh
Department of Botany, Gauhati University, Guwahati, Assam, India

P. Dwivedi
Department of Botany, Banaras Hindu University Plant Physiology, Varanasi, India

D. K. Pandey
Department of Biotechnology, Lovely Professional University, Punjab, India

N. K. Jha
Department of Biotechnology, School of Applied & Life Sciences (SALS), Uttaranchal University, 248007 Dehradun, India

T. Aftab (ed.), *Auxins, Cytokinins and Gibberellins Signaling in Plants*, Signaling and Communication in Plants,
https://doi.org/10.1007/978-3-031-05427-3_10

GA crosstalk with many other phytohormones like ethylene, jasmonic acid, brassinosteroids, strigolactones which further leads to increased tolerance to stresses such as drought, salinity, extreme temperature, etc. This chapter focuses in the signalling mechanism of GA along with its crosstalk with other related phytohormones in different plants.

1 Introduction

In past years, we have made significant advancements in our knowledge regarding transduction of Gibberellic acid (GA) signalling, which has resulted in alterations in GA-induced plant growth. The DELLA molecules, which act as suppressor of GA-induced responses, have been highlighted as playing a critical role in this area of research (Thomas and Sun 2004). The constantly growing population of human necessitates a significant rise in worldwide agricultural production. Nevertheless, adverse environmental circumstances have a significant impact on agricultural productivity. Drought, extreme temperature, heavy metal, salt stress, and other abiotic stressors cause biochemical, molecular, and physiological, damage within plants (Per et al. 2017; Tilman et al. 2011; Khan et al. 2012). Despite these limits, global output must improve by over 70% in upcoming years in order to feed an extra 2.3 billion of people (Tilman et al. 2011). Thus, innovative ways for creating the stress tolerance trait in vulnerable species of plants must be developed in order to increase agricultural yields albeit under inadequate circumstances. Plant hormones are a class of signalling molecules that enhance communication of cells at extremely low levels (Roychoudhury et al. 2015).

A recent intriguing investigation by Ueguchi-Tanaka et al. 2005 has recently resulted in the discovery of GA receptor in rice plants. This chapter focuses on the recently revealed phytohormonal crosstalk of GA with other hormones such as ethylene, strigolactones, brassinosteroids, and jasmonic acid during different stressed conditions.

2 Role of Gibberellic Acid During Drought

GA signalling, in addition to driving elongation and division of cells, interaction with several other pathways of phytohormone, it also leads to the transcription of certain stress-responsive genes which helps in establishing resistance to stressors like as drought (Jogawat 2019; Colebrook et al. 2014).

Lycopersicum esculenta (tomato) with low concentration of GA, deal with drought by raising levels of many amino acids including proline, allowing them to retain turgor of leaf for extended periods of time and enhancing drought tolerance (Omena-Garcia et al. 2019). The gid which is GA-receptor mutant variant of tomato, improves drought tolerance by delaying water loss and adversely impacting xylem growth.

Because GA regulates CK signalling, the SPINDLY (SPY) inhibitor of GA signalling when overexpressed diminishes tolerance to drought stress. The Spy mutants also exhibit better tolerance to drought by overexpressing few genes like LEA, CXX3, AREB1, and, RD20. In GA-regulated drought tolerance important role is played by GIBBERELLIN 2-OXIDASES (GAoxs) via lowering the concentration of GA. In rice, HEME OXYGENASE-1 activated by NO inhibits GA signalling induced cell death in the aleurone layer during drought. Remarkably, under drought stress, GA-regulated formation of vacuole are reduced by the activity of HEME OXYGENASE-1 along with (Wu et al. 2016). The ZFP185 which is a zinc-finger protein in rice, increases levels of GA while decreasing ABA levels, therefore it has a detrimental effect on tolerance to drought (Zhang et al. 2016). Altogether, research suggests that GA signalling is detrimental on tolerance to drought. A higher level of GA reduces tolerance to drought, while a low amount improves drought tolerance.

Crosstalk of gibberellic acid with other phytohormones have been briefly described below.

2.1 *Crosstalk Between Gibberellic Acid and Abscisic Acid (ABA)*

There exists an antagonistic relationship between abscisic acid (ABA) and gibberellins (GA), this relationship is a key regulator for the transition of development from embryogenesis to germination of seeds. In cereal aleurone layers the transcription of genes expressing hydrolytic enzymes required for development of seedling development, such as protease and a-amylases, is stimulated by GA but repressed by ABA.

Furthermore, ABA stimulates the gene transcription which may be involved in the development of tolerance against various kinds of stresses. The aleurone layers of cereal have been employed as a useful paradigm for researching effects of GA and ABA because to their well-defined molecular and biochemical markers. The RNAi-mediated knockdown of particular regulatory genes has accompanied both gain-of-function and loss-of-function approaches. The GAMyb is a transcription factor, which binds to a particular area in the promoter genes which are upregulated by GA, and SLN1 (SLR1) which is regulatory molecule which acts on upstream appears to be a functioning homolog of the GAI/RGA regulatory proteins in Arabidopsis. It has been proven that ABA induces and suppresses expression of genes via two separate signalling pathways, the upregulation of genes needs a transcription factor called ABI5, but being suppressed by a protein phosphatase 2C, and the downregulation of genes mediated by PKABA1 (a protein kinase). The suppression activity of ABA has been identified to be upstream of GAMyb, however it is localised downstream of SLN1 (SLR1) (HO et al. 2003).

Studies supporting the activity of a protein type called GRAS as a connecting hub between two phytohormones i.e., GA and ABA. Crosstalk between these two phytohormones has been growing, both physiologically and genetically, especially in the early stages of development of plant. In Arabidopsis, significant links between GA and ABA in the pathway of hormonal signalling has been observed via photo-reversibility of seed germination (Seo et al. 2006). In ABA-deficient mutant aba2-2, biosynthesis of GA was found to be increased which further caused better far red light dependent germination of seeds. This observation establishes an antagonistic relationship between biosynthesis of GA and ABA within seeds which are in germinating and growing stages (Seo et al. 2006). PHYTOCHROME-INTERACTING FACTOR3-LIKE5 (PIL5) which is a light sensitive protein mediates germination of seeds by regulating the transcription of GAI and RGA1 (both of them are DELLA proteins) along with this it also negatively regulates metabolism of ABA and GA (Oh et al. 2007). In rice, there are two WRKY factors; i.e., OsWRKY51 and OsWRKY71, these two factors are repressed by GA and induced via ABA. Therefore, these factors connect ABA and GA signalling within aleurone cells and allowing crosstalk between both the hormones (Xie et al. 2006). Furthermore, H_2O_2 has been identified as a major signalling molecule, in aleurone cells of barley it is synthesized by the activities GA which antagonises ABA signalling (Ishibashi et al. 2012). A phosphatidylethanolamine-binding protein called MFT (MOTHER OF FT AND TFL1) increases germination of seeds via a GA-mediated negative feedback control of ABA signalling (Xi et al. 2010). Studies confirmed the antagonistic crosstalk between GA and ABA signalling, and in response to environmental signals this crosstalk is a crucial process for modulating growth and development of plants (Xi et al. 2010). Moreover, seasonal modulation of seed dormancy provides more evidence for the importance of GA and ABA signalling interplay and antagonism in germination of seeds (Footitt et al. 2011). Surprisingly, synthesis of ABA and catabolism of GA elevated during dormancy, however when dormancy declined it was linked with a reduction in ABA and stimulation GA synthesis which is mediated by GA3ox1 (Footitt et al. 2011).

GRAS transcription factors of the SCR, DELLA, SCL and SHR types, have recently emerged as regulatory centres to incorporate development mediated by GA into environmental adaptability (Achard et al. 2008; Fode et al. 2008; Cui et al. 2012). Crosstalk between ABA-dependent and ABA-independent pathway with GA modulated tolerance of plants against many stresses such as drought, salinity, and cold establishing novel notions of convergence of hormonal signalling in plant cells. Interaction of the ABA and GA plant hormone pathways and convergence of signalling are expected to operate as a fundamental regulatory mechanism on response to environmental stimuli for cell-specific development. Understanding GRAS type TF (transcription factors) as essential connecting regulators for development of plants as well as in stress signalling will offer fresh and inventive methods for altering growth dynamics in response to stress within plants.

The understanding of the complexities of hormone signalling provides potential to uncover new notions for agricultural features and it will be a viable target for application in biotechnology to increase tolerance against environmental stresses in crops yet preserving development of plant and agricultural production. A diagrammatic representation of crosstalk between GA and ABA is shown in Fig. 1.

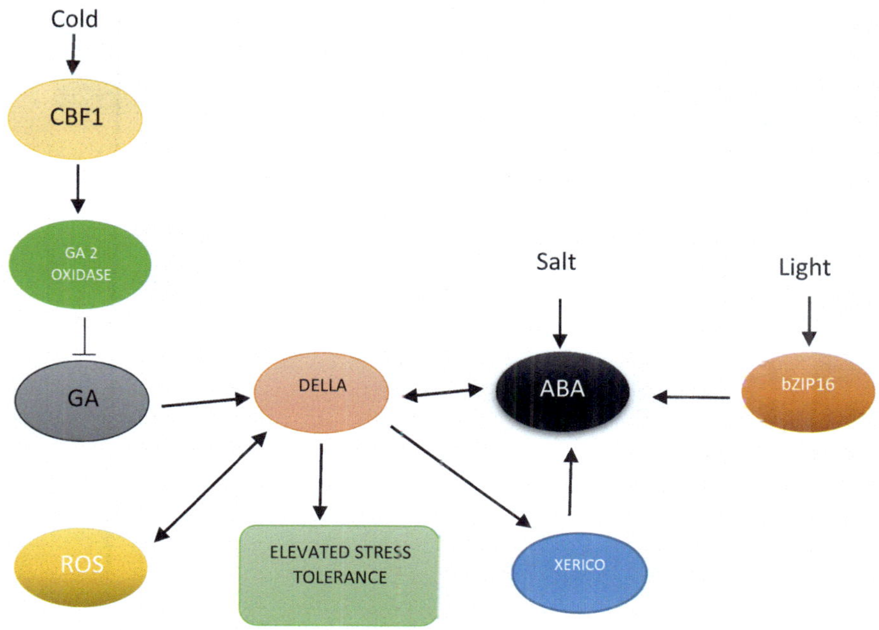

Fig. 1 Crosstalk between GA and ABA resulting in increased stress tolerance

2.2 Crosstalk Between Gibberellic Acid and Jasmonic Acid (JA)

DELLA proteins, which are important repressors of the GA signalling pathway, control Jasmonic acid signalling via competitive interaction of MYC2 with JAZ proteins (Hou et al. 2010). When GA is absent, stabilised DELLAs competes with the MYC2 for binding affinity to JAZs, therefore releasing MYC2, which then stimulates the transcription of JA-responsive genes. Moreover, in presence of elevated levels of GA, DELLA proteins are degraded along with release of JAZs, which then bind to MYC2 in the presence of elevated GA levels. Ultimately this results in the suppression of activity of MYC2 and the retardation of JA signalling. It's worth noting that the C-termini of JAZs along with domains of JA are required for connection between MYC2 and JAZs, as well as between DELLAs and JAZs (Chini et al. 2007; Hou et al. 2010). Furthermore, the C-termini have indeed been demonstrated to be crucial for the connection of JAZs with COI1 (Katsir et al. 2008). A beneficial connection between GA and JAs and GA has been proposed, because DELLA genes have been shown to interact with the transcription factors such as MYC4, MYC3, MYB24, and MYB21 (Lyons et al. 2013).

GA and JA also work together to control initiation of trichome, development of stamen, and biosynthesis of sesquiterpene (Fig. 2). Both DELLAs and JAZs interact with the similar downstream TFs, such as MYC2, and WD-repeat/bHLH/MYB

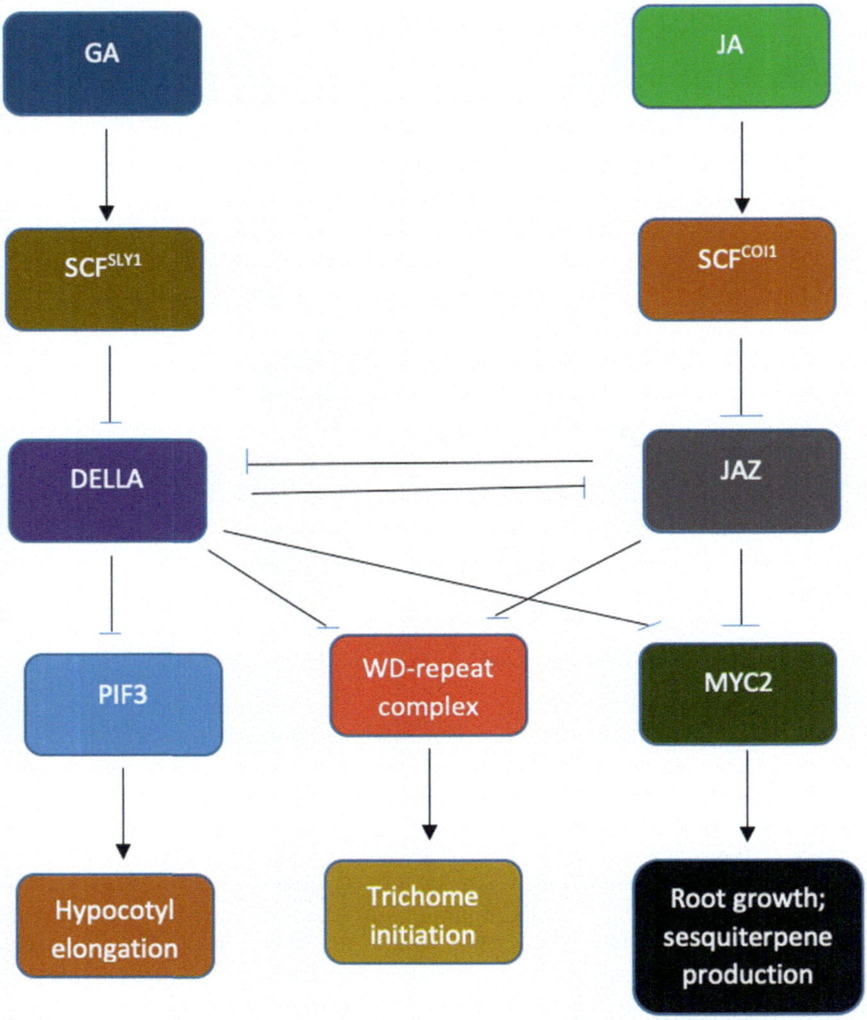

Fig. 2 Crosstalk between gibberellic acid and jasmonic acid leading to formation of trichomes

a, regulated the synergy between GA and JA in regulating formation of trichome and biosynthesis of sesquiterpene (Fig. 2). In response to signals of GA and JA, DELLAs and JAZs are degraded, and MYB or WD-repeat/bHLH/MYB complexes are released for initiation of formation of trichomes. Furthermore, GA was shown to control biosynthesis of JA by the action of DELLAs, which inhibit the production of LIPOXYGENASE (LOX) and DELAYED ANTHER DEHISCENCE1 (DAD1).

Furthermore, in case of cadmium stress, supplementing JA with GA3 showed to be a feasible strategy for allowing chickpea plants to flourish under Cd stress. The addition of JA and GA3 enhanced photosynthetic properties by protecting pigments and maintaining water balance. Aside from such activities, the combination of JA and GA3 demonstrated to be extremely effective in increasing osmoprotectants accumulation and effectively detoxifying ROS by boosting the capacities of both enzymatic and non-enzymatic antioxidants to prevent damage from oxidative stress (Ahmad et al. 2021). Therefore, there must exists some crosstalk between JA and GA3 which helps chickpea to overcome cadmium stress.

2.3 *Crosstalk Between Gibberellic Acid and Strigolactones*

The gibberellin regulates biosynthesis of Strigolactones which is revealed by in silico analysis (Marzec and Muszynska 2015), it was further confirmed to be controlled through a signal pathway of GID1-DELLA (Ito et al. 2017). Furthermore, from the reports of Gao et al. 2009, it was observed that expression of ORYZA SATIVA HOMEOBOX 1 (OSH1) is elevated in gibberellic acid-deficient mutants (Lo et al. 2008) as well as SL-signalling mutants (Gao et al. 2009), this suggests that GA and SL share a common molecular mechanism.

There have been several attempts to establish a direct molecular relationship between downstream genes of GA signalling and SLs. Moreover, D14, for example, interacts with SLR1 (SLENDER RICE1) which is a DELLA protein that functions as a signalling component downstream of GA and also acts as an inhibitor of GA signalling, in an SL-dependent way (Nakamura et al. 2013). However, degradation of SLR1 induced by SL has yet to be experimentally verified, and the operational effect(s) of this association remain mostly unclear. Modern efforts to identify a molecular link between signalling of GA and SL have generated evidence that support the functional independence of both hormones. For example, SL signalling had no effect on the deposition and stability of DELLA proteins (Bennett et al. 2016; Lantzouni et al. 2017).

2.4 *Crosstalk Between Gibberellic Acid and Brassinosteroids*

Brassinosteroids control GA production, and it has already been demonstrated that the BR-mediated TF BES1binds to promoters and controls biosynthesis of GA by binding to promoters. *Arabidopsis thaliana* mutant in BR signalling has been shown to have substantial impairment in biosynthesis of GA that has been associated with modified gene expression for biosynthesis of GA (Unterholzner et al. 2015). There is substantial reports that BRs have a role in GA-regulated control in growth under both normal and stressed circumstances. According to reports of De Vleesschauwer et al. (2012), it has been established that in rice an antagonistic relationship exists between GA and BR. Moreover, while examining the interacting role of GA and BRs, the BRs was used as a virulence factors to infect the roots of *Pythium graminicola*, and it was observed that BRs virulence factors take over the BR mechanisms in rice to fight against the disease. By the use of biosynthesis inhibitors such as uniconazole and brassinazole for GA and BR respectively, it has been revealed that BRs leads to inhibition of SA-regulated defences and it occurs downstream of biosynthesis of SA and upstream of the NPR1 and OsWRKY45 which are master regulators in defence. When BR was exogenously applied, GA-regulated immunity was upregulated because BR interferes with the metabolism of GA metabolism, resulting in stabilisation of DELLA which is a suppressor of GA and a protein called SLENDER RICE1 (SLR1), respectively. This further indicates that in *P. graminicola* by the uses of BR pathway it induces the antagonistic effect of GA and SA regulated defences. Tong et al. (2014) discovered that a mechanism driving BR–GA crosstalk becomes active based on levels of hormone and tissue. They discovered that, under different physiological circumstances, cell elongation is promoted by BR via upregulation of biosynthesis of GA. This upregulation in GA biosynthesis is mediated by enhancing the transcription of D18/GA3ox-2 which is a GA-synthetic gene in rice. Exogenously provided high quantities of BR reduced biosynthesis of GA via enhancing the transcription of the GA2-ox3 which is itself an inactivation gene for GA biosynthesis. GA inactivation gene GA2ox-3 concurrent with inhibition of BR, producing growth retardation by lowering endogenous levels of hormone. They hypothesised that GA, via a feedback mechanism, suppresses biosynthesis of BR. The interaction between DELLA and BZR1/BES1, results in regulated elongation of cell in Arabidopsis. Thereby establishing a direct signalling crosstalk between GA and BR. These cell elongations occur when DELLA proteins impact stability of protein, inhibiting BZR1 expressional activity, and GA releases DELLA-regulated suppression on BZR1, therefore promoting elongation of cells (Li and He 2013; Li et al. 2012a, b).

2.5 Gibberellic Acid and Ethylene Crosstalk in Salinity Stress

Most of the plant tissues expresses genes responsible for metabolism of GA and ethylene and this was confirmed by transcipt meta-analysis. Therefore, ACC which is a precursor for both the GAs and the ethylene is biosynthesised on a large scale (Dugardeyn et al. 2008). When ethephon along with GA3 is applied exogenously in *Amaranthus caudatus*, it ameliorate suppression of seeds germination under salt stress (Bialecka and Kepczynski 2009). When NaCl is present, the impact of ethephon was observed before GA3, and the ethephon was observed to be more potent than GA3. Furthermore, GA3 and ethylene was observed to alleviate the negative effects of salt on germination of seeds (Mohammed 2007; Kumar and Singh 1996; Khan and Huang 1988). In pea, Foo et al. (2006) observed an interaction between GA and ethylene, where the presence of phytochromes adversely inhibited ethylene biosynthesis, therefore lowering biosynthesis of GA. Moreover, GA and ethylene have a stimulatory effect in Arabidopsis on elongation of hypocotyl under light (De Grauwe et al. 2007). According to De Grauwe et al. (2008), increased biosynthesis of ethylene in mutant eto2-1 (ethylene overexpressing) is regulated by a GA responsive pathway because the gai eto2-1 double mutant (gibberellins insensitive; ethylene overexpressing) does not synthesis upregulated ethylene, implying that the GA regulates stability of the ACS5 protein. Furthermore, it was demonstrated that active ethylene signalling reduces GA levels, hence stabilising DELLA proteins (Vandenbussche et al. 2007). In Arabidopsis, GA treatment leads to degradation of five proteins of DELLA family by the action of 26S proteasome (Fu et al. 2002; Dill and Sun 2001). Furthermore, ethylene impacts stability of DELLA predominantly through variations in concertation of GA, allegedly through posttranscriptional modulation of few regulatory genes such as GA3ox, GA20ox, GA2ox (Vandenbussche et al. 2007; Achard et al. 2007).

The destabilisation of DELLA proteins by the action of GA is controlled by environmental cues (such as light and salt) as well as other plant phytohormone signalling (including ethylene and auxin), revealing the molecular underpinnings of this crosstalk (Achard et al. 2006). In *Fagus sylvatica*, the hormonal control of a gene called GA 20-oxidase has suggested that GA and ethylene crosstalk during the shift from dormancy of seed to seed germination (Calvo et al. 2004). Furthermore, treatment with GA causes expression of ACO in ga1-3 seeds (Ogawa et al. 2003). In the presence of light, an excess level of ethylene does circumvent the requirement of GA and stimulate seeds germination in ga1 mutant of Arabidopsis; however in the absence of light, the impact is considerably less (Koornneef and Karssen 1994; Karssen et al. 1989). The relationship between GA and ethylene appears to be antagonist since high concentrations of GA restore germination of seeds in etr1 mutant (Bleecker et al. 1988).

According to reports from Steffens et al. (2006), GA is not much effective on its own, however it works synergistically with ethylene to increase the number of penetrating roots and the rate of development of emergent roots. Because of the synergistic effects of ethylene and GA, they have common signalling component GA was practically ineffectual in stimulating root growth or development on its own, however when the roots were NBD (ethylene inhibitor) treated, neither of GA nor of ethylene could enhance root growth (Lorbiecke and Sauter 1999). As a result, the activity of GA on adventitious roots necessitated ethylene signalling and it was confirmed that GA functions downstream of the ethylene receptor, and activity of GA necessitated activated ethylene signalling via ethylene binding to its receptor. The biochemical characterisation of several GA-regulated mutants resulted in the identification of the DELLA and GID1, which are critical elements of pathway for the GA/GID1/DELLA pathway which allows plants to respond stimuli of GA (Harberd et al. 2009).

The pathway of GA–GID1–DELLA allows plants to endure temporary growth arrest and so withstand stress. Studies of the interaction among pathway of GA–GID1–DELLA and ethylene signalling revealed that in seedlings of DELLA-deficient mutant the growth of roots are inhibited by the action of ethylene signalling (Achard et al. 2003). Furthermore, the preservation of the enlarged structure of apical hook which is a characteristic of dark-grown seedlings which was treated with ethylene demonstrated to be reliant on the absence of inhibition of growth by the help of DELLA protein (Vriezen et al. 2004; Achard et al. 2003). According to the current research, GA and ethylene may function either antagonistically or synergistically. However, under stress, ethylene acts on DELLA protein and leads to decreased concentration of GA.

3 Conclusion and Future Prospective

GA is an important phytohormone in plants which promotes germination as well as growth of internodes during development of seedlings. Recent prevailing opinion is that different abiotic stress suppresses accumulation of GA along with the signalling pathway that go with it. Under stress conditions, there is an increase in cellular amount of DELLA proteins. These proteins helps to develop functional cross-talks with different phytohormones in plants such as ethylene, strigolactones, jasmonic acid, absicic acid, brassinosteroids, etc. Furthermore, GA controls the amount of reactive oxygen species within the cell. Under different types of stress conditions such as drought, salinity, high temperature, many key TFs such as DREBs, MYCs, JAZ, PIFs and CBFs engage in signalling pathway of GA. The final GA-regulated physiology under stress is complicated by such quick but nuanced interplay among various phytohormones.

Abiotic stressors such as salinity, cold, drought have been shown to cause epigenetic changes. Transgenic rice plants that upregulates the production of GA via overexpressing GA2ox, have produced high yielding variants with improved tolerance to stress. Furthermore, genome wide research must be carried out to uncover new catabolic locus of GA that may be successfully mapped for strict stress resistant cultivars, and high yielding variant. Effective field testing, accompanied by production of these transgenic crops, will assure worldwide food quality as well as a huge boost to the agrarian economy.

References

Achard P, Baghour M, Chapple A, Hedden P, Van Der Straeten D et al (2007) The plant stress hormone ethylene controls floral transition via DELLA-dependent regulation of floral meristem-identity genes. Proc Natl Acad Sci USA 104:6484–6489

Achard P, Cheng H, De Grauwe L, Decat J, Schoutteten H, Moritz T, Van Der Straeten D, Peng J, Harberd NP (2006) Integration of plant responses to environmentally activated phytohormonal signals. Science 311:91–94

Achard P, Gong F, Cheminant S, Alioua M, Hedden P, Genschik P (2008) The cold-inducible CBF1 factor-dependent signalling pathway modulates the accumulation of the growth-repressing DELLA proteins via its effect on gibberellin metabolism. Plant Cell 20:2117–2129

Achard P, Vriezen WH, van der Straeten D, Harberd NP (2003) Ethylene regulates Arabidopsis development via the modulation of DELLA protein growth repressor function. Plant Cell 15:612816–612825

Ahmad P, Raja V, Ashraf M, Wijaya L, Bajguz A, Alyemeni MN (2021) Jasmonic acid (JA) and gibberellic acid (GA3) mitigated Cd-toxicity in chickpea plants through restricted cd uptake and oxidative stress management. Sci Rep 11(1):1–7

Bennett T, Liang Y, Seale M, Ward S, Müller D, Leyser O (2016) Strigolactone regulates shoot development through a core signalling pathway. Biology Open 5:1806–1820

Bialecka B, Kepczynski J (2009) Effect of ethephon and gibberellin A3 on *Amaranthus caudatus* seed germination and a- and b-amylase activity under salinity stress. Acta Biol Cracov Ser Bot 51:119–125

Bleecker AB, Estelle MA, Somerville C, Kende H (1988) Insensitivity to ethylene conferred by a dominant mutation in *Arabidopsis thaliana*. Science 241:1086–1089

Calvo AP, Nicolas C, Nicolas G, Rodríguez D (2004) Evidence of a cross-talk regulation of a GA 20-oxidase (FsGA20ox1) by gibberellins and ethylene during the breaking of dormancy in *Fagus sylvatica* seeds. Physiol Plant 120:623–630

Chini A, Fonseca S, Fernandez G, Adie B, Chico JM, Lorenzo O, Garcia-Casado G, Lopez-Vidriero I, Lozano FM, Ponce MR, Micol JL (2007) The JAZ family of repressors is the missing link in jasmonate signalling. Nature 448:666–671

Colebrook EH, Thomas SG, Phillips AL, Hedden P (2014) The role of gibberellin signalling in plant responses to abiotic stress. J Exp Biol 217(1):67–75

Cui H, Hao Y, Kong D (2012) SCARECROW has a SHORT-ROOT-independent role in modulating the sugar response. Plant Physiol 158:1769–1778

De Grauwe L, Dugardeyn J, Van Der Straeten D (2008) Novel mechanisms of ethylene-gibberellin crosstalk revealed by the gai eto2-1 double mutant. Plant Signal Behav 3:1113–1115

De Grauwe L, Vriezen WH, Bertrand S, Phillips A, Vidal AM, Hedden P, Van Der Straeten D (2007) Reciprocal influence of ethylene and gibberellins on response-gene expression in *Arabidopsis thaliana*. Planta 226:485–498

De Vleesschauwer D, Van Buyten E, Satoh K, Balidion J, Mauleon R, Choi IR, Ver-Cruz C, Kikuchi S, Hofte M (2012) Brassinosteroids antagonize gibberellin- and salicylate-mediated root immunity in rice. Plant Physiol 158:1833–1846

Dill A, Sun T (2001) Synergistic derepression of gibberellin signaling by removing RGA and GAI function in *Arabidopsis thaliana*. Genetics 159:777–785

Dugardeyn J, Vandenbussche F, Van Der Straeten D (2008) To grow or not to grow: what can we learn on ethylene–gibberellin cross-talk by in silico gene expression analysis? J Exp Bot 59:1–16

Fode B, Siemsen T, Thurow C, Weigel R, Gatz C (2008) The Arabidopsis GRAS protein SCL14 interacts with class II TGA transcription factors and is essential for the activation of stress-inducible promoters. Plant Cell 20:3122–3135

Foo E, Ross JJ, Davies NW, Reid JB, Weller JL (2006) A role for ethylene in the phytochromemediated control of vegetative development. Plant J 46:911–921

Footitt S, Douterelo-Soler I, Clay H, Finch-Savage WE (2011) Dormancy cycling in Arabidopsis seeds is controlled by seasonally distinct hormone-signaling pathways. Proc Natl Acad Sci USA 108:20236–20241

Fu X, Richards DE, Ait-Ali T, Hynes LW, Ougham H, Peng J, Harberd NP (2002) Gibberellin mediated proteasome-dependent degradation of the barley DELLA protein SLN1 repressor. Plant Cell 14:3191–3200

Gao Z, Qian Q, Liu X et al (2009) Dwarf 88, a novel putative esterase gene affecting architecture of rice plant. Plant Mol Biol 71:265

Harberd NP, Belfield E, Yasumura Y (2009) The angiosperm gibberellin-GID1-DELLA growth regulatory mechanism: how an "inhibitor of an inhibitor" enables flexible response to fluctuating environments. Plant Cell 21:1328–1339

Ho TH, Gomez-Cadenas A, Zentella R, Casaretto J (2003) Crosstalk between gibberellin and abscisic acid in cereal aleurone. J Plant Growth Regul 22(2):185–194

Hou X, Lee LYC, Xia K, Yan Y, Yu H (2010) DELLAs modulate jasmonate signaling via competitive binding to JAZs. Dev Cell 19:884–894

Ishibashi Y, Tawaratsumida T, Kondo K, Kasa S, Sakamoto M, Aoki N, Zheng SH, Yuasa T, Iwaya-Inoue M (2012) Reactive oxygen species are involved in gibberellin/abscisic acid signaling in barley aleurone cells. Plant Physiol 158:1705–1714

Ito S, Yamagami D, Umehara M et al (2017) Regulation of strigolactone biosynthesis by gibberellin signaling. Plant Physiol 174:1250–1259

Jogawat A (2019) Crosstalk among phytohormone signaling pathways during abiotic stress. Molecular Plant Abiotic Stress Biol Biotechnol 12:209–220

Karssen CM, Zagorsky S, Kepczynski J, Groot SPC (1989) Key role for endogenous gibberellins in the control of seed germination. Ann Bot 63:71–80

Katsir L, Chung HS, Koo AJ, Howe GA (2008) Jasmonate signaling: a conserved mechanism of hormone sensing. Curr Opin Plant Biol 11:428–435

Khan AA, Huang XL (1988) Synergistic enhancement of ethylene production and germination with kinetin and 1-aminocyclopropane-1-carboxylic acid in lettuce seeds exposed to salinity stress. Plant Physiol 87:847–852

Khan MI, Iqbal N, Masood A, Khan NA (2012) Variation in salt tolerance of wheat cultivars: role of glycinebetaine and ethylene. Pedosphere 22(6):746–754

Koornneef M, Karssen CM (1994) Seed dormancy and germination. In: Meyerowitz EM, Somerville CR (eds) Arabidopsis. Cold Spring Harbor Laboratory, New York, pp 313–334

Kumar B, Singh B (1996) Effect of plant hormones on growth and yield of wheat irrigated with saline water. Ann Agric Res 17:209–212

Lantzouni O, Klermund C, Schwechheimer C (2017) Largely additive effects of gibberellin and strigolactone on gene expression in *Arabidopsis thaliana* seedlings. Plant J 92:924–938

Li QF, He JX (2013) Mechanisms of signaling crosstalk between brassinosteroids and gibberellins. Plant Sig Behav 8:e24686

Li QF, Wang C, Jiang L, Sun SS, He JX (2012a) An interaction between BZR1 and DELLAs mediates direct signaling crosstalk between brassinosteroids and gibberellins in Arabidopsis. Sci Signal 5(244):ra72. https://doi.org/10.1126/scisignal.2002908

Li YH, Liu YJ, Xu XL, Jin M, An LZ, Zhang H (2012b) Effect of 24-epibrassinolide on drought stress-induced changes in *Chorispora bungeana*. Bio Plant 56:192–196

Lo SF, Yang SY, Chen KT et al (2008) A novel class of gibberellin 2-oxidases control semidwarfism, tillering, and root development in rice. Plant Cell 20:2603–2618

Lorbiecke R, Sauter M (1999) Adventitious root growth and cell cycle induction in deepwater rice. Plant Physiol 119:21–29

Lyons R, Manners JM, Kazan K (2013) Jasmonate biosynthesis and signaling in monocots: a comparative overview. Plant Cell Rep 32:815–827

Marzec M, Muszynska A (2015) In silico analysis of the genes encoding proteins that are involved in the biosynthesis of the RMS/MAX/D pathway revealed new roles of strigolactones in plants. Int J Mol Sci 16:6757–6782

Mohammed AHMA (2007) Physiological aspects of mungbean plant (Vigna radiata L. Wilczek) in response to salt stress and gibberellic acid treatment. Res J Agr Biol Sci 3:200–213

Nakamura H, Xue YL, Miyakawa T et al (2013) Molecular mechanism of strigolactone perception by DWARF14. Nat Commun 4:2613

Ogawa M, Hanada A, Yamauchi Y, Kuwahara A, Kamiya Y, Yamaguchi S (2003) Gibberellin biosynthesis and response during Arabidopsis seed germination. Plant Cell 15:1591–1604

Oh E, Yamaguchi S, Hu J, Yusuke J, Jung B, Paik I, Lee HS, Sun TP, Kamiya Y, Choi G (2007) PIL5, a phytochrome-interacting bHLH protein, regulates gibberellin responsiveness by binding directly to the GAI and RGA promoters in Arabidopsis seeds. Plant Cell 19:1192–1208

Omena-Garcia RP, Martins AO, Medeiros DB, Vallarino JG, Ribeiro DM, Fernie AR, Araújo WL, Nunes-Nesi A (2019) Growth and metabolic adjustments in response to gibberellin deficiency in drought stressed tomato plants. Environ Exp Bot 1(159):95–107

Per TS, Khan NA, Reddy PS, Masood A, Hasanuzzaman M, Khan MI, Anjum NA (2017) Approaches in modulating proline metabolism in plants for salt and drought stress tolerance: phytohormones, mineral nutrients and transgenics. Plant Physiol Biochem 1(115):126–140

Roychoudhury A, Banerjee A, Lahiri V (2015) Metabolic and molecular-genetic regulation of proline signaling and itscross-talk with major effectors mediates abiotic stress tolerance in plants. Turk J Bot 39(6):887–910

Seo M, Hanada A, Kuwahara A, Endo A, Okamoto M, Yamauchi Y, North H, Marion-Poll A, Sun TP, Koshiba T, Kamiya Y, Yamaguchi S, Nambara E (2006) Regulation of hormone metabolism in Arabidopsis seeds: phytochrome regulation of abscisic acid metabolism and abscisic acid regulation of gibberellin metabolism. Plant J 48:354–366

Steffens B, Wang J, Sauter M (2006) Interactions between ethylene, gibberellin and abscisic acid regulate emergence and growth rate of adventitious roots in deepwater rice. Planta 223:604–612

Thomas SG, Sun TP (2004) Update on gibberellin signaling. A tale of the tall and the Short. Plant Physiol 135:668–676

Tilman D, Balzer C, Hill J, Befort BL (2011) Global food demand and the sustainable intensification of agriculture. Proc Natl Acad Sci 108(50):20260–20264

Tong H, Xiao Y, Liu D, Gao S, Liu L, Yin Y, Jin Y, Qian Q, Chu C (2014) Brassinosteroid regulates cell elongation by modulating gibberellin metabolism in rice. Plant Cell 26:4376–4393

Ueguchi-Tanaka M, Ashikari M, Nakajima M, Itoh H, Katoh E, Kobayashi M, Chow TY, Hsing YIC, Kitano H, Yamaguchi I, Matsuoka M (2005) GIBBERELLIN INSENSITIVE DWARF1 encodes a soluble receptor for gibberellin. Nature 437:693–698

Unterholzner SJ, Rozhon W, Papacek M, Ciomas J, Lange T, Kugler KG, Mayer KF, Sieberer T, Poppenberger B (2015) Brassinosteroids are master regulators of gibberellin biosynthesis in Arabidopsis. Plant Cell 27:2261–2272

Vandenbussche F, Vancompernolle B, Rieu I, Ahmad M, Phillips A, Moritz T, Hedden P, Van Der Straeten D (2007) Ethylene induced Arabidopsis hypocotyl elongation is dependent on but not mediated by gibberellins. J Exp Bot 58:4269–4281

Vriezen WH, Achard P, Harberd NP, Van Der Straeten D (2004) Ethylene-mediated enhancement of apical hook formation in etiolated *Arabidopsis thaliana* seedlings is gibberellin dependent. Plant J 37:505–516

Wu H, Zheng Y, Liu J, Zhang H, Chen H (2016) Heme oxygenase-1 delays gibberellin-induced programmed cell death of rice aleurone layers subjected to drought stress by interacting with nitric oxide. Front Plant Sci 19(6):1267

Xi W, Liu C, Hou X, Yu H (2010) MOTHER OF FT AND TFL1 regulates seed germination through a negative feedback loop modulating ABA signaling in Arabidopsis. Plant Cell 22:1733–1748

Xie Z, Zhang ZL, Zou X, Yang G, Komatsu S, Shen QJ (2006) Interactions of two abscisic-acid induced WRKY genes in repressing gibberellin signaling in aleurone cells. Plant J 46:231–242

Zhang Y, Lan H, Shao Q, Wang R, Chen H, Tang H, Zhang H, Huang J (2016) An A20/AN1-type zinc finger protein modulates gibberellins and abscisic acid contents and increases sensitivity to abiotic stress in rice (*Oryza sativa*). J Exp Bot 67(1):315–326

Crosstalk Between Salicylic Acid and Auxins, Cytokinins and Gibberellins Under Biotic Stress

Devendra Singh, Vinay Kumar Dhiman, Himanshu Pandey, Vivek Kumar Dhiman, and Devendra Pandey

Abstract Plants have evolved the defense systems to fight with the attacking pathogen. Phytohormones (plant hormones) like salicylic acid (SA) play as signals to mediate and trigger different plant resistance responses. It was also found that during the plant response against the abiotic stresses like heat, chilling, osmotic, and drought stress, SA plays a crucial role. On the other side, plant hormones like cytokinins, gibberellic acids (GA), and auxin, that were found connected with the abiotic stress and developmental responses, also perform a vital role in the plant defense signaling system against pathogens. These plant hormone pathways are interrelated either synergistically or antagonistically, giving plants greater control over their adaptation to their biotic environment and utilizing their limited resources to grow and develop cost-efficiently. In order to increase its virulence and to affect the plant signaling system, pathogens also start developing the strategies. This chapter provides detailed information regarding the signaling pathway in salicylic acid and SA-mediated interactions with other plant hormones. In addition, SA-regulated physiological functions were also discussed in this chapter.

D. Singh (✉)
Department of Biotechnology, B.N. College of Engineering And Technology, Lucknow 226201, Uttar Pradesh, India
e-mail: dev7600@gmail.com

V. K. Dhiman
Department of Microbiology, Dr. Yashwant Singh Parmar University of Horticulture and Forestry, Nauni, Solan 173230, India

H. Pandey
Department of Biotechnology, Dr. Yashwant Singh Parmar University of Horticulture and Forestry, Nauni, Solan 173230, India

V. K. Dhiman
Department of Biotechnology, Himachal Pradesh University Shimla, Shimla 171005, Himachal Pradesh, India

D. Pandey
Division of Crop Improvement and Biotechnology, Central Institute for Subtropical Horticulture, Lucknow 22610, Uttar Pradesh, India

T. Aftab (ed.), *Auxins, Cytokinins and Gibberellins Signaling in Plants*, Signaling and Communication in Plants,
https://doi.org/10.1007/978-3-031-05427-3_11

1 Introduction

Salicylic acid is a small aromatic phenol derivative structurally formed of phytohormone (ortho or 2-hydroxy benzoic acid) with hydroxyl derivative produced by plants (Mishra and Baek 2021). Salicin (ß-glucoside form of salicylic alcohol) compound was the first to be isolated hence named over as salicylic acid from willow tree bark (*Salix alba*). It can also be introduced exogenously as in a synthetically formed derivative (Mishra and Baek 2021). Salicylic acid is considered "an assisting" phytohormone with widely divergent, elaborated physiological regulatory mechanisms in plant metabolism, growth, disease resistance, and stresses (Dempsey et al. 2011). As SA is involved in regulating different metabolites, it acts as a hormone that directs the biosynthesis of other hormones or signaling hormones (jasmonic acid, ethylene, auxin, etc.) when required, thus affecting direct and indirect growth regulation in plants thereby is involved in stress-induced endurance in plants (Li et al. 2019). It is widely present in various plant species involved in structural and developmental growth in plants. Ion uptake, nutrient translocation, their transport, transpiration process in stomata, gas exchange, photosynthesis all are affected by salicylic acid presence. Structural changes in leaf and chloroplast, induced flowering, pathogen resistance proteins-enzyme activity, increasing antioxidant concentration in plants thus involved in defense against virus and fungal pathogens, etc. (Blokhina et al. 2003; El-Tayeb 2005). Developmental aspects such as germination, nodulation, the yield of the plant, senescence, etc., are also affected (Vlot et al. 2009).

SA is biosynthesized from two pathways that are distinct and use different precursors and routes for its synthesis (Chen et al. 2009). The phenylpropanoid which is produced in the cytoplasm via phenylalanine, whereas other pathways, i.e., isochorismate in the chloroplast. SA in plants is present in glucosylated and methylated forms (Chen et al. 2019). The glucose conjugate has a hydroxyl group and results in the formation of SA glucoside, which is the majority; on the other side, SA glucose conjugate with carboxyl group form SA ester of glucose in lesser or minor levels (Chen et al. 2009).

SA hormone in plants is a multipotent hormone that has a significant role in defense immunity stress responses (Vlot et al. 2009). In coordination with other hormones such as cytokinin, gibberellic acid, auxin, abscisic acid, etc., it contributes to regulating different aspects of growth stages and development of anatomy in plants, influencing the biochemical mediated responses still not as clear. Its dual contribution in defense and metabolism cannot be unnoticed as a balance and homeostasis in plants is achieved through the significant role of SA (An and Mou 2011).

Hormones-based studies and results are based on studies on *Arabidopsis thaliana*, signaling pathways, and interaction of hormones we know today (Allasia et al. 2018). In monocots such as rice, an additional role of defense and by these phytohormones is seen. Cross talking can be observed in rice plants, and the influence of SA can be seen in disease resistance, traits similar to a superior variety of rice can be achieved with phytohormones signaling at specific levels in plants (Vemanna et al. 2019). Identification of regulatory proteins their roles in transforming resources such as

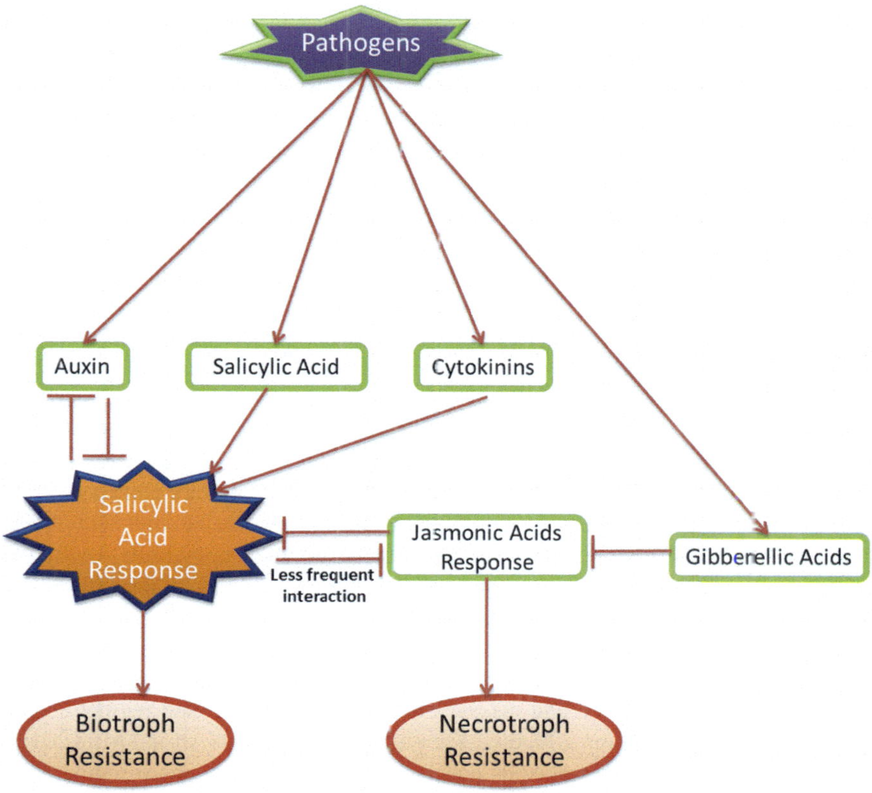

Fig. 1 A systematic overview of hormonal crosstalk that is involved in the plant defense against pathogens. Negative and positive interactions were indicated by lines with bars and arrows, respectively

plants are of significant importance. A systematic presentation of hormonal crosstalk is shown in Fig. 1.

2 Signaling Pathway in Salicylic Acid

There are receptors that sense the pathogens through patterns of molecules acting as receptors, i.e., pattern recognition receptors (PRRs) hence called pathogen-associated molecular patterns (PAMPs). All this leads to the initiation of PAMP-triggered immunity (PTI) as a defense against colonizing pathogens (Boller and Felix 2009; Schwessinger and Ronald 2012). Resistant proteins (R) identify molecules produced by pathogens as effector proteins, thereby initiating effector-triggered immunity (ETI) in plants. These responses are strong in hypersensitivity and lead to programmed cell death in infectious cells via ETI (Dodds and Rathjen 2010; Spoel and Dong 2012). Together PTI and ETI form systemic acquired resistance (SAR) so that uninfected cells remain prepared for such infections. PTI or ETI activation as defense signal pathways generate signals through infected tissues to tissues that require signaling at distal ends (Schwessinger and Ronald 2012; Spoel and Dong 2012). Hence it may be a way to retain immunity against many pathogens for a longer time via SAR. SA is found to be accumulated in leaves with systemic acquired resistance (SAR) due to pathogenic infections; over time, there is a development in SAR and increased expression of Pathogenesis related genes (PR) (Takatsuji and Jiang 2014). SA and its derivative analog compounds are directly involved in PR genes expression provide resistance against a range of viruses, bacteria, fungal infections in different monocot and dicot plants, e.g., 2,6-dichloroisonicotinic acid, benzothiadiazole S-methyl ester, probenazole (Takatsuji and Jiang 2014). Therefore, analogs of SA are also triggered for the development of systemic acquired resistance (SAR) and PR gene expression without worrying about drug resistance as with various pesticides as well as their absorption within plants. Hence these are named as Plant Activators because SA triggers the defense mechanisms for disease control in plants. Plants such as *Nicotiana tabacum* and *Arabidopsis* are able to degrade SA by the expression of enzymes within them, thereby preventing the SAR triggering (Lefevere et al. 2020).

Such conditions can be prevented by introducing endogenous SA. Therefore, SA plays an important role in introducing the defense mechanism, essentially becoming a regulator for plant defense and disease control. Various upstream regulators can control the SA accumulation affecting disease control in plants (An and Mou 2011). *Arabidopsis* and some other species consist of NPR1, a downstream regulator of SA acting as transcription coactivator of PR1. NPR1 (termed as NONEXPRESSER OF PR) and NPR3/NPR4 are two classes of receptors that support the SA defense hormone (Liu et al. 2020). The receptors are able to regulate gene expression induced by SA in two pathways, thus stimulating the genes involved in defense and immunity. NPR1 is a transcriptional activator, and its binding with SA promotes NPR1 activity (Ding et al. 2018).

On the other hand, NPR3/NPR4 act as a transcriptional repressor for SA-induced genes during no pathogenic infection. When SA is present, de-repression of NPR3/NPR4 takes place due to its binding to SA, thus activating defense genes (Ding et al. 2018; Liu et al. 2020). NPR1 and NPR4 require SA for the biosynthesis of N-hydroxypipecolic acid (NHP) essential for induced systemic acquired resistance (SAR). NPR protein family receptors NPR1 and NPR3/NPR4 are essential for plant defense and immune systems that are triggered by the positive feedback of SA and its modified forms with hydroxylation and glycosylation (Liu et al. 2020). When there is an increase in pathogenic infections, the rate level of SA biosynthesis increases and is produced mainly by the isochorismate pathway in the Arabidopsis plant. Isochorismate synthase 1 (ICS1) is a SA biosynthesizing enzyme encoded by SA-DEFICIENT 2 (SID2) expression when pathogens are detected. Both SID2 and ICS1 induction is due to the presence of transcriptional factor Systemic Acquired Resistance Deficient 1 (SARD1) and Cam binding protein (CBP60g), which leads to reduction of ICS1, thus assisting SA biosynthesis (Zhang et al. 2010; Wang et al. 2011). 2,5-Dihydroxybenzoic acid (DHBA) is a prominent form of benzoic acid that plays a major part in catabolizing SA and maintaining its level or homeostasis through hydroxylation. Another form of SA hydroxylates, i.e., SA 5-hydroxylase (S5H), is encoded by protein DMR6, which is further converted to 2,5-DHBA, finally forming SA also provides disease resistance (Zhang et al. 2017).

3 SA and Other Hormones

Plants consist of several hormones that interact within themselves in homeostasis for the proper functioning of plant metabolism. Some of these are also involved in immune and defense signaling pathways, unlike growth, maturation and other developmental, metabolic processes (Morgan and Connolly 2013). Salicylic acid initiates signaling with other phytohormones and proteins, thereby forming a cascade of responses and activating immune responses during pathogens invasion as well as differentiating them from damaged cells or foreign cells. Phytohormones such as jasmonic acid, abscisic acid. Auxin, cytokinin, gibberellic acid, peptides, brassinosteroid, etc. (Takatsuji and Jiang 2014).

4 SA Interaction with Auxin

Auxin is involved in overall plant growth, maturation as well as development. Pathogens can manipulate auxin biosynthesis within plants or even produce themselves, ceasing or preventing the hormone from functioning as in plant development (Chen et al. 2007; Robert-Seilaniantz et al. 2007). With the advancement in time, plants have evolved to repress the auxin signaling by pathogens. Plants that produce an excess of SA as defense signaling molecules act as resistant phenotype against auxin level; clearly, SA is responsible for interfering with auxin level in such plants (Mishra and Baek 2021). Hence, SA induces global repression of genes involved in auxin sensitivity via regulating Aux/IAA repressors mechanism. Therefore, blocking auxin sensitivity helped in increasing resistance against such pathogens (Wang et al. 2007), whereas introducing auxin via external sources causes the promotion of pathogens. Enzymes involved in crosstalk between SA and auxin GH3.5 that act as a conjugate between both phytohormones are key to switching the repression on and off (Westfall et al. 2016). Also, a low level of jasmonic acid indicates loss of arf6 and arf8 as auxin response factor genes required for the expression of auxin in plants due to mutation (Hentrich et al. 2013). Both SA signaling as well as auxin signaling is opposite to each other in an antagonistic way.

5 SA and Abscisic Acid

Abscisic acid is a key phytohormone in adapting abiotic stress in plants. It negatively regulates with SA as the accumulation of exogenous supply of ABA prevents accumulating SA, indirectly decreasing pathogens resistance in plants like *Arabidopsis* against *Pseudomonas syringae* (Mohr and Cahill 2003). ABA can also decrease SAR induction consecutively; it also interacts negatively with SA for molecular signaling (Yasuda et al. 2008). *Arabidopsis* and tomato have been shown to produce mutants that synthesize a lesser amount of phytohormone. ABA, conversely increasing resistance to pathogens and induced defense mechanisms compared to wild type. ABA is considered to have a negative impact on SA, which is directly related to SAR and other defense mechanisms decreasing pathogenic resistance in plants (Liu and Hou 2018). ABA affects callose deposit, increases reactive oxygen species, interference with genes involved in defense mechanisms.

6 SA and Cytokinins

Cytokinins are known to differentiate cells by proliferating them and multiplying their numbers during plant development. CKs have been found to be indulged in some pathogenic interaction in plants (Jameson 2000). In the case of *Agrobacterium tumefaciens* crown gall infection of dicotyledonous plants, overproduce CKs and indole acetic acid (Auxin) is due to genes (IPT for CKs, iaaM/H for auxin) involved in the production of enzymes isopentenyl transferase for CKs and enzymes tryptophan-2-monooxygenase and indoleacetamide hydrolase production for auxin present in bacterial DNA introduces in plants during infection (Jameson 2000). Cytokinins can be used by pathogens as a virulence factor in plants. CKs can also modulate SA signaling against hemibiotrophic bacteria (Pst DC3000) and biotrophic oomycete in *Arabidopsis* (Choi et al. 2010; Argueso et al. 2012). CKs form a complex (consisting of CK-activated transcriptional factor, *Arabidopsis* response regulator 2, TGA3 from a protein in *Arabidopsis*) with SA responsive transcriptional factor. PR1 and PR2 genes promoters bind to this complex for positive induction of defense response (Choi et al. 2010). In transgenic plants such as tomatoes, CKs were able to delay senescence weaken the Botrytis cinerea infection, whereas, in transgenic *Arabidopsis*, CK was able to endure more resistance from fungal pathogen *Alternaria brassicicola* KACC40036 (Choi et al. 2010).

7 SA and Gibberellic Acid

Phytohormone such as gibberellic acid (GA) has a primary role in growth promotion. It was first identified in *Gibberella fujikuroi,* a fungal pathogen of rice causing abnormal growth in rice plants (Cen et al 2020). However, GA was also observed to influence signaling during pathogen interference with plant activities. Bioactive GA hyperaccumulation in rice is due to a mutation in the *Eui1* gene, which is responsible for encoding GA degradation makes rice susceptible to pathogens *Magnaporthe oryzae*, but overexpressing the same gene *Eui1* resulted in resistance against pathogens (Yang et al. 2008). Thus, GA's role in resistance can be negative in hemibiotrophic pathogens. In another study, the gid1 mutant of rice showed resistance to fungal pathogen *Magnaporthe oryzae* due to its defectiveness for GA perception (Tanaka et al. 2006). Analyzing the quadruple (*rgl2-1*, *rga-t2*, *gai-6*, *rgl1-1*) infected *della* mutants genes revealed SA marked genes are induced earlier than JA (Chen et al. 2017). Hence it may be noted that there is a modulation in balancing SA and JA mediated in defense signaling. DELLA proteins which negatively regulate GA signaling due to mutation in *Arabidopsis*, DELLA proteins are responsible for susceptibility against biotrophs and resistance against necrotrophs (Navarro et al. 2008). As SA and JA genes marker patterns suggest that DELLA mutant proteins were able to resist disease in rice, it may be due to cross-talk between SA and JA in rice and Arabidopsis (Navarro et al. 2008).

Rice dwarf virus (RDV) is responsible for repressing GA biosynthesis, causing dwarfism in rice phenotypes similar to GA defective rice mutants, application of exogenous GA was able to restore normal phenotype. RDV modulation of GA metabolism for causing disease showed repression of *ent*-kaurene oxidase enzymes that is responsible for GA biosynthesis (An and Mou 2011). The *gid1* mutant showed resistance against blast fungus due to its defectitity for GA reception, thereby accumulating GA (Takatsuji and Jiang 2014).

8 SA-Regulated Physiological Functions

8.1 Effect on Seed Germination

Seed germination is regulated by different phytohormones such as gibberellins, auxins, cytokinins; SA's role in germination is ambiguous or unclear as it has been seen that SA affects the seed germination in a positive and negative way, it can increase seed vigor or sometimes can cease growth (Lee et al. 2010). In *Arabidopsis thaliana,* less than 1 mM concentration inhibits or slows the germination. Similarly, in barley, less than 0.250 mM SA showed a similar result impacting the negative effect of germination (Rivas-San and Plasencia 2011). In maize, germination is ceased with an SA level of 3 to 5 mM. SA regulation in the case of germination is negative due to its induction of oxidative stress. Oxidative stress was observed in the Arabidopsis plant when SA was treated up to 5 mM. An increase in H_2O_2 increases Cu and Zn activity when there is a lack of antioxidants such as catalases or peroxides etc. (Rivas-San and Plasencia 2011).

9 Photosynthesis

SA is responsible for affecting leaf, chloroplast structure, and RuBisCO activity due to its interaction with other phytohormones. SA regulates photosynthesis due to its effect on leaves, chloroplast structure, chlorophyll and carotenoid content, stomatal closure, enzymes such as RuBisCO carbonic anhydrase. Treating higher SA of 1–5 mM can affect the photosynthetic rate, RuBisCO activity in barley, reduction in chlorophyll content in wheat Arabidopsis plants. RuBisCO activity declination resulted in a 50% reduced protein level compared to control. Exogenous application of SA resulted in altered anatomy in leaves due to reduction in adaxial, abaxial epidermis, and mesophylls. Other changes include are increased chloroplast volume, grana thylakoids swelling, coagulated stroma. Therefore, indirect changes to plant structure lead to changes to its metabolic functioning and essential enzymes like RuBisCo. All resulted in a lower rate of photosynthesis, especially a higher level of SA effect on thylakoid membranes and stroma. Moharekar et al. have observed that SA was able to regulate the synthesis of carotenoids and xanthophylls but decreased the chlorophyll a and b ratio in wheat. Fariduddin et al. 2003 studied the foliage application of SA. Its derivative was able to increase transpiration rate, transpiration, and stomatal conductance in *Brassica juncea.* Similarly, in soybean foliage, the application of SA increased water efficiency, transpiration, and CO_2 level (Yusuf et al. 2013).

10 Respiration

SA has a regulation role in maintaining alternative oxidase signaling during stress in plants. SA can express the regulation of AOX in positive and negative depending on its concentration or level (Vanlerberghe 2013). The lower level of SA induces transcription proteins but not at a higher level. In thermogenic plants such as *Sauromatum guttatum* SA induces gene expression for AOX pathways regulation. In *Arabidopsis,* post transcription mechanisms involve abundance in the transcript and protein (Vanlerberghe 2013). As AOX plays a major role in providing metabolic homeostasis, its regulation in plants is important, especially during stress. Thus, SA is involved in regulating gene expression for AOX in both the thermogenic and non-thermogenic plant species. SA has a role in resisting major respiration choking chemicals such as cyanide as observed in tobacco cell culture suspension; increasing the amount of SA causes increased cyanide resistance and oxygen uptake measured by calorimetry as heat rate evolves in cells. Genes such as NtAOX1 are involved in expressing proteins abundance showed increased transcription when treated with SA (Rivas-San and Plasencia 2011).

11 Flowering

SA is known to influence the flowering in plants, as we know during the course of studies in flowering. It was first observed in tobacco callus as SA induces flowering in the callus when injected in μM concentration. It was found that aphid honeydew induces flowering in plants such as Lemnagibba kept under observation in a non-photoinductive cycle of light due to secretion of SA via phloem transmissible factor (Wada et al. 2010). SA was also observed to induce flowering in many plants genera, such as the Lemnaceae family, in both short as well as long-day plants and photoperiod sensitive plants. In species such as Pharbitisnil, stress conditions like poor nutrition induce flowering, amino-oxyacetic acid is used to treat flowering, but the application of SA restores similar conditions for flowering. Thus, concluding the importance of SA in inducing flowering in special conditions such as poor nutrient stress in plants (Wada et al. 2010; Yusuf et al. 2013).

12 Senescence

Senescence is when metabolic activity slows down in a plant. Declined photosynthesis rate, increased ROS level, and decreased antioxidants in plants are the signs of senescence. As SA is involved in regulating other processes such as homeostasis around the cells in plants, the photosynthesis rate is not surprising; it is certainly involved in regulating senescence in plants (Takatsuji and Jiang 2014). Accumulation of SA is also responsible for senescence up to some extent. Arabidopsis plants have shown an increase of SA utmost four times during the mid-stage of senescence. Interfering or abrupting biosynthesis of SA, introducing genes that slow or ceases its biosynthesis (transgenic NahG, mutant pad4), or abrupting signaling pathways such as in NPR1 can alter the stages of senescence, delaying its effect such as necrosis and yellowing leaves compared to wild type for same plants (Vogelmann et al. 2012; Takatsuji and Jiang 2014).

13 Growth

Growth and survival in plants are crucial for all life forms to exist. A plant species can exist only when it grows and flourish in the surroundings while maintaining its development. SA phytohormone affects growth depending upon plant species and their developmental stage. Soybean plants showed increases in root shoot growth with application of SA; as the level of SA was increased from 10 μM to 10 mM, there is an increase in growth up t0 45%, respectively, after a week compared to control (Takatsuji and Jiang 2014). Wheat also showed a similar result when 50 μM SA applications were given development in the apical meristem of roots was observed.

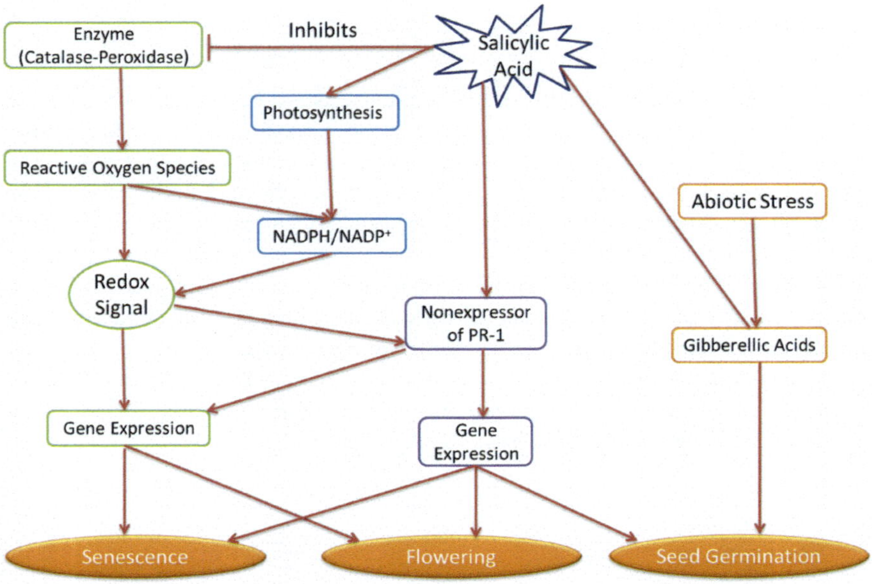

Fig. 2 A pictorial representation of salicylic acid (SA) functions in plant development and growth. Salicylic Acid is perceived by a transcriptional activator (Nonexpressor of PR-1) which regulates the expression of a gene that may involve seed germination, senescence, and also in flowering. Further, Salicylic Acid also inhibits the catalase-peroxidase enzyme activity, thus modulating the levels of ROS (reactive oxygen species). NPR1 (Nonexpressor of PR-1) oligomerization is also redox modulated

In chamomile plants, 50 μM SA stimulated 32% growth in leaf and 65% roots in growth, but with increasing the concentration as high as 250 μM ceases the growth (Kovacik et al. 2009). In *Arabidopsis, thaliana* trichome density and number is negatively affected the SA level compared to control (Takatsuji and Jiang 2014). Figure 2 shows the functions of SA in plant growth and development.

14 Concluding Remarks and Future Perspectives

Salicylic acid regulates signaling pathways that involve plant defense mechanisms with a varied cascade of receptors and defense action against pathogens. Many of the phytohormones are affected by SA, and crosstalk between SA and other hormones may influence the plant growth, development, homeostasis, maturity, responsiveness against different stresses. GA, CK, and auxin are the main phytohormones that regulate growth, cell division, development, and senescence in plants; governing these phytohormones need signaling and cross-talk for regulation of pathways and expression of genes responsible for phenotype characters required by the plant. These may be for survival, development, or establishment of the plant. SA is the junction that signals the other phytohormones directly or indirectly. It may regulate other hormones in positive or negative feedback. Crosstalking is required for adaptation to the environment by plants as signaling between hormones maintains the functionality of metabolism, and with changes in the environment, plants require sensing and adapting for survival. Crosstalk can be antagonistic depending on plant response to stress. Crosstalks between hormones are affected by different factors such as hormone level, age and development stage, environmental stress to the plants. Also, plant-pathogen interaction can stimulate hormones; differently, signaling can be a specific pathway that merges with another pathway depending upon the specificity of the requirement by the plant. Pathogenic infections influence integrated signaling in interconnected, complex coordination of hormones leading to activation of defense genes such as SAR and cascade of ETI and PTI mechanisms of PR genes. NPR1 and NPR3/NPR4 induction by SA in defense against pathogens. The same level of a hormone can have different effects on plants, negative or positive regulation in, especially in growth. Excess of SA inhibits growth in some plants but at an adequate level induces growth. SA can induce flowering, but at the same time, it can induce senescence by the accumulation of ROS in some parts of the plant. SA regulates other enzymes that are beneficial for plants, such as AOX required for the removal of scavengers that choke respiration in plants. SA also plays an important role in the resistance of diseases, as in the case of rice, SA interaction with JA regulates DELLA proteins required for resistance against fungal pathogens. Some interactions impact negatively the plants, such as ABA and SA, where ABA is antagonistic to SA.

Crosstalk between phytohormones can be experimentally tested for beneficial effects in plants as it plays an important role in the survival of plants; beneficial interactions provide more evolutionary adaptation in plants. Furthermore, these phytohormones can be exogenously introduced to plants that show positive results for growth development, as, for the economic value of crop application, part of hormones can be widely used. Disease resistance in plant type variety without using harmful chemicals can be induced based on requirement levels of phytohormones. Plant mutants can also sometimes alter the utilization of hormones in positive ways related to the development of plants or adaptation for newer conditions. Therefore, identifying the potential of phytohormones such as SA and inducing its application part is necessary by observing their role with changes in the environment. The same hormone can be potentially advantageous to certain plants but not to others.

As several processes that involve SA are not specifically known, SA's role in stress, such as both biotic and abiotic, is based on the plant sensitivity level of a SA during mitigation of stresses. Application of SA such as spraying, irrigating, and solutions have also induced its level in plants, thus involved in resistance to many stresses and diseases that may not be known to us. SA functionality can vary with plant species as a complex signaling pathway depends on the response of the plant to stress. Genes can be induced in the presence of SA as a protective response to the environment. Exogenous SA may not be linked directly to its endogenous level, but they surely affect plant metabolic and physiological behavior. Also, the plant's genetic nature is important and needs to be compatible with SA level in a positive way, such as in dicot plant of rice or tobacco or in Arabidopsis. The effect of SA in plant resistance to stresses and pathogens can be contradictory. The same treatment of SA level in plants can provide resistance to one stress but simultaneously can make plants vulnerable to other kinds of stresses that may not be as critical but surely can influence plants to cope with surroundings. SA exertion of different types of stresses is phenomenal but may be dependent or independent of plant species, pathways, and signaling, crosstalk with other hormones that may or may not be present at the required level.

References

Allasia V, Industri B, Ponchet M, Quentin M, Favery B, Keller H (2018) Quantification of salicylic acid (SA) and SA-glucosides in *Arabidopsis thaliana*. Bio-protocol 8(10):e2844. https://doi.org/10.21769/BioProtoc.2844

An C, Mou Z (2011) Salicylic acid and its function in plant immunity F. J Integr Plant Biol 53(6):412–428

Argueso CT, Ferreira FJ, Epple P, To JP, Hutchison CE, Schaller GE, Dangl JL, Kieber JJ (2012) Two-component elements mediate interactions between cytokinin and salicylic acid in plant immunity. PLoS Genet 8:1–13

Blokhina O, Virolainen E, Fagerstedt KV (2003) Antioxidants, oxidative damage and oxygen deprivation stress: a review. Ann Bot 91(2):179–194

Boller T, Felix G (2009) A renaissance of elicitors: perception of microbe-associated molecular patterns and danger signals by pattern-recognition receptors. Annu Rev Plant Biol 60:379–406

Cen YK, Lin JG, Wang YL, Wang JY, Liu ZQ, Zheng YG (2020) The Gibberellin producer *Fusarium fujikuroi*: Methods and technologies in the current toolkit. Front Bioeng Biotechnol 8:232. https://doi.org/10.3389/fbioe.2020.00232

Chen Z, Agnew JL, Cohen JD, He P, Shan P, Sheen J, Kunkel BN (2007) Pseudomonas syringae type III effector AvrRpt2 alters Arabidopsis thaliana auxin physiology. Proc Natl Acad Sci USA 104:20131–20136

Chen Z, Zheng Z, Huang J, Lai Z, Fan B (2009) Biosynthesis of salicylic acid in plants. Plant Signal Behav 4(6):493–496. https://doi.org/10.4161/psb.4.6.8392

Chen L, Xiang S, Chen Y, Li D, Yu D (2017) Arabidopsis WRKY45 interacts with the DELLA protein RGL1 to positively regulate age-triggered leaf senescence. Mol Plant 10(9):1174–1189. https://doi.org/10.1016/j.molp.2017.07.008

Chen L, Wang WS, Wang T, Meng XF, Chen TT, Huang XX, Li YJ, Hou BK (2019) Methyl salicylate glucosylation regulates plant defense signaling and systemic acquired resistance. Plant Physiol 180(4):2167–2181. https://doi.org/10.1104/pp.19.00091

Choi J, Huh SU, Kojima M, Sakakibara H, Paek KH, Hwang I (2010) The cytokinin-activated transcription factor ARR2 promotes plant immunity via TGA3/NPR1-dependent salicylic acid signaling in Arabidopsis. Dev Cell 19:284–295

Dempsey DA, Vlot AC, Wildermuth MC, Klessig DF (2011) Salicylic acid biosynthesis and metabolism. Arabidopsis Book 9:e0156. https://doi.org/10.1199/tab.0156

Ding Y, Sun T, Ao K, Peng Y, Zhang Y, Li X, Zhang Y (2018) Opposite roles of salicylic acid receptors NPR1 and NPR3/NPR4 in transcriptional regulation of plant immunity. Cell 173(6):1454–1467

Dodds PN, Rathjen JP (2010) Plant immunity: towards an integrated view of plant-pathogen interactions. Nat Rev Genet 11:539–548

El-Tayeb MA (2005) Response of barley grains to the interactive effect of salinity and salicylic acid. Plant Growth Regul 45(3):215–224

Hentrich M, Böttcher C, Düchting P, Cheng Y, Zhao Y, Berkowitz O, Masle J, Medina J, Pollmann S (2013) The jasmonic acid signaling pathway is linked to auxin homeostasis through the modulation of *YUCCA8* and *YUCCA9* gene expression. Plant j: Cell Mol Biol 74(4):626–637. https://doi.org/10.1111/tpj.12152

Jameson P (2000) Cytokinins and auxins in plant-pathogen interactions—an overview. Plant Growth Regul 32:369–380

Kovacik J, Grúz J, Bačkor M, Strnad M, Repčák M (2009) Salicylic acid-induced changes to growth and phenolic metabolism in Matricaria chamomilla plants. Plant Cell Rep 28(1):135–143

Lee S, Kim SG, Park CM (2010) Salicylic acid promotes seed germination under high salinity by modulating antioxidant activity in Arabidopsis. New Phytol 188(2):626–637. https://doi.org/10.1111/j.1469-8137.2010.03378.x

Lefevere H, Bauters L, Gheysen G (2020) Salicylic acid biosynthesis in plants. Front Plant Sci 11:338. https://doi.org/10.3389/fpls.2020.00338

Li N, Han X, Feng D, Yuan D, Huang LJ (2019) Signaling crosstalk between salicylic acid and ethylene/jasmonate in plant defense: do we understand what they are whispering? Int J Mol Sci 20(3):671. https://doi.org/10.3390/ijms20030671

Liu X, Hou X (2018) Antagonistic regulation of ABA and GA in metabolism and signaling pathways. Front Plant Sci 9:251. https://doi.org/10.3389/fpls.2018.00251

Liu Y, Sun T, Sun Y, Zhang Y, Radojičić A, Ding Y, Tian H, Huang X, Lan J, Chen S, Orduna AR (2020) Diverse roles of the salicylic acid receptors NPR1 and NPR3/NPR4 in plant immunity. Plant Cell 32(12)4:4002–4016

Mishra AK, Baek K-H (2021) Salicylic acid biosynthesis and metabolism: a divergent pathway for plants and bacteria. Biomolecules 11:705. https://doi.org/10.3390/biom11050705

Mohr PG, Cahill DM (2003) Abscisic acid influences the susceptibility of Arabidopsis thaliana to Pseudomonas syringae pv. tomato and Peronospora parasitica. Funct Plant Biol 30:461–469

Morgan JB, Connolly EL (2013) Plant-Soil Interactions: nutrient uptake. Nat Educ Knowl 4(8):2

Navarro L, Bari R, Achard P, Lison P, Nemri A, Harberd NP, Jones JD (2008) DELLAs control plant immune responses by modulating the balance of jasmonic acid and salicylic acid signaling. Curr Biol 18:650–655

Rivas-San Vicente M, Plasencia J (2011) Salicylic acid beyond defence: its role in plant growth and development. J Exp Bot 62(10):3321–3338

Robert-Seilaniantz A, Navarro L, Bari R, Jones JDG (2007) Pathological hormone imbalances. Curr Opin Plant Biol 10:372–379

Schwessinger B, Ronald PC (2012) Plant innate immunity: perception of conserved microbial signatures. Annu Rev Plant Biol 63:451–482

Spoel SH, Dong X (2012) How do plants achieve immunity? Defence without specialized immune cells. Nat Rev Immunol 12:89–100

Takatsuji H, Jiang CJ (2014) Plant hormone crosstalks under biotic stresses. In: Phytohormones: a window to metabolism, signaling and biotechnological applications, pp 323–350

Tanaka N, Matsuoka M, Kitano H, Asano T, Kaku H, Komatsu S (2006) Gid1, a gibberellin-insensitive dwarf mutant, shows altered regulation of probenazole-inducible protein (PBZ1) in response to cold stress and pathogen attack. Plant Cell Environ 29(4):619–631. https://doi.org/10.1111/j.1365-3040.2005.01441.x

Vanlerberghe GC (2013) Alternative oxidase: a mitochondrial respiratory pathway to maintain metabolic and signaling homeostasis during abiotic and biotic stress in plants. Int J Mol Sci 14(4):6805–6847

Vemanna RS, Bakade R, Bharti P, Kumar M, Sreeman SM, Senthil-Kumar M, Makarla U (2019) Cross-talk signaling in rice during combined drought and bacterial blight stress. Front Plant Sci 10:193. https://doi.org/10.3389/fpls.2019.00193

Vlot AC, Dempsey DMA, Klessig DF (2009) Salicylic acid, a multifaceted hormone to combat disease. Annu Rev Phytopathol 47:177–206

Vogelmann K, Drechsel G, Bergler J, Subert C, Philippar K, Soll J, Engelmann JC, Engelsdorf T, Voll LM, Hoth S (2012) Early senescence and cell death in Arabidopsis saul1 mutants involves the PAD4-dependent salicylic acid pathway. Plant Physiol 159(4):1477–1487
Wada KC, Yamada M, Shiraya T, Takeno K (2010) Salicylic acid and the flowering gene FLOWERING LOCUS T homolog are involved in poor-nutrition stress-induced flowering of Pharbitis nil. J Plant Physiol 167:447–452
Wang D, Pajerowska-Mukhtar K, Culler AH, Dong X (2007) Salicylic acid inhibits pathogen growth in plants through repression of the auxin signaling pathway. Curr Biol 17:1784–1790
Wang L, Tsuda K, Truman W, Sato M, Nguyen LV, Katagiri F, Glazebrook J (2011) CBP60g and SARD1 play partially redundant critical roles in salicylic acid signaling. Plant J 67(6):1029–1041
Westfall CS, Sherp AM, Zubieta C, Alvarez S, Schraft E, Marcellin R, Ramirez L, Jez JM (2016) *Arabidopsis thaliana* GH3.5 acyl acid amido synthetase mediates metabolic crosstalk in auxin and salicylic acid homeostasis. Proc Natl Acad Sci USA 113(48):13917–13922. https://doi.org/10.1073/pnas.1612635113
Yang DL, Li Q, Deng YW, Lou YG, Wang MY, Zhou GX et al (2008) Altered disease development in the eui mutants and Eui overexpressors indicates that gibberellins negatively regulate rice basal disease resistance. Mol Plant 1:528–537
Yasuda M, Ishikawa A, Jikumaru Y, Seki M, Umezawa T, Asami T, Maruyama-Nakashita A, Kudo T, Shinozaki K, Yoshida S, Nakashita H (2008) Antagonistic interaction between systemic acquired resistance and the abscisic acid-mediated abiotic stress response in Arabidopsis. Plant Cell 20(6):1678–1692. https://doi.org/10.1105/tpc.107.054296
Yusuf M, Hayat S, Alyemeni MN, Fariduddin Q, Ahmad A (2013) Salicylic acid: physiological roles in plants. In: Salicylic acid. Springer, Dordrecht, pp 15–30
Zhang Y, Xu S, Ding P, Wang D, Cheng YT, He J, Gao M, Xu F, Li Y, Zhu Z, Li X (2010) Control of salicylic acid synthesis and systemic acquired resistance by two members of a plant-specific family of transcription factors. Proc Natl Acad Sci 107(42):18220–18225
Zhang Y, Zhao L, Zhao J, Li Y, Wang J, Guo R, Gan S, Liu CJ, Zhang K (2017) S5H/DMR6 encodes a salicylic acid 5-hydroxylase that fine-tunes salicylic acid homeostasis. Plant Physiol 175(3):1082–1093

Understanding the Crosstalk Between Chromatin Remodeling Mechanism and Phytohormones Signaling for Maintenance of Plant Developmental Plasticity: An Insight

Samrat Banerjee, Pinaki Roy, and Sujit Roy

Abstract As inanimate in nature, plants exhibit a high level of developmental plasticity in their growth and development to combat environmental fluctuations. Plants have evolved highly efficient response mechanisms including phytohormones (auxin, cytokinin, gibberellin, abscisic acid, ethylene, brassinosteroids, salicylic acid, jasmonic acid) for maintaining growth and development in response to variable environmental stimuli. The tight control of the signaling network regulates the biosynthesis, degradation, and efficient transport of phytohormones at the site of their cellular response. Plants regulate the action of phytohormones through spatiotemporal distribution. Interestingly, it has been observed that phytohormones not only govern cell division, flowering, cell proliferation, seed germination but they also respond to several biotic and abiotic stress conditions. Recent studies revealed that the ATP-dependent chromatin remodelers (ACRs) also regulate the biosynthesis and signaling of phytohormones in plants. The dynamic nature of chromatin architecture determines transcriptional accessibility to DNA and gene expression levels in response to developmental and environmental stimuli. The single and double mutants of ACRs, particularly the SWI/SNF chromatin remodelers were found to be associated with complete impairment of the phytohormone signaling network. Moreover, epigenetic modifications also modulate the transport and signal transduction mechanisms of phytohormones. Interestingly, phytohormone signaling also affects the expression of many chromatin modifiers. So, the chromatin remodelers and phytohormones may interact at multiple levels to regulate plant growth. The complex crosstalk mechanism of phytohormone signaling and chromatin structure is still largely enigmatic. In this present book chapter, we have a specific focus on the function of chromatin modifiers in the modulation of chromatin structure and the interactions with the phytohormone biosynthesis and signaling to showcase their molecular crosstalk mechanism in the context of the multidimensional growth response in plants.

S. Banerjee · P. Roy · S. Roy (✉)
Department of Botany, UGC Centre for Advanced Studies, The University of Burdwan, Golapbag Campus, Burdwan 713104, West Bengal, India
e-mail: sujitroy2006@gmail.com

T. Aftab (ed.), *Auxins, Cytokinins and Gibberellins Signaling in Plants*, Signaling and Communication in Plants,
https://doi.org/10.1007/978-3-031-05427-3_12

1 Introduction

Being immobile, plants can't avoid various biotic and abiotic stresses which greatly affect their genomic stability. During evolution, plants evolve highly efficient physiological, biochemical and molecular mechanisms to attain desirable growth and development under environmental stress conditions. The developmental plasticity of plants helps them to combat and tolerate these environmental assaults throughout their life (Gratani 2014). Modulation of gene expression at the post-transcriptional or post-translational level regulates the cellular fate in plants. Under stress conditions, gene expression is often governed by modification of histone proteins, chromatin remodelers, and deposition of histone variants (Fan et al. 2005; Kim et al. 2019; Banerjee and Roy 2021). The chromatin ultrastructure is composed of DNA and histone proteins (Cedar and Bergman 2009). The higher-order structure of chromatin comprised of a tightly packed genome restricts the access of transcription factors and transcription machinery to genes. Thus, remodeling of chromatin facilitates the opening of the compact structure via conversion of chromatin from a transcriptionally inactive to a transcriptionally active state (Bannister and Kouzarides 2011; Wang et al. 2019).

For more than a decade, chromatin remodeling and dynamic changes in chromatin in the context of activation of DNA damage response (DDR) and repair have gained prime focus in the research of plants (House et al. 2014; Donà and Mittelsten 2015; Banerjee and Roy; 2021). To regulate the expression of stress-responsive genes, different classes of ATP-dependent chromatin remodelers (ACRs) such as SWI/SNF, ISWI, INO80, and CHD are reported to play a key role in plants (Bhadouriya et al. 2021). It was observed that mutants of any one of the subunits of ACRs exhibit impaired growth and development such as alteration of stem cell population in root and shoot apical meristem, defects in flower morphogenesis, repression of lateral root initiation, and leaf maturation (Fukaki et al. 2006; Sang et al. 2012; Wu et al. 2012; Efroni et al. 2013). Moreover, histone modifications (more specifically methylation) and DNA methylation are also involved in the reprogramming genome and gene silencing (Kim 2019).

The initial response of the plants following exposure to abiotic and biotic stresses includes modulation of intracellular calcium concentration, activation of kinase cascades, and production of reactive oxygen molecules (Verma et al. 2016). In addition to activation of the signaling cascades and production of ROS, phytohormones activate specific signal transduction pathways upon the perception of abiotic or biotic stress (Ku et al. 2018). The remodeling of chromatin also regulates the signaling and biosynthesis of major phytohormone genes. Phytohormones play a crucial role as a chemical messenger and regulate various plant physiological and developmental processes of plants (Kazan 2015). In very low concentration, they respond to both internal and external stimuli via involvement in the signal transduction pathways during environmental stress. Plant development broadly depends on biosynthesis and degradation, their cellular responses that control the developmental pattern and cell division to shape the plant body. Besides their crucial involvement in the development of shoot and root meristems, leaf senescence, cell division, it was also observed that phytohormones modulate the chromatin structure of major DNA repair proteins and facilitate genome stability (Donà et al. 2013). Phytohormones are composed of five major families, namely auxins (IAAs), cytokinins (CKs), abscisic acid (ABA), gibberellins (GAs), and ethylene (ET). Salicylates (SAs), jasmonates (JAs), brassinosteroids (BRs), strigolactones (SLs), and polyamines represent new families of phytohormones. The different phytohormones may crosstalk at different developmental stages in plants. It was observed that ABA, SA, JA, and ET play a crucial role in plant defense response against pathogens and abiotic stresses (Bari and Jones 2009; Nakashima and Yamaguchi-Shinozaki 2013).

Recently, it was observed that chromatin modifiers and phytohormones interact to regulate the developmental plasticity and genome stability in plants (Ojolo et al. 2018; Sarnowska et al. 2016). But the mechanism of their action remains unclear. It was observed that the SWI/SNF family of chromatin remodelers regulate the biosynthesis and activity of several phytohormones via interacting with their biosynthesis genes (Maury et al. 2019). Similarly, some phytohormones also modulate the chromatin structure of major plant developmental genes. The multilayered control of chromatin ultrastructure and mode of phytohormone action may unravel the transcriptional changes associated with developmental robustness and phenotypic plasticity of plants (Lachowiec et al. 2016). In this present book chapter, we have a specific focus on the function of chromatin modifiers in the modulation of chromatin structure and the interactions with the phytohormone biosynthesis and signaling to showcase their molecular crosstalk mechanism in the context of the multidimensional growth response in plants.

2 Chromatin Remodelers Associated with Regulation of Plant Growth and Development

Eukaryotic genome organization is achieved by the compaction of DNA into chromatin. Chromatin is composed of nucleosomes, a combination of DNA and proteins. A single nucleosome is made up of histone octamer which is wrapped by 147 base pairs of DNAs (Vincent et al. 2008; Yamamuro et al. 2016). Chromatin organization in the nucleus helps in the condensation of DNA into chromosomes, segregation of chromosomes, and transmission of genetic materials to the next generation (Nishioka et al. 2020). The DNA region that remains wrapped around the histone octamer is inaccessible to replication or transcription machinery (Han et al. 2015). Several additional proteins and modifiers facilitate the accession of DNA via altering chromatin ultrastructure (Kim 2019). Chromatins are physical packaging that regulates the expression and silencing of a gene (Hauk et al. 2010). Chromatin remodelers change the interaction of DNA and histone octamer non covalently but chromatin modifiers incorporate covalent changes by adding or removing the chemical group from histones or DNA (Han et al. 2015). ATP-dependent chromatin remodelers use the energy of ATP hydrolysis to exchange the nucleosome structure (Tsukiyama 2002; Banerjee and Roy 2021). Depending on their similarities and differences in their catalytic ATPases, chromatin remodeling ATPases have four subfamilies- Inositol auxotrophy 80 (INO80), SWI2/SNF2-related (SWR1), Chromodomain helicase DNA-binding (CHD), Imitation Switch 1 (ISW1), Switch/Sucrose non-fermentable (SWI/SNF).

2.1 *ISW1*

The ATPase domain of the ISWI subfamily has two RecA-like lobes, separated by a small insertion sequence and a C-terminal HAND-SANT-SLIDE (HSS) domain (Fig. 1) which helps in the movement of DNA along the surface of the nucleosome and is involved in transcriptional activation. On the other hand, the N-terminal flanking ATPase lobe containing the two domains is responsible for the regulation of the activity of the ATPase domain (Gentry and Hennig 2014; Clapier et al. 2017). In the

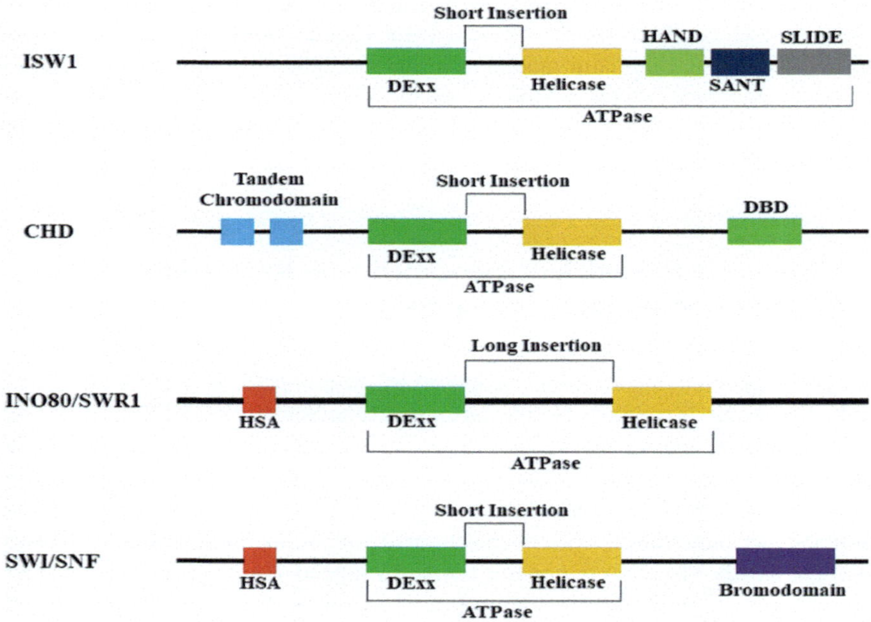

Fig. 1 The ATP-dependent chromatin remodelers (ACRs) are of four types (ISW1, CHD, INO80/SWR1, SWI/SNF). They are categorized based on their similarity and difference in the domain composition and architecture. The conserved ATPase domain is common in all the four ACR families and it helps to translocate the protein at the site of action

Arabidopsis thaliana, two genes encode the ISWI proteins CHROMATIN REMODELING11 (CHR11) and CHROMATIN REMODELING17 (CHR17) expressed mainly in young lateral organs and reproductive organs (Li et al. 2017).

2.2 CHD

The ATPase domain of the CHD subfamily shows similarity with the ISWI subfamily. They have two chromodomains to its N-terminus followed by a DNA binding domain (DBD) at C-terminus (Fig. 1) (Clapier et al. 2017; Gentry and Hennig 2014). The tandem chromodomain unit of the CHD subgroup of chromatin remodeler helps to distinguish between nucleosome and naked DNA (Hauk et al. 2010). Depending on structure and function this family has three groups CHD1, CHD3, and CHD7. CHD with histone chaperones play role in post replication chromatin assembly and nucleosome spacing.

2.3 INO80/SWR1

In INO80/SWR1 subfamily, a spacer is present in the ATPase core, which splits the ATPase domain (Fig. 1). The spacer helps to form an association with other core complex subunits (Han et al. 2015; Gentry and Hennig 2014). INO80 and SWR1 are involved in the genome-wide distribution of H2A.Z at the transcription start site and in *Arabidopsis*, INO80 regulates transcription and homologous recombinational repair (Han et al. 2015; Banerjee and Roy 2021). Loss of function of INO80 in *Arabidopsis* shows delay in floral development and retard root growth (Banerjee and Roy 2021).

2.4 SWI/SNF

The ATPase domain of this subfamily contains two RecA-like lobes, N-terminal helicase/SANT associated (HAS) are present in the ATPase domain and C-terminal bromodomain post HAS domain (Fig. 1). SWI/SNF complexes can bind to DNA directly with the help of DNA binding proteins and are recruited to the promoter regions of the DNA (Nishioka et al. 2020). BRM interacts with promoter and terminator regions of several genes and regulates their transcription (Archacki et al. 2017). The SWI/SNF subfamily is divided into three groups BRAHMA (BRM), SPLAYED (SYD), MINUSCULE (MINU). These subfamily members play role in transcriptional regulation, chromosome stability, and nuclear organization maintenance (Han et al. 2015; Banerjee and Roy 2021). There are several types of SWI/SNF complex are present in plants which are homolog of yeast and mammals. *Arabidopsis* have four SWI3 type proteins (AtSWI3A, AtSWI3B, AtSWI3C, AtSWI3D), Single copy of SNF5 (BUSHY), two SWP73 (AtSWP73A and AtSWP73B), eight classes of Actin related protein (ARP) and several other proteins (Jerzmanowski 2007). *Arabidopsis thaliana* Brahma (AtBRM) is involved in the regulation of vegetative and reproductive development (Farrona et al. 2004).

3 Phytohormones Regulate the Physiology and Development of Plants

In natural environmental conditions plants are exposed to several biotic (bacteria, virus, fungus, insect) and abiotic (drought, heat, cold, salinity) stress conditions. It is challenging for plants to grow and reproduce in such a harsh environment. Due to their sessile nature, they are unable to move in a favorable environment. So, to adapt to these adverse environmental stress conditions plants have evolved several mechanisms and altered developmental and physiological processes to grow and survive under these stress conditions (Waterworth et al. 2011; Verma et al. 2016). Phytohormones are produced in very low concentrations and are derived from various metabolic pathways and are structurally unrelated chemical compounds mainly involved in growth, development including pattern formation at the time of development, and regulate several plant processes (Santner et al. 2009). These phytohormones help the plants to cope up with changing and adverse environmental conditions in multiple ways (Verma et al. 2016; El-Esawi 2017). During a stress response, hormones are involved in the signal transduction pathway by activating phosphorylation cascade or second messenger and regulate several internal and external stimuli by crosstalk mechanism (El-Esawi 2017). Auxin (IAA), Cytokinin (CKs), Gibberellins (GAs), Abscisic acid (ABA), and Ethylene (ET) are the main five groups of phytohormone but there are some other phytohormones which include Jasmonates (JAs), Brassinosteroid (BRs), Salicylates (SAs) (El-Esawi 2017; Verma et al. 2016). Among these ABA, SA, JA, ET mediates defense response in plants against biotic and abiotic stress (Verma et al. 2016).

3.1 Auxin

Auxin the first discovered plant growth hormone is involved in the regulation of growth in plants in response to gravitation and light stimulation (Zhao 2010). Naturally occurring auxin is IAA (Indole-3-Acetic Acid) and synthetic auxin 2,4-dichlorophenoxyacetic acid, used as herbicides (Santner et al. 2009). IAA is produced in the young shoot, leaf primordial, young leaves and transported downward to the root tips and controls vascular differentiation in plants and this way IAA controls cell elongation and maintains apical dominance (Davies 2010; Fahad et al. 2015). IAA is involved in several cellular processes like cell enlargement, cell division, root initiation, apical dominance, flowering, growth of floral parts, etc. (Davies 2010) and plays an adaptive role in salinity, heavy metal stress, and regulate crosstalk mechanism in several biotic and abiotic stress (El-Esawi 2017).

3.2 Gibberellin

Gibberellins (GAs) are tetracyclic diterpenoids that play a crucial role in seed germination, stem and leaf elongation, and provide stress tolerance against osmotic stress (El-Esawi 2017; Fahad et al. 2014). GA was first isolated from the fungal pathogen *Gibberella fujikuroi* which causes disease in rice. This fungus is responsible for the excessive elongation of the stem leading to the lodging of the plant (Santner et al. 2009). GA3 is the most commonly found bioactive gibberellin but for stem elongation, GA1 is mainly responsible (Davies 2010). Besides stem elongation, there are several roles of GA exists such as induction of seed germination, bolting in long-day plants, fruit set, and growth (Davies 2010). GA increases plant photosynthetic efficiency by increasing leaf area index which helps in light perception (Fahad et al. 2014).

3.3 Cytokinins

Cytokinins (CKs) are adenine-based molecules where the N6-position is substituted (Santner et al. 2009) and Zeatin is the common CK present in plants (Davies 2010). CK play role in plant growth, development, cell division, chloroplast biogenesis, apical dominance, leaf senescence, anthocyanin production, and also respond to abiotic stress like salinity, high temperature, and drought (El-Esawi 2017; Fahad et al. 2014). Seed priming with CK increases salt tolerance (Fahad et al. 2014). CK acts as the ABA antagonist in water stress conditions. CK helps to break seed dormancy and inhibit leaf and fruit abscission via inhibiting ABA response (Fahad et al. 2014).

3.4 Abscisic Acid

Abscisic acid (ABA) is an isoprenoid compound produced by 2-C-methyl-D-erythritol-4-phosphate (MEP) pathway in plastid and play important role in seed dormancy, embryo formation, stomal opening, cell turgor maintenance, and stress response to survive in adverse environments (Santner et al. 2009; Fahad et al. 2014; El-Esawi 2017). ABA is well known as a stress hormone as ABA synthesis is upregulated in different types of stress (Fahad et al. 2014). ABA levels are increased in drought, salinity, cold, and heat stress (Verma et al. 2016). ABA biosynthetic genes, ZEP (ZEAXANTHIN EPOXIDASE), AAO3 (ALDEHYDE OXIDASE), NCED3 (9-CIS-EPOXYCAROTENOID DIOXYGENASE), and MCSU (MOLYBDENUM COFACTOR SULFURASE) expression is upregulated during osmotic stress (Verma et al. 2016). ABA functions as a secondary messenger in guard cells and helps plants to survive through the formation of dehydrins, osmoprotectants and induce cellular dehydration tolerance genes (Fahad et al. 2014; El-Esawi 2017). In high salinity and drought conditions, ABA is responsible for stomatal closure and maintaining water balance in plants (Verma et al. 2016).

3.5 Ethylene

Ethylene (ET) is a gaseous hormone synthesized from methionine via the Yang cycle and plays a crucial role in fruit ripening (Santner et al. 2009; Davies 2010). ET protects plants from herbivorous insects and necrotrophic pathogens (Verma et al. 2016). Moreover, ET also regulates flower senescence and stress response (El-Esawi 2017). Synthesis of ET is also enhanced due to wounding (Santner et al. 2009). ET, in association with JA and SA, activates a defense mechanism against biotic stress (El-Esawi 2017).

3.6 Brassinosteroids

Brassinosteroid (BR) is a type of steroidal hormone, have several activities in plants like seed germination, cell growth, reproductive growth, production of flower and fruit. BR also has a stress response against salinity, drought, heat, chilling (Fahad et al. 2014; El-Esawi 2017). BR was first isolated from *Brassica napus*. There are two bioactive BR, namely 24-epibrassinolide and 28-homobrassinolide (El-Esawi 2017). Under osmotic stress, seedling growth can be enhanced in *Sorghum* by BR application (Fahad et al. 2014).

3.7 Jasmonic Acid

Jasmonic acid (JA) is synthesized through the octadecanoid pathway (Santner et al. 2009) in several plant parts and cell organelles like leaves, roots, chloroplasts, and peroxisomes (Fahad et al. 2014). JA plays a crucial role in fruiting, flowering, senescence, and secondary metabolism and exhibits a defense response against drought, salinity, low-temperature, and heavy metal (El-Esawi 2017). JA protects plants from necrotrophs and herbivory (Verma et al. 2016). Herbivory and mechanical wounding are the main inductive signal for the increase of JA level (Santner et al. 2009).

3.8 Salicylic Acid

Salicylic acid (SA) is a phenolic compound, mainly involved in the expression of pathogenesis-associated proteins (El-Esawi 2017). SA is synthesized from the chorismate pathway and play important role in plant defense response against biotrophic and hemibiotrophic pathogens (Santner et al. 2009; Verma et al. 2016). SA not only plays a major role in the regulation of biotic stress but also shows the response in some abiotic stresses like drought, salinity, chilling, heavy metal tolerance, and heat (Fahad et al. 2014). In low concentration, SA shows antioxidant activity but in high concentration, SA is responsible for cell death (El-Esawi 2017).

4 Phytohormone and Chromatin Crosstalk Regulate Plant Growth

Chromatin remodelers regulate the phytohormone-mediated responses in plants under stress conditions. It was observed that the SWI/SNF family of chromatin remodelers play a crucial role in the regulation of several phytohormone responsive genes (Han et al. 2012; Yang et al. 2015). An SWI/SNF family gene, *Brm* regulates the expression of the key players of ABA response. In the absence of drought stress, BRM promotes the occupancy of nucleosomes at the transcription start site (TSS) of *ABA INSENSITIVE 5* (ABI5), one of the major transcriptional regulators of ABA response and inhibits its expression (Han et al. 2012). ABI5 is considered as a core signaling component of ABA signaling and is also acts as a signal integrator of crosstalk of ABA with other phytohormones. ABI5 also functions as a key regulator of the abiotic stress response (Skubacz et al. 2016). The mutants of *brm* exhibit increased drought tolerance similarly to the ABI5 overexpression lines. The double mutants of *brm/abi5* partially rescue the abnormal root growth phenotype (Han et al. 2012; Nishioka et al. 2020). Thus, BRM helps to keep the perfect balance between growth and stress responses in plants. Apart from these, BRM is also involved in the other major phytohormone signaling pathways, including auxin (Yang et al. 2015), cytokinin (Efroni et al. 2013), gibberellin (Archacki et al. 2013). In response to salt stress and ABA, the c-terminus of BRM physically interacts with ETHYLENE RESPONSE FACTOR VII (ERF VII) (Vicente et al. 2017). It was observed that BRM and ERF VII are binding with the same region of the *ABI5* cis-element containing double GCC motif and activating the *ABI5* promoter (Gibbs et al. 2014). Both the BRM and ERF VII compete for the same cis-element to control plant growth and development (Vicente et al. 2017). Moreover, it was observed that both ABI5 and ABI3 are negatively regulated by PICKLE (PKL), another SWI/SNF chromatin remodeling factor (Fig. 2). PKL promotes the histone methylation of the promoter of the *ABI5* gene leading to repression of chromatins and releasing the inhibition of germination of embryos (Perruc et al. 2007). HOOKLESS1 (HLS1), a histone acetyltransferase promotes histone H3 acetylation of ABI5 in association

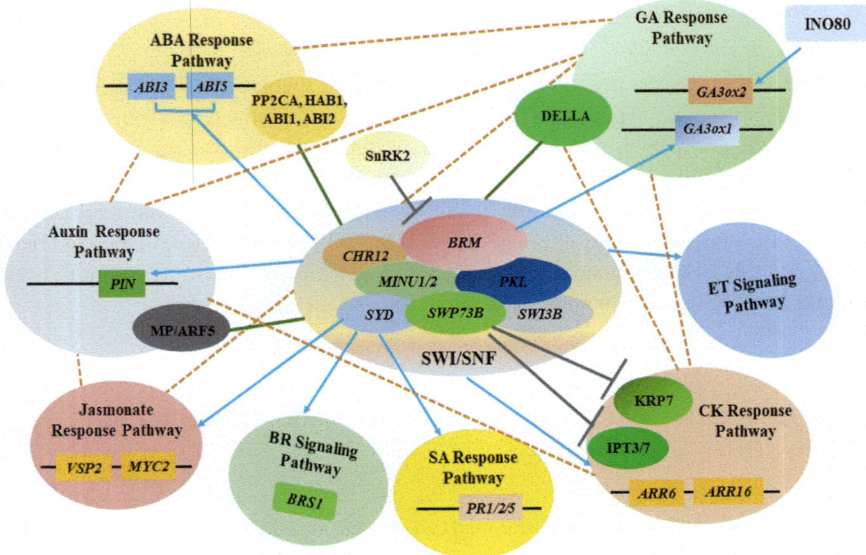

Fig. 2 Plant SWI/SNF family of chromatin remodelers play a crucial role in the regulation of phytohormone signaling and biosynthesis. Chromatin and phytohormone crosstalk mechanism regulate the plant growth at multiple developmental phases. Green lines represent physical interaction, blue arrows represent transcriptional regulation (positive or negative), grey lines with perpendicular bars indicate repression and red dashed lines indicate crosstalk

with MED18 (MEDIATOR18) and positively activates *ABI5* expression (Liao et al. 2016). In ambient growth conditions, BRM represses the expression of ABI5. But under environmental stress, ABA directly interacts with ABA receptors, PYR/PYL resulting in the phosphorylation of SnRK2. Subsequently, SnRK2 phosphorylates BRM and inhibits it from further repression of ABI5 expression (Fig. 2) (Weiner et al. 2010; Peirats-Llobet et al. 2016). Plants with partial loss of function of *brm3* exhibit enhanced ABI5 expression and increased growth arrest following the exposure to ABA and salt stress (Han et al. 2012).

From previous studies, it has been reported that chromatin remodeling factors such as BRM, SWI3c, PKL are actively involved in the GA pathway in *Arabidopsis* (Henderson et al. 2004; Archacki et al. 2013; Zhang et al. 2014). In rice, T-DNA insertion lines or RNAi lines of *osino80* exhibit GA deficient phenotype. These GA deficient phenotypic characteristics are recovered following the exogenous application of GA3. Moreover, transcriptomic analysis reveals that knockdown of *osino80* in rice leads to downregulation of two important GA biosynthesis genes namely *CPS1* and *GA3ox2* (Fig. 2) (Li et al. 2017). Moreover, in comparison to wild-type mice, the *osino80Ri-2* lines exhibit a decreased level of GA12, GA53, GA19, GA20, and GA1 (Li et al. 2017). Archacki et al. reported that BRM is a positive regulator of GA-mediated responses in *Arabidopsis*. The interaction of SWI3c with DELLA proteins is necessary for the regulation of GA biosynthesis and signaling (Sarnowska et al. 2013). Sarnowska et al., also revealed that the loss of function mutation of *swi3c* modulates the plant growth and affects the phytohormone signaling of ABA, brassinosteroids, and ethylene. In addition, mutation of *swi3c* inhibits the DELLA-mediated activation of the GA receptor gene, *GID1*, resulting in defects in GA signaling (Sarnowska et al. 2013). It was also observed that SWI3c directly interacts with DELLA proteins (Fig. 2) such as RGA-LIKE2 and RGA-LIKE3, affecting the activation of GID1 and GA3 oxidase which are involved in GA perception and GA biosynthesis (Sarnowska et al. 2013). The resemblance of *swi3c* and *brm* mutants (dark green leaf color and semi-dwarf nature) with those of GA deficient mutants indicate their role in GA biosynthesis (Fig. 2). The GA levels are decreased in swi3c plants and phenotypic traits such as curling leaves and expansion of leaf blades are not rescued following the exogenous application of GA (Sarnowska et al. 2013). Another DELLA interacting chromatin remodeler PKL plays a crucial function in GA-dependent responses in *Arabidopsis* (Zhang et al. 2014). Mutant seedlings of *pkl* are phenotypically similar to GA response mutants and exhibit a semi-dwarf phenotype. Interestingly, treatment of *pkl* mutants with external GA hinders the pickle root phenotype (Henderson et al. 2004). Leaves are the evolutionary decedents of shoots. It was observed that EARLY IN SHORT DAYS 1 (*ESD1*) encodes for an ACTIN-RELATED PROTEIN 6 (ARP6) and mutants of *esd1* depend on GA-mediated for early flowering phenotypes (Martin-Trillo et al. 2006). Moreover, CHR729 protein is a CHD3 chromatin-remodeling factor that is involved in the development of seedlings controlled by gibberellin in rice (Ma et al. 2015).

Auxin is considered one of the most important phytohormones in plants. Many genes associated with auxin signaling (*AXR3*), brassinosteroid signaling (*BRS1*), and gibberellic acid signaling (*RGL2*, *GA4*, and *GASA1*) are highly affected in *brm* and *syd* single mutants or *brm/syd* double mutants (Bezhani et al. 2007). The expression of auxin biosynthesis genes such as *YUC* is highly regulated by PIF transcription factors namely PIF4/5/7 (Hornitschek et al. 2012; Peng et al. 2018). Recently, it was reported that PIF7 physically interacts with the MORF RELATED GENE 2 (MRG2), an H3K4me3/H3K6me3-binding protein and as a result, the expression of YUC8 and IAA19 are upregulated under shade conditions (Peng et al. 2018). Peng and his colleagues have also revealed that PIF7 promotes histone acetylation (H4K5ac, H3K9ac, H3K27ac) at the *YUC8* gene locus. Recently, Lee and Seo have been reported that the AT-hook motif-containing nuclear-localized (AHL) proteins interact with *YUC9* locus and subsequently they recruit SWI2/SNF2-related 1 (SWR1) complex to enhance the histone exchange of canonical H2A with H2AZ containing nucleosome at the *YUC9* locus (Lee and Seo 2017). Ariel et al. proposed that chromatin loops modulate the expression pattern of an auxin-responsive gene *PID* (Ariel et al. 2014). BRM actively regulates the *PIN-FORMED* genes (Fig. 2) and auxin distribution in plant cells. Loss of function mutation of *brm* exhibits increased H3K27me3 levels in the promoters of *PIN1* and *PIN2* (Yang et al. 2015). Monopteros (MP)/Auxin Response Factor (ARF5) recruits two important chromatin remodelers, BRM and SYD in presence of auxin (Fig. 2). As a result, they change the chromatin dynamics to increase the DNA accessibility of auxin-responsive genes associated with flowering and leaf formation leading to their activation by transcription factors (Wu et al. 2015). However, in absence of auxin, MP/ARF5 is suppressed by AUX/IAA proteins and thus BRM/SYD can't help to facilitate DNA accessibility (Wu et al. 2015). The expression of auxin response factor 3 (ARF3) and PIN1 is altered during embryogenesis in *met1* mutant lines resulting in modulation of auxin gradient (Li et al. 2011). Recently, it was observed that PKL represses the deposition of H3K27me3 in IAA19 and IAA29 auxin-related genes and enhances their expression (Luo et al. 2018).

Cytokinins (CKs) are generally associated with mitotic cell cycle progression in shoots. However, their over-production results in inhibition of root elongation and lateral root development (Kuderova et al. 2008). Root growth is positively regulated by SWI/SNF ATP-dependent chromatin remodeling protein BAF60. BAF60 regulates the formation of chromatin loops and deposition of active histone marks in the major CK biosynthesis genes, *IPT3* and *IPT7*. From chromosome conformation capture (3C) experiments, it was observed that BAF60 (also known as SWP73B) plays a negative role in gene loop formation in CK biosynthesis gene and thus they remain transcriptionally inactive (Jégu et al. 2015). Moreover, BAF60 also hinders the deposition of active histone marks (H3K4me3) and the recruitment of RNA Pol II in the *KRP7* gene, a CK-regulated cell cycle inhibitor (Fig. 2) (Jégu et al. 2015). Moreover, DNA methylation is also associated with the repression of the CK biosynthesis gene, IPT5b (Feng et al. 2017). Cytokinin also promotes the expression of the *MET1* gene via the regulation of the cell cycle (Liu et al. 2018). The determinate growth of leaves is also modulated by SWI/SNF family gene *Brm*. It was observed that BRM physically interacts with bHLH transcription factor TEOSINTE BRANCHED 1, CYCLOIDEA, PROLIFERATING CELL FACTOR1/2 4 (TCP4). Both these protein partners bind with the promoter of a cytokinin inhibitor gene, ARR16 (Efroni et al. 2013; Nishioka et al. 2020). In this way, BRM regulates the leaf growth via modulation of cytokinin biosynthesis (Fig. 2) in a developmental phase-specific manner (Efroni et al. 2013). It was also observed that the suppression of CIN-TCP4 activity results in delayed leaf maturation along with prolonged leaf cell proliferation (Ori et al. 2007; Efroni et al. 2008). Both the BRM and SWI3c interacted with TCP transcription factors as evident from bimolecular fluorescence complementation studies. Moreover, BRM and TCP4 regulate the activity of ARR6, other than ARR16, in young leaves. ARR6 expression level is relatively higher at proliferating stages of leaf development (Efroni et al. 2013).

It was observed that in *Arabidopsis* AGO1 physically interacts with SWI/SNF chromatin remodeling complexes and binds with stress-responsive genes in response to cold and phytohormones (Liu et al. 2018). A plethora of studies revealed that the PLANT HOMEODOMAIN (PHD) proteins are involved in the regulation of housekeeping processes of plant life such as germination, flowering time, root development, meiotic and post-meiotic development, and embryo meristem formation (Wu et al. 2021). Coexpression studies revealed that in *Gossypium hirsutum GhPHD* genes regulate the phytohormone signaling network to improve abiotic stress tolerance (Wu et al. 2021). A PHD protein related to auxin-mediated genetic network, GSR1 (Germostatin Resistance locus 1) interacts with ARF16 to control seed germination (Ye et al. 2016). Moreover, GhPHD5 regulates the auxin homeostasis in plants which may improve tolerance to drought, salt, and heat stress. In silico studies revealed that GhPHD34 and GhPHD107 may be involved in the development of heat tolerance in plants via modulating auxin and ethylene signal transduction pathways (Wu et al. 2021). In dehydration stress, ATX1, an H3K4 methyltransferase promotes the transcriptional activity of ABA biosynthetic genes (Ding et al. 2011).

5 Regulation of Phytohormone Signaling via Modulation of Chromatin Under Stress Conditions

Chromatin modifications play an important role in the transcriptional regulation of various stress-responsive genes. It was observed that HDAcs respond to different plant hormones such as ABA, JA, and ethylene (Sridha and Wu 2006; Zhou et al. 2005). Moreover, brassinosteroid signaling and gene expression also involve dynamic changes in chromatin structure facilitated by NAP1 protein and other chromatin remodeling complexes (Shigeta et al. 2011). Histone acetylation and deacetylation are mainly affecting the ABA-responsive genes (Sridha and Wu 2006). It was reported that the RNA interfering lines of Histone deacetylase 6 (HDA6) exhibits hypersensitive responses to ABA (Chen et al. 2010). The expression of ABI1, ABI2, KAT1, KAT2, and RD29B is greatly reduced in *hda19-1* mutants (Chen and Wu 2010). Transgenic lines overexpressing AtHD2C affects several ABA-related genes (Sridha and Wu 2006). Another core histone deacetylase, HDA9 interacts with PWR and HOS15, and this protein complex is directly involved in the repression of many stress-responsive genes, including the ethylene response factor (ERF4/5/6/11). Moreover, HDA9 also interacts with the ABI4 transcription factor and repress the expression of CYP707s (ABA catabolism related gene) under drought stress condition (Baek et al. 2020). Meanwhile, overexpression lines of HDA19 exhibit increased expression of jasmonic acid and ethylene-regulated pathogenesis-related genes such as β-1, 3 glucanase, ERF1, and basic chitinase. Taken together, HDA19 mainly interconnects the hormone response pathway with biotic stress response (Zhou et al. 2005).

From initial studies, it was observed that histone modifications play a crucial function in the regulation of salicylic acid (SA) biosynthetic gene and SA responsive genes. Plants infected with *Pseudomonas syringae* exhibit enhanced expression of SA-related genes via inactivation of SRT2, an HDAc protein that generally suppress the expression of SA biosynthetic genes namely *PAD4*, *EDS5,* and *SID2* (Wang et al. 2010). Interestingly, null mutation of *hda19* results in enhanced expression of SA responsive genes (Tian et al. 2005). Application of salicylic acid or its synthetic analog acibenzolar s-methyl induces chromatin modification of the promoters of plant defense-related genes. The modification of histones results in gene priming of pathogen defense genes which leads to the enhancement of plant stress response. It has been observed that during systemic acquired resistance, the histone H2A is replaced by H2AZ in defense gene promoters (van den Burg and Takken 2009).

Moreover, chromatin modifications such as H3/H4 acetylation and H3K4 methylation might develop a memory of the previous infection in plants (Jaskiewicz et al. 2011). It was also observed that both BRM and SYD regulate several SA-responsive genes (Bezhani et al. 2007). SYD has further been involved in wound stress-mediated expression of downstream genes related to JA or ET signaling pathway (Walley et al. 2008). Following the wound stress, SYD was recruited to the promoter of the *MYC2* gene. The mutants of syd are highly susceptible to *Botrytis cinerea* but resistant to *Pseudomonas syringae* (Walley et al. 2008). Plants with reduced expression of BRM, MSI1, HDA19 exhibit increased drought tolerance and ABA-dependent growth defects (Maury et al. 2019). PsSNF5 plays a crucial role in chromatin remodeling and is accumulated under drought stress and abscisic acid in germinating embryos specifically at the later stage of embryo development in *Pisum sativum* which is an indication of ABA-mediated chromatin remodeling (Ríos et al. 2007). Under abiotic stress at the time of seed maturation ATP-dependent chromatin remodeling occurs by the regulation of abscisic acid (Chinnusumy et al. 2008).

6 Phytohormone-Chromatin Crosstalk Modulates Developmental Plasticity in SAM

Being sessile, plants need to constantly respond to a wide range of environmental cues to maintain their growth and developmental plasticity (Gaillochet and Lohmann 2015). Sometimes plants integrate the signal of environmental fluctuations without reflecting any changes in phenotype. Plant growth is governed by a complex interplay of phytohormone signaling, remodeling of chromatin structure, and modulating gene expression. It is speculated that the meristems are the central places for this cellular crosstalk though regulatory mechanisms remain largely enigmatic. During plant development, phytohormone signaling pathways regulate some key chromatin modifiers such as PRC1 and PRC2 with histone methyltransferase (HMT) activity playing a key role in transcriptional regulation (Mozgová et al. 2017; Ikeuchi et al. 2015). A brassinosteroid signaling transcription factor, BZR1 recruits ELF6 which represses H3K27me3 activity of PRC2 at Flowering Locus C (FLC) hindering precocious floral transition (Li et al. 2018a, b). Alteration of DNA methylation pattern in *Arabidopsis* is generally associated with the changed phytohormone response of JA, SA, and ethylene (Lafon-Placette et al. 2018). Moreover, recent evidence suggests that fertilization-dependent auxin downregulates the activity of PRC2 in the seed coat of *Arabidopsis* (Figueiredo et al. 2015; Figueiredo and Köhler 2018). Auxin biosynthesis and signaling genes in the SAM and leaves of *Arabidopsis* are also regulated by PRC2 (Lafos et al. 2011). Taken together, it can be concluded that the phytohormone signaling cascades direct the activity of chromatin modifiers.

Epigenetic modification and phytohormone signaling pathways have a complementary role in the maintenance of stem cell pluripotency, differentiation, and reprogramming (Cao et al. 2015; Ojolo et al. 2018). Chromatin modifiers regulate the

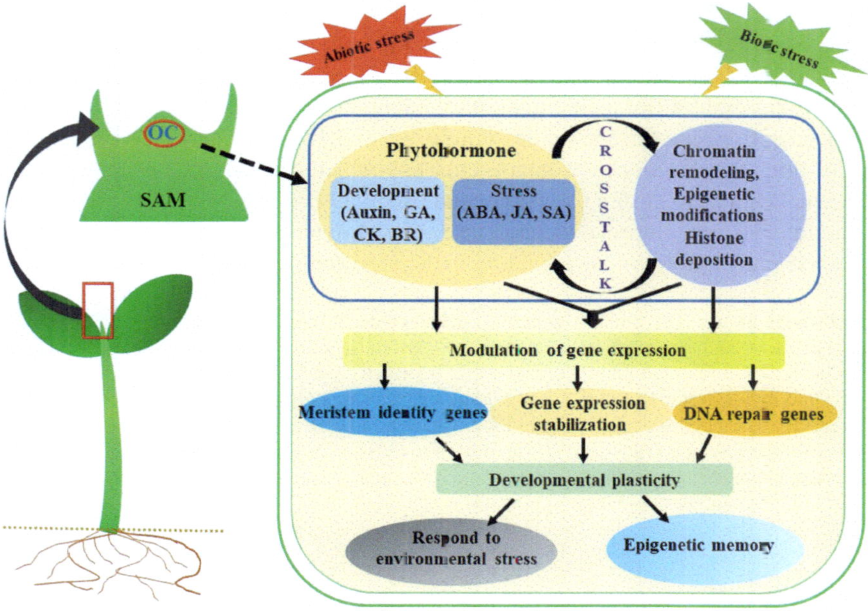

Fig. 3 Schematic representation showing chromatin and phytohormone crosstalk mechanism in shoot apical meristem (SAM) of plants. Meristems are the major location of this crosstalk mechanism under abiotic and biotic stress conditions. The cooperative action of phytohormones and chromatin regulates the developmental plasticity of plants

activity of meristem cell identity genes which are the key targets of phytohormones (Galinha et al. 2007). Interestingly, DNA methylation, cytokinin signaling, H3K27me3, or chromatin remodeling control the expression of WUSCHEL (WUS), a SAM-organizing homeobox gene (Kwon 2005; Dodsworth 2009; Cao et al. 2015; Liu et al. 2018). Moreover, genes of stem cell niche maintaining transcription factors such as WOX4, WOX5, PLT1, PLT2 are the potential targets of PRC2 and phytohormones (Lafos et al. 2011; Maury et al. 2019). Thus, meristems are the central region of chromatin and phytohormone crosstalk in plants under both normal and stress conditions to maintain developmental plasticity (Fig. 3).

7 Conclusion and Future Perspective

Chromatin remodelers and modifiers play a crucial role in the regulation of plant development, genome stability, stress tolerance, and adaptation (Kim 2019; Banerjee and Roy 2021). Interestingly, phytohormones also modulate the physiological and developmental patterns of plants (Sarnowska et al. 2016). Plants' response to stress requires a well-organized and accurate regulation. It was observed that SWI/SNF chromatin remodeling complexes are involved in the fine-tuning of phytohormonal responses through various feedback loops. Mutants of several chromatin remodelers and modifiers further confirm the existence of crosstalk between chromatin and phytohormones. But their extent of interaction to which they perform jointly or independently needs further studies (Ojolo et al. 2018). Meristems are the central part of this phytohormone-chromatin crosstalk which can integrate different environmental signals to ensure developmental plasticity (Fig. 3). Moreover, single-cell methodologies may improve the understanding of dynamics of chromatin structure in response to phytohormone signaling in the meristems of plants.

Acknowledgements The authors gratefully acknowledge Council of Scientific and Industrial Research, Govt. of India, (Ref. No. 38(1587)/16/EMR-II, dated: 17/05/2016 to SR), UGC, Govt. of India (Start-Up research grant No.F.30-158/2015 (BSR) and DST & BT, Govt. 4 of WB (Ref. No: ST/P/S & T/1G-5/2018 to SR) for providing financial supports performing research related to the topic discussed in this review. SB and PR are thankful to CSIR, Govt. of India, and UGC, Govt. of India for the research fellowship. We apologize to those authors whose work could not be cited due to space limitations.

References

Archacki R, Buszewicz D, Sarnowski TJ, Sarnowska E, Rolicka AT, Tohge T et al (2013) Brahma ATPase of the SWI/SNF chromatin remodeling complex acts as a positive regulator of gibberellin-mediated responses in *Arabidopsis*. PLoS ONE 8:e58588

Archacki R, Yatusevich R, Buszewicz D, Krzyczmonik K, Patryn J, Iwanicka-Nowicka R, Biecek P, Wilczynski B, Koblowska M, Jerzmanowski A, Swiezewski S (2017) Arabidopsis SWI/SNF chromatin remodeling complex binds both promoters and terminators to regulate gene expression. Nucleic Acids Res 45(6):3116–3129

Ariel F, Jegu T, Latrasse D, Romero-Barrios N, Christ A, Benhamed M, Crespi M (2014) Noncoding transcription by alternative RNA polymerases dynamically regulates an auxin-driven chromatin loop. Mol Cell 55(3):383–396

Baek D, Shin G, Kim MC, Shen M, Lee SY, Yun DJ (2020) Histone deacetylase HDA9 with ABI4 contributes to abscisic acid homeostasis in drought stress response. Front Plant Sci 11:143

Banerjee S, Roy S (2021) An insight into understanding the coupling between homologous recombination mediated DNA repair and chromatin remodeling mechanisms in plant genome: an update. Cell Cycle 20(18):1760–1784

Bannister AJ, Kouzarides T (2011) Regulation of chromatin by histone modifications. Cell Res 3:381–395

Bari R, Jones JD (2009) Role of plant hormones in plant defence responses. Plant Mol Biol 69(4):473–488

Bezhani S, Winter C, Hershman S, Wagner JD, Kennedy JF, Kwon CS, Pfluger J, Su Y, Wagner D (2007) Unique, shared, and redundant roles for the arabidopsis SWI/SNF chromatin remodeling ATPases BRAHMA and SPLAYED. Plant Cell Online 19:403–416

Bhadouriya SL, Mehrotra S, Basantani MK, Loake GJ, Mehrotra R (2021) Role of chromatin architecture in plant stress responses: an update. Front Plant Sci 11:603380

Cao X, He Z, Guo L, Liu X (2015) Epigenetic mechanisms are critical for the regulation of WUSCHEL expression in floral meristems: figure 1. Plant Physiol 168:1189–1196

Cedar H, Bergman Y (2009) Linking DNA methylation and histone modification: patterns and paradigms. Nat Rev Genet 10:295–304

Chen LT, Wu K (2010) Role of histone deacetylases HDA6 and HDA19 in ABA and abiotic stress response. Plant Signal Behav 5(10):1318–1320
Chen LT, Luo M, Wang YY, Wu K (2010) Involvement of Arabidopsis histone deacetylase HDA6 in ABA and salt stress response. J Exp Bot 61(12):3345–3353
Chinnusamy V, Gong Z, Zhu JK (2008) Abscisic acid-mediated epigenetic processes in plant development and stress responses. J Integr Plant Biol 50(10):1187–1195
Clapier CR, Iwasa J, Cairns BR, Peterson CL (2017) Mechanisms of action and regulation of ATP-dependent chromatin-remodelling complexes. Nat Rev Mol Cell Biol 18(7):407–422
Davies, P (2010) The plant hormones: their nature, occurrence, and functions. https://doi.org/10.1007/978-1-4020-2686-7_1
Ding Y, Avramova Z, Fromm M (2011) The Arabidopsis trithorax-like factor ATX1 functions in dehydration stress responses via ABA-dependent and ABA-independent pathways. Plant J 66:735–744
Dodsworth S (2009) A diverse and intricate signalling network regulates stem cell fate in the shoot apical meristem. Dev Biol 336:1–9
Donà M, Mittelsten SO (2015) DNA damage repair in the context of plant chromatin Plant Physiol 168(4):1206–1218
Donà M, Macovei A, Faè M, Carbonera D, Balestrazzi A (2013) Plant hormone signaling and modulation of DNA repair under stressful conditions. Plant Cell Rep 32(7):1043–1052
Efroni I, Blum E, Goldshmidt A, Eshed Y (2008) A protracted and dynamic maturation schedule underlies Arabidopsis leaf development. Plant Cell 20:2293–2306
Efroni I, Han SK, Kim HJ, Wu MF, Steiner E, Birnbaum KD et al (2013) Regulation of leaf maturation by chromatin-mediated modulation of cytokinin responses. Dev Cell 24:438–445
El-Esawi M (2017) Hormonal regulation in plant development and stress tolerance. https://doi.org/10.5772/intechopen.69806
Fahad S, Hussain S, Bano A (2015) Potential role of phytohormones and plant growth-promoting rhizobacteria in abiotic stresses: consequences for changing environment. Environ Sci Pollut Res 22:4907–4921
Fahad S, Hussain S, Matloob A, Khan F et al (2014) Phytohormones and plant responses to salinity stress: a review. Plant Growth Regul 75:391–404
Fan Y, Nikitina T, Zhao J, Fleury TJ, Bhattacharyya R, Bouhassira EE et al (2005) Histone H1 depletion in mammals alters global chromatin structure but causes specific changes in gene regulation. Cell 123:1199–1212
Farrona S, Hurtado L, Bowman JL, Reyes JC (2004) The Arabidopsis thaliana SNF2 homolog AtBRM controls shoot development and flowering. Development 131(20):4965–4975
Feng Y, Zhang X, Wu T, Xu X, Han Z, Wang Y (2017) Methylation effect on IPT5b gene expression determines cytokinin biosynthesis in apple rootstock. Biochem Biophys Res Commun 482(4):604–609
Figueiredo DD, Köhler C (2018) Auxin: a molecular trigger of seed development. Genes Dev 32:479–490
Figueiredo DD, Batista RA, Roszak PJ, Köhler C (2015) Auxin production couples endosperm development to fertilization. Nat Plants 1:15184
Fukaki H, Taniguchi N, Tasaka M (2006) PICKLE is required for SOLITARY-ROOT/IAA14 mediated repression of ARF7 and ARF19 activity during *Arabidopsis* lateral root initiation. Plant J 48(2006):380–389
Gaillochet C, Lohmann JU (2015) The never-ending story: from pluripotency to plant developmental plasticity. Development 142:2237–2249
Galinha C, Hofhuis H, Luijten M, Willemsen V, Blilou I, Heidstra R et al (2007) PLETHORA proteins as dose-dependent master regulators of Arabidopsis root development. Nature 449:1053–1057
Gentry M, Hennig L (2014) Remodelling chromatin to shape development of plants. Exp Cell Res 321(1):40–46

Gibbs DJ, Md Isa N, Movahedi M, Lozano-Juste J, Mendiondo GM, Berckhan S, Marín-de la Rosa N, Vicente Conde J, Sousa Correia C, Pearce SP et al (2014) Nitric oxide sensing in plants is mediated by proteolytic control of group VII ERF transcription factors. Mol Cell 53:369–379
Gratani L (2014) Plant phenotypic plasticity in response to environmental factors. Adv Bot 2014:208747
Han SK, Sang Y, Rodrigues A, Wu MF, Rodriguez PL et al (2012) The SWI2/SNF2 chromatin remodeling ATPase BRAHMA represses abscisic acid responses in the absence of the stress stimulus in Arabidopsis. Plant Cell 24:4892–4906
Han SK, Wu MF, Cui S, Wagner D (2015) Roles and activities of chromatin remodeling ATPases in plants. Plant J 83(1):62–77
Hauk G, McKnight JN, Nodelman IM, Bowman GD (2010) The chromodomains of the Chd1 chromatin remodeler regulate DNA access to the ATPase motor. Mol Cell 39(5):711–723
Henderson JT, Li HC, Rider SD, Mordhorst AP, Romero-Severson J, Cheng JC et al (2004) Pickle acts throughout the plant to repress expression of embryonic traits and may play a role in gibberellin-dependent responses. Plant Physiol 134:995–1005
Hornitschek P et al (2012) Phytochrome interacting factors 4 and 5 control seedling growth in changing light conditions by directly controlling auxin signaling. Plant J 71:699–711
House NCM, Koch MR, Freudenreich CH (2014) Chromatin modifications and DNA repair: beyond double-strand breaks. Front Genet 4:296
Ikeuchi M, Iwase A, Rymen B, Harashima H, Shibata M, Ohnuma M et al (2015) PRC2 represses dedifferentiation of mature somatic cells in Arabidopsis. Nat Plants 1:15089
Jaskiewicz M, Conrath U, Peterhänsel C (2011) Chromatin modification acts as a memory for systemic acquired resistance in the plant stress response. EMBO Rep 12(1):50–55
Jégu T, Domenichini S, Blein T, Ariel F, Christ A et al (2015) A SWI/SNF chromatin remodeling protein controls cytokinin production through the regulation of chromatin architecture. PLoS ONE 10(10):e0138276
Jerzmanowski A (2007) SWI/SNF chromatin remodeling and linker histones in plants. Biochim Biophys Acta 1769(5–6):330–345
Kazan K (2015) Diverse roles of jasmonates and ethylene in abiotic stress tolerance. Trends in Plant Sci 20:219–229
Kim JH (2019) Chromatin remodeling and epigenetic regulation in plant DNA damage repair. Int J Mol Sci 20(17):4093
Kim JH, Ryu TH, Lee SS et al (2019) Ionizing radiation manifesting DNA damage response in plants: an overview of DNA damage signaling and repair mechanisms in plants. Plant Sci 278:44–53
Ku YS, Sintaha M, Cheung MY, Lam HM (2018) Plant hormone signaling crosstalks between biotic and abiotic stress responses. Int J Mol Sci 19:3206
Kuderova A, Urbankova I, Valkova M, Malbeck J, Brzobohaty B, Nemethova D et al (2008) Effects of conditional IPT-dependent cytokinin overproduction on root architecture of *Arabidopsis* seedlings. Plant Cell Physiol 49:1001
Kwon CS (2005) WUSCHEL is a primary target for transcriptional regulation by SPLAYED in dynamic control of stem cell fate in Arabidopsis. Genes Dev 992–1003
Lachowiec J, Queitsch C, Kliebenstein DJ (2016) Molecular mechanisms governing differential robustness of development and environmental responses in plants. Ann Botany 117:795–809
Lafon-Placette C, Le Gac A-L, Chauveau D, Segura V, Delaunay A, Lesage-Descauses M-C et al (2018) Changes in the epigenome and transcriptome of the poplar shoot apical meristem in response to water availability affect preferentially hormone pathways. J Exp Botany 69:537–551
Lafos M, Kroll P, Hohenstatt ML, Thorpe FL, Clarenz O, Schubert D (2011) Dynamic regulation of H3K27 trimethylation during Arabidopsis differentiation. PLoS Genet 7:e1002040
Lee K, Seo PJ (2017) Coordination of matrix attachment and ATPdependent chromatin remodeling regulate auxin biosynthesis and *Arabidopsis* hypocotyl elongation. PLoS ONE 12(7):e0181804
Li D, Liu J, Liu W (2017) The ISWI remodeler in plants: protein complexes, biochemical functions, and developmental roles. Chromosoma 126:365–373

Li C, Liu Y, Shen WH, Yu Y, Dong A (2018a) Chromatin-remodeling factor OsINO80 is involved in regulation of gibberellin biosynthesis and is crucial for rice plant growth and development. J Integr Plant Biol 60:144–159

Li Q-F, Lu J, Yu J-W, Zhang C-Q, He J-X, Liu Q-Q (2018b) The brassinosteroid-regulated transcription factors BZR1/BES1 function as a coordinator in multisignal-regulated plant growth. Biochim Biophys Acta Gene Regul Mech 1861:561–571

Li W, Liu H, Cheng ZJ, Su YH, Han HN, Zhang Y et al (2011) DNA methylation and histone modifications regulate de novo shoot regeneration in Arabidopsis by modulating WUSCHEL expression and auxin signaling. PLoS Genet 7:e1002243

Liao CJ, Lai Z, Lee S, Yun DJ, Mengiste T (2016) *Arabidopsis* HOOKLESS1 regulates responses to pathogens and abscisic acid through interaction with MED18 and acetylation of WRKY33 and ABI5 chromatin. Plant Cell 28:1662–1681

Liu HC, Lämke J, Lin S, Hung MJ, Liu KM, Charng Y, Bäurle I (2018) Distinct heat shock factors and chromatin modifications mediate the organ-autonomous transcriptional memory of heat stress. Plant J 95:401–413

Luo J, Zhou JJ, Zhang JZ (2018) Aux/IAA gene family in plants: molecular structure, regulation, and function. Int J Mol Sci 19(1):259

Ma X, Ma J, Zhai H, Xin P, Chu J, Qiao Y, Han L (2015) CHR729 is a CHD3 protein that controls seedling development in rice. PLoS ONE 10(9):e0138934

Martin-Trillo M, Lázaro A, Poethig RS, Gómez-Mena C, Piñeiro MA, Martinez-Zapater JM, Jarillo JA (2006) *EARLY IN SHORT DAYS 1* (*ESD1*) encodes ACTIN-RELATED PROTEIN 6 (AtARP6), a putative component of chromatin remodeling complexes that positively regulates *FLC* accumulation in *Arabidopsis*. Development 133(7):1241–1252

Maury S, Sow MD, Le Gac AL, Genitoni J, Lafon-Placette C, Mozgova I (2019) Phytohormone and chromatin crosstalk: the missing link for developmental plasticity? Front Plant Sci 10:395

Mozgová I, Mu-oz-Viana R, Hennig L (2017) PRC2 represses hormone-induced somatic embryogenesis in vegetative tissue of *Arabidopsis thaliana*. PLoS Genet 13:e1006562

Nakashima K, Yamaguchi-Shinozaki K (2013) ABA signaling in stress-response and seed development. Plant Cell Rep 32(7):959–970

Nishioka S, Sakamoto T, Matsunaga S (2020) Roles of BRAHMA and its interacting partners in plant chromatin remodeling. Cytologia 85(4):263–267

Ojolo SP, Cao S, Priyadarshani SVGN, Li W, Yan M, Aslam M et al (2018) Regulation of plant growth and development: a review from a chromatin remodeling perspective. Front Plant Sci 9:1232

Ori N, Cohen AR, Etzioni A, Brand A, Yanai O, Shleizer S, Menda N, Amsellem Z, Efroni I, Pekker I et al (2007) Regulation of LANCEOLATE by miR319 is required for compound-leaf development in tomato. Nat Genet 39:787–791

Peirats-Llobet M, Han SK, Gonzalez-Guzman M, Jeong CW, Rodriguez L, Belda-Palazon B et al (2016) A direct link between abscisic acid sensing and the chromatin-remodeling ATPase BRAHMA via core ABA signaling pathway components. Mol Plant 9:136–147

Peng ML et al (2018) Linking phytochrome-interacting factor to histone modification in plant shade avoidance. Plant Physiol 176:1341–1351

Perruc E, Kinoshita N, Lopez-Molina L (2007) The role of chromatin-remodeling factor PKL in balancing osmotic stress responses during Arabidopsis seed germination. Plant J 52:927–936

Ríos G, Gagete AP, Castillo J, Berbel A, Franco L, Rodrigo MI (2007) Abscisic acid and desiccation-dependent expression of a novel putative SNF5-type chromatin-remodeling gene in Pisum sativum. Plant Physiol Biochem 45(6–7):427–435

Sang Y, Silva-Ortega CO, Wu S, Yamaguchi N, Wu MF, Pfluger J et al (2012) Mutations in two non-canonical *Arabidopsis* SWI2/SNF2 chromatin remodeling ATPases cause embryogenesis and stem cell maintenance defects. Plant J 72:1000–1014

Santner A, Calderon-Villalobos L, Estelle M (2009) Plant hormones are versatile chemical regulators of plant growth. Nat Chem Biol 5:301–307

Sarnowska EA, Rolicka AT, Bucior E, Cwiek P, Tohge T, Fernie AR et al (2013) DELLA-interacting SWI3C core subunit of switch/sucrose nonfermenting chromatin remodeling complex modulates gibberellin responses and hormonal cross talk in *Arabidopsis*. Plant Physiol 163:305–317

Sarnowska E, Gratkowska DM, Sacharowski SP, Cwiek P, Tohge T, Fernie AR et al (2016) The role of SWI/SNF chromatin remodeling complexes in hormone crosstalk. Trends Plant Sci 21:594–608

Shigeta T, Yoshimitsu Y, Nakamura Y, Okamoto S, Matsuo T (2011) Does brassinosteroid function require chromatin remodeling? Plant Signal Behav 6(11):1824–1827

Skubacz A, Daszkowska-Golec A, Szarejko I (2016) The role and regulation of ABI5 (ABA-insensitive 5) in plant development, abiotic stress responses and phytohormone crosstalk. Front Plant Sci 7:1884

Sridha S, Wu K (2006) Identification of *AtHD2C* as a novel regulator of abscisic acid responses in *Arabidopsis*. Plant J 46:124–133

Tian L, Fong MP, Wang JJ, Wei NE, Jiang H, Doerge RW, Chen ZJ (2005) Reversible histone acetylation and deacetylation mediate genome-wide, promoter-dependent and locus-specific changes in gene expression during plant development. Genetics 169:337–345

Tsukiyama T (2002) The in vivo functions of ATP-dependent chromatin-remodelling factors. Nat Rev Mol Cell Biol 3(6):422–429

van den Burg HA, Takken FLW (2009) Does chromatin remodeling mark systemic acquired resistance? Trends Plant Sci 14:286–294

Verma V, Ravindran P, Kumar PP (2016) Plant hormone-mediated regulation of stress responses. BMC Plant Biol 16:86

Vicente J, Mendiondo GM, Movahedi M, Peirats-Llobet M, Juan Y, Shen Y, Dambire C, Smart K, Rodriguez PL, Charng Y, Gray JE, Holdsworth MJ (2017) The CysArg/N-end rule pathway is a general sensor of abiotic stress in flowering plants. Curr Biol 27:3183–3190

Vincent JA, Kwong TJ, Tsukiyama T (2008) ATP-dependent chromatin remodeling shapes the DNA replication landscape. Nat Struct Mol Biol 15(5):477–484

Walley JW, Rowe HC, Xiao Y, Chehab EW, Kliebenstein DJ, Wagner D, Dehesh K (2008) The chromatin remodeler SPLAYED regulates specific stress signaling pathways. PLoS Pathog 4(12):e1000237

Wang C, Gao F, Wu J, Dai J, Wei C, Li Y (2010) Arabidopsis putative deacetylase AtSRT2 regulates basal defense by suppressing PAD4, EDS5 and SID2 expression. Plant Cell Physiol 51(8):1291–1299

Wang J, Gao S, Peng X et al (2019) Roles of the INO80 and SWR1 chromatin remodeling complexes in plants. Int J Mol Sci 20(18):4591

Waterworth WM, Drury GE, Bray CM, West CE (2011) Repairing breaks in the plant genome: the importance of keeping it together. New Phytol 192(4):805–822

Weiner JJ, Peterson FC, Volkman BF, Cutler SR (2010) Structural and functional insights into core ABA signaling. Curr Opin Plant Biol 13:495–502

Wu MF, Sang Y, Bezhani S, Yamaguchi N, Han SK, Li Z et al (2012) SWI2/SNF2 chromatin remodeling ATPases overcome polycomb repression and control floral organ identity with the Leafy and Sepallata3 transcription factors. PNAS 109:3576–3581

Wu M-F, Yamaguchi N, Xiao J, Bargmann B, Estelle M, Sang Y et al (2015) Auxin-regulated chromatin switch directs acquisition of flower primordium founder fate. eLife 4:e09269

Wu H, Zheng L, Qanmber G, Guo M, Wang Z, Yang Z (2021) Response of phytohormone mediated plant homeodomain (PHD) family to abiotic stress in upland cotton (*Gossypium hirsutum* spp.). BMC Plant Biol 21(1):1–20

Yamamuro C, Zhu JK, Yang Z (2016) Epigenetic modifications and plant hormone action. Mol Plant 9(1):57–70

Yang S, Li C, Zhao L, Gao S, Lu J, Zhao M et al (2015) The Arabidopsis SWI2/SNF2 chromatin remodeling ATPase BRAHMA targets directly to PINs and is required for root stem cell niche maintenance. Plant Cell 27:1670–1680

Ye Y, Gong Z, Lu X, Miao D, Shi J, Lu J, Zhao Y (2016) Germostatin resistance locus 1 encodes a PHD finger protein involved in auxin-mediated seed dormancy and germination. Plant J 85(1):3–15
Zhang D, Jing Y, Jiang Z, Lin R (2014) The chromatin-remodeling factor PICKLE integrates brassinosteroid and gibberellin signaling during skotomorphogenic growth in Arabidopsis. Plant Cell 26:2472–2485
Zhao Y (2010) Auxin biosynthesis and its role in plant development. Annu Rev Plant Biol 61:49–64
Zhou C, Zhang L, Duan J, Miki B, Wu K (2005) *HISTONE DEACETYLASE19* is involved in jasmonic acid and ethylene signaling of pathogen response in *Arabidopsis*. Plant Cell 17:1196–1204

Phytohormone-Mediated Regulation of Sprouting in Tuber and Storage Root Crops

Kirtikumar R. Kondhare

Abstract Tuber and storage root crops are staple food in several tropical countries. Potato and yam are the two widely cultivated tuber crops, and sweet potato is the most important storage root crop after cassava. The belowground storage organs of these crops are delicious, nutritionally rich and have medicinal applications. Potato and yam tubers as well as sweet potato storage roots have unique abilities to sprout under favorable conditions to form new plants. However, the short dormancy period of these storage organs, especially sweet potato, causes significant losses in the quality for human consumption, and is a major limitation for their global cultivation. The sprouting phenomenon is widely studied in potato, and numerous genes related to metabolism, transport and signaling pathways of phytohormones and sugars are identified that could act as crucial regulators of the sprouting process. However, the literature is scarce regarding sprouting of yam tubers and sweet potato storage roots. Despite the enormous importance of the tuber and storage root crops, knowledge about the molecular mechanism governing the sprouting process is limited. In this chapter, the roles of various molecular factors, phytohormones and their signaling crosstalk are discussed during the sprouting of tubers in potato and yam, and sweet potato storage roots.

1 Introduction

Several tuber and storage root crops serve as a staple food in many tropical and temperate countries (Chandrasekara and Kumar 2016). Two widely cultivated tuber crops include potato (*Solanum tuberosum*) and yam (*Dioscorea alata*), whereas the major storage root crops are sweet potato (*Ipomoea batatas*), cassava (*Manihot esculenta*), beetroot (*Beta vulgaris*), carrot (*Daucus carota*), radish (*Raphanus sativus*),

K. R. Kondhare (✉)
Plant Molecular Biology Group, Biochemical Sciences Division, CSIR-National Chemical Laboratory, Dr. Homi Bhabha Road, Pune, Maharashtra 411008, India
e-mail: kr.kondhare@ncl.res.in

Academy of Scientific and Innovative Research (AcSIR), Ghaziabad 201002, India

T. Aftab (ed.), *Auxins, Cytokinins and Gibberellins Signaling in Plants*, Signaling and Communication in Plants,
https://doi.org/10.1007/978-3-031-05427-3_13

turnip (*Brassica rapa*), etc. (Chandrasekara and Kumar 2016). These crops are nutritionally rich with diverse medicinal and industrial applications, and are considered as ideal crops for overcoming the global food security challenges (Chandrasekara and Kumar 2016). Among these crops, the belowground storage organs (i.e. tubers of potato and yam as well as storage roots of sweet potato) exhibit a unique ability to sprout under favorable conditions to produce new plants (Fig. 1). Although sprouting is necessary for vegetative propagation, the short dormancy period of the edible tubers and storage roots leads to the significant losses in their quality for human consumption (Cheema et al. 2010; More et al. 2019). Apart from the growth practices employed and the environmental conditions prevalent during the tuber or storage root formation and maturation (More et al. 2019), numerous molecular, biochemical and genetic factors are known to affect the dormancy and sprouting of tubers in potato (Aksenova et al. 2013; Sonnewald and Sonnewald 2014).

Over the past decades, the advancement in the transcriptomics, proteomics, metabolomics and plant genetic engineering has provided some insights regarding the tuber dormancy and the sprout growth in potato (reviewed in Gong et al. 2021). However, the specific genes and the molecular mechanisms controlling the sprout initiation are not well understood in the tuber and storage root crops like potato, yam and sweet potato. This chapter aims to provide the current overview of the tuber dormancy phenomenon and the factors influencing the sprouting process in potato and the underlying genetic regulatory network. Moreover, the information available

Fig. 1 Dormancy to sprouting transitions in tubers of potato (*Solanum tuberosum*) and storage roots of sweet potato (*Ipomoea batatas*). Dormant (**a**) and sprouted (**b**, **c**) tubers of potato are shown along with dormant (**d**) and sprouted (**e**, **f**) storage roots of sweet potato. Arrows (white color) in (**b**) and (**e**) indicate sprout initials from potato tubers and sweet potato storage roots, respectively. Scale bar = 1 cm

regarding the sprouting of yam tubers and sweet potato storage roots is also summarized. The overall role of the molecular factors, phytohormones as well as their signaling crosstalk during potato, yam and sweet potato sprouting is discussed. The potential of microRNAs (miRNAs) and long non-coding RNAs (lncRNAs) in the regulation of phytohormone metabolism and signaling during potato tuber sprouting is also described. The special emphasis of this chapter is to highlight the role of key phytohormones in the regulation of the sprouting process in tubers of potato and yam, and storage roots of sweet potato.

2 Molecular Control of Tuber Sprouting in Potato

Potato (*Solanum tuberosum* L.) belongs to the Solanaceae family and is the most essential food crop after cereals like wheat, maize and rice (www.fao.org). It is a succulent plant grown worldwide for the belowground storage organ—tuber, which is used as a staple food, raw material for industrial food products and starch production (Muthoni et al. 2014). As potatoes have high market demand throughout the year, it is imperative to find the ways for effective control of the tuber dormancy period and the sprout initiation process (Suttle 2004b; Kloosterman et al. 2007). Tuber dormancy and sprouting are the two crucial processes of the potato tuber life cycle that determine its growth and yield (Sonnewald and Sonnewald 2014). These complex processes are influenced by the genetic background of varieties, the stage of tuber development, the crop management practices, environmental conditions like photoperiod (daylength), light intensity, humidity, temperature, carbon dioxide levels in the field, and soil conditions (texture and moisture) (Biemelt et al. 2000; Sonnewald 2001; Claassens 2002; Agrimonti and Marmiroli 2008; Aksenova et al. 2013; Sonnewald and Sonnewald 2014). Different environmental, biochemical and genetic factors and the associated morphological/anatomical changes governing the tuber dormancy, sprout initiation and its subsequent growth are mentioned in Table 1.

The tuber sprouting process in potato involves three physiological stages: (i) dormancy, (ii) sprout initiation, and (iii) sprout growth (Fig. 1). During the sprouting, the tuber buds and the tuber tissues exhibit peculiar growth characteristics as well as specific metabolic and cellular gene expression changes (Ronning et al. 2015). In the dormant potatoes, the tuber buds show arrested growth due to the blockage of meristematic cells in the G1/S-phase transition and the metabolism rate is minimal, whereas the tuber tissues show the absence of growth and morphogenesis, and the preservation of storage carbohydrates and patatins (Table 1). In fact, dormant buds are known to have approximately 77% of the cells' nuclei in the growth phase (G1), whereas about 13% of the cells' nuclei remain in the preparative phase for mitosis (G2) leading to a highly reduced growth of tuber buds (Haider et al. 2021). During the sprout initiation stage, the changes in the tuber buds include the removal of blockage of meristematic cells in the G1/S-phase transition, the initiation of growth and morphogenesis, and the activation of metabolism (Table 1). The tuber tissue exhibits the absence of growth and morphogenesis, a transition from storage to source function, and

Table 1 Factors contributing to the dormancy and sprouting of potato and yam tubers, and sweet potato storage roots and the associated changes (morphological, anatomical, biochemical, and genetic)

Crop	Factors	Characteristics features	References
Potato	Environmental	Tubers produced from long-day plants have longer dormancy period than short-day plants. High temperatures (>35 °C) during tuber formation cause early sprouting. Reduced soil moisture and nutrient availability during the tuber maturation reduces the dormancy period. The presence or absence of light during post-harvest storage has no effect on the dormancy duration, but dramatically affects the morphology of emerging sprouts. Hypoxic and anoxic conditions during the tuber storage can break tuber dormancy. Immature tubers have a greater sprouting capacity compared to mature tubers stored under the same conditions	Hutchingson (1978) Burton (1989) Levy and Veilleux (2007) Suttle (2007) Haider et al. (2021)
	Morphological	Growth of the tuber eyes, especially the transition of meristematic cells from G1-to-S phase or G2-to-M2 phase, is arrested in dormancy; however, growth and morphogenesis is enhanced during the sprout initiation	Muthoni et al. (2014)
	Biochemical	Levels of ABA decrease, whereas that of auxin, CK and GA increase during the dormancy release and the sprout initiation. ABA and ethylene extend the tuber dormancy and suppress tuber sprouting. CK and GA are required for the sprout initiation and the sprout growth, respectively. Auxin levels increase at the sprout initiation and is responsible for the vascular development between the sprout initials and the tuber pith. Nitric oxide (NO) treatment induces sprouting. Metabolism rate remains low during the dormancy, but it increases during the sprout growth. Storage carbohydrates and patatins are preserved during dormancy, but the carbohydrate metabolism increases during the sprout initiation	Hartmann et al. (2011) Mani et al. (2014) Muthoni et al. (2014)
	Genetic	Wild potato species have a longer tuber dormancy, whereas majority of the cultivated potato species have a short dormancy	Suttle (2007)
Yam	Environmental	Environmental factors have no effect on the endo-dormancy (Phase I) of tubers. Temperature and relative humidity can influence the endo-/eco-dormancy (Phase II) and eco-dormancy (Phase III) of tubers. When dormant tubers are grown on high humidity, it enhances the shoot apical meristem formation and sprout growth. Constant warm temperatures with high humidity or complete dark conditions promote tuber sprouting. Tuber storage at low temperatures (15–16 °C) or treating dormant tubers with gamma irradiation (5–12 krad) from 1 to 8 months can delay sprouting. Storage of tubers in complete dark conditions also influences tuber sprouting. Higher respiration and carbohydrate loss are responsible for the dormancy breakage. Better aeration (ventilation) can reduce the weight loss of tubers during the storage and can delay the sprout initiation. The application of nitrogen fertilizer in the field at higher rates increases sprouting in stored tubers	Mozie (1975), Adesuyi (1982), Craufurd et al. (2001), Ile et al. 2006

(continued)

Table 1 (continued)

Crop	Factors	Characteristics features	References
	Biochemical	Application of auxins (indole acetic acid; naphthalene acetic acid) or cytokinin (benzyladenine) delays sprouting to varying degrees. Application of GA3 (100 μM) to the dormant tubers inhibits sprouting unlike potato and sweet potato, whereas applying GA biosynthesis inhibitors stimulates sprouting. Application of ABA biosynthesis inhibitor—fluridone, and ABA + fluridone treatments induce tuber sprouting on new underground tubers, whereas ABA alone prolongs the tuber dormancy. Ethephon is known to break dormancy and enhances sprouting. Jasmonic acid treatment at a lower concentrations (0.1–1 μM) enhances sprouting, whereas sprouting is inhibited at higher concentrations (30–100 μM). The respiration rate and metabolic activity of the tubers is negligible during the tuber dormancy, but it increases significantly during the dormancy breakage and the sprout initiation. Alpha-amylase activity increases in the tubers during the sprouting, which coincides with the higher sugar content. The activities of enzymes, such as hexokinase, alcohol dehydrogenase, phosphorylase, and glucose-6-phosphate dehydrogenase increase during the sprouting, whereas polyphenol oxidase activity and the starch content decrease. L-ascorbate, total polyphenols, carotenoids and lipids increase during the storage, and their levels peak during the sprout initiation. Reduced level of glutathione is associated with an increase in the tuber dormancy. During the tuber maturation, activated ABA signaling helps to maintain the dormancy by arresting the growth of cells in a tuber. Starch, discorin, heat-shock proteins, chaperon, and various inhibitors of proteases, trypsin and cysteine accumulate during the dormancy phase. Auxin helps to break dormancy and mobilize starch. Cell cycle proliferation is initiated and discorin proteins start to deplete during sprouting. The sprout initiation is accompanied by the active cell growth in the bud region; the starch level reduces and discorin protein continues to deplete	Gregory (1968), Igwilo (1982), Ikediobi and Oti (1983), Okagami and Tanno (1993), Shiwachi et al. (2003), Ile et al. (2006), Jaleel et al. (2007), Adu-Gyamfi and Blay (2009), Braide and Hamadina (2018), Sharma and Deswal (2021)
	Morphological and anatomical	Dormant tuber is characterised by four regions: (i) the protective region consisting an outer layer of primary suberized cells, and an inner layer of the radially arranged suberized cork cells; (ii) the cortex region comprising parenchyma cells beneath the protective region; (iii) the meristematic region containing 2–4 layers of small, flattened and stretched-out undifferentiated cells that lie beneath the cortex, and (iv) the storage parenchyma region filled with starch grains and scattered vascular tissues. In the dormant tubers, there is no sign of any active meristematic activity (Phase I: endo-dormancy). Shoot apical bud development involves active cell division and differentiation in the meristematic cells, and leads to the formation of a localized mass of cells, called the primary thickening meristem that subsequently give rise to the tuber germinating meristem. The cells in the primary thickening meristem are small, either irregular or oblong in shape and arranged in a horizontal array, whereas the shape of the cells in the tuber germinating meristem changes from horizontal to a more vertical array and these cells have widespread activity in the meristematic layer (typically 10–40 cell layers thick depending on the developmental stage) (Phase II: endo-/eco-dormancy). The cells at the apex of the tuber germinating meristem get organized and form a shoot apical meristem tangentially to the tuber germinating meristem. Foliar primordia initiates from the peripheral cells of the shoot apical meristem and subsequently complete apical shoot buds are formed (Phase III: eco-dormancy). Tubers with thick peels tolerate rougher handling during the storage and such tubers sprout less	Wickham et al. (1981), Akoroda (1993), Ile et al. (2006)

(continued)

Table 1 (continued)

Crop	Factors	Characteristics features	References
Sweet potato	Environmental	Adequate curing of storage root can delay sprouting, whereas inadequate or excessive curing can induce sprouting. Providing adequate ventilation to storage roots during storage delays the sprout initiation. Gamma radiation (10–20 krad) or isopropyl *N*-(3-chlorophenyl) carbamate (CIPC) treatment can inhibit sprouting. Hot water treatment delays the sprout initiation. Incubating storage roots at temperatures, in the range 12–16 °C, can delay the sprouting process, whereas incubation at ≥25 °C stimulates it. Optimum relative humidity (>95%) delays sprouting, whereas that in the range of 70–90% can induce sprouting. Maleic hydrazide treatment during the storage controls sprouting	Jenkins (1982), Cantwell and Suslow (2001), Cheema et al. (2010), Chakraborty et al. (2017)
	Morphological	Unlike the eyes on the potato tubers, the sprout buds are not visible prior to the sprout growth in sweet potato storage roots	Cheema et al. (2010)
	Biochemical	Changes in the levels of various phytohormones in storage roots of sweet potato during dormancy to sprout initiation are not yet studied. However, application of GA stimulates, whereas its biosynthesis inhibitors supresses sprouting. Ethylene application (at ≥5 ppm) inhibits the sprout growth. Application of auxin (NAA; naphthaleneacetic acid) reduces sprouting. A decline in the starch content accompanied by an increment in activity of α-amylases is observed during the sprout initiation	Paton and Scriven, (1989), Zhang et al. (2002), Cheema et al. (2010)

enhanced activity of carbohydrate metabolism. During the sprout growth stage, the tuber buds show the highest growth activity and metabolism rate, whereas the tuber tissues exhibit the absence of growth and morphogenesis and metabolic changes to provide an active supply of energy to the growing bud (Aksenova et al. 2013).

The phytohormones, mostly cytokinin (CK), abscisic acid (ABA), gibberellin (GA) and auxin modulate various cellular and molecular changes during the dormancy and sprouting of potato tubers (Saidi and Hajibarat 2021). The cell cycle and division in the meristematic tissues of the tuber buds are arrested or significantly reduced during the dormancy, whereas they are re-activated or induced during the sprout initiation. These changes are mediated through the coordinated synthesis and action of cyclin-dependent kinases (CDKs), their downstream targets and regulatory partners. CDKs are activated by D-type cyclins (CYCD) and activated CDKs catalyze the transfer of G1 cells into the S phase (Lipavská et al. 2011). A number of reports indicate that CK induces the expression of CYCD3 and also helps in binding of CYCD3 to CDKs; thereby promotes the resting cells of G0/G1 to enter in the cell cycle and sprouting is initiated (Francis and Sorrell 2001; Werner et al. 2001; Tang et al. 2004; Lipavská et al. 2011; Velappan et al. 2017; Gao et al. 2019; Skalák et al. 2019). When the tuber buds are dormant, ABA acts as a mitotic inhibitor and it inactivates the CDK/CYCD complex by inducing the expression of CDK inhibitors (CKIs), which results in the cell cycle arrest at the G1/S check-point and subsequent accumulation of cells at G1 (Velappan et al. 2017). It is proposed that GA can induce the level of CYCD-REGULATED Cdc 2 KINASE that is involved in triggering the G2-to-M transition of the cell cycle and enhances the rate at which cells are produced (Francis and Sorrell 2001). In the dormant potato tubers, auxin is accumulated in the meristem of tuber buds. Increased auxin levels in the tuber buds have been associated with the dormancy maintenance; however, the reduced auxin levels lead to the dormancy breakage (Aksenova et al. 2013). In potato, a calcium-dependent protein kinase 1 (StCDPK1) phosphorylates and activates the auxin transporter StPIN4. *StCDPK1* is found to be expressed in the vascular tissues in the dormant tubers, whereas its expression further enhances in the tuber buds during the sprouting, and this differential gene expression pattern of *StCDPK1* is governed by a microRNA, miR390 (Santin et al. 2017). Auxin signaling has been shown to regulate the cell cycle either directly or through the crosstalk with other phytohormones. Low auxin concentration can promote the sprout growth following the dormancy breakage (Muthoni et al. 2014). Another report showed that auxin inhibits CK synthesis and thereby reduces the levels of both *CYCD3* and *CDK3* expression leading to a prolonged maintenance of the tuber dormancy (Cheng et al. 2015). These findings suggest the crucial role of phytohormones in the tuber dormancy maintenance, dormancy breakage as well as in the sprout initiation and its subsequent growth.

Considering the physiological changes happening at the onset of the sprout initiation and subsequent growth of the sprout buds, it appears that several genes associated with phytohormone, sugar metabolism and signaling pathways could govern these changes during the sprouting process (described in Sects. 2.1 and 2.2). In a study by Liu et al. (2012), the authors used suppression subtractive hybridization, and identified a gene encoding ADP RIBOSYLATION FACTOR in potato that was associated with the tuber dormancy release. RT-qPCR analysis of potato tuber eyes revealed that the expression of *ADP RIBOSYLATION FACTOR* was significantly higher at the sprout initiation compared to the tuber dormancy, suggesting its crucial role in the tuber sprout initiation. Recently, Morris et al. (2019) identified several quantitative trait loci (QTLs) that are associated with the tuber sprout growth. Underlying one such QTL is a gene that encodes for a specific Flowering Locus T (FT) family member protein, known as TERMINAL FLOWER 1/CENTRORADIALIS (StCEN). The authors found that the manipulation of *StCEN* expression influenced the overall sprout growth, but the tuber dormancy break time was unaffected in these lines compared to controls. Furthermore, RNA interference (RNA*i*) lines of *StCEN* led to an increased rate of the tuber sprout growth accompanied by reduced ABA levels and enhanced levels of CK. In contrast, its overexpression lines displayed the reduced rate of the tuber sprout growth that was associated with enhanced ABA levels, but reduced CK levels (Morris et al. 2019). Additionally, in a recent review by Gong et al. (2021), the sprout growth that involves the active cell division and meristem development in the eyes of potato tubers has also been linked with DNA methylation and histone acetylation, implicating that the epigenetic regulation could also be one of the mechanism regulating the tuber dormancy maintenance and the sprout initiation.

2.1 Role of Phytohormones and Their Signaling During Potato Tuber Sprouting

Environmental, biochemical and genetic factors play a crucial role in determining the potato tuber dormancy period and the sprouting process (Table 1). Various phytohormones and their crosstalk with other molecular factors have been shown to be important for regulating the tuber dormancy, sprout initiation and its subsequent growth. Exogenous application of phytohormones to tubers (Table 1) and their effect on the dormancy duration change and the sprout initiation have suggested that they could serve as key regulators of the tuber dormancy and sprouting in potato (Aksenova et al. 2013). Phytohormones like ABA, ethylene and brassinosteroid (BR) control the tuber dormancy stage, whereas auxin, CK and GA regulate the sprout initiation phase (Table 1). CK and GA are also known to regulate the sprout growth process (Aksenova et al. 2013).

In the potato tubers, the endogenous levels of phytohormones exhibit dynamic changes from the tuber dormancy to the sprout initiation and the subsequent sprout growth (Sukhova et al. 1993; Suttle 2004a; Morris et al. 2019). As shown in Fig. 2, the auxin (IAA; indole-3-acetic acid) level is low in dormant tubers; however, its level peaks at the sprout initiation phase and it is drastically reduced during the sprout growth process. GA level increases gradually from dormancy to sprouting transitions, with its level being highest during the sprout growth. In contrast, ABA exhibit an opposite pattern compared to GA levels, with the highest level being

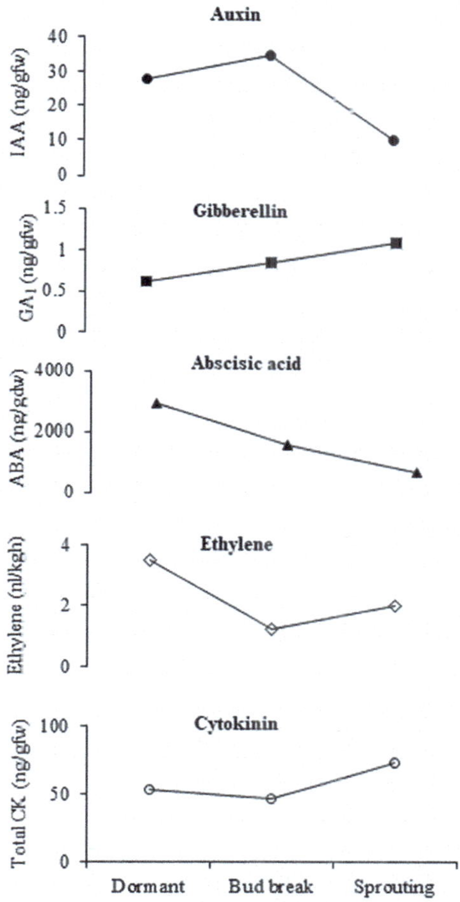

Fig. 2 Phytohormone changes during dormancy to sprouting transitions in the tuber buds of potato. The endogenous levels of various phytohormones—auxin (IAA; indole-3-acetic acid), gibberellin (GA_1; gibberellin 1), abscisic acid (ABA), ethylene and cytokinin (CK) in potato tuber buds are shown as per the reports of Sukhova et al. (1993), Suttle (2004a) and Morris et al. (2019). Total CK represents a combined level of different CK forms: zeatin (Z), zeatin riboside (ZR), isopentenyladenine (iP) and isopentenyladenosine (iPA)

in the dormant tubers and the lowest during the sprout growth process. Ethylene shows a pattern opposite to auxin, wherein its level is highest in the dormant tubers, which is then reduced at the sprout initiation and the subsequent growth stages. The level of CK remains low during the dormancy and sprout initiation stages, but it increases during the sprout growth stage (Fig. 2). The in silico gene expression analysis revealed that apart from the biosynthesis, the genes involved in the transport, catabolism and signaling of various phytohormones exhibit differential expression in a mature tuber versus a tuber sprout, suggesting their importance in the potato tuber sprouting process (Fig. 3a). A recent report by Zhang et al. (2021) found that the heat stress treatment reduced the tuber dormancy, but induced sprouting. Through a comparative transcriptomics of heat stressed potato tubers, authors further revealed that the dormancy-associated genes, such as *DOG1* and *SLP* were downregulated, whereas the genes related to the ABA catabolism (*ABA 8'-HYDROXYLASE*), GA

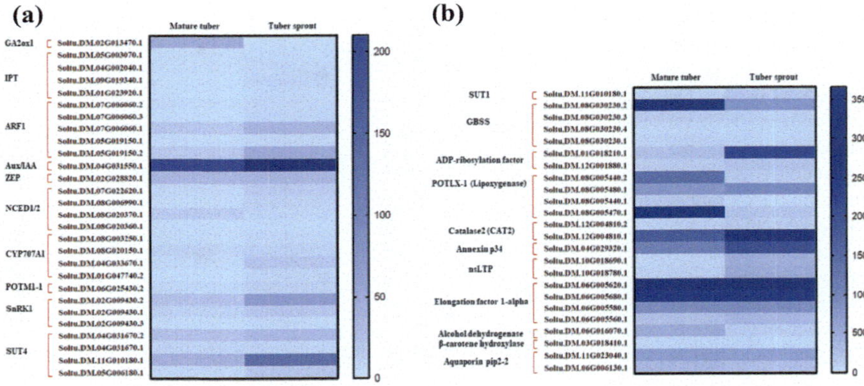

Fig. 3 Heat map depicting the expression of key genes involved in the potato tuber sprouting. The expression of genes related to biosynthesis and signaling of various phytohormones (**a**) and other categories (**b**) are shown in a mature tuber versus a tuber sprout. The color code represents the gene transcript abundance (mean values) in the range mentioned (right side of the heat-map) based on the TPM (transcript per million) values retrieved from the transcriptomics data of the potato genotype—RH89-039-16 using the Solanaceae Genomics Resource SpudDB website (http://spuddb.uga.edu). The genes included in the heat maps are as per the report of Gong et al. (2021) and other references cited in the text. Abbreviations: *GA2OX1 = GIBBERELLIN 2-OXIDASE; IPT = ISOPENTENYL TRANSFERASE; ARF1 = AUXIN RESPONSE FACTOR 1; Aux/IAA = AUXIN/INDOLE-3-ACETIC ACID (auxin repressor); ZEP = ZEAXANTHIN EPOXIDASE; NCED1/2 = NINE-CIS-EPOXYCAROTENOID DEHYDROGENASES; CYP707A1 = ABSCISIC ACID 8'-HYDROXYLASE; POTM1-1 = POTATO MADS-BOX TRANSCRIPTION FACTOR; SnRK1 = SUCROSE TRANSPORTER; StSUT 4 = SUCROSE TRANSPORTER 4; GBSS = GRANULE-BOUND STARCH SYNTHASE; nsLTP = PUTATIVE NON-SPECIFIC LIPID TRANSFER PROTEIN*

biosynthesis (*ENT-KAURENOIC ACID OXIDASE*) and auxin signaling (*AUXIN-RESPONSIVE PROTEIN IAA16*) were upregulated during the heat stress-induced tuber sprouting. These findings proposed that genes like *DOG1* and *SLP* could serve as dormancy markers, and the *ABA 8'-HYDROXYLASE* gene as a sprouting marker during the heat stress-induced sprouting process.

Numerous reports have explored the physiological roles of key phytohormones related genes in the potato tuber sprouting process. For example, a constitutive overexpression or the leaf-specific overexpression of a GA biosynthesis gene, *GA20OXIDASE* (*StGA20OX1*) in potato led to a reduction in the tuber dormancy period and also displayed an early sprouting phenotype, whereas these phenotypes remained comparable to controls in *StGA20OX1* antisense lines (Table 2) (Carrera et al. 2000). Using the constitutive overexpression or RNA suppression lines of a GA catabolism gene *StGA2OX1* in potato, Kloosterman et al. (2007) observed no change in duration of the tuber dormancy period in both types of lines compared to tubers from wild-type plants (Table 2). However, as expected, the authors found a reduced inter-nodal length of the sprout in the overexpression lines, suggesting that *StGA2OX1* does not regulate the tuber dormancy period, but it could be involved in the sprout elongation process. In an investigation by Hartmann et al. (2011), the authors

Table 2 List of transgenic studies highlighting the role of key phytohormone metabolism, transport and signalling related genes in controlling the sprouting phenomena in tubers of potato and yam

Crop	Phytohormones	Gene expression analysis or transgenic manipulation	Associated phenotype	References
Potato	Auxin	qRT-PCR analysis as well as *in-situ* hybridization experiments revealed enhanced (threefold) expression of an *AUXIN RESPONSE FACTOR* (*StARF6*) in the sprouting buds, especially in the meristematic, procambial and early vascular tissues compared to the dormant buds *AUXIN RESPONSE FACTOR ARF1*, *AUX/IAA* and related genes (*StIAA2* and *StIAA*) are up-regulated during the tuber dormancy breaking	Not yet known	Faivre-Rampant et al. (2004), Liu et al. (2012)
	CK	Constitutive overexpression of CK biosynthesis gene *ISOPENTENYLTRANSFERASE* (*IPT*) or CK catabolism gene *CYTOKININ OXIDASE/DEHYDROGENASE 1* gene (*CKX1*)	Constitutive overexpression of *IPT* did not influence the tuber sprout induction; however, these lines exhibited earliness for sprout induction when treated with GA_3 under in vitro conditions. *CKX1* overexpressing lines showed a prolonged dormancy period	Hartmann et al. (2011)
	GA	Overexpression or RNA*i* of GA biosynthesis gene *GA20OXIDASE* (*StGA20ox1*) Overexpression or RNA*i* lines of a GA catabolism gene *GA2OXIDASE* (*StGA2ox1*)	Constitutive or the leaf-specific overexpression of *StGA20OX1* caused a reduction in tuber dormancy and exhibited early sprouting phenotype, whereas these phenotypes remained comparable to controls in *StGA20OX1* RNA*i* lines No change in the duration of the tuber dormancy period was observed in constitutive overexpression or RNA*i* lines of *StGA2OX1* compared to wild-type	Carrera et al. (2000), Kloosterman et al. (2007)
	ABA	Although ABA biosynthesis or signalling genes are not yet characterized during sprouting process, it is quite established that reduced ABA levels in the meristem and periderm of tubers are associated with the dormancy break and the sprout initiation. This is catalysed by the enhanced expression ABA catabolism genes like *ABA 8′-HYDROXYLASE* (*StCYP707A*) and *SUCROSE TRANSPORTER SnRK1*, and a decreased expression of ABA biosynthesis genes *9-CISEPOXYCAROTENOID DIOXYGENASE* (*StNCED1/2*). At the sprout initiation, known ABA-inducible genes are also downregulated	Not yet known	Destefano-Beltran et al. (2006), Campbell et al. (2008), Lin et al. (2015)

(continued)

Table 2 (continued)

Crop	Phytohormones	Gene expression analysis or transgenic manipulation	Associated phenotype	References
	Ethylene	Ethylene responsive factor 5 and ethylene responsive element binding protein 6 are upregulated during the tuber dormancy breakage	Not yet known	Liu et al. (2012)
	SL	RNA*i* of SL biosynthetic pathway gene *CAROTENOID CLEAVAGE DIOXYGENASE8* (*StCCD8*)	RNA*i* lines of *StCCD8* showed a higher rate of tuber sprouting during storage compared to wild-type	Pasare et al. (2013)
Yam	Auxin	Enhanced expression of an auxin-induced protein—PCNT115 and auxin-dependent transcription factor at the onset of tuber sprouting. ADP ribosylation factor 1 helps in breaking the dormancy and mobilization of starch during the dormancy breakage	Not yet known	Sharma and Deswal (2021)
	GA	The expression of a GA receptor gene *GIBBERELLIN INSENSITIVE DWARF 1 (DoGID1A)* increases gradually during yam bulbil sprouting	Not yet known	Long et al. (2019)
	ABA	Increased expression of *PYL9* (ABA receptor) and transcription factors like DEAD-BOX and glycine-rich RNA-binding proteins help to maintain dormancy	Not yet known	Sharma and Deswal (2021)

used transgenic approach to manipulate the endogenous levels of CK; wherein a gene encoding CK biosynthesis enzyme ISOPENTENYLTRANSFERASE (IPT) from *Agrobacterium tumefaciens* and the gene encoding CK catabolism enzyme CYTOKININ OXIDASE/DEHYDROGENASE 1 (CKX1) from Arabidopsis were overexpressed independently in potato, and the tuber sprouting phenotype of transgenic plants was observed (Table 2). The constitutive overexpression lines of *IPT* had enhanced accumulation CK levels, yet they did not influence the tuber sprout induction compared to wild-type potato tubers. However, *IPT* overexpression lines exhibited earliness for the sprout induction when treated with GA_3 under in vitro conditions. In contrast, *CKX1* overexpressing lines had reduced CK levels and their tubers showed a prolonged dormancy period. Moreover, the tuber buds from *CKX1* overexpressing lines were non-responsive to GA_3 application. These results support that CKs are essential for terminating the tuber dormancy and stimulating the cell division to initiate sprouting in potato tubers. Also, it appears that higher CK levels increase the GA-responsiveness of the tuber buds and induce sprout initiation in potato.

Gene expression analysis as well as *in-situ* hybridization experiments revealed that the expression level of a gene involved in auxin signaling - *AUXIN RESPONSE FACTOR* (*StARF6*) increased by nearly threefold in the sprouting buds, especially in the meristematic, procambial and early vascular tissues compared to the dormant buds (Table 2). The authors proposed that *StARF6* could serve as an important marker to govern the meristem activation in potato tubers (Faivre-Rampant et al., 2004). RNA*i* lines of the key strigolactone (SL) biosynthetic pathway gene *CAROTENOID CLEAVAGE DIOXYGENASE8* (*CCD8*) in potato lead to an approximately threefold decrease in root SL levels compared to wild-type plants, and the tubers from the RNA*i* lines showed a higher rate of sprouting during their storage (Table 2) (Pasare et al., 2013). Interestingly, during twelve weeks of storage at room temperature, all the tubers of the *CCD8* RNA*i* lines sprouted, whereas no sprouting was observed in wild-type tubers, implying that SL delays the tuber sprouting rate and efficiency. Ethylene is known to suppress the tuber sprouting; however, the exact role of ethylene remains to be elucidated (Table 2).

Tuber dormancy is believed to be genetically inherited. An investigation by Bisognin et al. (2018) found a quantitative trait locus (QTL), which can control the dormancy and sprouting of potato tubers. This QTL was located on chromosomes 2, 3, and 7, and contained genes involved in signaling of phytohormones, such as ABA, IAA and GA. A recent report by Sharma et al. (2021) employed a dual approach of conventional QTL analysis coupled with a combined bulk-segregant analysis (BSA), and numerous BSA-QTLs responsible for the tuber sprout elongation and the dormancy release were identified. Among them, many QTLs harbored genes involved in various phytohormone biosynthesis and signaling pathways, such as GA (*GA20OXIDASE*, PGSC0003DMG400000170; *GA20OXIDASE 4*, PGSC0003DMG400000011; GA receptor *GID1*, PGSC0003DMG400000139, PGSC0003DMG400012756; DELLA protein *RGL2*, PGSC0003DMG400007285; *YABBY1*, PGSC0003DMG400025969), CK (two-component system sensor histidine kinase/response regulator, PGSC0003DMG400038579), auxin (AUXIN-INDUCED

PROTEIN X10A; PGSC0003DMG400026010), BR (*BRASSINOSTEROID INSENSITIVE 1-ASSOCIATED RECEPTOR KINASE 1*, PGSC0003DMG401000056) and ethylene (*1-AMINOCYCLOPROPANE-1-CARBOXYLATE SYNTHASE 2*, PGSC0003DMG400000193). The candidate QTLs could be further explored for developing new potato cultivars. Useful manipulation of the tuber sprout initiation (delay or earliness) can be achieved based on the purpose of the tubers- food source or plant propagation. Apart from phytohormone related genes, several other genes, such as *SUCROSE TRANSPORTER 1* (*SUT1*), *GRANULE BOUND STARCH SYNTHASE* (*GBSS*), *ADP RIBOSYLATION FACTOR*, *POTLX-1* (*LIPOXYGENASE*), *CATALASE 2* (*CAT2*), *ANNEXIN P34*, *ALCOHOL DEHYDROGENASE*, *β-CAROTENE HYDROXYLASE*, *AQUAPORIN PIP2-2*, etc. were differentially expressed between a mature tuber and the tuber sprout, implying their role in the initiation of the potato tuber sprout (Fig. 3b).

2.2 Sugar Metabolism and Its Crosstalk with Plant Hormones During Potato Tuber Sprouting

The sprout initiation and its subsequent growth is dependent on the supply of sucrose from the tuber pith cells to the growing bud. The starch stored in the tubers is hydrolyzed by the coordinated action of α- and β-AMYLASES, and the debranching enzymes like ISOAMYLASE and LIMIT DEXTRINASE to form glucose and maltose. These sugars are further exported to the cytosol by the maltose and glucose transporters, respectively. In the cytosol, INORGANIC PYROPHOSPHATASE (PPase) catalyze the conversion of these sugars into sucrose, which is then transported through the vascular strands to the sprout bud as energy source. Earlier, Hajirezaei and Sonnewald (1999) developed the transgenic potato plants that expressed a *PPase* gene under the control of a strong and constitutive chimeric *ST-LS1/35S* promoter. Wild-type tubers began to sprout after four months of storage at room temperature, whereas transgenic tubers did not sprout even after a prolonged storage period of two years, suggesting that constitutive expression of *PPase* inhibits sprout growth. As the tubers of these transgenic lines had enhanced levels of glucose, fructose and sucrose, early sprout growth was expected in them compared to the wild-type tubers. Additionally, the transgenic lines contained reduced starch content in tubers, yet the rate of respiration of tubers was unaltered in transgenic lines compared to wild-type. Based on these observations, the authors proposed that reduced transport of sucrose to the sprout bud could have inhibited the sprout initiation from transgenic tubers. Subsequently, a pioneering study by Farré et al. (2001) specifically expressed the *PPase* gene from *E. coli* in the cytosol of tuber cells using a tuber-specific *patatin* promoter. Interestingly, the tubers from the transgenic lines displayed an earliness for sprouting compared to wild-type tubers, and they had increased levels of sucrose and glucose along with decreased starch content, which could have stimulated the early sprouting phenotype. These studies suggest that manipulating the cytosolic inorganic pyrophosphate (PPi) content represents a cost-effective and environment-friendly strategy for controlling the tuber sprouting in potato.

Potato transgenic lines overexpressing bacterial *1-DEOXY-D-XYLULOSE 5-PHOSPHATE SYNTHASE* gene (*DXS*) showed an earliness for the tuber sprout initiation. The tubers from these overexpression lines were found to be sprouted at the time of harvest, whereas the wild-type tubers were dormant at the same time (Morris et al. 2006). The authors further noticed ~sixfold increase in the levels of a CK—*trans*-zeatin riboside in the tubers of these overexpression lines at harvest; however, the levels of other MEP (2-C-methyl-D-erythritol 4-phosphate) pathway-derived hormones, such as GA and ABA remained unchanged compared to the wild-type tubers, suggesting that increased levels of CK contribute to earliness in tuber sprouting (Morris et al. 2006). In order to modulate the trehalose-6-phosphate (T6P) content specifically in the tubers of transgenic potato plants, Debast et al. (2011) generated two constructs that either express the *E. coli OtsA* gene encoding a TREHALOSE-6-PHOSPHATE SYNTHASE (*TPS*) or the *OtsB* gene encoding a TREHALOSE-6-PHOSPHATE PHOSPHATASE (TPP) under the control of a tuber-specific *B33* promoter. The analysis showed that the tubers of *TPS* lines had five-fold increase in T6P levels, whereas the tubers from *TPP* overexpression lines had 50% reduction in T6P levels compared to the wild-type tubers. The authors further reported that the sprout formation process was significantly accelerated in the tubers of *B33-TPP* lines, whereas it was delayed in *B33-TPS* lines compared to controls. Further, it was observed that the tubers of *B33-TPP* lines had threefold less ABA levels accompanied with enhanced expression of the ABA catabolism gene—*ABA-8′-HYDROXYLASE*, whereas *B33-TPS* lines had slightly higher ABA levels that were associated with reduced expression of *ABA-8′-HYDROXYLASE* gene. These findings support the positive role of T6P and ABA in tuber dormancy maintenance and delaying sprout initiation, and their negative role in potato tuber sprouting.

2.3 Potential Role of MicroRNAs and Long Non-coding RNAs in the Potato Tuber Sprouting

Previously, hundreds of microRNAs (miRNAs) involved in stolon-to-tuber development have been identified in potato (Zhang et al. 2013; Lakhotia et al. 2014; Kondhare et al. 2018). However, only three miRNAs, such as miR156, miR172 and miR390 are studied for their role in tuber development (Martin et al. 2009; Bhogale et al. 2014; Santin et al. 2017). An investigation by Ou et al. (2015) identified putative miRNAs and their target genes that were differentially expressed during potato tubers' adaptive responses to the cold stress, dormancy transition, and carbohydrate metabolism. Numerous genes involved in phytohormone metabolism, transport and signaling were found as targets of these differentially expressed miRNAs, indicative of their potential role in potato tuber sprouting (Table 3).

Table 3 List of miRNAs and their targets including phytohormone-related genes that were differentially expressed during the tuber sprouting in potato. The list is prepared as per the report of Ou et al. (2015)

miRNA	Target gene(s)
stu-miR160a/b	*AUXIN RESPONSE FACTORS* (*ARFs*)
stu-miR393a/b	Auxin receptor—*TRANSPORT INHIBITOR RESPONSE PROTEIN 1* (*TIR1*)
miR159/miR319a	GA responsive MYB transcription factors—*GAMYB1* and *-2*
miR473 and miR477	GA signaling component—DELLA protein *RGL1*
novel-miRNA052	Two-component sensor protein *HISTIDINE PROTEIN KINASE*
novel-miRNA068 and -102	*AUXIN EFFLUX CARRIERs* (*AECs*)
novel-miRNA088	ABA receptor—*ABSCISIC INSENSITIVE 1B*
novel-miRNA104	Ethylene receptor—*ETHYLENE INSENSITIVE3 (EIN3)*
stu-miR166	*CLASS III HOMEODOMAIN LEUCINE ZIPPER TRANSCRIPTION FACTOR* (*HD-ZIP III*)
stu-miR172	*APETALA2-LIKE PROTEIN (RAP2)*
stu-miR396	*GLUTAMATE DECARBOXYLASE (GAD)*
stu-miR396	*SUCROSE NON-FERMENTING-LIKE KINASE (SnRK2.4)*
stu-miR396a-3p	*GLUCURONOSYLTRANSFERASE IRREGULAR XYLEM 7-LIKE (IRX7)*
stu-miR482f-5p	*BAH DOMAIN-CONTAINING PROTEIN SUO-LIKE* (*SUO*)

Stu = *Solanum tuberosum*; miR = MicroRNA

Long non-coding RNAs (lncRNAs) are widely present in mammals and plants. The average lengths of these lncRNAs are >200 nucleotides. They harbor structural characteristics of mRNA like the 5′ cap and poly-A tail, but do not code for proteins (Mercer et al. 2009; Hung and Chang 2010). In the past decade, several studies have ascertained their role in regulating the expression of target genes by modulating DNA methylation, histone modification and chromatin re-modelling (Chekanova 2015). Recently, Hou et al. (2018) performed a genome-wide analysis of lncRNAs from the apical meristems during the dormancy release and the sprouting stages of potato tubers. Authors identified 723 differentially expressed lncRNAs, which were enriched for functions in cellular components of the potato apical buds and cellular metabolic processes. Furthermore, 386 differentially expressed lncRNAs (out of 723) were also found as putative targets of 235 potato miRNAs. The transcription factor prediction of differentially expressed lncRNAs' target genes included several auxin responsive TFs like *ARF* (10), *ARR-B* (5) and *AUX/IAA* (8), abscisic acid signaling gene *ABAI3VP1* (13), and a gene involved in brassinosteroid-mediated signaling pathway—*BRASSINAZOLE-RESISTANT 1* protein (1). In summary, the studies described above suggest the putative role of miRNAs and lncRNAs in the regulation of the tuber sprouting process. However, none of the miRNAs or lncRNAs are functionally characterized for their role in the process of the potato tuber sprouting initiation, and it certainly demands further research.

3 Tuber Spouting in Yam and the Role of Phytohormones

Yam (*Dioscorea alata* L.) belongs to the Dioscoreaceae family, and the genus *Dioscorea* contains more than 600 plant species. Seven *Dioscorea* species are most widely cultivated for human consumption, which include *D. rotundata* Poir., *D. cayenensis* Lam., *D. dumetorum* (Knuth) Pax, *D. trifida* L., *D. alata* L. and *D. esculenta* Lour. Burkill, and *D. bulbifera* L. Selective medicinally important yam species, such as *D. floribunda* Mart. and Gal., *D. spiculiflora* Hemsl, and *D. composita* Hemsl are also grown in several countries (Akoroda 1993). Yam tubers are rich in nutrients, but also contain bioactive metabolites, such as resistant starch, steroidal sapogenins like diosgenin, the storage protein dioscorin, and mucilage polysaccharides. These health-promoting products can help in preventing cardiovascular disease, diabetes, and disorders of the gut microbiome (Epping and Laibach 2020). Yams are generally grown in humid tropical countries and is a major food in West Africa after cassava. It is also produced in Latin America, Asia and Oceania. An average yield of yam (9973 kg per ha) is highest among all the tuber and storage root crops (www.fao.org). Only 16% portion of the tubers are non-edible, which is also the lowest compared to cassava (26%) and sweet potato (21%). A typical growth cycle of yam crop is 6–10 months, and the tubers after harvest remain dormant for the next 2–4 months. The growth phase of the crop corresponds with the wet season, whereas the tuber dormancy phase coincides with the dry season. Long growing season, the requirement of high labor cost along with the relatively large amount of planting material, and the large storage space with a huge cost for its storage are the major problems faced by the yam growers. Despite several nutritional benefits offered by yam tubers, it remains as one of the most neglected and underutilized crops throughout the world (www.fao.org), and there is an urgent need for its crop improvement.

The tuber shape and size can vary greatly due to genetic and environmental factors. The yam tuber grows from a corm-like structure located at the base of the vine. Occasionally, this corm remains attached to the tuber after harvest and sprouts will develop from it. When the corm separates from the tuber, sprouting occurs from the tuber near to the point at which the corm was attached (Diop 1998). A transverse section of a mature yam tuber shows the presence of four concentric layers, namely corky periderm, cortex, meristematic layer and ground tissue. The corky periderm (the outer portion of the yam tuber comprising the cork cells) provides an effective barrier against a water loss and invasion by pathogens. Cortex is a layer located immediately beneath the cork, comprising thin-walled cells with very little stored starch. Meristematic layer includes elongated thin-walled cells under the cortex, and sprouts are initiated from this layer. Ground tissue represents the central portion of the tuber, and is composed of thick-walled starchy cells, with vascular bundles separating throughout the mass. Most yams are essentially composed of water, starch, small quantities of protein and other minor constituents (Diop 1998).

Tuber dormancy is an important mechanism for the adaptation of yams to their natural environments (Craufurd et al. 2001). As fresh tubers are purchased for the propagation purposes, the prolonged dormancy is crucial to determine the shelf life of tubers. Therefore, long dormancy serves as a desirable attribute in yam breeding and selection programs (Shiwachi et al. 2003). It is indispensable for farmers to have a clear idea about the sprout planting period i.e. when the tubers lose their dormancy and start to sprout so that the next planting cycle can be started with uniformity in their growth. The striking difference between potato and yam is the absence of pre-formed meristems on yam tubers (Wickham et al. 1984), and as a result, known sprout suppressants of potato are not effective on yam tubers as they mostly act on preformed meristems to suppress their growth. An interesting report by Ile et al. (2006) proposed three phases of tuber dormancy in yam. Phase I starts from the tuber initiation and lasts until the appearance of the tuber germinating meristems; Phase II then continues until the initiation of foliar primordia, and Phase III lasts up to the appearance of the shoot bud on the surface of a tuber (Table 1). Phase I is classified as endo-dormancy; lasts for 200–220 days after initiation of tubers (or ~285 days after planting), and is governed by endogenous factors only. Phase II is known as endo-/eco-dormancy; lasts for ~35 days after endo-dormancy, and is controlled by phytohormones, physiological and environmental factors. Moreover, the initiation of phase II is recognized by the appearance of the tuber germinating meristems and this phase ends with the emergence of foliar primordia. Phase III is called as eco-dormancy; continues for ~10 days after phase II. This phase is regulated by phytohormones and environmental factors, and leads to the formation of a shoot bud (Table 1).

In past, unique dormancy-inducing phenolic substances, called as batatasins I, II and III, were detected in dormant yam bulbils and their application could inhibit sprouting of yam bulbils (Hashimoto et al. 1972). Similar to potato tubers, the endogenous levels of ABA increase in yam tubers in favor of dormancy (Hasegawa and Hashimoto 1973). In an experiment, ABA biosynthesis inhibitor—fluridone (30 μM) was applied to the developing tubers of the two yam cultivars having the prolonged tuber dormancy in the hydroponics system containing a nutrient medium with or without the inhibitor, and the authors observed the induction of sprouting within 30 days of tuber formation (Awologbi and Hamadina 2016). This signifies the positive role of ABA in the induction and maintenance of yam tuber dormancy.

During tuber maturation in yam, activated ABA signaling and the abundance of transcription factors like DEAD-BOX and GLYCINE-RICH RNA-BINDING proteins help to maintain dormancy (Table 2; Fig. 4). A storage protein—DISCORIN (DIO 1/2/3/4/5/A/B and their precursors), heat-shock proteins (HSP70 and HSP81-88), chaperon (Clpcl), and various inhibitors of proteases, trypsin and cysteine start to accumulate. Starch synthesis is enhanced due to the upregulation of sucrose synthase and downregulation of genes encoding α-AMYLASE and INVERTASE. Cell growth is arrested. However, during the dormancy breakage, auxin helps in the mobilization

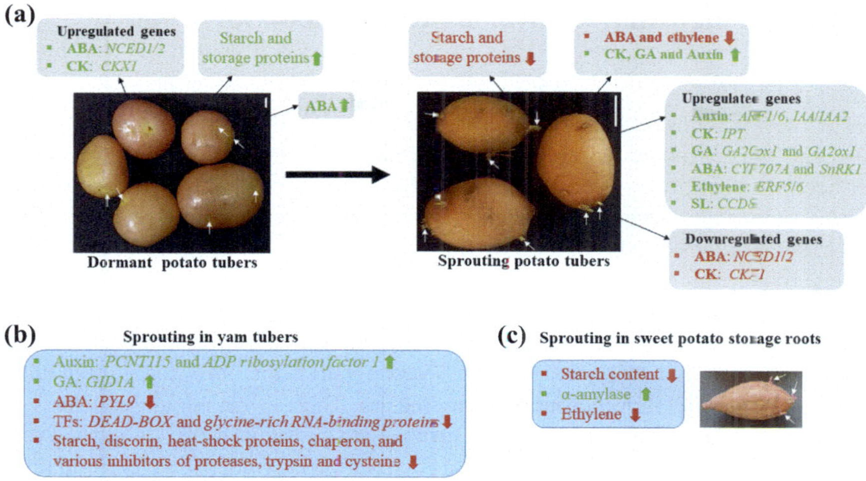

Fig. 4 Schematics summarizing the role of various phytohormones and molecular factors during the sprouting of tubers in potato and yam, and storage roots of sweet potato. **a** The levels of phytohormones and expression of their metabolism and signaling related genes in the dormant versus sprouting tubers of potato. Arrows (upward; green color) represent an increase in levels of phytohormones, gene expression or sugar/storage protein content, whereas arrows (downward, red color) indicate a decrease in the respective levels. The abbreviations are defined in the Fig. 3 legend and the text of this article. Factors involved in sprouting of yam tubers (**b**) and storage roots of sweet potato (**c**) are also summarized. Arrows (white) indicate 'dormant tuber eyes' or 'sprouts' in potato (panel A) and sprouts in sweet potato storage roots (panel C)

of starch by increasing the expression of an auxin-induced protein—PCNT115 and auxin-dependent transcription factor—ADP RIBOSYLATION FACTOR 1 (Table 2; Fig. 4). Cell cycle proliferation is initiated and DISCORIN protein starts to deplete. Further, the sprout initiation is accompanied by active cell growth in the bud region, downregulation of a gene encoding SUCROSE SYNTHASE and upregulation of α-AMYLASE and INVERTASE. DISCORIN continues to deplete (Table 1; Fig. 4) (Sharma and Deswal 2021). However, none of the genes is functionally characterized for its role in the sprouting process.

A study by Suja et al. (2003) demonstrated that white yam tubers when treated with indole-3-acetic acid (IAA; 1000 ppm) exhibited a delay in the sprout initiation, but the sprout numbers were comparable to controls, suggesting a negative role for auxin in the yam tuber sprout initiation. Unlike potato and sweet potato, GA is known to inhibit the tuber sprouting in yams (Okagami and Tanno 1993). Consistently, Ile et al. (2006) showed that soaking yam tubers in gibberellic acid (GA_3; at 1000 mg/L) solution for 2 h delayed the formation of the tuber germinating meristems. The expression of a GA receptor gene *GIBBERELLIN INSENSITIVE DWARF 1* (*DoGID1A*) increases gradually during the yam bulbil sprouting (Table 2; Fig. 4) (Long et al. 2019). However, the limited information is available regarding the tuber dormancy as well as the mechanism of sprout initiation in yam.

4 Sweet Potato and Control of Storage Root Sprouting

Sweet potato (*Ipomoea batatas* L.) belongs to the Convolvulaceae family and is the world's seventh most essential food crop. It is also the third most preferred amongst tuber and storage root crops after potato and cassava. In terms of biomass production per hector, sweet potato is better than all other food crops, and it is best suited for tropical soils with minimum fertilizers and irrigation (Loebenstein et al. 2003). One of the unique characteristics of sweet potato is the production of nutritious and edible storage roots, which are also a rich source of different vitamins (like A, C, B2, B5, B6, and B9), minerals (potassium, copper, etc.) and storage proteins (Cheema et al. 2010). The storage roots of sweet potato, following the short dormancy period, can sprout and generate new plants (Fig. 1). Sweet potato is the only storage root crop, whose storage roots show sprouting phenomenon. Thus, sweet potato represents an ideal system to understand the sprouting mechanism in storage roots. Owing to a very short shelf life of storage roots, sweet potato cultivation is avoided in several countries (Doku 1989; Kurup and Balgopalan 1991; Rees et al. 2001). Injuries to storage roots during harvest, improper or inadequate curing methods and the storage conditions, and sprouting are the major reasons for the quality loss of sweet potato storage roots (Ravi and Aked 1996). Sprouting leads to a significant loss of weight and moisture of the storage roots, ultimately resulting in a shrinkage of the storage organ. A number of factors are responsible for sprouting of sweet potato storage roots (Table 1). These include delayed harvest of storage roots, the presence of excessive soil moisture during the storage root maturation stage, storage root anatomy, dormancy period, storage conditions like optimum temperature and humidity, and the increased respiration by storage roots after harvest (Bourke 1982; Jana 1982; Winaro 1982; Wills et al. 2007; Edmunds et al. 2008). Sprouting occurs most efficiently when storage roots are stored at ambient temperatures, whereas it is inhibited by a constant incubation at 15 °C (Table 1). As per the guidelines by the United States Department of Agriculture (2005), sweet potato storage roots with 10% or more sprouts longer than 19 mm are considered as defective roots and are not considered for human consumption.

In countries, where sweet potato is grown seasonally, the extension of shelf life becomes critical to fulfill the market demand. Curing and proper storage of sweet potato is important to mitigate postharvest physiological disorders and to increase the shelf life of storage roots (Ahn et al. 1980; Chang and Kays 1981). Curing or wound healing involves the desiccation of cells on the upper surface, the accompanying lignification of underlying cells, followed by wound periderm formation. Cured sweet potatoes can be stored at 13 to 16 °C for several weeks (Table 1) (Bartz and Brecht 2005). Although the implementation of ideal curing ways and storage conditions are effective ways for increasing the shelf life (Ravi et al. 1996), the limited access to electricity and the higher capital costs involved makes it the foremost challenge for farmers to store storage roots at controlled temperatures in developing countries. The bulkiness of storage roots is another challenge for their economic transport to distal places. A number of other ways have been employed to control the sprouting process in sweet potato storage roots e.g. the use of plant growth regulators like IAA (10–100 ppm) (Paton and Scriven 1989), chemicals (e.g. Sodium hypochlorite; 0.3–9% by volume) (Lewthwaite and Triggs 1995), irradiation (gamma radiation; 0.2–1 kGy) (McGuire and Sharp 1995), hot water treatment (50 °C dipping for 30 min) (Hu and Tanaka 2007), controlled atmospheric storage, and maleic hydrazide treatment (Table 1).

The total sugar content slowly increases during the storage of sweet potatoes (Huang et al. 2014). Zhang et al. (2002) studied the biochemical changes in sweet potatoes during the storage, wherein the most genotypes exhibited a decline in the starch content during the storage (from 0 to 180 days). An increment in α-amylase activity was observed during the first two months of storage; however, the activity decreased gradually thereafter (Table 1; Fig. 4). Nevertheless, there are no reports that describe the role of molecular and hormonal factors in sprouting of sweet potato storage roots. Thus, there is an urgency to identify the genes involved in the sprouting process, and accordingly devise the biotechnological strategies for increasing the shelf life of sweet potato storage roots.

5 Conclusions and Future Prospects

The sprouting process is somewhat studied in the tubers of potato and yam; however, it is yet to be explored in sweet potato storage roots (Table 2; Fig. 4). Considering the differences in the origin of belowground storage organs (i.e. modified stem in case of potato and yam, and storage root in case of sweet potato), the presence of preformed meristems on potato tubers and their absence on sweet potato storage roots, and the variable composition of tuber versus storage roots, it is expected that the regulatory mechanisms of sprouting control could be mostly unique in potato versus sweet potato along with few common pathways. Only future research will shed light on this. The way forward approach for controlling the sprouting process and to enhance the shelf life of tubers or storage roots of these crops could be employing the storage organ specific manipulation of the phytohormone signaling related genes depending on their roles in the sprout initiation process. This could be achieved using the tuber-specific promoters of genes, such as *PATATIN B33* and *GRANULE-BOUND STARCH SYNTAHSE* (*GBSS*) in potato (Debast et al. 2011; Miroshnichenko et al. 2020) or the storage root-specific promoters of *SPORAMIN* genes in sweet potato. As proposed by Rees et al. (1997), it is also necessary to utilize the existing sweet potato varieties or develop new varieties with delayed sprouting phenotype in order to enhance the shelf life of tubers and storage roots that would boost the farmers' income improving their socio-economic conditions. Moreover, a combined approach utilizing the transcriptomics, genomics and metabolite profiling data of dormant versus sprout initiating tubers/storage roots with advanced genome editing tools like CRISPR/Cas9 would be useful to generate new varieties of these commercially important tuber and storage root crops with delayed sprouting and enhanced shelf life.

Acknowledgements The author acknowledges the research fellowship and grant (IFA18-LSPA 123) from the Department of Science and Technology- Inspire Faculty Program, Government of India. Thanks to the support from the Council of Scientific and Industrial Research-National Chemical Laboratory (CSIR-NCL), Pune, India.

Conflict of Interest The author declare that there is no conflict of interest.

References

Adesuyi SA (1982) The application of advanced technology to the improvement of yam storage. In: Yams. Ignames. Oxford, Clarendon Press, pp 312–319

Adu-Gyamfi R, Blay E (2009) Influence of storage duration and growth regulators on sprouting of yam minisett influence of storage duration and growth regulators on sprouting and field performance of yam (*Dioscorea rotundata* Poir.) Minisett. J Root Crops 35(2):219–225

Agrimonti C, Marmiroli N (2008) Gene expression during transition from dormancy to sprouting in potato tubers. Fruit Veg Cereal Sci Biotechnol 2(1):95–109

Ahn K, Collins WW, Pharr DM (1980) Influence of pre-harvest temperature and flooding on sweet potato roots in storage. Hort Sci 15:261

Akoroda MO (1993) 49 - Yams: Dioscorea spp. In: Kalloo G, Bergh BO (eds) Genetic improvement of vegetable crops. Pergamon, ISBN 9780080408262, pp 717–733

Aksenova NP, Sergeeva LI, Konstantinova TN, Golyanovskaya SA, Kolachevskaya OO, Romanov GA (2013) Regulation of potato tuber dormancy and sprouting. Russ J Plant Physiol 60:301–312

Awologbi E, Hamadina E (2016) Early induction of sprouting on seed tubers of yam (dioscorea spp.) soon after tuber initiation in a hydroponics system. Exp Agric 52(3):405–417

Bartz J, Brecht J (2005) Postharvest physiology and pathology of vegetables, 2nd edn. Marcel Dekker INC, New York

Bhogale S, Mahajan AS, Natarajan B, Rajabhoj M, Thulasiram HV, Banerjee AK (2014) MicroRNA156: a potential graft-transmissible microRNA that modulates plant architecture and tuberization in *Solanum tuberosum* ssp. andigena. Plant Physiol 164:1011–1027

Biemelt S, Hajirezaei M, Hentschel E, Sonnewald U (2000) Comparative analysis of abscisic acid content and starch degradation during storage of tubers harvested from different potato varieties. Potato Res 43:371–382

Bisognin DA, Manrique-Carpintero NC, Douches DS (2018) QTL analysis of tuber dormancy and sprouting in potato. Am J Potato Res 95:374–382

Bourke RM (1982) Sweet potato in Papua New Guinea. 45. International: review. In: Griggs TD (eds) Sweet potato L Villareal. Proceedings of 1st international symposium. Asian vegetable research developmental centre (AVRDC), Shanhua, Tainan, Taiwan

Braide S, Hamadina EI (2018) A role of abscisic acid in the induction of tuber dormancy in yam (*Dioscorea alata* L.). Adv Life Sci 8(1):32–38

Burton WG (1989) Dormancy and sprout growth. In: The potato, 3rd edn, Longman scientific and technical. Longman Group Ltd., UK, pp 471–504

Campbell M, Segear E, Beers L, Knauber D, Suttle J (2008) Dormancy in potato tuber meristems: chemically induced cessation in dormancy matches the natural process based on transcript profiles. Funct Integr Genomic 8:317–328

Cantwell M, Suslow T (2001) Sweet potato. Crisphead: Recommendations for maintaining postharvest quality. http://postharvest.ucdavis.edu/Commodity_Resources/Fact_Sheets/Datastores/Vegetables_English/?uid=34&ds=799. Accessed 17 Sept 2021

Carrera E, Bou J, García-Martínez JL, Prat S (2000) Changes in GA20-oxidase gene expression strongly affect stem length, tuber induction and tuber yield of potato plants. Plant J 22:247–256

Chakraborty C, Roychowdhury R, Chakraborty S, Chakravorty P, Ghosh D (2017) A review on post-harvest profile of sweet potato. Int J Curr Microbiol Appl Sci 6:1894–1903
Chandrasekara A, Kumar TJ (2016) Roots and tuber crops as functional foods: a review on phytochemical constituents and their potential health benefits. Int J Food Sci 3631647
Chang LA, Kays SJ (1981) Effect of low oxygen storage on sweet potato roots. J Am Soc Hortic Sci 106:481
Cheema M, Rees D, Westby A, Taylor M (2010) Hormonal control of sprouting of sweetpotatoes in storage. Acta Hortic 858:173–177
Chekanova JA (2015) Long non-coding RNAs and their functions in plants. Curr Opin Plant Biol 27:207–216
Cheng Y, Liu H, Cao L, Wang S, Li Y, Zhang Y, Jiang W, Zhou Y, Wang H (2015) Down-regulation of multiple CDK inhibitor ICK/KRP genes promotes cell proliferation, callus induction and plant regeneration in Arabidopsis. Front Plant Sci 6:825
Claassens MMJ (2002) Carbohydrate metabolism during potato tuber dormancy and sprouting. PhD thesis (123101), Wageningen University
Craufurd PQ, Summerfield RJ, Asiedu R, Vara Prasad PV (2001) Dormancy in yams. Exp Agric 37:75–109
Debast S, Nunes-Nesi A, Hajirezaei MR, Hofmann J, Sonnewald U, Fernie AR, Börnke F (2011) Altering trehalose-6-phosphate content in transgenic potato tubers affects tuber growth and alters responsiveness to hormones during sprouting. Plant Physiol 156:1754–1771
Destefano-Beltran L, Knauber D, Huckle L, Suttle JC (2006) Effects of postharvest storage and dormancy status on ABA content, metabolism, and expression of genes involved in ABA biosynthesis and metabolism in potato tuber tissues. Plant Mol Biol 61:687–697
Diop A (1998) Storage and processing of roots and tubers in the tropics. In: Calverley DJB (eds) Post-harvest technology. Food and Agriculture Organization of the United Nations
Doku EV (1989) Root crops in Ethiopia. Root crops and low-input agriculture: Helping to meet food self sufficiency goals in eastern and southern Africa. In: Proceedings of 3rd eastern and southern Africa regional workshop on roots and tuber crops. Int. Inst. Trop. Agric., Ibadan, Nigeria
Edmunds BA, Boyette N, Clark CA, Ferrin D, Smith TP, Holmes G (2008) Postharvest handling of sweet potatoes. North Carolina Cooperative Extension Services, Raleigh, North Carolina
Epping J, Laibach N (2020) An underutilized orphan tuber crop—Chinese yam: a review. Planta 252:58
Faivre-Rampant O, Cardle L, Marshall D, Viola R, Taylor MA (2004) Changes in gene expression during meristem activation processes in Solanum tuberosum with a focus on the regulation of an auxin response factor gene. J Exp Bot 55(397):613–622
Farré E, Bachmann A, Willmitzer L, Trethewey RN (2001) Acceleration of potato tuber sprouting by the expression of a bacterial pyrophosphatase. Nat Biotechnol 19:268–272
Francis D, Sorrell DA (2001) The interface between the cell cycle and plant growth regulators: a mini review. Plant Growth Regul 33(1):1–12
Gao Y, Zhao M, Wu XH, Li D, Borthakur D, Ye JH, Zheng XQ, Lu JL (2019) Analysis of differentially expressed genes in tissues of Camellia sinensis during dedifferentiation and root re-differentiation. Sci Rep 9:2935
Gong HL, Dusengemungu L, Igiraneza C, Rukundo P (2021) Molecular regulation of potato tuber dormancy and sprouting: a mini-review. Plant Biotechnol. Rep. 15:417–434
Gregory LE (1968) Effect of Ethephon on breaking of dormancy in yarn tubers. J Agric Univ Puerto Rico 56:46–56
Haider MW, Nafeesa M, Amina M, Asadb HU, Ahmad I (2021) Physiology of tuber dormancy and its mechanism of release in potato. J Hortic Sci Technol 4(1):13–21
Hajirezaei M, Sonnewald U (1999) Inhibition of potato tuber sprouting: low levels of cytosolic pyrophosphate lead to non-sprouting tubers harvested from transgenic potato plants. Potato Res 42:353–372

Hartmann A, Senning M, Hedden P, Sonnewald U, Sonnewald S (2011) Reactivation of meristem activity and sprout growth in potato tubers require both cytokinin and gibberellin. Plant Physiol 155(2):776–796

Hasegawa K, Hashimoto T (1973) Quantitative changes of batatasins and abscisic acid in relation to the development of dormancy in yam bulbils. Plant Cell Physiol 14(2):369–377

Hashimoto T, Hasegawa K, Kawarada A (1972) Batatasins: New dormancy-inducing substances of yam bulbils. Planta 108:369–374

Hou X, Du Y, Liu X, Zhang H, Liu Y, Yan N, Zhang Z (2018) Genome-wide analysis of long non-coding RNAs in potato and their potential role in tuber sprouting process. Int J Mol Sci 19(1):101

Hu W, Tanaka SI (2007) Effect of heat treatments on the quality and storage life of sweet potato. J Sci Food Agric 87:313–319

Huang CL, Liao WC, Chan CF, Lai YC (2014) Storage performance of Taiwanese sweet potato cultivars. J Food Sci Technol 51:4019–4025

Hung T, Chang HY (2010) Long noncoding RNA in genome regulation: prospects and mechanisms. RNA Biol 7:582–585

Hutchingson RW (1978) The dormancy of seed potatoes. 2. The effect of storage temperature. Potato Res 21:267–275

Igwilo NH (1982) Evaluation of yam cultivars for seed yam production using minisett technique. NCCRI, Annual Report (Technical report) vol I

Ikediobi CO, Oti E (1983) Some biochemical changes associated with post-harvest storage of white yam (*Dioscorea rotundata*) tubers. J Sci Food Agric 34(10):1123–1129

Ile EI, Craufurd PQ, Battey NH, Asiedu R (2006) Phases of dormancy in yam tubers (*Dioscorea rotundata*). Ann Bot 97(4):497–504

Jaleel CA, Gopi R, Manivannan P, Kishorekumar A, Gomathinayagam M, Panneerselvam R (2007) Changes in biochemical constituents and induction of early sprouting by triadimefon treatment in white yam (*Dioscorea rotundata* Poir.) tubers during storage. J Zhejiang Uni Sci B 8(4):283–288

Jana RK (1982) Status of sweet potato cultivation in East Africa and its future. In: Villareal RL, Griggs TD (eds) Proceedings of 1st international sweet potato symposium. Asian Vegetable Research Development Center (AVRDC), Shanhua, Tainan, Taiwan, p 63

Jenkins PD (1982) Losses in sweet potatoes (*Ipomoea batatas*) stored under traditional conditions in Bangladesh. Trop Sci 24:17

Kloosterman B, Navarro C, Bijsterbosch G, Lange T, Prat S, Visser RG, Bachem CW (2007) StGA2ox1 is induced prior to stolon swelling and controls GA levels during potato tuber development. Plant J 52:362–373

Kondhare KR, Malankar NN, Devani RS, Banerjee AK (2018) Genome-wide transcriptome analysis reveals small RNA profiles involved in early stages of stolon-to-tuber transitions in potato under photoperiodic conditions. BMC Plant Biol 18:284

Kurup GT, Balgopalan C (1991) Sweet potato production, post harvest handling and utilization in India. In: Dayal TR, Scott G, Kurup GT, Balgopalan C (eds) Sweet potato in south asia: post-harvest handling storage, processing and uses. CIP - CANADIAN TOBACCO CONTROL RESEARCH INITIATIVE, Trivandrum, India, p 33

Lakhotia N, Joshi G, Bhardwaj AR, Katiyar-Agarwal S, Agarwal M, Jagannath A, Goel S, Kumar A (2014) Identification and characterization of miRNAome in root stem leaf and tuber developmental stages of potato (*Solanum tuberosum* L.) by high-throughput sequencing. BMC Plant Biol 14:6

Levy D, Veilleux RE (2007) Adaption of potato to high temperatures and salinity-a review. Am J Potato Res 84:487–506

Lewthwaite SL, Triggs CM (1995) Sprout suppression in sweet potato roots following immersion in sodium hypochlorite solutions. New Zealand J Crop Hortic Sci 23:283–287

Lin Y, Liu T, Liu J, Liu X, Ou Y, Zhang H, Li M, Sonnewald U, Song B, Xie C (2015) Subtle regulation of potato acid invertase activity by a protein complex of invertase, invertase inhibitor, and sucrose nonfermenting1-related protein kinase. Plant Physiol 168:1807–1819

Lipavská H, Mašková P, Vojvodová P (2011) Regulatory dephosphorylation of CDK at G2/M in plants: yeast mitotic phosphatase cdc25 induces cytokinin like effects in transgenic tobacco morphogenesis. Ann Bot 107(7):1071–1086

Liu B, Zhang N, Wen Y, Si H, Wang D (2012) Identification of differentially expressed genes in potato associated with tuber dormancy release. Mol Biol Rep 39:11277–11287

Loebenstein G, Fuentes S, Cohen J, Salazar LF (2003) Sweet potato. In: Loebenstein G, Thottappilly G (eds) Virus and virus-like diseases of major crops in developing countries. Springer, Dordrecht, pp 223–248

Long W, Meng J, Xu S, Zhang X, Duan Y, Yang R (2019) Cloning and expression analysis of gibberellin receptor gene DoGID1A in *Dioscorea opposita*. J Agric Biotechnol 27(11):1933–1941

Mani F, Bettaieb T, Doudech N, Hannachi C (2014) Physiological mechanisms for potato dormancy release and sprouting: a review. African Crop Sci J. 22:155–174

Martin A, Adam H, Díaz-Mendoza M, Zurczak M, González-Schain ND, Suárez-López P (2009) Graft-transmissible induction of potato tuberization by the microRNA miR172. Development 136:2873–2881

McGuire RG, Sharp JL (1995) Market quality of sweet potatoes after gamma-irradiation for weevil control. HortScience 30:1049–1051

Mercer TR, Dinger ME, Mattick JS (2009) Long non-coding RNAs: insights into functions. Nat Rev Genet 10:155–159

Miroshnichenko D, Firsov A, Timerbaev V, Kozlov O, Klementyeva A, Shaloiko L, Dolgov S (2020) Evaluation of plant-derived promoters for constitutive and tissue-specific gene expression in potato. Plants 9(11):1520

More SJ, Ravi V, Raju S (2019) Tropical tuber crops. In: Postharvest physiological disorders in fruits and vegetables, 1st edn. CRC Press, pp 40

Morris WL, Ducreux LJM, Hedden P, Millam S, Taylor MA (2006) Overexpression of a bacterial 1-deoxy-D-xylulose 5-phosphate synthase gene in potato tubers perturbs the isoprenoid metabolic network: implications for the control of the tuber life cycle. J Exp Bot 57(12):3007–3018

Morris WL, Alamar MC, Lopez-Cobollo RM, Cañete JC, Bennett M, Kaay JV, Stevens J, Sharma SK, McLean K, Thompson AJ, Terry LA, Turnbull CGN, Bryan GJ, Taylor MA (2019) A member of the TERMINAL FLOWER 1/CENTRORADIALIS gene family controls sprout growth in potato tubers. J Exp Bot 70(3):835–843

Mozie O (1975) Sprout growth in stored white yams (*Dioscorea rotundata* Poir.). Trop Sci 17:45–46

Muthoni J, Kabira J, Shimelis H, Melis R (2014) Regulation of potato tuber dormancy: a review. Aust J Crop Sci 8:754–759

Okagami N, Tanno N (1993) Gibberellic acid-induced prolongation of the dormancy in tubers or rhizomes of several species of East Asian Dioscorea. Plant Growth Regul 12:119–123

Ou Y, Liu X, Xie C, Zhang H, Lin Y, Li M, Song B, Liu J (2015) Genome-wide identification of microRNAs and their targets in cold-stored potato tubers by deep sequencing and degradome analysis. Plant Mol Biol Rep 33:584–597

Pasare SA, Ducreux LJ, Morris WL, Campbell R, Sharma SK, Roumeliotis E, Kohlen W, van der Krol S, Bramley PM, Roberts AG, Fraser PD, Taylor MA (2013) The role of the potato (*Solanum tuberosum*) CCD8 gene in stolon and tuber development. New Phyt 198:1108–1120

Paton J, Scriven F (1989) Use of NAA to inhibit sprouting in sweet potatoes (*Ipomoea batatas*). J Sci Food Agric 48:421–427

Ravi V, Aked J (1996) Review on tropical root and tuber crops. II. Physiological disorders in freshly stored roots and tubers. Crit Rev Food Sci Nutr 36:711–731

Ravi V, Aked J, Balagopalan C (1996) Review on tropical root and tuber crops. I. Storage methods and quality changes. Crit Rev Food Sci Nutr 36:661–709

Rees D, Kapinga R, Matunda K (2001) Effect of damage on market value and shelf-life of sweet potato in urban markets of Tanzania. Trop Sci 41:142

Rees D, Pollard A, Matters D, Carey E (1997) Relating post harvest physiology of sweet potato storage roots with storability for a range of East African varieties. In: Altman A, Waisel Y (eds) Biology of root formation and development. Springer, US

Ronning CM, Stegalkina SS, Ascenzi RA, Bougri O, Hart AL, Utterbach TR, Vanaken SE, Riedmuller SB, White JA, Cho J, Pertea GM, Lee Y, Karamycheva S, Sultana R, Tsai J, Quackenbush J, Griffiths HM, Restrepo S, Smart CD, Fry WE, Van Der Hoeven R, Tanksley S, Zhang P, Jin H, Yamamoto ML, Baker BJ, Buell CR (2015) Comparative analyses of potato expressed sequence. Plant Physiol 131:419–429

Saidi A, Hajibarat Z (2021) Phytohormones: plant switchers in developmental and growth stages in potato. J Genet Eng Biotechnol 19(1):89

Santin F, Bhogale S, Fantino E, Grandellis C, Banerjee AK, Ulloa RM (2017) Solanum tuberosum StCDPK1 is regulated by miR390 at the posttranscriptional level and phosphorylates the auxin efflux carrier StPIN4 in vitro, a potential downstream target in potato development. Physiol Plant 159:244–261

Sharma S, Deswal R (2021) *Dioscorea alata* tuber proteome analysis uncovers differentially regulated growth-associated pathways of tuber development. Plant Cell Physiol 62(1):191–204

Sharma SK, McLean K, Colgan RJ, Rees D, Young S, Sønderkær M, Terry LA, Turnbull C, Taylor MA, Bryan GJ (2021) Combining conventional QTL analysis and whole-exome capture-based bulk-segregant analysis provides new genetic insights into tuber sprout elongation and dormancy release in a diploid potato population. Heredity 127:253–265

Shiwachi H, Ayankanmi T, Asiedu R, Onjo M (2003) Induction of sprouting in dormant yam (Dioscorea spp.) tubers with inhibitors of gibberellins. Exp Agric 39(2):209–217

Skalák J, Vercruyssen L, Claeys H, Hradilová J, Černý M, Novák O, Plačková L, Saiz-Fernández I, Skaláková P, Coppens F, Dhondt S (2019) Multifaceted activity of cytokinin in leaf development shapes its size and structure in Arabidopsis. Plant J 97(5):805–824

Sonnewald U (2001) Control of potato tuber sprouting. Trends Plant Sci 6:333–335

Sonnewald S, Sonnewald U (2014) Regulation of potato tuber sprouting. Planta 239(1):27–38

Suja G, Nair VM, Saraswathy P, Geetha Kumari VL (2003) Influence of growth-promoting substances on sprouting of white yam (*Dioscorea rotundata*) setts. Trop Sci 43(4):170–173

Sukhova LS, Macháčková I, Eder J, Bibik ND, Korableva NP (1993) Changes in the levels of free IAA and cytokinins in potato tubers during dormancy and sprouting. Biologia Plant 35(3):387–391

Suttle JC (2004a) Involvement of endogenous gibberellins in potato tuber dormancy and early sprout growth: a critical assessment. J Plant Physiol 161:157–164

Suttle JC (2004b) Physiological regulation of potato tuber dormancy. Am J Potato Res 81:253–262

Suttle JC (2007) Dormancy and sprouting. In: Vreugdenhil D (eds) Potato biology and biotechnology. Advances and perspectives, Elsevier, pp. 287–309

Tang W, Harris L, Newton RJ (2004) The biochemical control of the cell cycle by growth regulators in higher plants. J for Res 15(3):227–232

United States Department of Agriculture (2005) United States standards for grades of sweet potatoes. www.ams.usda.gov/sites/default/files/media/Sweetpotato_Standard%5B1%5D.pdf. Accessed 20 March 2018

Velappan Y, Signorelli S, Considine MJ (2017) Cell cycle arrest in plants: what distinguishes quiescence, dormancy and differentiated G1? Ann Bot 120(4):495–509

Werner T, Motyka V, Strnad M, Schmülling T (2001) Regulation of plant growth by cytokinin. Proc Natl Acad Sci USA 98(10):487–492

Wickham LD, Wilson LA, Passam HC (1981) Tuber germination and early growth in four edible Dioscorea species. Ann Bot 47:87–95

Wickham LD, Passam DHC, Wilson LA (1984) Dormancy responses to post-harvest application of growth regulators in Dioscorea species. 2. Dormancy responses in ware tubers of *D. alata* and *D. esculenta.* J Agric Sci 102:433–436

Wills RBH, McGlasson B, Grahm D, Joyce D (2007) Postharvest: An introduction to physiology and handling of fruits, vegetables and ornamentals, 5th edn. University of New South Wales Press, Sydney

Winaro FG (1982) Sweet potato processing and by-product utilization in the tropics—international review. In: Villareal L, Griggs TD (eds) Sweet potato. Proceedings of 1st international symposium. Asian Vegetable Research Developmental Centre (AVRDC), Shanhua, Tainan, Taiwan, p 373
Zhang Z, Wheatley CC, Corke H (2002) Biochemical changes during storage of sweet potato roots differing in dry matter content. Postharvest Biol Technol 24:317–325
Zhang R, Marshall D, Bryan GJ, Hornyik C (2013) Identification and characterization of miRNA transcriptome in potato by high-throughput sequencing. PLoS ONE 8(2):e57233
Zhang G, Tang R, Niu S, Si H, Yang Q, Rajora OP, Li X-Q (2021) Heat-stress-induced sprouting and differential gene expression in growing potato tubers: Comparative transcriptomics with that induced by postharvest sprouting. Hortic Res 8:226

Role of Phytohormones in Plant-Microbial Interaction

Nikhilesh Dhar, N. S. Raja Gopalan, P. T. Nikhil, and Sridev Mohapatra

Abstract Plants in their lifetime interact with a wide variety of microbes that can be pathogenic or beneficial. While beneficial microbes can be either endophytes or rhizospheric, pathogenic microbes are mostly free-living and colonize either the phyllosphere or the rhizosphere. These microbes can be either fungi or bacteria and have co-evolved to interact with their host plants using specific mechanisms. Plant-associated microbes release several chemicals onto the surfaces that they colonize, which are known to modulate the biology of the colonized plants directly or indirectly. Plants have evolved to specifically respond to the presence of such microbes, either by gearing up their defence responses (in case of pathogenic microbes) or in a mutually beneficial manner (in case of beneficial microbes). It is but obvious that this relationship between plants and their interacting microbes involves a variety of signalling and metabolic networks, both, from the microbe, as well as the plant side. In this cross-talk between microbes and plants, a very important role is played by phytohormones such as auxins, cytokinins, gibberellins, abscisic acid, jasmonic acid, salicylic acid, ethylene, brassinosteroids etc. Some of these phytohormones are commonly synthesized in the plants and also the plant-associated microbes are known to release some of these phytohormones into their habitat, consequently influencing plant responses. These microbes are also known to impact signalling mechanisms in their host plants by modulating the metabolism of important phytohormones in them. Such modulations in plant phytohormone metabolism have a pleotropic impact on a wide array of metabolic and signalling networks in them, thus affecting, not only their specific responses to the microbes, but also their growth, development and general stress response mechanisms. This chapter highlights the importance of phytohormones in plant–microbe interaction, both in case of pathogenic as well as beneficial microbes.

N. Dhar
Department of Plant Pathology, University of California, UC Davis Extension at USDA-Salinas, Davis, CA, USA

N. S. R. Gopalan · P. T. Nikhil · S. Mohapatra (✉)
Department of Biological Sciences, Birla Institute of Technology and Science (Pilani), Hyderabad Campus, Hyderabad, Telangana, India
e-mail: sridev.mohapatra@hyderabad.bits-pilani.ac.in

T. Aftab (ed.), *Auxins, Cytokinins and Gibberellins Signaling in Plants*, Signaling and Communication in Plants,
https://doi.org/10.1007/978-3-031-05427-3_14

1 Introduction

The role of phytohormones in growth, development and stress tolerance/resistance in plants is a very well-studied area of research with a strong history. The metabolism and signalling mechanisms of auxin, cytokinin and gibberellins in plants have been elaborated in the other chapters of this book. In this chapter, we highlight several aspects of phytohormone signalling involved in the interaction between plants and microbes, with special emphasis on the above three phytohormones. Plants, in their lifetime, encounter both, pathogenic as well as beneficial microbes that have co-evolved with their hosts for mutual interaction. These microbes can either be fungi or bacteria. While pathogenic microbes can be both, phyllospheric (colonizing leaf surfaces) and rhizospheric (colonizing rhizosphere/roots), beneficial microbes typically live in the soil and are, therefore, rhizospheric. Rhizospheric microbes can either be free-living or endo-colonizers and are collectively termed as plant growth promoting microbes (PGPM; PGPR for rhizobacteria and PGPF for fungi). Pathogenic microbes can either be biotrophs (need live plant tissues for nutrition) or necrothrophs (live on dead plant tissues), and in either case, can cause extensive damage to plants. Plants have evolved to recognize the presence of pathogens on their surface and respond by recruiting specific defence-related metabolic and signalling pathways that provide them resistance against these pathogens. Beneficial, soil microbes on the other hand co-exist with plants in a mutually beneficial manner, as obligatory or facultative symbionts. They are known to positively impact plant growth, development and stress tolerance/resistance by modulating several metabolic and signalling pathways in plants. It has been well established that there is a substantial amount of cross-talk between plants and the microbes they interact with. Much of this cross-talk is mediated by phytohormones such as auxins, cytokinins, gibberellic acid (GA), abscisic acid (ABA), salicylic acid (SA), jasmonic acid (JA), ethylene, brassinosteroids etc. Some of these phytohormones (particularly auxinss, cytokinins and GA) are common growth regulators in both, plants and microbes and are biosynthesized via similar (in some cases, identical) pathways. They are metabolized according to their specific requirements in growth, development and stress mitigation in plants versus bacteria. Plant–microbe interaction with respect to phytohormone signalling works in two ways:

(a) The phytohormones biosynthesized by several of these microbes are released exogenously. For example, several strains of PGPM release these growth hormones into the rhizosphere where they positively modulate the root structure architecture, thus enabling the microbes to efficiently colonize the rhizosphere.
(b) Many of these microbes are known to modulate the endogenous phytohormone metabolism and signaling processes in plants, which enables several aspects of plant responses to these microbes.

Both these phenomena have been discussed in this chapter.

2 Biosynthesis and Metabolism of Phytohormones in Microbes

When cultured in the growth media, many rhizospheric bacteria (harmful and beneficial) are known to produce different phytohormones like auxins, cytokinins, GA and ABA (reviewed by Spaepen and Vanderleyden 2011). Among these microbial-produced phytohormones, auxin is more extensively studied than others. Most soil bacteria produce auxin in the form of indole-3-acetic-acid (IAA) in a tryptophan (precursor for IAA) dependent biosynthetic pathway. The first step involves an enzyme tryptophan mono-oxygenase which converts the amino acid tryptophan into indole-3-pyruvic acid (IPA). Indole-3-pyruvate decarboxylase (IPDC) then converts IPA to IAA (Duca et al. 2014; Patten et al. 2013; Shao et al. 2015). An alternate pathway has also been suggested which involves the conversion of tryptophan to indole-3-acetamide (IA) via the enzyme tryptophan mono-oxygenase followed by conversion of IA to IAA via IA dehydrogenase (Patten et al. 2013; Zupan and Zambryski 1995). Most beneficial bacteria produce auxin via the IPA pathway, while pathogenic (gall/knot forming) soil bacteria commonly use the IA pathway for biosynthesis of IAA. Many bacteria including *Azospirillium* sp., *Enterobacter* sp., and *Pseudomonas* sp., encode the *ipdC* gene responsible for converting indole-3-pyruvic acid to indole-3-acetic acid (Koga et al. 1991; Patten and Glick 2002b; Xie et al. 2005). It has been reported that *iaaM* and *iaaH* genes are involved in IA mediated biosynthesis of IAA in *Pseudomonas fluorescens*, *Erwinia herbicola*, *Erwinia crysanthemi* 3937 and *Ralstonia solanacearum* (Kochar et al. 2011; Yang et al. 2007). A PGPR strain, *Azospirillum brasilense* Yu62 encodes a gene, *aldA* that converts indole-3-acetaldehyde to IAA (suggested as an alternative pathway for IAA biosynthesis in some bacteria). The Pathogenic bacterium *Pseudomonas savastanoi* produces conjugated forms of auxin (responsible for auxin homeostasis in plants) via the *iaaL* gene which encodes the enzyme IAA-lysine synthase, responsible for conjugation of auxin (Glass and Kosuge 1986). These pathways have been summarized in Fig. 1.

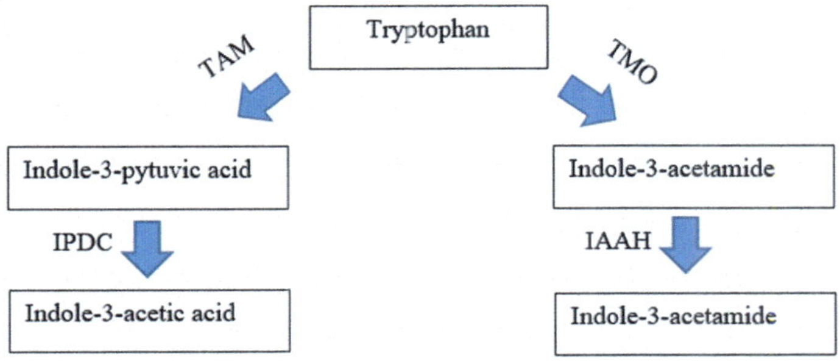

Fig. 1 Commonly used IAA biosynthesis pathway in bacteria. TAM: tryptophan aminotransferase, IPDC: indole-3-pyruvic acid decarboxylase, TMO: tryptophan monooxygenase, IAAH: indole-3-acetamide hydrolase

Apart from auxin, many plant-associated bacteria such as *Agrobacterium tumifaciens*, *Rhodococcus fascians*, *Ralstonia solanacearum* and *Pseudomonas savastonii* etc., are also known to produce the phytohormone cytokinin (Akiyoshi et al. 1989; Barry et al. 1984; MacDonald et al. 1986). The biosynthesis of cytokinin in bacteria generally begins with two isopentenyl transferases (IPT): adenylate-IPT and tRNA-IPT. The transfer of isopentyl group to N6-amino group (prenylation) of the free adenosine nucleotides such as AMP, ADP and ATP or tRNA-bound adenosine phosphate catalyses the first step in cytokinin biosynthesis (Sakakibara 2006; Takei et al. 2001). After prenylation, hydroxylation of isopentyl adenosine nucleotides by cytochrome P450 monooxygenase forms trans-zeatin (tZ). The adenylate, IPT is more commonly found in pathogenic bacteria, and it typically uses adenosine monophosphate (AMP) as a prenyl chain acceptor and dimethylallyl pyrophosphate (DMAPP) and 4-hydroxy-3-methyl-2-enyl pyrophosphate (HMBPP) as prenyl chain donor. The cytokinins are only available as nucleosides incorporated into the tRNA sequence in many bacteria. In most bacteria, tRNA IPT catalyses the formation of isopentyl-ribosides, and the miaA gene encodes the tRNA IPT is present in all bacterial species (except *Mycoplasma* sp.). The tRNA IPT uses both DMAPP and HMBPP as side-chain donors to produce cytokinin (Sakakibara 2006; Takei et al. 2001). While auxin and cytokinin are well studied and characterised in microbes, there are also reports of the production of gibberellic acid (GA) in beneficial growth-promoting bacteria. Most bacteria use geranyl–geranyl diphosphate (GGDP) as the precursor for GA synthesis. Studies conducted by Freiberg et al. (1997) show the presence of GA biosynthetic gene cluster in *Rhizobium* NGR234, a legume colonising bacterium that codes for three cytochrome P450, one geranyl–geranyl diphosphate synthase (GGDPS) and two di-terpene synthases. Also, in *Bradyrhizobium japonicum*, three genes CYP112, CYP114 and CYP117, which encode cytochrome P450 have been identified (Tully et al. 1998).

Similar to bacteria, plant associated fungi also produce phytohormones during their interaction with the host. Many fungi are known to produce phytohormones like auxin, GA and cytokinin (Liao et al. 2018; Pons et al. 2020; Takeda et al. 2015). Similar to bacteria, many fungal species utilize the IPA pathway for biosynthesis of auxin (Kumla et al. 2020). It is well known that one GA was first identified in fungi such as *Fusarium fujikuroi* (Sawada 1912). The biosynthetic pathway for GA in fungi is different from the plants in that it involves the cytochrome P450 gene cluster and GGDP as a precursor similar to the pathway seen earlier in bacteria (Bömke and Tudzynski 2009).

3 Phytohormone-Releasing PGPM and Their Benefits to Plants

It is a well-established fact that PGPM not only biosynthesize phytohormones for their growth and development, but also release them into the rhizosphere. These hormones include auxin, cytokinin, GA, ABA etc. either in free or conjugated forms (Saharan and Nehra 2011; Vacheron et al. 2013; Vejan et al. 2016; Gouda et al. 2018; Numan et al. 2018). This property of PGPM is considered to be a key mechanism in improving plant growth and stress tolerance. For example, it is well established that IAA-producing PGPR are able to increase root surface area in inoculated plants. (reviewed by Mantelin and Touraine 2003). Auxin proding *Bacillus subtilis* LK14 increased plant biomass and chlorophyll content in tomato plants (Khan et al. 2016). Auxin released from the PGPR, *Azospirillum* has has been known to help in imparting abiotic stress tolerance to a variety of crops such as legumes and graminaceous plants (Arzanesh et al. 2011; Cassán et al. 2014). An auxin-overproducing strain of *Pseudomonas putida* GR12-2 has been shown to promot root elongation in canola seedlings (Xie et al. 1996). Similar observations have been reported by Patten and Glick (2002a) in the same strain. *Micrococcus luteus* chp37, a cytokinin producing PGPR improved, both, root and shoot biomass in maize, helping the plants overcome water-stress (Raza and Faisal 2013). Similar observations have been made in the in cytokinin-producing *Bacillus subtilis* strains when inoculated with lettuce (Arkhipova et al. 2007) and *Platycladus orientalis* (Liu et al. 2013). Selvakumar et al. (2018) reported that cytokinin-producing *Citricoccus zhacaiensis* and *Bacillus amyloliquefaciens* had similar impacts on tomato plants. A GA-producing PGPR, *Pseudomonas putida* H-2-3 improved drought tolerance in soyabean (Kang et al. 2014a, b, c). GA and ABA-producing *Azospirillum lipoferum* conferred drought tolerance to maize (Cohen et al. 2009) and GA-producing *Azospirillum brasilense* to wheat (Creus et al. 2004).

In a study by Salomon et al. (2014), inoculation of ABA producing *Bacillus licheniformis* and *Pseudomonas fluorescens* improved growth of grapevine under drought conditions. Sandhya et al. (2009, 2010) have reported the production of a variety of phytohormones in the drought-tolerant PGPR, *Pseudomonas putida* GAP-P45 and imparts drought tolerance to maize and sunflower. Ghosh et al. (2019) later reported that the same strain produces different concentrations of auxin, cytokinin, GA and ABA in the growth medium and imparts drought tolerance to *Arabidopsis thaliana.*

Much like rhizobacteria, several strains of soil fungi are also known to release phytohormones into the culture medium. Species such as *Aspergillus fumigatus* and *Penicillium citrinum* have been known to secrete GA in culture medium (Waqas et al. 2012). Cytokinin produced by fungi is involved in drought and salinity stress through cross-talk with ABA (Ansari et al. 2013). Inoculation with *Penicillium* species during salt stress, promoted shoot growth due to the enhanced secretion of active GAs in cucumber plants (Gupta et al. 2021; Waqas et al. 2012).

4 Precise Roles of Microbe-Derived Phytohormones

What happens to the phytohormones that are released by the soil microbes and how exactly do they help in PGPM-induced growth, development and stress tolerance in plants? Considering the pleotropic role of phytohormones in almost every aspect of plant life-history, it can be hypothesized that the mechanisms involving PGPM-mediated improvement in plant health would involve the uptake and utilization of microbe-derived phytohormones. Some studies involving bacterial mutants have established the importance of microbe-derived phytohormones in such beneficial interactions which is described in greater details later in this chapter. However, precise studies on the uptake and utilization of these phytohormones are missing. We can broadly categorize the effects of these microbe-derived phytohormones on (a) modulation of the rhizosphere/roots and (b) modulation of endogenous signalling networks in the host plants.

5 Role of Microbe-Derived Phytohormones in Modulation of Rhizosphere/Roots

It is believed that phytohormones released by PGPM modulate the rhizosphere which helps in their colonization, however, there are not enough reports to corroborate this hypothesis. There are more reports on the role of microbe-produced exopolysaccharide on root colonization. However, the impact of these phytohormone-producing microbes on modulation of root structure architecture is established (Kudoyarova et al. 2019), pointing to a possible role of these phytohormones on bacterial colonization. Since the roots are the first points of contact of any bacterial exudate and the plant, therefore, the impact of these microbe-derived chemicals on parameters such as primary root length, lateral root length and branching, root hair density etc. is but expected. The role of plant-derived phytohormones on root growth and development is well established. It is a well-known fact that the two primary hormones responsible for regulating root growth are auxin and cytokinin. Classical studies have proven that auxin induces lateral root growth in plants, while, both auxin and cytokinin can become inhibitory towards root growth at relatively higher concentrations (Taiz et al. 2015). There are several reports on the impact of phytohormone-producing bacteria on root growth and development, however, there are very few reports establishing a direct correlation between microbe derived phytohormones and root growth or root structure architecture. Nonetheless, based on the roles of phytohormones such as auxin and cytokinin on root growth and architecture, some speculative correlations can be made. For example, it has been seen that auxin or cytokinin-releasing PGPR induce increase in root hair density and lateral root length while, in some cases, inhibition of primary root length has been observed (Shah et al. 2016; Ghosh et al. 2017; Kudoyarova et al. 2019). There are also reports on enhancement of the total root surface by these microbes. It is obvious that many of these changes can lead to increased mineral uptake by roots, thereby increasing root exudation and, possibly, facilitating PGPM colonization The role of auxin and cytokinin producing bacteria on several aspects of root development has been comprehensively reviewed by Kudoyarova et al. (2019). According to this review, the impact of these bacteria on root development is two-fold; a positive impact by some strains and a negative impact by others. The negative impact has been correlated with the inhibitory roles of higher concentrations of auxin and cytokinin on root growth. For example, an increase in root biomass and endogenous root auxin levels were observed in wheat plants inoculated with auxin-producing *Paenibacillus illinoisensis* IB 1087 and *Pseudomonas extremaustralis* IB-K13-1A (Kudoyarova et al. 2017). Similar observations have been reported in wheat inoculated with the salt tolerant PGPR *Pseudomomas moraviensis* (Ul Hassan and Bano 2019) and *Azospirillum* (Dobbelaere et al. 1999). Similarly, drought mitigation in wheat due to enhanced root growth and lateral root formation by auxin producing *Azospirillum* was reported by Arzanesh et al. (2011) Changes in the root architecture due to the production of IAA by PGPR was found in the studies of Mantelin and Touraine (2003), helping the plants in rapid uptake of water from the soil under drought conditions. A beneficial species of fungi, *Trichoderma virens*, inoculated onto *Arabidopsis thaliana*, changed root architecture by modulating IAA concentration (Contreras-Cornejo et al. 2009). There are reports that mutant strains deficient in IAA production were impaired in the type of root growth promotion that the wild type strain could impart. For example, IAA-deficient mutants of *Pseudomomas putida* whose impact was studied on root growth in canola (*Brassica napus*) (Patten and Glick 2002b). Studies with mutants altered in IAA production and their role in modulating root structure architecture in plants has been reviewed by Spaepen and Vanderleyden (2011).

In case of cytokinin producing microbes, *Bacillus amyloliquefaciens* UCMB5113 inhibited primary root growth in *Arabidopsis thaliana* (Asari et al. 2017). This observation was correlated with, both, bacterial cytokinin production which increased root cytokinin levels or the increased auxin levels that were also detected in colonized roots (Asari et al. 2017). Similar changes in root growth in *Arabidopsis thaliana* by phytohormone producing, abiotic stress mitigating soil bacteria has been reported with *Pseudomonas putida* GAP-P45 (Ghosh et al. 2017) and with *Pseudomonas putida* AKMP7 (Shah et al. 2016). These studies reported a reduction in primary root length of Arabidopsis plants under water-stress conditions, when inoculated with the aforementioned strains of bacteria. While, in the former study (i.e., Ghosh et al. 2017), bacterial inoculation was positively corelated with water-stress tolerance, in the later (Shah et al. 2016), water-stress mediated deterioration was observed in plants inoculated with AKMP7.

6 Impact of PGPM on Plant Phytohormone Metabolism and Signalling

An important aspect of plant-PGPM interaction is the widely reported phenomenon of modulation in plant phytohormone levels and metabolism by PGPM. However, the precise mechanism leading to such modulations of endogenous phytohormone levels in plants by PGPM is not clearly understood. Form the current status of research, it can be hypothesized that this may be a key mechanism in PGPM-mediated stress amelioration in plants (Barnawal et al. 2017; Dodd et al. 2010). However, whether these modulations occur due to uptake of bacterial phytohormones by plants or due to alteration in plant's endogenous hormone metabolism induced by bacteria, or a combination of both, remains unclear. Modulations in plant phytohormone levels have been seen in general growth and development as well as stress amelioration processes when inoculated with PGPM. For example, Marulanda et al. (2009) reported an increase in IAA concentration in clover, inoculated with *Pseudomonas putida* and *Bacillus megaterium*. This was positively correlated with enhanced plant biomass. In another study, Kang et al. (2014a) reported increased levels of endogenous GAs in cucumber plants inoculated with PGPR strains like *Burkholdera cepacia SE4*, *Promicromonospora* spp. SE188 and *Acinetobacter calcoaceticus* SE370. *Aeromonas punctata* PNS-1, *Serratia marcescens* 90–166 and *Azospirillum brasilense* Sp245 are PGPR strains capable of producing auxin and inoculation of these bacteria increased the endogenous levels of auxin (Iqbal and Hasnain 2013; Shi et al. 2010). Endogenous levels of IAA in the roots of plants inoculated with *Phyllobacterium brassicacearum* STM196 and *Bacillus* sp. LZR216 were higher than the control plants with concomitant increase in expression of genes involved in IAA biosynthesis (Contesto et al. 2010). Apart from its effects on biosynthesis of phytohormones, PGPR can also affect the transport of endogenous auxin by altering the expression of auxin transporters which can also regulate the growth stimulating activities. For example, inoculation with *Bacillus* sp. LZR216 decreased the synthesis of PIN and AUX (auxin transporters). On the other hand, PIN2 and PIN3 expression is elevated in plants inoculated with *Bacillus phytofirmans* PsJN and mutation in pin2 impacts negatively on the growth promotion by these *Bacillus* species (Poupin et al. 2016; Wang et al. 2015). Certain substances with auxin-like activity produced by PGPR can also affect the endogenous auxin levels. For example, cyclopeptides produced by mutant *Pseudomonas aeruginosa* exhibit auxin-like activity weakly which enhance the lateral root formation and growth compared to the wild-type *Pseudomonas* that are not capable of producing the cyclopeptides (Ortiz-Castro et al. 2011). In-vitro studies using *Arabidopsis thaliana* have shown that volatiles such as acetoin and 3,4-butanediol produced by *Bacillus subtilis* GB03and *Bacillus amyloliquefaciens* IN937a alter endogenous auxin levels with increased expression of auxin biosynthetic genes and they also influence the expression levels of IAA transporter genes (Ryu et al. 2003; Zhang et al. 2007).

Several PGPR strains are capable of synthesizing cytokinin as well as altering the endogenous levels of cytokinin in their host plants (Tsukanova et al. 2017). For instance, elevated levels of endogenous cytokinin were observed in six-week-old *Arabidopsis thaliana* upon seed bacterization with *Burkholderia phytofirmans* PsJN (Su et al. 2016). Exposure of tomato plants to volatiles emitted by *Bacillus subtilis* SYST2 enhanced the expression of cytokinin biosynthetic gene (*Sl*CKX1) and endogenous cytokinin levels. Inoculation of *Bacillus subtilis* (AE016877) increased the endogenous cytokinin levels of *Platycladus orientalis* plants by 97.10% compared to its respective control (Liu et al. 2013). Similarly, PGPM inoculation can affect the endogenous levels of GAs in plants (Tsukanova et al. 2017). Studies using mutant rice plants impaired in biosynthesis of gibberellins shows that PGPR *Leifsonia soli* SE134 and *Enterococcus faecium* LKE12 capable of producing gibberellins can compensate the shoot growth compared to its control (Kang et al. 2014b; Lee et al. 2015). There are certain strains which do not synthesize gibberellins but are capable of elevating the gene expression of GA biosynthetic genes For instance, *Burkholderia phytofirmans* PsJN increases the expression of *At*GA3ox1which is involved gibberellin synthesis in *Arabidopsis thaliana* (Poupin et al. 2013).

It has been reported that several species of stress-mitigating PGPM are known to modulate phytohormone metabolism and homeostasis in the plants exposed to one or more of these stresses. Environmental stress on plants can broadly be classified into abiotic and biotic stresses. Stress tolerant PGPM are known to help plants in amelioration of both these stresses. There is a substantial body of work on the positive impact of stress-tolerant PGPR on abiotic stress (drought, salinity, thermal stress etc.) tolerance in plants. As far as the role of PGPR in amelioration of biotic stress is concerned, they are mostly helpful in limiting the growth of soil pathogens by a variety of mechanisms, thus earning them the title of “biocontrol” agents. With respect to phythormone signalling, auxin, GA, cytokinin and ABA are considered important for PGPM-mediated abiotic stress tolerance, while SA, JA and ethylene signalling are considered to be important for the phenomenon of biocontrol.

Pereyra et al. (2012) have observed that, on inoculation of wheat seedlings with *Azospirillum* under osmotic stress, there were some morphological changes in the xylem architecture which was due to the upregulation of indole-3-pyruvate decarboxylase gene and increased IAA production in the plants. Enhanced tomerance of cucumber plants to salinity and drought waas positively correlated with endogenous GA levels when inoculated with *Burkholdera cepacia SE4*, *Promicromonospora* spp. *SE188* and *Acinetobacter calcoaceticus SE370* (Kang et al. 2014a) *Phyllobacterium brassicacearum* STM196 has been reported to modulate internal hormonal signaling in *Arabidopsis thaliana* (Contesto et al. 2010; Galland et al. 2012; Bresson et al. 2013). Ghosh et al. (2019) observed that when *Arabidopsis thaliana* plants were inoculated with the IAA and cytokinin producing PGPR *Pseudomonas putida* GAP-P45, the levels of the four major phytohormones-auxin (IAA), cytokinin (Tz), GA and ABA were differently modulated in the roots versus shoots. *P. putida* GAP-P45 downregulated endogenous ABA levels in *Arabidopsis thaliana* under water -stress, while elevating IAA and tZ accumulation in shoots and roots. This bacterium caused an increase in endogenous GA content in shoot but decrease of the same in root tissue of the plants under water-stress was observed. A beneficial species of fungus, *Trichoderma virens* caused alterations in root structure architecture of *Arabidopsis thaliana*, by modulating IAA concentration (Contreras-Cornejo et al. 2009). The beneficial fungus, *Piriformospora indica* has been reported to alleviate drought and salinity stress in several plants by secretion of the phytohormone cytokinin and its cross-talk with plant ABA levels (Ansari et al. 2013).

7 Hormonal Signaling in Plant-Pathogen Interaction

In the last two decades, the role of phytohormone signalling in plant-pathogen interaction has been widely explored. Apart from the classical growth hormones elaborated above, exciting research around the world has extended that repertoire to include salicylic acid (SA), jasmonic acid (JA), nitric oxide (NO), strigolactones, karrikins etc. Some pathogens may directly secrete hormones or cause alteration of hormone levels and signalling components to overcome host defence responses with a combination of effectors and interference mechanisms to manipulate available host resources to their advantage (Bari and Jones 2009; Berens et al. 2017). Thus, the nature and magnitude of these interactions determine the overall outcome of host-microbial interactions. On the other hand, application of the hormones can be used to alter plant–microbe interactions directly or by interference in signalling to assess synergistic responses that enhance plant productivity as well as defense responses (Delaney et al. 1994; Li et al. 1996; Tjamos et al. 2005; Johansson et al. 2006). Therefore, a better understanding of plant hormonal responses to pathogens during susceptible and resistant interactions will aid in the effective management of plant diseases while maintaining crop yields. This part of the chapter examines historical and novel developments on plant hormonal changes in response to various pathogens in general. We present a brief overview of the known host genetic components, defense strategies, and molecular mechanisms underlying defense responses about hormone signalling pathways and identify research areas for the future. While the main focus of this chapter is on the roles of auxin, cytokinin and GA signalling, the major hormones impacting plant -pathogen interaction are SA, JA and ethylene. Hence, we first review the roles of these hormones and then move on to the others.

8 Role of Salicylic Acid (SA), Jasmonic Acid (JA), and Ethylene (ET) Signaling in Plant Defense

Salicylic acid (SA) is a plant hormone that plays a central role in plant defense response to pathogens, not only locally, but also during systemic defense response (SAR) in the face of a challenge from multiple pathogens (Delaney et al. 1994; Cao et al. 1997). SA-mediated defense responses are most effective against biotrophic and hemibiotrophic pathogens with a diminished role against necrotrophic pathogens (Glazebrook 2005). In plants, pathogen infection induces biosynthesis of SA mainly through the upregulation of the SA biosynthetic *Isochroismate Synthase 1/Salicyclic Acid Induction Deficient* (*SID2*) gene from the precursor chorismate, which is synthesized in the chloroplast. This induces downstream responses like the production of reactive oxygen species (ROS), hypersensitive response (HR), lignification of the cell wall, production of secondary metabolites like antimicrobials, etc. Interestingly, pretreatment of plants with SA induces resistance to various pathogens (Gong et al. 2017), while plants expressing the *Pseudomonas putida NahG* (encoding the enzyme salicylate hydroxylase that converts SA to catechol) resulted in increased susceptibility to pathogens (Liu et al. 2014; Zheng et al. 2019). Activation of SA mediated defense responses by applying beneficial microbes, including the rhizosphere endophytic PGPR strain *Paenibacillus alvei* K165 to increase resistance in the host plants (Tjamos et al. 2005). Furthermore, transcriptional reprogramming following pathogen infection has been shown to involve SA signaling with various pathogens employing effector molecules to infect the host plant by altering hormone signaling as an infection strategy (Bari and Jones 2009). *Verticillium dahliae Isochorismate synthase* (*VdIsc1*) is one such effector protein employed by the pathogen to suppress SA levels in the host during early infection stages (Liu et al. 2014). When cotton plants were inoculated with a *VdIsc1* deletion mutant, the host SA and SAG levels and the SA marker, *PR1*, were significantly up-regulated. Other pathogen effectors like *VdSCP41* target the master immune regulators *calmodulin-binding protein 60-like g* (*CBP60g*) and *SAR Deficient 1* (*SARD1*), which in turn regulate the defense responses in multiple plant species by binding directly to the promoters of SA signaling components (Qin et al. 2018). Thus, SA signaling plays a significant role during plant-microbial interactions with SA perception, and microbe-mediated alterations of the host SA levels determine the host resistance mechanisms to various pathogenic and beneficial microbes in general.

Another class of plant hormones that plays an important role in multiple processes like biotic stress, abiotic stresses, and plant development includes the jasmonates and jasmonic acid (JA), structurally similar to the metazoan prostaglandins (Chini et al. 2007). In plants, JA is synthesized in chloroplasts through the octadecanoid pathway and is indispensable for resistance against various pests, including herbivory, insects, and necrotrophs, as well as in plant reproduction, including pollen fertility (Xie et al. 1998; Berens et al. 2017). Though the JA signaling pathway genes are activated quickly in response to this pathogen, the levels of active JA and JA-Ile are noticeable at a later stage in the progression of the disease, indicating an active role for the hormone in fungal defense response Moreover, JA pathway mutants (like *jar1*, *coi1*, and *cyp94B3*) exhibit increased fungal resistance with reduced tissue colonization and lower fungal biomass in roots (Scholz et al. 2018). Lignin polymerization, which is involved in such resistance responses during fungal infection, has been reported to be induced by pathogen infection in a manner involving JA signalling leading to increased resistance in mutants defective in the lignin polymerization enzyme laccase1 (Hu et al. 2018, 2019). Enhancement of JA signalling through involving a negative regulator of JA signaling *Homeodomain Transcription Factor 1* (*HDTF1*) has been linked to increased resistance of cotton to *V. dahliae* and *Botrytis cinerea* (Gao et al. 2016). Other transcriptional regulators like the plant-specific homeodomain-leucine zipper (HD-ZIP) family protein HB12 and the plant-specific NAC transcription family member ATAF1 are involved in the negative regulation of JA mediated defense responses (He et al. 2016, 2018a, b), underscoring the fact that transcriptional reprogramming during defense responses to pathogen actively involves JA mediated signal responses. JA signaling pathway is activated upon fungal infection in a manner that requires stearoyl acyl-carrier-protein desaturase (*SSI2*), leading to activation of defense responses and heightened resistance (Gao et al. 2013a, b). JA signaling pathway has also been linked to pathogen resistance responses through the mediator complex that comprises the conserved multiprotein cofactor of RNA polymerase II and regulates transcription through 20–30 subunits that form four mediator subcomplexes. This complex communicates with hormone signaling to influence multiple plant processes like development, flowering, non-coding RNA processing, secondary metabolism, and defense response to various abiotic and biotic stresses (An and Mou 2013; Li et al. 2018). Multiple studies have shown that the absence of JA leads to the activation of SA-mediated response for resistance against pathogenic fungal infection (Johansson et al. 2006), while pretreatment of plants with JA before subsequent infection leads to resistance against this fungal pathogen by activating basal defense responses (Johansson et al. 2006; Gao et al. 2013a). Though the JA signaling pathway genes are activated quickly in response to this pathogen, the active JA and JA-Ile levels show noticeable changes in only the later part of pathogen infection. Thus, pretreatment of plants with JA before subsequent infection leads to resistance against this fungal pathogen by activating basal defense responses (Johansson et al. 2006; Gao et al. 2013a). Reciprocal regulation of JA-SA mediated defense responses has been observed during the Verticillium wilt of plants (Li et al. 2014).

Ethylene (ET) is a gaseous plant hormone usually associated with senescence and cell death processes in plants which manifest as chlorosis/yellowing of the foliage from the loss of chlorophyll. Necrotrophic pathogen infection in plants results in similar cell death processes leading to the establishment and successful propagation of the pathogen. Thus, it is not surprising that plant defense response to necrotrophic pathogens has been shown to extensively involve ethylene signaling (Broekaert et al. 2006), with evidence of increased ET levels coinciding with the onset of disease symptoms (Cronshaw and Pegg 1976). The ethylene biosynthetic enzyme 1-aminocyclopropane-1-carboxylate synthase (*ACS6*) has been reported to be induced upon pathogen infection, further confirming the role of ethylene during pathogen infection (Wang et al. 2004; Zhou et al. 2012). Increased ethylene production has been observed in susceptible cultivars than tolerant potato cultivars upon exposure to fungal culture filtrate and toxin. At the same time, inhibition of the host ET signaling abrogated this toxin-induced and ET-mediated symptom development in the host (Mansoori and Smith 2005). Ethylene production followed pathogen infection, and ET signaling components are required for defense response against fungal pathogens. In contrast, treatment of the pathogen-challenged host with the ethylene precursor molecule 1-aminocyclopropane-1-carboxylic acid (ACC) resulted in increased host resistance underlined by increased fresh weight of the inoculated plants (Johansson et al. 2006). Interestingly, silencing of the ET receptor genes *Never*

Ripe (*SlNr*) and *SlETR4* resulted in reduced disease incidence, severity, and reduced fungal biomass indicative of enhanced resistance to fungal infection (Pantelides et al. 2010a). Pathogen resistance through lignification of the cell-wall also involves ET signaling with *Ethylene Response Factor-like* gene (*ERF1 like*) and *Ethylene Response Factor 6* (*GhERF6*) transcription factors shown to bind to the GCC-box element in the promoters of the defense-related genes and positively regulate defense against fungal pathogens (Guo et al. 2016; Yang et al. 2015). However, the role of ET during fungal infection has been contradictory, with some studies indicating a dual role for ET in resistance and the promotion of wilt. In this context, it is interesting to note that impaired perception of ET in the *Arabidopsis etr1-1* mutant led to a significant reduction in pathogen growth and increased resistance during Verticillium wilt ET, with a similar result achieved upon application of the ET inhibitor aminoethoxy vinyl glycine (AVG) either before or at the time of inoculation (Pantelides et al. 2010b; Robison et al. 2001a). On the other hand, plants transformed with a catabolic enzyme—the bacterial *ACC deaminase* gene under root-specific to inhibit ET synthesis, significantly reduced or delayed disease symptoms by targeted degradation of the ET precursor ACC, resulting in an overall reduction of disease symptoms (Robison et al. 2001b). Such dual nature of ET signaling can be explained by the fact that initial perception of the pathogen may be activating ET signaling and ET biosynthesis at the site of infection to limit the spread of the pathogen, but once the pathogen has already been established in a host, it might aid in the establishment of the necrotrophic phase of the pathogen.

9 Role of Cytokinin (CK), Auxin (AUX), Gibberellic Acid (GA), Brassinosteroid (BR), Nitric Oxide (NO), and ß-Aminobutyric Acid (BABA) in Pathogen Resistance

A decrease in the cytokinin levels in the tracheal fluid and the above-ground tissue in cotton plants and tomato plants treated with a pathogenic strain of *Verticillium* spp. in symptom development has been reported (Misaghi et al. 1972; Patrick et al. 1977). A decrease in water potential in the root leading to a reduction in CK levels has been proposed to underline the visible yellowing of the leaves due to chlorosis and loss of pigments during Verticillium wilt (Patrick et al. 1977). Up-regulation of cytokinin degrading enzymes leading to a decrease in the host cytokinin levels, particularly trans-zeatin [tZ], has been documented to concur with *Verticillium*-induced premature senescence (Reusche et al. 2013). Taken together, these results indicate that reduced cytokinin levels most likely promote infection caused by a necrotrophic pathogen or during the necrotrophic phase in the life cycle of a pathogen laying the framework for efficient colonization of the host leaves by active induction of senescence (Reusche et al. 2013). This is supported by the fact that external application of synthetic cytokinins and inhibition of the cytokinin degrading enzymatic activity led to reduced symptoms and proliferation of the fungus on the host (Reusche et al. 2013). Thus, cytokinin levels seem to play a role during plant defense responses to specific pathogens that promote senescence as a strategy for enhancing pathogenicity on susceptible hosts.

Many microbes produce/secrete bioactive molecules, including hormones as secondary metabolites, including the model bacterial pathogen *Pseudomonas syringae* and the fungal pathogen *Verticillium dahliae.* Recent work has shown that volatile compounds (VCs) from *Verticillium* spp. cause preferential allocation of resources driving the root growth over shoot growth by manipulating auxin (AUX) signaling pathways in the host plant. Various approaches, including chemical inhibition of the AUX signaling pathway using an auxin efflux inhibitor, compromised this change in growth pattern, further underscoring the role of AUX during pathogen infection. Indeed, several components of the AUX signaling pathway, including *TIR1*, *TIR3*, *AUX1*, and *AXR1*, were subsequently shown to be involved in the regulation of resistance to both fungal and bacterial pathogens in the host plants (Li et al. 2018).

GA signalling has been shown to be involved in a few instances of disease pathology, including the bakanae disease on rice (Studt et al. 2013) and during the infection by rice dwarf virus (Zhu et al. 2005). In this regard, the DELLA proteins that negatively regulate the GA signaling pathway involved in plant growth are fast emerging as plant defense response regulators that balance plant growth during biotic and abiotic stress (Hou et al. 2010). Recent work has shown that increased bioactive GA through induction of GA biosynthetic genes combined with suppression of *DELLA* genes contribute to hyper GA signaling seems to increase the susceptibility of the defense compromised *ndr1-1* mutant in *Arabidopsis* while promoting a faster transition to flowering in response to pathogen infection (Dhar et al. 2019).

A class of steroid hormones that are found in both plants and animals includes the recently discovered brassinosteroids (BR) is perceived by cell surface receptors in plants in contrast to that found in the animal counterpart and is primarily involved in plant growth, development, cell differentiation, and photomorphogenesis (Wang et al. 2012). However, a role for BR signaling in plant defense is starting to emerge, with its role in balancing defense and development attracting further attention. Epibrassinolide treatment in longer-term seemed to positively regulate resistance to fungal pathogen in tomato and cotton while activating the JA signaling pathway (Krishna 2003; Gao et al. 2013b; Roos et al. 2014; Bibi et al. 2017). Studies have shown that both pathogen infection and epibrassinolide significantly elevated the level of the enzymes involved in carbohydrate metabolism, including sucrose phosphate synthase (SPS), vacuolar/cell wall-bound acid invertase (AI), and cytosolic sucrose synthase (SuSy), helping to negate the pathogen-induced osmotic stress in the host, thus contributing to host resistance (Goicoechea et al. 2000; Bibi et al. 2014). Various components of BR signaling, including *BAK1* that encodes an LRR-RLK associated with the BR receptor *Brassinosteroid Insensitive 1* (*BRI1*), is required for resistance against the fungus in multiple species (Fradin et al. 2009; Gao et al. 2013a; Roos et al. 2014). Yet another RLK *Suppressor of BIR1* (*SOBIR1*) that is associated *BAK1-interacting receptor-like-kinase 1* (*BIR1*) and hence an integral component of BR signaling in plants is required for resistance against various pathogens and regulates plant defense responses through its interaction with multiple receptor-like proteins (RLPs) (Liebrand et al. 2013; Zhou et al. 2019). Taken together, these findings suggest that BR signalling might play a more significant role in plant defense response than previously expected for a hormone that initially came to prominence for its primary involvement in plant growth and development.

Recently the gaseous hormone nitric oxide (NO) has been implicated in ROS and defense signalling in plants. *Verticillium* infection and *Vd* toxins have been known to induce cell death with the active involvement of ROS (Jia et al. 2007), while *Vd* toxin alters hormone balances during *Verticillium* infection (Pegg and Brady 2002). However, both NO and H_2O_2 are produced in cotton suspension cells treated with *Vd* toxin have been known to produce both NO and H_2O_2, with the upregulation of defense associated redox proteins glutathione S-transferases (GSTs) proteins associated with the NO signaling pathway (Jia et al. 2007). Leaves of *Arabidopsis* plants treated with the *Vd* toxin produce NO with peak activity around an hour post-treatment which is abrogated in the NR deficient *nia1 x nia2* mutant (Shi and Li 2008). Additionally, the *Vd* toxin-induced NO production results in the depolymerization and destabilization of the cortical microtubules rather than the actin microfilaments inside the cell, resulting in HR-like cell death along with the activation of the host defense responses (Shi et al. 2009). Yet another study with *Vd* toxins and mutant analysis has shown that H_2O_2 functions upstream of NO as a result of modulation of dynamic microtubule cytoskeleton through the blockage of NO production, mediated by nitrate reductase, leading to activation of defense against the fungal pathogen *V. dahliae* infection (Yao et al. 2012). Further work has shown that the downregulation of defense-related SA and NO hormone levels in the VIGS-mediated silencing of the coiled-coil (CC)–NBS–LRR-type gene (*GbRVd*) in cotton predisposes them to be susceptibility against *Verticillium* infection (Yang et al. 2016).

Yet another active chemical, ß-aminobutyric acid (BABA), emerging as a novel plant growth regulator, has been shown to induce resistance to various pathogens. It prevents disease symptoms, including stunting of plants during Verticillium wilt of the oilseed rape by activating higher synthesis and accumulation of phenylpropanoids (Kamble et al. 2013). This heightened resistance is underlined by a change in vascular architecture and storage of resistance-enhancing phenolics, leading to the containment of the pathogen by inhibiting colonization of the shoot (Kamble et al. 2013).

10 How Indispensable Are Microbe-Derived Phytohormones for Plant Responses to Microbes?

As described above, phytohormones play a key role in the interaction between plants and associated microbes, whether beneficial or pathogenic. Both plants and microbes biosynthesize phytohormones and several microbes release them into their colonizing habitats. These phytohormones and other chemicals released by the microbes modulate the endogenous metabolic and signaling networks within the plant, thus impacting the way they respond to the presence of microbes in their vicinity. As described above, several studies involving plant and microbial mutants have identified the importance of microbe-derived phytohormones in plant responses to their associated microbiota. While these studies help establish the connection between plant and microbial phytohormone signaling networks, further studies need to be done to understand the precise mechanisms underlying the responses of plants to microbe-derived phytohormones. However, it is clear now that the different classes of plant hormones not only have specific roles in impacting plant physiology, development and stress tolerance, but with accumulating evidence, they also seem to help plants in interacting with their biotic environment. Thus, phytohormone based signalling mechanisms between plants and associated microbiota are intricately interconnected, sharing multiple pathway components with overlapping functions to maintain a well-oiled cellular machinery that helps the sessile organisms in deriving benefit from soil microbiota and safe-guarding them from the continuous barrage of omnipresent pathogenic microbes in nature.

References

Akiyoshi DE, Regier DA, Gordon MP (1989) Nucleotide sequence of the tzs gene from *Pseudomonas solanacearum* strain K60. Nucleic Acids Res. https://doi.org/10.1093/nar/17.21.8886

An C, Mou Z (2013) The function of the Mediator complex in plant immunity. Plant Signal Behav 8(3):e23182. https://doi.org/10.4161/psb.23182

Ansari MW, Trivedi DK, Sahoo RK, Gill SS, Tuteja N (2013) A critical review on fungi mediated plant responses with special emphasis to *Piriformospora indica* on improved production and protection of crops. Plant Physiol Biochem 70:403–410. https://doi.org/10.1016/J.PLAPHY.2013.06.005

Arkhipova TN, Prinsen E, Veselov SU, Martinenko EV, Melentiev AI, Kudoyarova GR (2007) Cytokinin producing bacteria enhance plant growth in drying soil. Plant Soil 292(1–2):305–315. https://doi.org/10.1007/s11104-007-9233-5

Arzanesh MH, Alikhani HA, Khavazi K, Rahimian HA, Miransari M (2011) Wheat (*Triticum aestivum* L.) growth enhancement by *Azospirillum* sp. under drought stress. World J Microbiol Biotechnol 27(2):197–205. https://doi.org/10.1007/s11274-010-0444-1

Asari S, Tarkowská D, Rolčík J, Novák O, Palmero DV, Bejai S, Meijer J (2017) Analysis of plant growth-promoting properties of Bacillus amyloliquefaciens UCMB5113 using *Arabidopsis thaliana* as host plant. Planta 245(1). https://doi.org/10.1007/s00425-016-2580-9

Bari R, Jones JD (2009) Role of plant hormones in plant defense responses. Plant Mol Biol 53:412–428. https://doi.org/10.1007/s11103-008-9435-0

Barnawal D, Bharti N, Pandey SS, Pandey A, Chanotiya CS, Kalra A (2017) Plant growth-promoting rhizobacteria enhance wheat salt and drought stress tolerance by altering endogenous phytohormone levels and *TaCTR1* / *TaDREB2* expression. Physiol Plant 161(4):502–514. https://doi.org/10.1111/ppl.12614

Barry GF, Rogers SG, Fraley RT, Brand L (1984) Identification of a cloned cytokinin biosynthetic gene. Proc Natl Acad Sci. https://doi.org/10.1073/pnas.81.15.4776

Berens ML, Berry HM, Mine A, Argueso CT, Tsuda K (2017) Evolution of hormone signaling networks in plant defense. Ann Rev Phytopathol 55:401–425. https://doi.org/10.1146/annurev-phyto-080516-035544

Bibi N, Fan K, Dawood M, Nawaz G, Yuan SN, Wang XD (2014) Exogenous application of epibrassinolide attenuated *Verticillium* wilt in upland cotton by modulating the carbohydrates metabolism, plasma membrane ATPases and intracellular osmolytes. Plant Growth Regul 73:155–164. https://doi.org/10.1007/s10725-013-9877-5

Bibi N, Ahmed IM, Fan K, Dawood M, Li F, Yuan S et al (2017) Role of brassinosteroids in alleviating toxin-induced stress of *Verticillium dahliae* on cotton callus growth. Environ Sci Pollut Res Int 24:12281–12292. https://doi.org/10.1007/s11356-017-8738-6

Bömke C, Tudzynski B (2009) Diversity, regulation, and evolution of the gibberellin biosynthetic pathway in fungi compared to plants and bacteria. Phytochemistry. https://doi.org/10.1016/j.phytochem.2009.05.020

Bresson J, Varoquaux F, Bontpart T, Touraine B, Vile D (2013) The PGPR strain *Phyllobacterium brassicacearum* STM196 induces a reproductive delay and physiological changes that result in improved drought tolerance in *Arabidopsis*. New Phytol 200(2):558–569. https://doi.org/10.1111/nph.12383

Broekaert WF, Delaure SL, De Bolle MF, Cammue BP (2006) The role of ethylene in host pathogen interactions. Ann Rev Phytopathol 44:393–416. https://doi.org/10.1146/annurev-phyto.44.070505.143440

Cao H, Glazebrook J, Clarke JD, Volko S, Dong X (1997) The *Arabidopsis NPR1* gene that controls systemic acquired resistance encodes a novel protein containing ankyrin repeats. Cell 88:57–63. https://doi.org/10.1016/S0092-8674(00)81858-9

Cassán F, Vanderleyden J, Spaepen S (2014) Physiological and agronomical aspects of phytohormone production by model plant-growth-promoting rhizobacteria (PGPR) belonging to the genus *Azospirillum*. J Plant Growth Regul 33(2). https://doi.org/10.1007/s00344-013-9362-4

Chini A, Fonseca S, Fernandez G, Adie B, Chico JM, Lorenzo O et al (2007) The JAZ family of repressors is the missing link in JA signaling. Nature 448:666–671. https://doi.org/10.1038/nature06006

Cohen AC, Travaglia CN, Bottini R, Piccoli PN (2009) Participation of abscisic acid and gibberellins produced by endophytic *Azospirillum* in the alleviation of drought effects in maize. Botany 87(5):455–462. https://doi.org/10.1139/B09-023

Contesto C, Milesi S, Mantelin S, Zancarini A, Desbrosses G, Varoquaux F, Bellini C, Kowalczyk M, Touraine B (2010) The auxin-signaling pathway is required for the lateral root response of *Arabidopsis* to the rhizobacterium *Phyllobacterium brassicacearum*. Planta 232(6):1455–1470. https://doi.org/10.1007/s00425-010-1264-0

Contreras-Cornejo HA, Macías-Rodríguez L, Cortés-Penagos C, López-Bucio J (2009) Trichoderma virens, a plant beneficial fungus, enhances biomass production and promotes lateral root growth through an auxin-dependent mechanism in *Arabidopsis*. Plant Physiol 149(3):1579–1592. https://doi.org/10.1104/pp.108.130369

Creus CM, Sueldo RJ, Barassi CA (2004) Water relations and yield in *Azospirillum*-inoculated wheat exposed to drought in the field. Can J Bot 82(2):273–281. https://doi.org/10 1139/b03-119

Cronshaw DK, Pegg GF (1976) Ethylene as a toxin synergist in *Verticillium* wilt of tomato. Physiol Plant Pathol 9:33–34. https://doi.org/10.1016/0048-4059(76)90073-4

de Jonge R, van Esse HP, Maruthachalam K, Bolton MD, Santhanam P, Saber MK et al (2012) Tomato immune receptor *Ve1* recognizes effector of multiple fungal pathogens uncovered by genome and RNA sequencing. Proc Natl Acad Sci USA 109:5110–5115. https://doi.org/10.1073/pnas.1119623109

Delaney TP, Uknes S, Vernooij B, Friedrich L, Weymann K, Negrotto D et al (1994) A central role of salicylic acid in plant disease resistance. Science 266:1247–1250. https://doi.org/10.1126/science.266.5188.1247

Dhar N, Short DPG, Mamo BE, Corrion AJ, Wai CM, Anchieta A et al (2019) *Arabidopsis* defense mutant *ndr1-1* displays accelerated development and early flowering mediated by the hormone gibberellic acid. Plant Sci 285:200–213. https://doi.org/10.1016/j.plantsci.2019.04.006

Dobbelaere S, Croonenborghs A, Thys A et al (1999) Phytostimulatory effect of *Azospirillum brasilense* wild type and mutant strains altered in IAA production on wheat. Plant Soil 212:153–162

Dodd IC, Zinovkina NY, Safronova VI, Belimov AA (2010) Rhizobacterial mediation of plant hormone status. Ann Appl Biol 157(3):361–379. https://doi.org/10.1111/j.1744-7348.2010.00439.x

Duca D, Lorv J, Patten CL, Rose D, Glick BR (2014) Indole-3-acetic acid in plant-microbe interactions. Antonie Van Leeuwenhoek Int J Gen Mol Microbiol. https://doi.org/10.1007/s10482-013-0095-y

Fradin EF, Zhang Z, Ayala JCJ, Castroverde CDM, Nazar RN, Robb J et al (2009) Genetic dissection of *Verticillium* wilt resistance mediated by tomato *Ve1*. Plant Physiol 150:320–332. https://doi.org/10.1111/j.1364-3703.2006.00323.x

Freiberg C, Fellay R, Bairoch A, Broughton WJ, Rosenthal A, Perret X (1997) Molecular basis of symbiosis between *Rhizobium* and legumes. Nature. https://doi.org/10.1038/387394a0

Galland M, Gamet L, Varoquaux F, Touraine B, Touraine B, Desbrosses G (2012) The ethylene pathway contributes to root hair elongation induced by the beneficial bacteria *Phyllobacterium brassicacearum* STM196. Plant Sci 190:74–81. https://doi.org/10.1016/j.plantsci.2012.03.008

Gao W, Long L, Zhu LF, Xu L, Gao WH, Sun LQ et al (2013a) Proteomic and virus-induced gene silencing (VIGS) analyses reveal that gossypol, brassinosteroids, and jasmonic acid contribute to the resistance of cotton to *Verticillium dahliae*. Mol Cell Proteomics 12:3690–3703. https://doi.org/10.1074/mcp.M113.031013

Gao X, Li F, Li M, Kianinejad AS, Dever JK, Wheeler TA et al (2013b) Cotton *GhBAK1* mediates *Verticillium* wilt resistance and cell death. J Integr Plant Biol 55:586–596. https://doi.org/10.1111/jipb.12064

Gao W, Long L, Xu L, Lindsey K, Zhang X, Zhu L (2016) Suppression of the homeobox gene *HDTF1* enhances resistance to *Verticillium dahliae* and *Botrytis cinerea* in cotton. J Integr Plant Biol 58:503–513. https://doi.org/10.1111/jipb.12432

Ghosh D, Sen S, Mohapatra S (2017) Modulation of proline metabolic gene expression in *Arabidopsis thaliana* under water-stressed conditions by a drought-mitigating *Pseudomonas putida* strain. Ann Microbiol. https://doi.org/10.1007/s13213-017-1294-y

Ghosh D, Gupta A, Mohapatra S (2019) A comparative analysis of exopolysaccharide and phytohormone secretions by four drought-tolerant rhizobacterial strains and their impact on osmotic-stress mitigation in *Arabidopsis thaliana*. World J Microbiol Biotechnol. https://doi.org/10.1007/s11274-019-2659-0

Glass NL, Kosuge T (1986) Cloning of the gene for indoleacetic acid-lysine synthetase from *Pseudomonas syringae* subsp. savastanoi. J Bacteriol. https://doi.org/10.1128/jb.166.2.598-603.1986

Glazebrook J (2005) Contrasting mechanisms of defense against biotrophic and necrotrophic pathogens. Annu Rev Phytopathol 43:220–227. https://doi.org/10.1146/annurev.phyto.43.040204.135923

Goicoechea N, Aguirreolea J, Cenoz S, Garcia-Mina JM (2000) *Verticillium dahliae* modifies the concentration of proline, soluble sugars, starch, soluble protein and abscisic acid in pepper plants. Eur J Plant Pathol 106:19–25. https://doi.org/10.1023/A:1008724816041

Gong Q, Yang Z, Wang X, Butt HI, Chen E, He S et al (2017) Salicylic acid-related cotton (*Gossypium arboreum*) ribosomal protein GaRPL18 contributes to resistance to *Verticillium dahliae*. BMC Plant Biol 3:17–59. https://doi.org/10.1186/s12870-017-1007-5

Gouda S, Kerry RG, Das G, Paramithiotis S, Shin H-S, Patra JK (2018) Revitalization of plant growth promoting rhizobacteria for sustainable development in agriculture. Microbiol Res 206:131–140. https://doi.org/10.1016/J.MICRES.2017.08.016

Guo W, Jin L, Miao Y, He X, Hu Q, Guo K et al (2016) An ethylene response-related factor, *GbERF1-like*, from *Gossypium barbadense* improves resistance to *Verticillium dahliae* via activating lignin synthesis. Plant Mol Biol 91:305–318. https://doi.org/10.1007/s11103-016-0467-6

Gupta S, Schillaci M, Walker R, Smith PMC, Watt M, Roessner U (2021) Alleviation of salinity stress in plants by endophytic plant-fungal symbiosis: current knowledge, perspectives and future directions. Plant Soil 461(1–2):219–244. https://doi.org/10.1007/S11104-020-04618-W

He X, Zhu L, Xu L, Guo W, Zhang X (2016) *GhATAF1*, a NAC transcription factor, confers abiotic and biotic stress responses by regulating phytohormonal signaling networks. Plant Cell Rep 35:2167–2179. https://doi.org/10.1007/s00299-016-2027-6

He X, Wang T, Zhu W, Wang Y, Zhu L (2018a) *GhHB12*, a HD-ZIP I transcription factor, negatively regulates the cotton resistance to *Verticillium dahliae*. Int J Mol Sci 19:3997. https://doi.org/10.3390/ijms19123997

He X, Zhu L, Wassan GM, Wang Y, Miao Y et al (2018b) *GhJAZ2* attenuates cotton resistance to biotic stresses via the inhibition of the transcriptional activity of *GhbHLH171*. Mol Plant Pathol 19:896–908. https://doi.org/10.1111/mpp.12575

Hou X, Lee LYC, Xia K, Yan Y, Yu H (2010) DELLAs modulate jasmonate signaling via competitive binding to JAZs. Dev Cell 19:884–894. https://doi.org/10.1016/j.devcel.2010.10.024

Hu Q, Min L, Yang X, Jin S, Zhang L, Li Y, Ma Y, Qi X, Li D, Liu H, Lindsey K, Zhu L, Zhang X (2018) Laccase GhLac1 modulates broad-spectrum biotic stress tolerance via manipulating phenylpropanoid pathway and jasmonic acid synthesis. Plant Physiol 176:1808–1823. https://doi.org/10.1104/pp.17.01628

Hu X, Puri KD, Gurung S, Klosterman SJ, Wallis CM, Britton M, Durbin-Johnson B, Phinney B, Salemi M, Short DPG, Subbarao KV (2019) Proteome and metabolome analyses reveal differential responses in tomato-*Verticillium dahliae*-interactions. J Proteom 207:103449. https://doi.org/10.1016/j.jprot.2019.103449

Iqbal A, Hasnain S (2013) Aeromonas punctata PNS-1: A promising candidate to change the root morphogenesis of *Arabidopsis thaliana* in MS and sand system. Acta Physiol Plant 35(3):657–665. https://doi.org/10.1007/S11738-012-1106-8

Jia ZQ, Yuan HY, Li YZ (2007) NO and H_2O_2 induced by *Verticillium dahliae* toxins and its influence on the expression of *GST* gene in cotton suspension cells. Chin Sci Bull 52:1347–1354. https://doi.org/10.1007/s11434-007-0211-3

Johansson A, Staal J, Dixelius C (2006) Early responses in the *Arabidopsis-Verticillium longisporum* pathosystem are dependent on NDR1, JA- and ET-associated signals via cytosolic NPR1 and RFO1. Mol Plant Microbe Interact 19:958–969. https://doi.org/10.1094/MPMI-19-0958

Kamble A, Koopmann B, von Tiedemann A (2013) Induced resistance to *Verticillium longisporum* in *Brassica napus* by b-aminobutyric acid. Plant Pathol 62:552–561. https://doi.org/10.1111/j.1365-3059.2012.02669.x

Kang S-M, Khan AL, Waqas M, You Y-H, Kim J-H, Kim J-G, Hamayun M, Lee I-J (2014a) Plant growth-promoting rhizobacteria reduce adverse effects of salinity and osmotic stress by regulating phytohormones and antioxidants in *Cucumis sativus*. J Plant Interact 9(1):673–682. https://doi.org/10.1080/17429145.2014.894587

Kang S-M, Latif Khan A, You Y-H, Kim J-G, Kamran M, Lee I-J (2014b) Gibberellin production by newly isolated strain *Leifsonia soli* SE134 and its potential to promote plant growth. J Microbiol Biotechnol 24(1):106–112. https://doi.org/10.4014/jmb.1304.04015

Kang S-M, Radhakrishnan R, Khan AL, Kim M-J, Park J-M, Kim B-R, Shin D-H, Lee I-J (2014c) Gibberellin secreting rhizobacterium, *Pseudomonas putida* H-2-3 modulates the hormonal and stress physiology of soybean to improve the plant growth under saline and drought conditions. Plant Physiol Biochem 84:115–124. https://doi.org/10.1016/j.plaphy.2014.09.001

Khan AL, Halo BA, Elyassi A, Ali S, Al-Hosni K, Hussain J, Al-Harrasi A, Lee I-J (2016) Indole acetic acid and ACC deaminase from endophytic bacteria improves the growth of *Solanum lycopersicum*. Electron J Biotechnol 21:58–64. https://doi.org/10.1016/J.EJBT.2016.02.001

Kochar M, Upadhyay A, Srivastava S (2011) Indole-3-acetic acid biosynthesis in the biocontrol strain *Pseudomonas fluorescens* Psd and plant growth regulation by hormone overexpression. Res Microbiol. https://doi.org/10.1016/j.resmic.2011.03.006

Koga J, Adachi T, Hidaka H (1991) Molecular cloning of the gene for indolepyruvate decarboxylase from *Enterobacter cloacae*. MGG Mol Gen Genet. https://doi.org/10.1007/BF00273581

Krishna P (2003) Brassinosteroid-mediated stress responses. J Plant Growth Regul 22:289–297. https://doi.org/10.1007/s00344-003-0058-z

Kudoyarova GR, Vysotskaya LB, Arkhipova TN, Kuzmina LY, Galimsyanova NF, Sidorova LV, Gabbasova IM, Melentiev AI, Veselov SY (2017) Effect of auxin producing and phosphate solubilizing bacteria on mobility of soil phosphorus, growth rate, and P acquisition by wheat plants. Acta Physiol Plant 39(11). https://doi.org/10.1007/s11738-017-2556-9

Kudoyarova G, Arkhipova T, Korshunova T, Bakaeva M, Loginov O, Dodd IC (2019) Phytohormone mediation of interactions between plants and non-symbiotic growth promoting bacteria under edaphic stresses. Front Plant Sci 10. https://doi.org/10.3389/fpls.2019.01368

Kumla J, Suwannarach N, Matsui K, Lumyong S (2020) Biosynthetic pathway of indole-3-acetic acid in ectomycorrhizal fungi collected from northern Thailand. PLoS One. https://doi.org/10.1371/journal.pone.0227478

Lee K-E, Radhakrishnan R, Kang S-M, You Y-H, Joo G-J, Lee I-J, Ko J-H, Kim J-H (2015) *Enterococcus faecium* LKE12 cell-free extract accelerates host plant growth via gibberellin and indole-3-acetic acid secretion. J Microbiol Biotechnol 25(9):1467–1475. https://doi.org/10.4014/jmb.1502.02011

Li C, He X, Luo X, Xu L, Liu L, Min L et al (2014) Cotton *WRKY1* mediates the plant defense-to-development transition during infection of cotton by *Verticillium dahliae* by activating *JASMONATE ZIM-DOMAIN1* expression. Plant Physiol 166:2179–2194. https://doi.org/10.1104/pp.114.246694

Li N, Wang W, Bitas V, Subbarao K, Liu X, Kang S (2018) Volatile compounds emitted by diverse *Verticillium* species enhance plant growth by manipulating auxin signaling. Mol Plant Microbe Interact 31:102–1031. https://doi.org/10.1094/MPMI-11-17-0263-R

Li J, Zingen-Sell I, Buchenauer H (1996) Induction of resistance of cotton plants to *Verticillium* wilt and of tomato plants to *Fusarium* wilt by 3-aminobutyric acid and methyl jasmonate. J Plant Dis Protect 103(3):288–299. https://www.jstor.org/stable/43216097

Liao D, Wang S, Cui M, Liu J, Chen A, Xu G (2018) Phytohormones regulate the development of arbuscular mycorrhizal symbiosis. Int J Mol Sci. https://doi.org/10.3390/ijms19103146

Liebrand TWH, van den Berg GCM, Zhang Z, Smit P, Cordewener JHG, America AHP et al (2013) Receptor-like kinase *SOBIR1/EVR* interacts with receptor-like proteins in plant immunity against fungal infection. Proc Natl Acad Sci USA 110:10010–10015. https://doi.org/10.1073/pnas.1220015110

Liu F, Xing S, Ma H, Du Z, Ma B (2013) Cytokinin-producing, plant growth-promoting rhizobacteria that confer resistance to drought stress in *Platycladus orientalis* container seedlings. Appl Microbiol Biotechnol 97(20):9155–9164. https://doi.org/10.1007/s00253-013-5193-2

Liu T, Song T, Zhang X, Yuan H, Su L, Li W et al (2014) Unconventionally secreted effectors of two filamentous pathogens target plant salicylate biosynthesis. Nat Commun 5:4686. https://doi.org/10.1038/ncomms5686

MacDonald E, Powell G, Regier D, Glass NL, Roberto F, Kosuge T, Morris R (1986) Secretion of zeatin, ribosylzeatin, and ribosyl-1″-methylzeatin by *Pseudomonas savastanoi*: plasmid-coded cytokinin biosynthesis. Plant Physiology (bethesda). https://doi.org/10.1104/pp.82.3.742

Mansoori B, Smith CJ (2005) Elicitation of ethylene by *Verticillium albo-atrum* phytotoxins in potato. J Phytopathol 153:143–149. https://doi.org/10.1111/j.1439-0434.2005.00943.x

Mantelin S, Touraine B (2003) Plant growth-promoting bacteria and nitrate availability: impacts on root development and nitrate uptake. J Exp Bot 55(394):27–34. https://doi.org/10.1093/jxb/erh010

Marulanda A, Barea J-M, Azcón R (2009) Stimulation of plant growth and drought tolerance by native microorganisms (AM fungi and bacteria) from dry environments: mechanisms related to bacterial effectiveness. J Plant Growth Regul 28(2):115–124. https://doi.org/10.1007/s00344-009-9079-6

Misaghi I, Kosuge T, Devay JE (1972) Changes in cytokinin activity associated with development of *Verticillium* wilt and water stress in cotton plants. Physiol Plant Pathol 2:187–196. https://doi.org/10.1016/0048-4059(72)90001-X

Numan M, Bashir S, Khan Y, Mumtaz R, Shinwari ZK, Khan AL, Khan A, AL-Harrasi A (2018) Plant growth promoting bacteria as an alternative strategy for salt tolerance in plants: a review. Microbiol Res 209:21–32. https://doi.org/10.1016/J.MICRES.2018.02.003

Ortiz-Castro R, Díaz-Pérez C, Martínez-Trujillo M, del Río RE, Campos-García J, López-Bucio J (2011) Transkingdom signaling based on bacterial cyclodipeptides with auxin activity in plants. Natl Acad Sci 108.https://doi.org/10.1073/pnas.1006740108

Pantelides IS, Tjamos SE, Paplomatas EJ (2010a) Insights into the role of ethylene perception in tomato resistance to vascular infection by *Verticillium dahliae*. Plant Path 59:130–138. https://doi.org/10.1111/j.1365-3059.2009.02206.x

Pantelides IS, Tjamos SE, Paplomatas E. J. (2010b). Ethylene perception via *ETR1* is required in *Arabidopsis* infection by *Verticillium dahliae*. Mol Plant Pathol 11:191–202. https://doi.org/10.1111/j.1364-3703.2009.00592.x

Patrick TW, Hall R, Fletcher RA (1977) Cytokinin levels in healthy and *Verticillium*-infected tomato plants. Can J Bot 55:377–382. https://doi.org/10.1139/b77-048

Patten CL, Glick BR (2002b) Role of *Pseudomonas putida* indoleacetic acid in development of the host plant root system. Appl Environ Microbiol. https://doi.org/10.1128/AEM.68.8.3795-3801.2002b

Patten CL, Blakney AJC, Coulson TJD (2013) Activity distribution and function of indole-3-acetic acid biosynthetic pathways in bacteria. Crit Rev Microbiol.https://doi.org/10.3109/1040841X.2012.716819

Patten CL, Glick BR (2002a) Regulation of indoleacetic acid production in *Pseudomonas putida* GR12-2 by tryptophan and the stationary-phase sigma factor RpoS. Can J Microbiol. https://doi.org/10.1139/w02-053

Pegg GF (1976) The response of ethylene-treated tomato plants to infection by *Verticillium albo-atrum*. Physiol Plant Path 9:215–226. https://doi.org/10.1016/0048-4059(76)90055-2

Pegg GF, Brady BL (2002) *Verticillium* wilts. CABI Publishing, New York

Pegg GF, Cronshaw DK (1976) Ethylene production in tomato plants infected with *Verticillium albo-atrum*. Physiol Plant Path 8:279–295. https://doi.org/10.1016/0048-4059(76)90022-9

Pereyra MA, García P, Colabelli MN, Barassi CA, Creus CM (2012) A better water status in wheat seedlings induced by *Azospirillum* under osmotic stress is related to morphological changes in xylem vessels of the coleoptile. Appl Soil Ecol 53:94–97. https://doi.org/10.1016/J.APSOIL.2011.11.007

Pons S, Fournier S, Chervin C, Bécard G, Rochange S, Dit Frey NF, Pagès VP (2020) Phytohormone production by the arbuscular mycorrhizal fungus *Rhizophagus irregularis*. PLoS One.https://doi.org/10.1371/journal.pone.0240886

Poupin MJ, Timmermann T, Vega A, Zuñiga A, González B (2013) Effects of the plant growth-promoting bacterium *Burkholderia phytofirmans* PsJN throughout the life cycle of *Arabidopsis thaliana*. PLoS One 8(7). https://doi.org/10.1371/journal.pone.0069435

Poupin MJ, Greve M, Carmona V, Pinedo I (2016) A complex molecular interplay of auxin and ethylene signaling pathways is involved in Arabidopsis growth promotion by *Burkholderia phytofirmans* PsJN. Front Plant Sci 7(APR2016). https://doi.org/10.3389/FPLS.2016.00492/FULL

Qin J, Wang K, Sun L, Xing H, Wang S, Li L et al (2018) The plant-specific transcription factors CBP60g and SARD1 are targeted by a *Verticillium* secretory protein VdSCP41 to modulate immunity. eLife 7:e34902. https://doi.org/10.7554/eLife.34902

Raza FFA, Faisal M (2013) Growth promotion of maize by desiccation tolerant *Micrococcus luteus*-chp37 isolated from Cholistan desert, Pakistan. AJCS 7(Issue 11)

Reusche M, Klaskova J, Thole K, Truskina J, Novak O, Janz D et al (2013) Stabilization of cytokinin levels enhances *Arabidopsis* resistance against *Verticillium longisporum*. Mol Plant Microbe Interact 26:850–860. https://doi.org/10.1094/MPMI-12-12-0287-R

Robison MM, Griffith M, Pauls KP, Glick BR (2001a) Dual role for ethylene in susceptibility of tomato to *Verticillium* wilt. J Phytopath 149:385–388. https://doi.org/10.1111/j.1439-0434.2001.tb03867.x

Robison MM, Shah S, Tamot B, Pauls KP, Moffatt BA, Glick BR (2001b) Reduced symptoms of *Verticillium* wilt in transgenic tomato expressing a bacterial ACC deaminase. Mol Plant Path 2:135–145. https://doi.org/10.1046/j.1364-3703.2001.00060.x

Roos J, Bejai S, Oide S, Dixelius C (2014) *RabGAP22* is required for defense to the vascular pathogen *Verticillium longisporum* and contributes to stomata immunity. PLoS One 4, 9(2):e88187. https://doi.org/10.1371/journal.pone.0088187

Ryu CM, Faragt MA, Hu CH, Reddy MS, Wei HX, Paré PW, Kloepper JW (2003) Bacterial volatiles promote growth in *Arabidopsis*. Proc Natl Acad Sci USA. https://doi.org/10.1073/pnas.0730845100

Saharan BS, Nehra V (2011) Plant growth promoting rhizobacteria: a critical review. Life Sci Med Res 2011

Sakakibara H (2006) Cytokinins: activity, biosynthesis, and translocation. Annu Rev Plant Biol. https://doi.org/10.1146/annurev.arplant.57.032905.105231

Salomon MV, Bottini R, de Souza Filho GA, Cohen AC, Moreno D, Gil M, Piccoli P (2014) Bacteria isolated from roots and rhizosphere of *Vitis vinifera* retard water losses, induce abscisic acid accumulation and synthesis of defense-related terpenes in in vitro cultured grapevine. Physiol Plant 151(4):359–374. https://doi.org/10.1111/ppl.12117

Sandhya V, Ali SZ, Venkateswarlu B, Reddy G, Grover M (2010) Effect of osmotic stress on plant growth promoting *Pseudomonas* spp. Arch Microbiol 192(10):867–876. https://doi.org/10.1007/s00203-010-0613-5

Sandhya V, Ali SKZ, Grover M, Reddy G, Venkateswarlu B (2009) Alleviation of drought stress effects in sunflower seedlings by the exopolysaccharides producing *Pseudomonas putida* strain GAP-P45. Biol Fertil Soils 46(1):17–26. https://doi.org/10.1007/s00374-009-0401-z

Sawada K (1912) Diseases of agricultural products in Japan. Form Agric Rev 63:10–16

Scholz SS, Schmidt-Heck W, Guthke R, Furch ACU, Reichelt M, Gershenzon J et al (2018) *Verticillium* d*ahliae-Arabidopsis* interaction causes changes in gene expression profiles and jasmonate levels on different time scales. Front Microbiol 9:217. https://doi.org/10.3389/fmicb.2018.00217

Selvakumar G, Bindu GH, Bhatt RM, Upreti KK, Paul AM, Asha A, Shweta K, Sharma M (2018) Osmotolerant cytokinin producing microbes enhance tomato growth in deficit irrigation conditions. Proc Natl Acad Sci India Sect B Biol Sci 88(2):459–465. https://doi.org/10.1007/s40011-016-0766-3

Shah DA, Sen S, Shalini A, Ghosh D, Grover M, Mohapatra S (2016) An auxin secreting *Pseudomonas putida* rhizobacterial strain that negatively impacts water-stress tolerance in *Arabidopsis thaliana*. Rhizosphere 3(P1). https://doi.org/10.1016/j.rhisph.2016.11.002

Shao J, Li S, Zhang N, Cui X, Zhou X, Zhang G, Shen Q, Zhang R (2015) Analysis and cloning of the synthetic pathway of the phytohormone indole-3-acetic acid in the plant-beneficial *Bacillus amyloliquefaciens* SQR9. Microb Cell Factories. https://doi.org/10.1186/s12934-015-0323-4

Shi FM, Li YZ (2008) *Verticillium dahliae* toxins-induced nitric oxide production in *Arabidopsis* is major dependent on nitrate reductase. BMB Rep 41:79–85. https://doi.org/10.5483/bmbrep.2008.41.1.079

Shi FM, Yao LL, Pei BL, Zhou Q, Li XL, Li Y et al (2009) Cortical microtubule as a sensor and target of nitric oxide signal during the defense responses to *Verticillium dahliae* toxins in *Arabidopsis*. Plant Cell Environ 32:428–438. https://doi.org/10.1111/j.1365-3040.2009.01939.x

Shi CL, Park HB, Lee JS, Ryu S, Ryu CM (2010) Inhibition of primary roots and stimulation of lateral root development in *Arabidopsis thaliana* by the rhizobacterium *Serratia marcescens* 90-166 is through both auxin-dependent and -independent signaling pathways. Mol Cells 29(3):251–258. https://doi.org/10.1007/s10059-010-0032-0

Spaepen S, Vanderleyden J (2011) Auxin and plant-microbe interactions. Cold Spring Harb Perspect Biol 3(4). https://doi.org/10.1101/cshperspect.a001438

Studt L, Schmidt FJ, Jahn L, Sieber CMK, Connolly LR, Niehaus E-M et al (2013) Two histone deacetylases, FfHda1 and FfHda2, are important for *Fusarium fujikuroi* secondary metabolism and virulence. Appl Environ Microbiol 79:7719–7734. https://doi.org/10.1128/AEM.01557-13

Su F, Gilard F, Guérard F, Citerne S, Clément C, Vaillant-Gaveau N, Dhondt-Cordelier S (2016) Spatio-temporal responses of *Arabidopsis* leaves in photosynthetic performance and metabolite contents to *Burkholderia phytofirmans* PsJN. Front Plant Sci 7(MAR2016). https://doi.org/10.3389/FPLS.2016.00403/FULL

Taiz L, Zeiger E, Møller IM, Murphy A (2015) Plant physiology and development, 6th edn. In: Sinauer Associates (vol 1)

Takeda N, Handa Y, Tsuzuki S, Kojima M, Sakakibara H, Kawaguchi M (2015) Gibberellins interfere with symbiosis signaling and gene expression and alter colonization by Arbuscular Mycorrhizal fungi in *Lotus Japonicus*. Plant Physiol. https://doi.org/10.1104/pp.114.247700

Takei K, Sakakibara H, Sugiyama T (2001) Identification of genes encoding adenylate Isopentenyl-transferase, a cytokinin biosynthesis enzyme, in *Arabidopsis thaliana*. J Biol Chem. https://doi.org/10.1074/jbc.M102130200

Tjamos SE, Flemetakis E, Paplomatas EJ, Katinakis P (2005) Induction of resistance to *Verticillium dahliae* in *Arabidopsis thaliana* by the biocontrol agent K-165 and pathogenesis-related proteins gene expression. Mol Plant Microbe Interact 18:555–561. https://doi.org/10.1094/MPMI-18-0555

Tsukanova KA, Chebotar V, Meyer JJM, Bibikova TN (2017) Effect of plant growth-promoting Rhizobacteria on plant hormone homeostasis. S Afr J Bot 113:91–102. https://doi.org/10.1016/J.SAJB.2017.07.007

Tully RE, Van Berkum P, Lovins KW, Keister DL (1998) Identification and sequencing of a cytochrome P450 gene cluster from *Bradyrhizobium japonicum*. Biochim Biophys Acta Gene Struct Expr. https://doi.org/10.1016/S0167-4781(98)00069-4

Ul Hassan T, Bano A (2019) Construction of IAA-Deficient Mutants of Pseudomonas moraviensis and their comparative effects with wild type strains as bio-inoculant on wheat in saline sodic soil. Geomicrobiol J 36(4). https://doi.org/10.1080/01490451.2018.1562498

Vacheron J, Desbrosses G, Bouffaud M-L, Touraine B, Moënne-Loccoz Y, Muller D, Legendre L, Wisniewski-Dyé F, Prigent-Combaret C (2013) Plant growth-promoting rhizobacteria and root system functioning. Front Plant Sci 4:356. https://doi.org/10.3389/fpls.2013.00356

Vejan P, Abdullah R, Khadiran T, Ismail S, Nasrulhaq Boyce A (2016) Role of plant growth promoting rhizobacteria in agricultural sustainability—a review. Molecules 21(5):573. https://doi.org/10.3390/molecules21050573

Wang JY, Cai Y, Gou JY, Mao YB, Xu YH, Jiang WH et al (2004) *Vd*NEP, an elicitor from *Verticillium dahliae*, induces cotton plant wilting. Appl Environ Microbiol 70:4989–4995. https://doi.org/10.1128/AEM.70.8.4989-4995.2004

Wang Z-Y, Bai M-Y, Oh E, Zhu J-Y (2012) Brassinosteroid signaling network and regulation of photomorphogenesis. Ann Rev Genet 46:701–724. https://doi.org/10.1146/annurev-genet-102209-163450

Wang J, Zhang Y, Li Y, Wang X, Nan W, Hu Y, Zhang H, Zhao C, Wang F, Li P, Shi H, Bi Y (2015) Endophytic microbes *Bacillus* sp. LZR216-regulated root development is dependent on polar auxin transport in *Arabidopsis* seedlings. Plant Cell Rep 34(6):1075–1087. https://doi.org/10.1007/S00299-015-1766-0

Waqas M, Khan A, Kamran M, Hamayun M, Kang S (2012) Endophytic fungi produce gibberellins and indoleacetic acid and promotes host-plant growth during stress. Molecules 17 10754–10773. https://doi.org/10.3390/molecules170910754

Xie H, Pasternak JJ, Glick BR (1996) Isolation and characterization of mutants of the plant growth-promoting rhizobacterium *Pseudomonas putida* GR12-2 that overproduce indoleacetic acid. Curr Microbiol. https://doi.org/10.1007/s002849900012

Xie DX, Feys BF, James S, Nieto-Rostro M, Turner JG (1998) COI1: an *Arabidopsis* gene required for jasmonate-regulated defense and fertility. Science 280:1091–1094. https://doi.org/10.1126/science.280.5366.1091

Xie B, Xu K, Hong XZ, San FC (2005) Isolation of transposon mutants from *Azospirillum brasilense* Yu62 and characterization of genes involved in indole-3-acetic acid biosynthesis. FEMS Microbiol Lett. https://doi.org/10.1016/j.femsle.2005.05.020

Yang S, Zhang Q, Guo J, Charkowski AO, Glick BR, Ibekwe AM, Cooksey DA, Yang CH (2007) Global effect of indole-3-acetic acid biosynthesis on multiple virulence factors of *Erwinia chrysanthemi* 3937. Appl Environ Microbiol. https://doi.org/10.1128/AEM.01770-06

Yang CL, Liang S, Wang HY, Han LB, Wang FX, Cheng HQ et al (2015) Cotton major latex protein 28 functions as a positive regulator of the *ethylene responsive factor 6* in defense against *Verticillium dahliae*. Mol Plant 8:399–411. https://doi.org/10.1016/j.molp.2014.11.023

Yang J, Ma Q, Zhang Y, Wang X, Zhang G, Ma Z (2016) Molecular cloning and functional analysis of *GbRVd*, a gene in *Gossypium barbadense* that plays an important role in conferring resistance to Verticillium wilt. Gene 575:687–694. https://doi.org/10.1016/j.gene.2015.09.046

Yao LL, Pei BL, Zhou Q, Li YZ (2012) NO serves as a signaling intermediate downstream of H_2O_2 to modulate dynamic microtubule cytoskeleton during responses to VD-toxins in *Arabidopsis*. Plant Signal Behav 7:174–177. https://doi.org/10.4161/psb.18768

Zhang H, Kim MS, Krishnamachari V, Payton P, Sun Y, Grimson M, Farag MA, Ryu CM, Allen R, Melo IS, Paré PW (2007) Rhizobacterial volatile emissions regulate auxin homeostasis and cell expansion in *Arabidopsis*. Planta 226(4):839–851. https://doi.org/10.1007/S00425-007-0530-2

Zheng X, Koopmann B, von Tiedemann A (2019) Role of salicylic acid and components of the phenylpropanoid pathway in basal and cultivar-related resistance of oilseed rape (*Brassica napus*) to *Verticillium longisporum*. Plants 8(11):491. https://doi.org/10.3390/plants8110491

Zhou BJ, Jia PS, Gao F, Guo HS (2012) Molecular characterization and functional analysis of a necrosis- and ethylene-inducing, protein-encoding gene family from *V. dahliae*. Mol Plant Microbe Interact 25:964–975. https://doi.org/10.1094/MPMI-12-11-0319

Zhou Y, Sun L, Wassan GM, He X, Shaban M, Zhang L et al (2019) *GbSOBIR1* confers *Verticillium* wilt resistance by phosphorylating the transcriptional factor *GbbHLH171* in *Gossypium barbadense*. Plant Biotechnol J 17:152–163. https://doi.org/10.1111/pbi.12954

Zhu S, Gao F, Cao X, Chen M, Ye G, Wei C et al (2005) The rice dwarf virus P2 protein interacts with ent-kaurene oxidases in vivo, leading to reduced biosynthesis of gibberellins and rice dwarf symptoms. Plant Physiol 139:1935–1945. https://doi.org/10.1104/pp.105.072306

Zupan JR, Zambryski P (1995) Transfer of T-DNA from agrobacterium to the plant cell. Plant Physiol.https://doi.org/10.1104/pp.107.4.1041

Iron Toxicity Tolerance in Rice: Roles of Auxins and Gibberellins

Olumide Samuel Daramola, Abraham Attah Shaibu, and Vimal Kumar Semwal

Abstract Rice is subjected to high iron (Fe) levels in acidic lowland soils which results in disturbance of basic metabolism, several changes in physiological processes and as a consequence reduction in productivity. In response to Fe toxicity in soils, rice like other plants produce a number of hormones (also known as phytohormones), including auxins, gibberellic acids and cytokinins. These hormones are organic substances that regulates plant growth and development, and play important role in rice defence against Fe toxicity. These hormones are part of signal-transduction pathway that stimulates reactions for Fe toxicity responses. The biosynthesis, transport, redistribution and conjugation of these plant hormones in rice has been shown not only to reduce high Fe inside rice plant tissue, but also to alleviate the adverse effect of Fe toxicity. In the present review, we discuss the conditions that enhances Fe toxicity in rice, effects of Fe toxicity in rice and tolerance strategies to Fe toxicity in rice. A special attention has been paid on the role and mechanism of phytohormones in enhancing tolerance and overcoming Fe toxicity-induced adverse effects.

1 Introduction

Rice is one of the most important staple food crops for more than 4 billon people worldwide (Jaggard et al. 2010). It is an indispensable crop for food security, providing significant amount of daily caloric intake, and major source of employment for billions of households in Asia, Africa and Americas (Ma et al. 2007). As the threat to food security continue to increase due to the increasing human population and climate change, rice is a highly strategic and priority crop that must increase in production and yield to feed the estimated 9.1 billion people by 2050 (Jaggard et al. 2010).

O. S. Daramola
West Florida Research and Education Center, University of Florida, Jay, FL, USA

A. A. Shaibu · V. K. Semwal (✉)
Africa Rice Center Regional Station, Oyo Road, P.M.B 5320, Ibadan, Nigeria
e-mail: v.semwal@cgiar.org

T. Aftab (ed.), *Auxins, Cytokinins and Gibberellins Signaling in Plants*,
Signaling and Communication in Plants,
https://doi.org/10.1007/978-3-031-05427-3_15

Although an important crop for food security worldwide, rice is currently facing several biotic and abiotic stresses which has limited its yield and productivity in all rice ecosystems. Among the different abiotic stresses, soil nutrient toxicities is one of the principal factors that have limited the growth and productivity of rice (Sikirou et al. 2015; Swamy et al. 2016; Sahebi et al. 2018; Melandri et al. 2021; Kirk et al. 2021). About 100 million hectares of the 128 million hectares cultivated to lowland rice worldwide is constrained by some sort of nutrient toxicity (Brady 1982; Becker and Asch 2005). Among the nutrient toxicities, iron toxicity is known as the most widely distributed nutritional disorder in rice production, and the most important constraint to rice production on acid soils (Neue et al. 1998; Dobermann and Fairhurst 2000; Kirk et al. 2021). Although there are several approaches and interventions including application of mineral nutrients and development or selection of tolerance cultivars, the role of phytohormones is considered an important factor in rice defense system under iron toxicity stress.

Plant hormones such as salicylic acid (SA) plays a role in rice defense; jasmonates (JAs), and ethylene (ET) reported to act as signals to trigger and mediate a diverse array of defense responses (Spoel and Dong 2012). Plants encounter various abiotic and biotic environmental stresses during the different growth stages of their life cycle and under stresses, plants have developed some sophisticated mechanisms to sense changes in environmental conditions and adapt their growth and development to adjust to different stress conditions (Dong-Lei et al. 2013). In the past, researchers found that growth-controlling hormones, such as auxin, gibberellic acids (GAs), brassinosteroids (BRs), and abscisic acid (ABA), are actively involved in plant immunity and thereby fine-tune immunity and growth and development in plants (Bari and Jones 2009; Grant and Jones 2009). It is known that a dozen mutants with constitutive activation of defense often reduce growth, whereas mutations in genes that function in growth and development often alter disease resistance. Therefore, activation of plant defense responses could generally utilize the growth and development primarily due to hormone pathways. This is often called defense primarily due to hormone pathways.

Current knowledge about the hormone-based defense signaling pathways and the interaction between the immunity and growth largely rely on scientific studies. For example, *Arabidopsis* as a monocotyledonous plant and rice one of the most important staple food crops for which the entire genome has been sequenced, are considered model plants to study general biological processes in cereal crops and other plants. In contrast to the extensive studies on phytohormones signaling mechanisms in *Arabidopsis*, relatively limited information is available on molecular mechanisms of immune responses and roles of hormones in rice, although several rice resistance genes have been cloned and functionally characterized. Researchers working on rice are increasingly making efforts to understand roles of phytohormones in acquisition of Fe-toxicity tolerance. In present review we focus on briefly describing the Fe-toxicity responses and its effects on rice productivity with emphasis on roles of major phytohormones in acquisition of Fe Toxicity tolerance in rice.

2 Soil and Environmental Conditions that Enhances Fe Toxicity in Rice

The occurrence of Fe toxicity in rice depends on the prevailing soil and environmental conditions. Although Fe toxicity in rice can occur in wide range of soil types including acid clay soils (Alaily et al. 1998), peat soils (Deturck 1994), sulfate soils (Tinh 1999), valley-bottom soils (Sahrawat and Diatta 1995), podzols, arisols, gleysols, ferralsols and fluvisols (Cherif et al. 2009; Yang et al. 2018), it is most common in acid sulfate soils and waterlogged soil conditions (Becker and Asch, 2005). Depending on the soil type, Fe content can range between 20,000 and 550,000 mg kg^{-1} (0.2 to 55%), with highest concentration at 2–15 cm soil depth (Audebert and Sahrawat 2000; Audebert and Fofana 2009). Also, about 1500 ppm Fe can be added to exposed soil sites through the interflow of Fe from the upland slopes, thereby increasing soil soluble Fe content (Yoshida 1981; Zahra et al. 2021). However, the concentration of soil soluble Fe that is accessible to, and can affect the rice plant ranges between 10 and >2000 mg L^{-1} depending on the soil and environmental conditions, (Benckiser et al. 1983; Singh et al. 2009).

Several soil conditions including conditions for reduction of Fe^{3+} to Fe^{2+} particularly in waterlogged paddy rice conditions (Prade et al. 1990), increased amount of extractable and exchangeable soil Fe^{2+} (Ponnamperuma 1972; Nugraha et al. 2016), reduced content of soil clay material (Das et al. 1997; Rajkumar et al. 1997; Sharma and Dubey 2004), low soil pH, low soil fertility, increased soil organic matter, soil aeration, temperature and redox buffer content (Ponnamperuma 1972; Onaga et al. 2013; Nugraha et al. 2016), reduced soil cation exchange capacity, the presence of stress factors (Becker and Asch 2005; Sahrawat 2010; Bashir et al. 2010), increased population and activities of soil fungi and microorganisms (Bonneville et al. 2004), and edaphic soil characteristics such as Fe:Mn, K:Fe and Fe:Zn ratio (Zancani et al. 2007; Abhilash et al. 2009) have been reported to enhance the severity of Fe toxicity in rice. High concentration (100–1000 mg L^{-1}) of soluble Fe^{2+} are found in acid soils, with up to 5000 Mg Kg^{-1} in acid sulfate soils (Ponnamperuma 1972; Harmsen and Van Breemen 1975). Fe is reported to be freely available in large amount (2000 Mg Kg^{-1}) at pH less than 5, causing Fe toxicity to the rice plant (Da Silveira et al. 2007; Onaga et al. 2013). Similarly, the activities of soil microorganisms such as the production of organic acids, regulation of Fe oxidation, mobilization of Fe oxides, and reduction process in the root zone influences Fe solubility and availability (Frei et al. 2016; Vejchasarn et al. 2016). Some facultative soil microbial populations including *Pseudomonas, Bacillus megaterium, B. pumilus , Geobacter, Clostridium,* and *Bacillus* sp., play a major role in the conversion of Fe^{3+} oxides in the soil thereby enhancing Fe toxicity (Ponnamperuma 1972; Bonneville et al. 2004).

Poor water management and other environmental factors such as poor drainage (Audebert and Fofana 2009) and increased industrial discharge (Deka and Sarma 2012) results in deterioration of soil properties which promotes the uptake of Fe^{2+} in soil, thereby enhancing Fe toxicity. Substantial amount of Fe is released into the environment through several anthropogenic activities including textile and steel industrial activities, sludge disposal in water treatment plants, laundry bluing, pigment manufacturing, and tanneries (Jayaweera et al. 2008; Xing and Zhang 2010).

3 Symptoms of Fe Toxicity in Rice

The typical visual symptom of Fe toxicity in rice is the copper coloration of the plant leaves regarded as 'leaf bronzing' and stunted overall vegetative growth (Fig. 1,

Fig. 1 Typical Fe toxicity effects on rice leaves stunted growth severe leaf bronzing at late vegetative stage in Suakoko, Liberia in 2020 wet season (**a**), irrecoverable damage and death of rice plants due to artificially induced Fe toxicity stress in a pot-based study in Ibadan, Nigeria in 2020 dry season (**b**) and severe chlorophyll loss and bronzing symptoms at reproductive stage at Edozighi, Nigeria in 2021 wet season (**c**)

Yamanouchi and Yoshida 1981; Bode et al. 1995). This symptom is usually first noticed as tiny brown spots spreading from the leaf tip to the base of older rice leaves with higher transpiration rates, and eventually becomes a rusty brown dry spot on the entire leaves as the severity of toxicity increases (Yamanouchi and Yoshida 1981; Fairhurst et al. 2007). Leaf bronzing is usually associated with other symptoms and different growth defect depending on the growth stage of the rice plant. Fe toxicity during the seedling stage results in stunted rice plant with reduced tillering (Fig. 1a) Abraham and Pandey 1989), while toxicity during the vegetative stage results in reduced plant height and dry matter accumulation (Abu et al. 1989). During the reproductive stage Fe toxicity in rice is usually associated with delayed flowering (Ayotade 1979), reduced number of panicle (Singh et al. 1992) and increased spikelet sterility (Virmani 1977). Although leaf bronzing is often used to indicate Fe toxicity stress level in rice (Lantin and Neue 1989; Bode et al. 1995), some studies, have however, reported significant reduction in rice yield due to Fe toxicity without the occurrence of leaf bronzing (Li et al. 2001; Sahrawat 2005; Sikirou et al. 2018; Onaga et al. 2013).

4 Effect of Fe Toxicity on Cellular Damage and Rice Yield

Fe toxicity is responsible for significant yield reduction in rice, and sometimes under severe condition can result in complete crop failure (Audebert and Sahrawat 2000). Between 10 and 90% reduction in rice yield was reported due to Fe toxicity in rice (Audebert and Fofana 2009; Cherif et al. 2009). Excess Fe in the soil causes oxidative burst which is toxic to the root of the rice plant, and can affect the uptake of other nutrients, such as phosphorus, zinc and copper (De Dorlodot et al. 2005). Hence, Fe toxicity- induced yield reduction in rice is often associated with poor soil nutrient status (Ottow et al. 1983). Absorption of excess Fe by the rice plant have also been reported to reduce the root and shoot length (Verma and Pandey 2017). Excessive uptake of Fe by the root of the rice plant, and it subsequent translocation to the shoot and leaves causes irreversible damage of different cellular components such as proteins, DNA, nucleic acids, and membrane lipids, which can induce cell death (Thongbai and Goodman 2000; Blokhina et al., 2003). Also, excess Fe^{2+} act as a catalyst in the Fenton reaction and generates ROS which is harmful to the rice plant and can result in cellular damage (Fang et al., 2001; Onaga et al. 2016). The free radicals generated due to Fe toxicity can also oxidize chlorophyll, thereby reducing the chlorophyll content (Monteiro and Winterbourn 1988). In addition, the oxidative burst occasioned by excess accumulation of Fe result in the disruption of the energization of the thylakoid membrane (Verma and Pandey 2017), thereby reducing the efficiency of photosystem II (Li et al. 2019). Combined effects of free radicals on cellular macro molecules and membranes leads to accelerated cell/ tissue and in some cases whole plant death leading to loss of biomass and grain yield in rice crop.

5 Adaptation and Tolerance Strategies to Fe Toxicity in Rice

Rice plant employs different strategies to cope with the constraint of Fe toxicity. These include physiological and morphological mechanisms to avoid, survive, and/or tolerate the adverse effect of excess Fe in the soil and in the plant (Tanaka et al. 1966; Kabayashi et al. 2014). For adaptation to Fe toxicity, rice plant use tolerance mechanisms including avoidance of excess Fe^{2+} by creating a physical barrier through enzymatic oxidation or the release of oxygen in the root, internal storage and distribution Fe^{2+} in shoot (Gross et al. 2003; Curie and Briat 2003), tolerance of excess Fe uptake involving antioxidants and free radicals through Fenton reaction (Li et al. 2019; Imam et al. 2017). Rice plant can also employ physiological root-based tolerance mechanism by retaining excess Fe in roots (Becker and Asch 2005). Oxidative stress under Fe toxicity condition can also be reduced in rice through biochemical and physiological processes including oxidation of chlorophyll; catalyzing hydroxyl radicals; releasing of root exudates, ROS and oxidant enzymes such as dehydroascorbate reductase, ascorbate peroxide, catalase and peroxidase (Muller et al. 2015; Audebert and Sahrawat 2000; Wu et al. 2014; Drame et al. 2010). Adaptation to excess Fe in rice can also be through molecular mechanism via different genes and transporter such as *OsA1 to OsA10*, *OsZIP1* to *OsZIP10* to control uptake, absorption, transportation, and translocation of Fe (Rout et al. 2015; Dos Santos et al. 2017; Kim and Guerinot 2007; Kar et al. 2021). Hence, physiological, and biochemical processes exist that can mitigate stress effects and confer Fe toxicity tolerance in rice.

6 Management Strategies of Fe Toxicity in Rice

The influx of Fe^{2+} and mobility of Fe in the rhizosphere can be reduced by management strategies at the landscape level of the lowland rice field or through cultivar selection. The negative effects of Fe toxicity in rice leaf tissue can also be reduced by management strategies that employs the mineral nutrient application and adaptive germplasm selection (Becker and Asch 2005). Fe toxicity caused by the influx of reduced Fe from adjacent slopes can be managed at the landscape level by engaging strategies that reduces the amount and the solute charge of interflow water, which can be achieved by planting deep-rooting upland vegetation rather than bare fallow (Bognonkpe and Becker 2003; Van de Giesen et al. 2005). In West Africa for example, the cultivation of banana in the hydromorphic fringe of a valley was reported to efficiently intercept the inflow of water and the nutrients it contained (Bognonkpe and Becker 2000). The construction of irrigation and drainage canals on the valley fringe has also been reported as an efficient way of intercepting interflow. Regardless of the origin of Fe toxicity, several crop and soil–water management strategies including measures that prevent rapid drop of pH in both soil and rhizosphere (Ottow et al. 1983; Becker and Asch 2005), re-oxidation or removal of Fe in both soil and rhizosphere

(Beyrouty et al., 1994; Baggie and Bah 2001; Becker and Asch 2005), application of mineral nutrients that strengthens the rice plant (Mitra et al. 1993; Sahrawat 2000), and avoidance of excessive influx of Fe into the root of the rice plant (Audebert and Saharawat 2000; Becker and Asch 2005) can effectively prevent the build of Fe^{2+} and minimizes the adverse effect of Fe toxicity in rice (Becker and Asch 2005). At the plant level, the use of tolerant rice cultivars is one of the most common strategies adopted to address the problem of Fe toxicity. Several rice cultivars with different degrees of adaptation to Fe toxicity have been developed by breeders (WARDA 1993; Wissuwa 2005; Wan et al. 2005).

7 Hormonal Responses to Fe Toxicity

Phytohormones play critical roles in abiotic stress tolerance in crops and other plants including Fe toxicity tolerance in rice. The major roles of phytohormones Fe toxicity tolerance in rice and other crops are summarized in sections below and Fig. 2 and Table1.

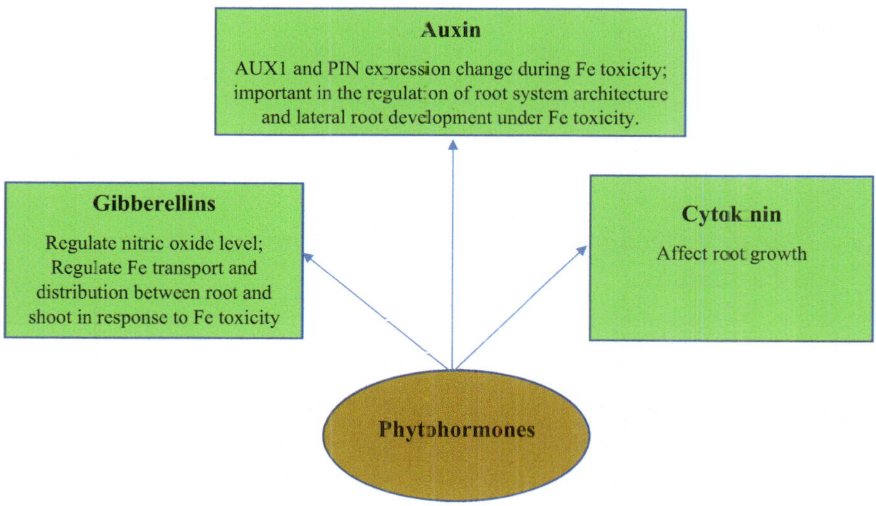

Fig. 2 Summary of the roles of phytohormones in iron toxicity tolerance in rice and other crop plants

Table 1 Phytohormones involved in Fe toxicity tolerance in crops and other plants and summary of their mechanisms in improving Fe toxicity tolerance

Phytohormones	Plants	Adaptive mechanism/reaction	References
Auxin			
	Wheat	Regulation of phytosiderophore which limits iron acquisition in root via auxin signalling derived from shoot with expression of *TaSAMS* and *TADMASI* genes	Kabir et al. (2016)
	Arabidopsis	AUX1 function in ethylene-mediated tolerance of lateral root to Fe toxicity	Li et al. (2015)
	Arabidopsis	Auxin NAA promoted lateral root initiation and alleviated inhibition of lateral root formation due to Fe toxicity condition	Li et al. (2015)
	Arabidopsis	Auxin IAA promoted the frequency of lateral root initiation under Fe toxicity condition	Li et al. (2015)
Gibberellins			
	Rice	Regulation of Fe transport and translocation between roots and shoot of seedlings through the inhibition of *OSYSL2* gene expression	Wang et al. (2017)
	Rice	*Bacillus pumilus*-produced gibberellin affected iron partitioning in lowland rice by increasing iron concentration in the roots and decreasing iron concentration in the shoot	Torries (2009)

7.1 Role and Mechanism of Auxin in Enhancing Tolerance to Fe Toxicity

Auxin is one of the most important plant growth hormones because it is required for the induction of cell division and growth of plant tissue (Asgher et al. 2015). It is regarded as one of the major regulators of plant development, with essential roles in plant developmental processes like cell elongation, leaf expansion, formation and growth of plant root, flower and fruit development, formation of auxiliary bud, abscission, apical dominance, phototropism and differentiation of vascular tissue (Taiz and Zeiger 2002; Zhao et al. 2010). Research studies (Tyburski et al. 2009; Krishnamurthy and Rathinasabapathi 2013) has shown that auxin signaling and transport plays significant role in plant tolerance to abiotic stresses.

The role of auxin under conditions of iron toxicity is one of the most important auxin stress signaling functions described (Fukaki and Tasaka 2009; Peret et al. 2009). Studies (Kobayashi and Nishizowa 2012; Gayomba et al. 2015) have reported the involvement of Auxin in the complex signaling cascades that regulates plant response to iron toxicity (Fig. 2, Table 1). The study of Kabir et al. (2016) showed that auxin group IBA and IAA significantly increased in the root of wheat plant in response to iron stress. Auxin modulate gene expression in response to iron availability and function as a positive regulator of iron acquisition gene (Fukaki and Tasaka 2009). Auxin signaling is important in the regulation of root system architecture and lateral root development under iron toxicity (Casimiro et al. 2003; Peret et al. 2009). Auxin formed the basic machinery for basipetal auxin transport in *Arabidopsis* which is critical for lateral root initiation (De smet et al. 2007). This was made possible by the reduction in expression of PIN2 protein in root tips with consequence arrest of lateral root initiation near the growing tip of the primary root in early response to excess iron. These modulation in hormone homeostasis help root system architecture to adjust rapidly to resist excessive iron absorption and avoid serious iron toxicity (Li et al. 2015; De smet et al. 2007). The promotion of lateral root development and alleviation of excess iron-mediated inhibition of lateral root formation in *Arabidopsis* was linked to the increase of auxin in the root tip (Casimiro et al. 2003). Auxin resistant 1 (AUX1)- an auxin influx carrier regulates lateral root initiation under the condition of iron toxicity stress. AUX1 was reported to function in ethylene-mediated tolerance of lateral root to iron toxicity in *Arabidopsis* (Li et al. 2015). According to the study of Lewis et al. (2011), ethylene-mediated lateral root formation is dependent on auxin pathway Auxin signaling was also reported to drive iron toxicity tolerance in wheat (Kabir et al. 2016). AUX1 expression and accumulation in the lateral root apex of *Arabidopsis* stimulate lateral root elongation for adaptation to iron toxicity condition (Giehl et al. 2012). Auxin effect is possibly through ethylene because ethylene production is promoted in the presence of high level auxin via the synthesis of ACC synthase (Kim et al. 1992).

7.2 Role and Mechanism of Gibberellic Acids in Enhancing Tolerance to Fe Toxicity in Rice

Gibberellins are carboxylic acids component synthesized from acetyl coenzyme A. Gibberellins regulate various plant growth and developmental processes including flowering, germination, dormancy, expansion of leaves, cell elongation, chlorophyll biosynthesis and fruit senescence (Jiang et al. 2007). Gibberellins is also responsible for the regulation of the activities of nitrogen assimilation enzymes and metal transport and translocation (Wang et al. 2017). Few studies has reported the role of gibberellins in the regulation of plant responses to Iron toxicity (Table 1; Fu and Harberd 2003; Gayomba et al. 2015). The study of Guo et al. (2015) showed that the application of exogenous gibberellic acids decreased iron plauque. *Bacillus pumilus*-produced gibberellin affected iron partitioning in lowland rice by increasing iron concentration in the roots and decreasing iron concentration in the shoot. Similarly, in a recent study, exogenous application of gibberellic acid decreased translocation of iron to shoot through the inhibition of *OSYSL2* gene expression involved in iron transport in rice, this was made possible through the amendment of iron homeostasis by negative regulation of iron translocation from rice root to the shoot (Wang et al. 2017).

8 Conclusion and Prospects

Fe toxicity is a major constrains to lowland rice production globally. It is one of the major challenges for sustainable rice production particularly in highly weathered soils in inland valley of Sub Saharan Africa. Large genetic variation exists in available rice germplasm for Fe toxicity tolerance. Several QTLs related to tolerance to iron toxicity stress are known for both vegetative and reproductive growth stages of rice. Some of these QTL co-localize with previously reported QTL that were mapped under more chronic iron stress, suggesting that they were associated with 'universal' defense mechanisms. However, most QTLs had rather small effects and were distributed throughout the genome, confirming the complexity of the genetics behind adaptation to varying iron toxic conditions. Further, QTLs were associated with either exclusion or inclusion mechanisms of iron tolerance. As demonstrated in few scientific literatures cited, iron exclusion via oxidation at the root surface is an important adaptive trait under iron toxicity stress. The trait appears to be favored by root architecture and can be genetically dissected within the IR29/Pokkali mapping population reported. Pyramiding this trait with further shoot based adaptive traits may be effective in the breeding for iron toxicity tolerance. The progress in conventional rice breeding under abiotic stresses coupled with the characterization of QTL mapping populations and their subsequent use in marker-assisted breeding is seen to accelerate the process of developing appropriate germplasm that will provide the quantitative traits linked with the genetic loci are related to well-understood and clearly described adaptation mechanisms (exclusion, avoidance, tolerance). The understanding of improved crop adaptation for a range of iron stress situations needs to be translated into repeatable and robust tools for the screening of improved rice cultivars. Phytohormones like auxins, gibberellins and cytokinin play critical roles in Fe toxicity tolerance and downstream signal transduction pathways regulating expression of favorable genes contributing to stress avoidance and tolerance. More studies are needed to elucidate protective roles of these phytohormones in acquisition of Fe toxicity tolerance in rice. Integrated approaches by using highly tolerant varieties supported by best agronomic approaches described above at farmers field could mitigate stress effects and contribute towards sustainable lowland rice production in Fe toxic soils.

References

Abhilash P, Pandey VC, Srivastava P, Rakesh P, Chandran S, Singh N (2009) Phytofiltration of cadmium from water by *Limnocharis flava* (L.) Buchenau grown in free-floating culture system. J Hazardous Mater 170(2–3):791–797

Abraham MJ, Pandey DK (1989) Performance of selected varieties and advanced generation genotypes in rainfed lowland iron toxic soil. Int Rice Res Newsl 14:21–21

Abu MB, Tucker ES, Harding SS, Sesay JS (1989) Cultural practices to reduce iron toxicity in rice. Int Rice Res Newsl 14:19–19

Alaily F, Sib J, Madrid K, Soubeiga Z, Roth C (1998) Paddy soils from basaltic colluvium in the Highland of Madagascar. Z Pflanzenernahr Bodenk 161:633–638

Asgher M, Khan MIR, Anjum NA, Khan NA (2015) Minimising toxicity of cadmium in plants—role of plant growth regulators. Protoplasma 252(2):399–413

Audebert A, Fofana M (2009) Rice yield gap due to iron toxicity in West Africa. J Agron Crop Sci 195(1):66–76

Audebert A, Sahrawat KL (2000) Mechanisms for iron toxicity tolerance in lowland rice. J Plant Nutr 23(11–12):1877–1885

Ayotade KA (1979) Reaction of some rice varieties to iron toxicity in flooded strongly acid ferralitic soil in Nigeria. WARDA (West Africa Rice Dev Assoc) Techn Newsl 1(1)

Baggie I, Bah AR (2001) Low-cost management of iron toxicity in farmers' fields in Sierra Leone. Int Rice Res Notes 26(1):1–1

Bari R, Jones JD (2009) Role of plant hormones in plant defense responses. Plant Mol Biol 69(4):473–488

Bashir K, Ishimaru Y, Nishizawa NK (2010) Iron uptake and loading into rice grains. Rice 3(2):122–130

Becker M, Asch F (2005) Iron toxicity in rice: conditions and management concepts. J Plant Nutr Soil Sci 168(4):558–573

Benckiser G, Ottow JCG, Santiago S, Watanabe I (1983) Iron toxicity—effect of phosphorus, potassium, calcium and magnesium fertilization on rhizoflora, redox potential and iron uptake by different rice varieties (*Oryza sativa* L.). Landw Forschung 36:285–299

Beyrouty CA, Grigg BC, Norman RJ, Wells BR (1994) Nutrient uptake by rice in response to water management. J Plant Nutr 17:39–55

Blokhina O, Virolainen E, Fagerstedt KV (2003) Antioxidants, oxidative damage and oxygen deprivation stress: a review. Ann Bot 91(2):179–194

Bode K, Döring O, Lüthje S, Neue HU, Böttger M (1995) The role of active oxygen in iron tolerance of rice (*Oryza sauva* L.). Protoplasma 184(1):249–255

Bognonkpe JP, Becker M (2003) Effect of land use on soil N dynamics at watershed scale in West Africa. University of Goettingen and ATSAF: Technological and Institutional Innovations for Sustainable Rural Development. Klartext GmbH, Göttingen, Germany, p 334

Bognonkpe JPI, Becker M (2000) Native soil N dynamics and use efficiency by lowland rice as a function of slope management. Deutscher Tropentag. Müllerbader Press. University of Hohenheim

Bonneville S, Van Cappellen P, Behrends T (2004) Microbial reduction of iron (III) oxyhydroxides: effects of mineral solubility and availability. Chem Geol 212(3–4):255–268

Brady NC (1982) Potential of increasing production and cropping intensity of rain-fed wetland rice fields in Asia. In: Report of a workshop on cropping systems research in Asia. International Rice Research Institute, pp 11–14

Casimiro I, Beeckman T, Graham N, Bhalerao R, Zhang H, Casero P, Bennett MJ (2003) Dissecting Arabidopsis lateral root development. Trends Plant Sci 8(4):165–171

Chérif M, Audebert A, Fofana M, Zouzou M (2009) Evaluation of iron toxicity on lowland irrigated rice in West Africa

Curie C, Briat JF (2003) Iron transport and signaling in plants. Annu Rev Plant Biol 54(1):183–206

Da Silveira VC, Fadanelli C, Sperotto RA, Stein RJ, Basso LA, Santos DS, Vaz IDS, Dias JF, Fett JP (2007) Role of ferritin in the rice tolerance to iron overload. Scientia Agricola 66(4):549–555
Das P, Samantaray S, Rout GR (1997) Studies on cadmium toxicity in plants: a review. Environ Pollut 98(1):29–36
De Dorlodot S, Lutts S, Bertin P (2005) Effects of ferrous iron toxicity on the growth and mineral composition of an interspecific rice. J Plant Nutr 28(1):1–20
Deka J, Sarma HP (2012) Heavy metal contamination in soil in an industrial zone and its relation with some soil properties. Arch Appl Sci Res 4(2):831–836
Deturck P (1994) Iron toxicity to rainfed lowland rice in Sri Lanka. Dissertationes de Agricultura (Belgium). no. 259
Dobermann A, Fairhurst T (2000) Rice: nutrient disorders and nutrient management. Potash & Phosphate Institute, Potash & Phosphate Institute of Canada, and International. Rice Research Institute, Singapore and Los Baños, p 191
Dos Santos RS, de Araujo Júnior AT, Pegoraro C, de Oliveira AC (2017) Dealing with iron metabolism in rice: from breeding for stress tolerance to biofortification. Genetic Mol Biol 40:312–325
Dramé KN, Saito K, Koné B, Chabi A, Dakouo D, Ebenezer Annan-Afful E, Monh S, Abo E, Sié M (2010) Coping with iron toxicity in the lowlands of sub-Saharan Africa: experience from Africa Rice Center. In: AfricaRice (ed) Innovation and partnerships to realize Africa's rice potential. Proceedings of the 2nd Africa Rice Congress, Bamako, pp 1.9.1–1.9.8
Fairhurst T, Witt C, Buresh R, Dobermann A, Fairhurst T (2007) Rice: a practical guide to nutrient management. International Rice Research Institute
Fang S, Yu J, Feng J, Han C, Li D, Liu Y (2001) Identification of rice black-streaked dwarf fijivirus in maize with rough dwarf disease in China. Adv Virol 146(1):167–170
Frei M, Tetteh RN, Razafindrazaka AL, Fuh MA, Wu LB, Becker M (2016) Responses of rice to chronic and acute iron toxicity: genotypic differences and biofortification aspects. Plant Soil 408(1):149–161
Fu X, Harberd NP (2003) Auxin promotes Arabidopsis root growth by modulating gibberellin response. Nature 421(6924):740–743
Fukaki H, Tasaka M (2009) Hormone interactions during lateral root formation. Plant Mol Biol 69(4):437–449
Gayomba SR, Zhai Z, Jung HI, Vatamaniuk OK (2015) Local and systemic signaling of iron status and its interactions with homeostasis of other essential elements. Front Plant Sci 6:716
Giehl RF, Lima JE, von Wirén N (2012) Localized iron supply triggers lateral root elongation in Arabidopsis by altering the AUX1-mediated auxin distribution. Plant Cell 24(1):33–49
van de Giesen N, Stomph TJ, Ridder N (2005) Surface runoff scale effects in West African watersheds: modeling and management options. Agric Water Manage 72(2):109–130
Grant MR, Jones JD (2009) Hormone (dis) harmony moulds plant health and disease. Science 324(5928):750–752
Gross J, Stein RJ, Fett-Neto AG, Fett JP (2003) Iron homeostasis related genes in rice. Genet Mol Biol 26(4):477–497
Guo Y, Zhu C, Gan L, Ng D, Xia K (2015) Effects of exogenous gibberellic acid3 on iron and manganese plaque amounts and iron and manganese uptake in rice. PLoS ONE 10(2):e0118177
Harmsen K, Van Breemen N (1975) Translocation of iron in acid sulfate soils: II. Production and diffusion of dissolved ferrous iron. Soil Sci Soc Am J 39(6):1148–1153
Imam MU, Zhang S, Ma J, Wang H, Wang F (2017) Antioxidants mediate both iron homeostasis and oxidative stress. Nutrients 9(7):671
Jaggard KW, Qi A, Ober ES (2010) Possible changes to arable crop yields by 2050. Philos Trans Royal Biol Sci 365:2835–2851
Jayaweera MW, Kasturiarachchi JC, Kularatne RK, Wijeyekoon SL (2008) Contribution of water hyacinth (Eichhornia crassipes (Mart.) Solms) grown under different nutrient conditions to Fe-removal mechanisms in constructed wetlands. J Environ Manage 87(3):450–460

Jiang C, Gao X, Liao L, Harberd NP, Fu X (2007) Phosphate starvation root architecture and anthocyanin accumulation responses are modulated by the gibberellin-DELLA signaling pathway in Arabidopsis. Plant Physiol 145(4):1460–1470
Kabir AH, Begum MC, Haque A, Amin R, Swaraz AM, Haider SA, Hossain MM (2016) Genetic variation in Fe toxicity tolerance is associated with the regulation of translocation and chelation of iron along with antioxidant defence in shoots of rice. Funct Plant Biol 43(11):1070–1081
Kar S, Mai H-J, Khalouf H, Abdallah HB, Flachbart S, Fink-Straube C, Bräutigam A et al (2021) Comparative transcriptomics of lowland rice varieties uncovers novel candidate genes for adaptive iron excess tolerance. Plant Cell Physiol 62(4):624–640
Kim SA, Guerinot ML (2007) Mining iron: iron uptake and transport in plants. FEBS Lett 581(12):2273–2280
Kim WT, Silverstone A, Yip WK, Dong JG, Yang SF (1992) Induction of 1-aminocyclopropane-1-carboxylate synthase mRNA by auxin in mung bean hypocotyls and cultured apple shoots. Plant Physiol 98(2):465–471
Kirk GJ, Manwaring HR, Ueda Y, Semwal VK, Wissuwa M (2021) Below-ground plant–soil interactions affecting adaptations of rice to iron toxicity. Plant Cell Environ
Kobayashi T, Nishizawa NK (2012) Iron uptake, translocation, and regulation in higher plants. Annu Rev Plant Biol 63:131–152
Kobayashi T, Nakanishi Itai R, Nishizawa NK (2014) Iron deficiency responses in rice roots. Rice 7(1):1–11
Krishnamurthy A, Rathinasabapathi B (2013) Auxin and its transport play a role in plant tolerance to arsenite-induced oxidative stress in Arabidopsis thaliana. Plant, Cell Environ 36(10):1838–1849
Lantin RS, Neue HU (1989) Iron toxicity: a nutritional disorder in wetland rice. In: Irrigated Rice Meeting, Brazil, 26-30 Sept 1989
Lewis DR, Negi S, Sukumar P, Muday GK (2011) Ethylene inhibits lateral root development, increases IAA transport and expression of PIN3 and PIN7 auxin efflux carriers. Development 138(16):3485–3495
Li H, Yang X, Luo A (2001) Ameliorating effect of potassium on iron toxicity in hybrid rice. J Plant Nutr 24(12):1849–1860
Li H, Li Y, Xiang L, Huang Q, Qiu J, Zhang H, Valange S (2015) Heterogeneous photo-Fenton decolorization of Orange II over Al-pillared Fe-smectite: response surface approach, degradation pathway, and toxicity evaluation. J Hazard Mater 287:32–41
Li L, Ye L, Kong Q, Shou H (2019) A vacuolar membrane ferric-chelate reductase, OsFRO1, alleviates Fe toxicity in rice (*Oryza sativa* L.). Front Plant Sci 10:700
Ma JF, Yamaji N, Mitani N, Tamai K, Konishi S, Fujiwara T (2007) An efflux transporter of silicon in rice. Nature 448:209–212
Melandri G, Sikirou M, Arbelaez JD, Shittu A, Semwal VK, Konaté KA, Maji AT, Ngaujah SA, Akintayo I, Govindaraj V, Shi X, Agosto-Peréz FJ, Greenberg AJ, Atlin G, Venuprasad R, McCouch SR (2021) Multiple small-effect alleles of Indica origin enhance high iron-associated stress tolerance in rice under field conditions in West Africa. Front Plant Sci 11:604938. https://doi.org/10.3389/fpls.2020.604938
Mitra GN, Sahu SK, Ali MH (1993) Effect of potassium application on alleviating iron toxicity and nutrient uptake by rice in lateritic soils. In: Proceedings of the workshop on micronutrients, Bhubaneswar, India, pp 269–272
Monteiro HP, Winterbourn CC (1988) The superoxidedependent transfer of iron from ferritin to transferrin. Biochemistry Journal 256:923–928
Müller C, Kuki KN, Pinheiro DT, de Souza LR, Siqueira Silva AI, Loureiro ME, Almeida AM (2015) Differential physiological responses in rice upon exposure to excess distinct iron forms. Plant Soil 391(1):123–138
Neue HU, Quijano C, Senadhira D, Setter T (1998) Strategies for dealing with micronutrient disorders and salinity in lowland rice systems. Field Crop Res 56(1–2):139–155
Nugraha AA, Liutkus A, Vincent E (2016) Multichannel audio source separation with deep neural networks. IEEE/ACM Trans Audio Speech Lang Process 24(9):1652–1664

Onaga G, Dramé KN, Ismail AM (2016) Understanding the regulation of iron nutrition: can it contribute to improving iron toxicity tolerance in rice? Funct Plant Biol 43(8):709–726

Onaga G, Egdane J, Edema R, Abdelbagi I (2013) Morphological and genetic diversity analysis of rice accessions (*Oryza sativa* L.) differing in iron toxicity tolerance. J Crop Sci Biotechnol 16(1):53–62

Ottow JCG, Benckiser G, Watanabe I, Santiago S (1983) Multiple nutritional soil stress as the prerequisite for iron toxicity of wetland rice (*Oryza sativa* L.). Trop Agric 60(2)

Péret B, De Rybel B, Casimiro I, Benková E, Swarup R, Laplaze L, Bennett MJ (2009) Arabidopsis lateral root development: an emerging story. Trends Plant Sci 14(7):399–408

Ponnamperuma FN (1972) The chemistry of submerged soils. In: Advances in agronomy, vol 24. Academic Press, pp 29–96

Prade K, Ottow JCG, Jacq VA, Malouf G, Loyer JY (1990) Relationships between the properties of flooded rice soils and iron toxicity in Lower Casamance (Senegal). Studies, review and summary of previous work. Cahiers ORSTOM, Serie Pedologie, IRD, Montpelliers, France, pp 453–474

Rajkumar GR, Patil CV, Prakash SS, Yeledhalli NA (1997) Distribution of forms of iron in relation to the properties of rice soils. J Maharashtra Agric Univ 22:267–270

Rout GR, Sahoo S (2015) Role of iron in plant growth and metabolism. Rev Agric Sci 3:1–24

Sahebi M, Hanafi MM, Rafii MY, Mahmud TMM, Azizi P, Osman M, Atabaki N (2018) Improvement of drought tolerance in rice (*Oryza sativa* L.): genetics, genomic tools, and the WRKY gene family. BioMed Research International

Sahrawat KL (2000) Elemental composition of the rice plant as affected by iron toxicity under field conditions. Commun Soil Sci Plant Anal 31:2819–2827

Sahrawat KÁ (2005) Iron toxicity in wetland rice and the role of other nutrients. J Plant Nutr 27(8):1471–1504

Sahrawat KL (2010) Reducing iron toxicity in lowland rice with tolerant genotypes and plant nutrition. Plant Stress 4(2):70–75

Sahrawat KL, Jones MP, Diatta S (1995) Response of upland rice to phosphorus in an Ultisol in the humid forest zone of West Africa. Fertilizer Res 41(1):11–17

Sharma P, Dubey RS (2004) Ascorbate peroxidase from rice seedlings: properties of enzyme isoforms, effects of stresses and protective roles of osmolytes. Plant Sci 167:541–550

Sikirou M, Saito K, Achigan-Dako EG, Dramé KN, Ahanchédé A, Venuprasad R, Adam A (2015) Genetic improvement of iron toxicity tolerance in rice: progress, challenges and prospects in West Africa. Plant Prod Sci 18:423–434

Sikirou M, Shittu A, Konaté KA, Maji AT, Ngaujah AS, Sanni KA, Venuprasad R (2018) Screening African rice (*Oryza glaberrima*) for tolerance to abiotic stresses: I Fe Toxicity. Field Crops Res 220:3–9

Singh BP, Das M, Prasad RN, Ram M (1992) Characteristics of Fe-toxic soils and affected plants and their correction in acid Haplaquents of Meghalaya. Int Rice Res Newsl 17:18–19

Singh HP, Kaur S, Batish DR, Sharma VP, Sharma N, Kohli RK (2009) Nitric oxide alleviates arsenic toxicity by reducing oxidative damage in the roots of *Oryza sativa* (rice). Nitric Oxide 20(4):289–297

de Smet I, Tetsumura T, De Rybel B, Frey NFD, Laplaze L, Casimiro I, Beeckman T (2007) Auxin-dependent regulation of lateral root positioning in the basal meristem of Arabidopsis

Spoel SH, Dong X (2012) How do plants achieve immunity? Defence without specialized immune cells. Nat Rev Immunol 12(2):89–100

Swamy BPM, Rahman MA, Inabangan-Asilo MA, Amparado A, Manito C, Chadha-Mohanty P, Reinke R, Slamet-Loedin IH (2016) Advances in breeding for high grain zinc in rice. Rice 9:49

Taiz L, Zeiger E (2002) Photosynthesis: physiological and ecological considerations. Plant Physiol 9:172–174

Tanaka A, Loe R, Navasero SA (1966) Some mechanisms involved in the development of iron toxicity symptoms in the rice plant. Soil Sci Plant Nutr 12(4):32–38

Thongbai P, Goodman BA (2000) Free radical generation and post-anoxic injury in rice grown in an iron-toxic soil. J Plant Nutr 23(11–12):1887–1900

Tinh TK (1999) Reduction chemistry of acid sulphate soils: reduction rates and influence of rice cropping

Tyburski J, Dunajska K, Mazurek P, Piotrowska B, Tretyn A (2009) Exogenous auxin regulates H_2O_2 metabolism in roots of tomato (*Lycopersicon esculentum* Mill.) seedlings affecting the expression and activity of CuZn-superoxide dismutase, catalase, and peroxidase. Acta Physiologiae Plantarum 31(2):249–260

Vejchasarn P, Lynch JP, Brown KM (2016) Genetic variability in phosphorus responses of rice root phenotypes. Rice 9(1):1–16

Verma L, Pandey N (2017) The effect of iron toxicity on seed germination and early seedling growth of green gram (*Vigna radiata* L. Wilczek). Int 6(8):1427–1430

Virmani SS (1977) Varietal tolerance of rice to iron toxicity in Liberia. Int Rice Res Newsl

Wan JL, Zhai HQ, Wan JM (2005) Mapping of QTLs for ferrous iron toxicity tolerance in rice (*Oryza sativa* L.). Yi Chuan Xue Bao 32:1156–1166

Wang Y, Zhao J, Lu W, Deng D (2017) Gibberellin in plant height control: old player, new story. Plant Cell Rep 36(3):391–398

WARDA (1993) Development of lands. Training in rice production. Instructor's manual. West Africa Rice Develop. Assoc, Hong Kong, pp 29–46

Wissuwa M (2005) Mapping nutritional traits in crop plants. In: Broadley M, White P (eds) Plant nutritional genomics. Blackwell Publishing, UK, in press

Wu LB, Shhadi MY, Gregorio G, Matthus E, Becker M, Frei M (2014) Genetic and physiological analysis of tolerance to acute iron toxicity in rice. Rice 7(1):1–12

Xing Y, Zhang Q (2010) Genetic and molecular bases of rice yield. Annu Rev Plant Biol 61:421–442

Yamanouchi M, Tominaga Y, Fujiyama H, Nagai T (1989) Iron oxidizing mechanisms in nutrient solution during cultivation of rice plant. J Faculty Agric Tottori Univ 25:9–14

Yang DL, Yang Y, He Z (2013) Roles of plant hormones and their interplay in rice immunity. Mol Plant 6(3):675–685

Yang YP, Zhang HM, Yuan HY, Duan GL, Jin DC, Zhao FJ, Zhu YG (2018) Microbe mediated arsenic release from iron minerals and arsenic methylation in rhizosphere controls arsenic fate in soil-rice system after straw incorporation. Environ Pollut 236:598–608

Yoshida S (1981) Fundamentals of rice crop science. International Rice Research Institute

Zahra N, Hafeez MB, Shaukat K, Wahid A, Hasanuzzaman M (2021) Fe toxicity in plants: impacts and remediation. Physiol Plant 173(1):201–222

Zancani M, Peresson C, Patui S, Tubaro F, Vianello A, Macrì F (2007) Mitochondrial ferritin distribution among plant organs and its involvement in ascorbate-mediated iron uptake and release. Plant Sci 173(2):182–189

Zhao K, Liu X, Xu J, Selim HM (2010) Heavy metal contaminations in a soil–rice system: identification of spatial dependence in relation to soil properties of paddy fields. J Hazard Mater 181(1–3):778–787

New Auxin and Cytokinin Related Compounds Based on Synthetic Low Molecular Weight Heterocycles

V. A. Tsygankova, Ya. V. Andrusevich, O. I. Shtompel, R. M. Solomyanny, A. O. Hurenko, M. S. Frasinyuk, G. P. Mrug, O. V. Shablykin, S. G. Pilyo, A. M. Kornienko, and V. S. Brovarets

Abstract Our chapter is devoted to the biological screening of new effective substitutes of plant hormones auxins and cytokinins among synthetic low molecular weight heterocyclic compounds, derivatives of pyridine, pyrimidines, pyrazolotriazinones, oxazoles, oxazolopyrimidines and isoflavonoids. The auxin-like and cytokinin-like activity of chemical low molecular weight heterocyclic compounds, derivatives of pyridine, pyrimidine, pyrazole, isoflavones, oxazolopyrimidine and oxazole was studied. A specific bioassay on auxin-like activity showed a high stimulating effect of the chemical heterocyclic compounds, derivatives of pyridine, pyrimidine, pyrazole and isoflavones used at the concentration 10^{-8} M on the formation of adventitious roots on the 14th-day-old leaf petioles isolated from seedlings of haricot bean (*Phaseolus vulgaris* L.) cultivar Belozernaya, which was similar or higher of the effect of plant hormones auxins IAA and NAA used at the same concentration 10^{-8} M. A specific bioassay on cytokinin-like activity showed a high stimulating effect of the chemical heterocyclic compounds, derivatives of pyridine, pyrimidine, pyrazole, isoflavones, oxazolopyrimidine and oxazole used at the concentrations 10^{-8} M and 10^{-9} M on the growth of biomass of 16th-day-old cotyledons isolated from seeds of muscat pumpkin (*Cucurbita moschata* Duch. et Poir.) cultivar Gilea, which was similar or higher of the effect of plant hormone cytokinin Kinetin used at the same concentrations 10^{-8} M and 10^{-9} M. The results obtained confirmed the inducing auxin-like and cytokinin-like effect of chemical low molecular weight heterocyclic compounds, derivatives of pyridine, pyrimidine, pyrazole, isoflavones, oxazolopyrimidine and oxazole on plant cell elongation, division, and differentiation that are the basic processes of plant growth. The practical application of derivatives of pyrimidine, pyrazole, isoflavones, pyridine, oxazolopyrimidine and oxazole as new plant growth regulators was proposed.

V. A. Tsygankova (✉) · Ya. V. Andrusevich · O. I. Shtompel · R. M. Solomyanny · A. O. Hurenko · M. S. Frasinyuk · G. P. Mrug · O. V. Shablykin · S. G. Pilyo · A. M. Kornienko · V. S. Brovarets
Department for Chemistry of Bioactive Nitrogen-Containing Heterocyclic Compounds, V.P. Kukhar Institute of Bioorganic Chemistry and Petrochemistry, National Academy of Sciences of Ukraine, 1, Murmanskaya str, Kyiv 02660, Ukraine
e-mail: vTsygankova@ukr.net

T. Aftab (ed.), *Auxins, Cytokinins and Gibberellins Signaling in Plants*, Signaling and Communication in Plants,
https://doi.org/10.1007/978-3-031-05427-3_16

1 Introduction

The successful development of modern agriculture is based on practical application of high-intensive technologies of plant growing. Today, natural plant growth regulators, such as plant hormones or their synthetic analogs having phytohormone-like activity are widely used in the agricultural practice to accelerate plant growth, increase plant productivity, and to protect plant against biotic and biotic stress factors causing adverse effects on plant growth and yield (Basra 2000; Rademacher 2015; Meena 2015; Haggag et al. 2015; Wania et al. 2016; Denancé et al. 2013).

The very promising approach is the development of new classes of plant growth regulating substances having high physiological activity at low concentrations, broad specificity of action in various agricultural crops and lack of toxic effect for environment, animal and human health. The great theoretical and practical interest for plant biologists is study of specificity of plant growth regulating activity of new bioactive compounds of chemical or natural origin, for this purpose, various specific for plant hormone-like activity bioassays are used.

The best known auxin-like and cytokinin-like bioassays are based on the key role of plant hormones auxins and cytokinins in control of plant cell division, elongation, and differentiation that are basic processes of plant organogenesis, i.e. the formation of the plant vegetative and reproductive organs such as leaf, stem, root, flower, fruit and seed, as well as the formation of the adventitious roots on the isolated stem and leaf cuttings, increase in the biomass of cotyledons isolated from plant seeds, delaying of leaf senescence (Lam-Son and Sikander 2014; Zhao 2010; Enders and Strader 2015; Mok and Mok 2001; Gyulai and Heszky 1995; Basu 1972; Chen and Leisner 1985; Pop et al. 2011; Pandey et al. 2011; Takatsuka and Umeda 2014).

Considerable attention is currently being given to the study of the plant growth regulatory activity of synthetic low molecular weight heterocyclic compounds, derivatives of pyridine, pyrimidine, pyrazole, isoflavones, oxazolopyrimidine and oxazole, which have already found practical application in the agriculture as effective substitutes of traditional plant growth regulators, herbicides, fungicides and antimicrobial agents (Minn et al. 2008; Cansev et al. 2016; Sergiev et al. 2004; Corsi et al. 2011; Whittingham et al. 2010; Baum and Chen 1987; Chang and Baum 1990; Newton and Waldeck 2000).

The advantage of application of synthetic low molecular weight heterocyclic compounds is their high efficiency at their application at very low concentrations and ecological safety due to lack of toxic effect on the human, animal and plant cells; in addition, they are widely used in medical practice as therapeutic agents for treatment of nervous, allergic, gastroesophageal, cancer, bacterial, viral, fungal, infectious, and inflammatory diseases (Jain et al. 2006, 2016; Kumar et al. 2014; Quin and Tyrell 2010).

Today the new classes of the plant growth regulating substances are elaborated on the base of synthetic low molecular weight heterocyclic compounds synthesized in the V.P. Kukhar Institute of Bioorganic Chemistry and Petrochemistry of National Academy of Sciences of Ukraine. Our numerous researchers showed that synthetic low molecular weight heterocyclic compounds, derivatives of pyridine, pyrimidine, pyrazole, isoflavones, oxazolopyrimidine and oxazole revealed a high stimulating auxin-like and cytokinin-like effect on seed germination and vegetative growth of various crops (Tsygankova et al. 2017a, b, 2018a, b, c, d, 2019). Since synthetic low molecular weight heterocyclic compounds are applied at very low non-toxic for human, animal and plant concentrations, it is possible to prevent the negative effects on environmental pollution of pesticides used in high concentrations and with a long half-life (Nicolopoulou-Stamati et al. 2016).

The main task of our present work was study of auxin-like and cytokinin-like activity of synthetic low molecular weight heterocyclic compounds, derivatives of pyridine, pyrimidine, pyrazole, isoflavones, oxazolopyrimidine and oxazole using specific bioassays on the isolated organs of haricot bean and pumpkin plants (Tsygankova et al. 2018e).

2 Materials and Methods

2.1 *Bioassay on Auxin-Like Activity*

As is known, the major plant hormones auxins are involved in control of plant embryogenesis, seed germination, cell elongation and division in plant hypocotyls and coleoptiles, apical dominance, cambium cell division, plant tropisms, growth and development of root system, promotion of fruit setting and prevention of leaf abscission (Zhao 2010; Enders and Strader 2015; Gyulai and Heszky 1995; Basu 1972; Pop et al. 2011; Pandey et al. 2011; Takatsuka and Umeda 2014).

In our work to study auxin-like activity of chemical heterocyclic compounds, derivatives of pyridine, pyrimidine, pyrazole, and isoflavones we used specific bioassay conducted on the leaf petioles isolated from seedlings of haricot bean (*Phaseolus vulgaris* L.) cultivar Belozernaya (Tsygankova et al. 2018e). As is known, this bioassay is based on key role of auxins in regulation of formation of adventitious roots on the stem and leaf cuttings (Basu 1972; Pop et al. 2011; Pandey et al. 2011). The activity of chemical low molecular weight heterocyclic compounds, derivatives of pyridine, pyrimidine, pyrazole and isoflavones was compared with the activity of plant hormones auxins IAA and NAA.

The chemical structure, name and molecular mass (MM) of plant hormones auxins IAA (1*H*-Indol-3-ylacetic acid) and NAA (1-Naphthylacetic acid), and tested chemical heterocyclic compounds, derivatives of pyrimidine (compounds № 1–3), pyrazole (compounds № 4–6), isoflavones (compounds № 7–9), and pyridine (compound № 10) are shown in the Table 1.

To study auxin-like activity of synthetic low molecular weight heterocyclic compounds, seeds of haricot bean (*Phaseolus vulgaris* L.) cultivar Belozernaya were sterilized in 1% $KMnO_4$ solution for 3 min and 96% ethanol solution for 1 min and washed three times in the sterilized distilled water. After this procedure seeds were

Table 1 Chemical structure of plant hormones auxins and chemical heterocyclic compounds, derivatives of pyrimidine, pyrazole, isoflavones and pyridine

№	Chemical structure of compounds	Chemical name and relative molecular mass of compounds
IAA	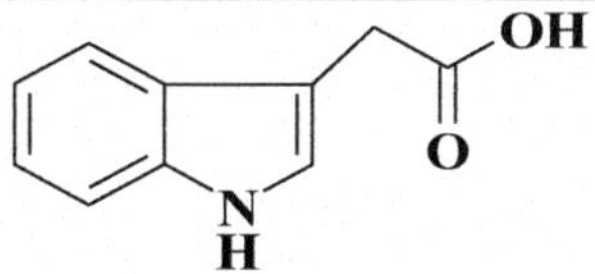	IAA (1*H*-Indol-3-ylacetic acid), MM = 175.19
NAA	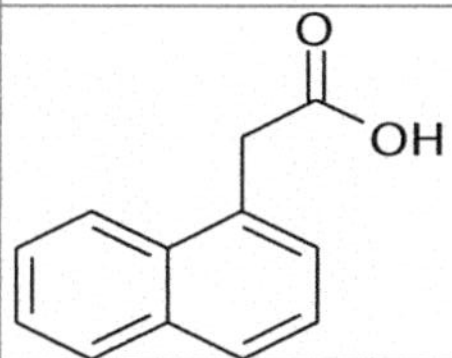	NAA (1-Naphthylacetic acid), MM = 186.21
1	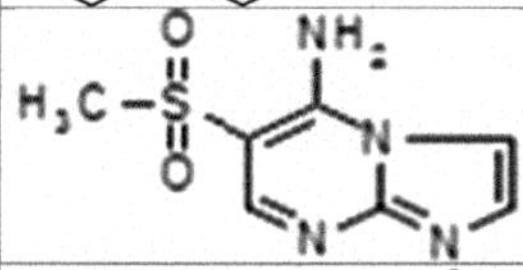	6-Methanesulfonyl-imidazo[1,2-*a*]pyrimidine-5-ylamine, MM = 212.23
2	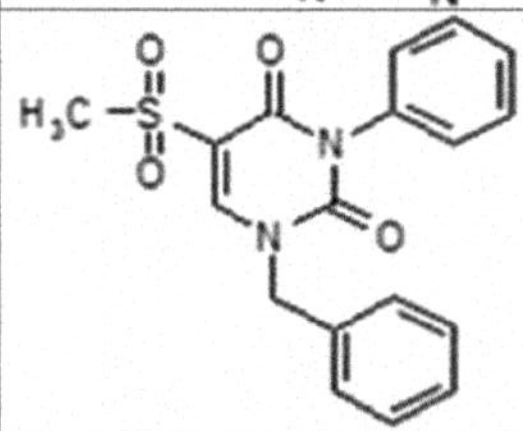	1-Benzyl-5-methanesulfonyl-3-phenyl-1*H*-pyrimidine-2,4-dione, MM = 356.48

(continued)

Table 1 (continued)

№	Chemical structure of compounds	Chemical name and relative molecular mass of compounds
3	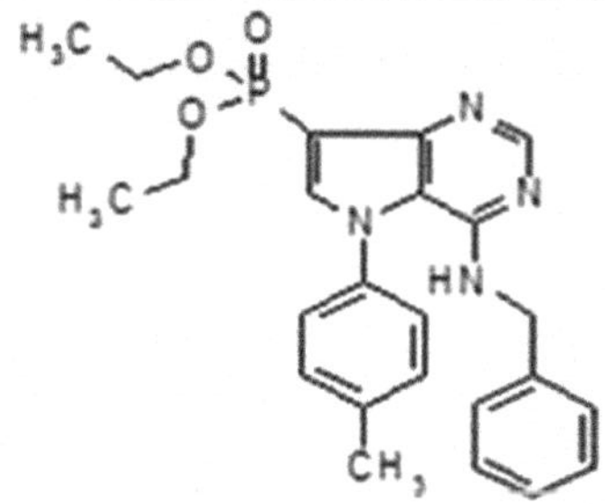	4-Benzylamino-5-*p*-tolyl-5*H*-pyrrolo-[3,2-*d*]pyrimidin-7-yl)-phosphonic acid diethyl ester, MM = 450.48
4	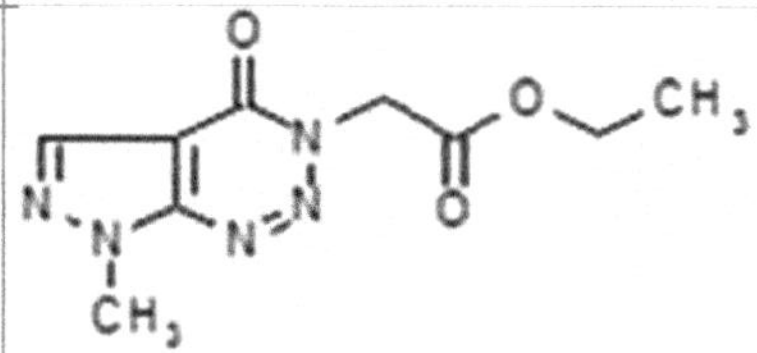	Ethyl 2-(4-oxo-7-methyl-4,7-dihydro-3*H*-pyrazolo[3,4-*d*][1,2,3]triazin-3-yl)acetate, MM = 237.22
5	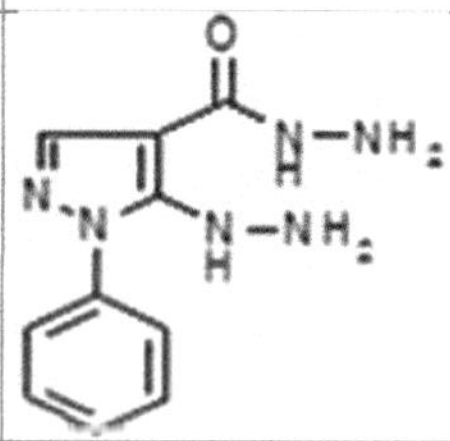	5-Hydrazino-1-phenyl-1*H*-pyrazole-4-carbohydrazide, MM = 232.25

(continued)

Table 1 (continued)

№	Chemical structure of compounds	Chemical name and relative molecular mass of compounds
6	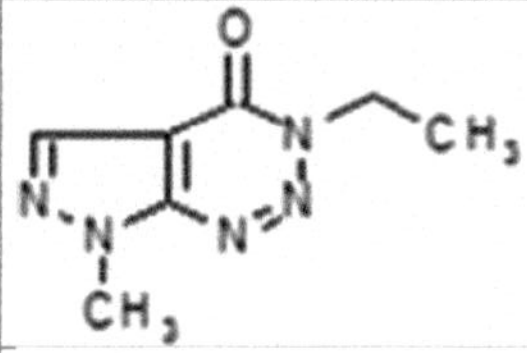	3-Ethyl-7-methyl-3,7-dihydro-4*H*-pyrazolo[3,4-*d*] [1,2,3]triazin-4-one, MM = 179.18
7	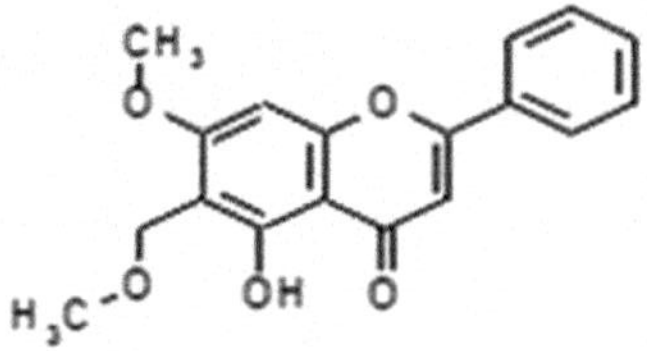	5-Hydroxy-7-methoxy-6-(methoxymethyl)-2-phenyl-4*H*-chromen-4-one, MM = 312.32
8	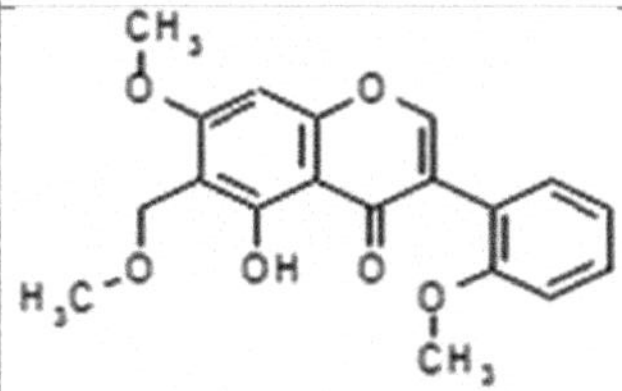	5-Hydroxy-7-methoxy-6-(methoxymethyl)-3-(2-methoxyphenyl)-4*H*-chromen-4-one, MM = 342.35
9	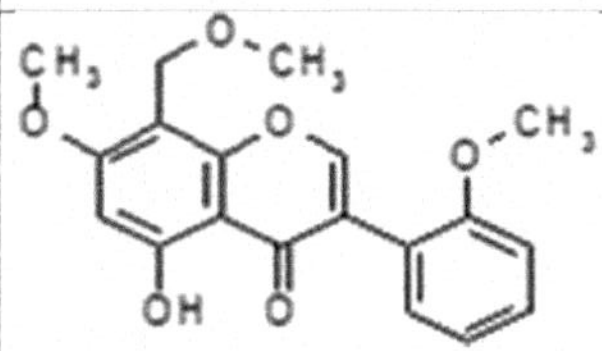	5-Hydroxy-7-methoxy-8-(methoxymethyl)-3-(4-methoxyphenyl)-4*H*-chromen-4-one, MM = 342.35

(continued)

Table 1 (continued)

№	Chemical structure of compounds	Chemical name and relative molecular mass of compounds
10	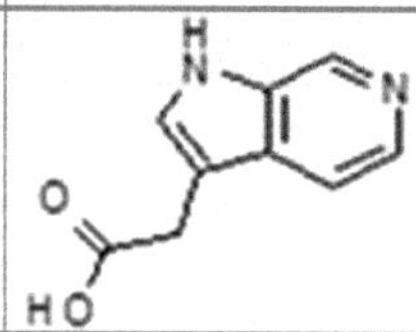	((1*H*-pyrrolo[2,3-*c*]pyridin-3-yl)-acetic acid), MM = 176.175

placed in the cuvettes (each containing 15–20 seeds) on the perlite moistened with distilled water. Then seeds were placed in the thermostat for their germination in darkness at the temperature 23 °C during 48 h. Sprouted seedlings were placed in the plant growth chamber in which seedlings were grown for 10 days at the 16/8 h light/dark conditions, at the temperature 23–25 °C, light intensity 3000 lx and air humidity 60–80%. To stimulate the formation of roots on the leaf petioles isolated from haricot bean seedlings they were cut at a distance of 4.3 mm from their base and then were placed immediately to a depth of 3 cm in separate glass test-tubes 30 cm in length containing either distilled water (control), or water solution of chemical heterocyclic compounds used at the concentration 10^{-8} M, or water solution of plant hormones auxins IAA and NAA used at the same concentration 10^{-8} M. After 14th days the indices of total roots number (pcs) and total length of roots (mm) calculated per one experimental haricot bean leaf petiole were determined and compared with the analogical indices of control leaf petiole on which the formation of adventitious roots should not be observed.

2.2 *Bioassay on Cytokinin-Like Activity*

Plant hormones cytokinins take an important part in control of embryo patterning, seed germination, de-etiolation, cell cycle control and protein synthesis, chloroplast differentiation, overcoming of apical dominance, releasing of lateral buds from dormancy, flower and fruit development and delaying of leaf senescence (Wania et al. 2016; Lam-Son and Sikander 2014; Mok and Mok 2001; Chen and Leisner 1985).

In our work to study cytokinin-like activity of chemical heterocyclic compounds, derivatives of pyridine, pyrimidine, pyrazole, isoflavones, oxazolopyrimidine and oxazole we used specific bioassay conducted on the cotyledons (i.e. food-storage organs) isolated from seeds of muscat pumpkin (*Cucurbita moschata* Duch. et Poir.) cultivar Gilea (Tsygankova et al. 2018e). As is known, this bioassay is based on key role of cytokinins in regulation of cell division in isolated plant organs, which leads to an increase in their biomass (Mok and Mok 2001; Chen and Leisner 1985). The activity of chemical heterocyclic compounds was compared with the activity of plant hormone cytokinin Kinetin.

The chemical structure, name and molecular mass (MM) of tested chemical heterocyclic compounds, derivatives of pyridine, pyrimidine, pyrazole, isoflavones are shown in the Table 1.

The chemical structure, name and molecular mass (MM) of plant hormone cytokinin Kinetin (*N*-(2-Furylmethyl)-7*H*-purin-6-amine), and tested chemical heterocyclic compounds, derivatives of oxazolopyrimidine (compounds № 1–4) and oxazole (compounds № 5 and 6) are shown in the Table 2.

To study cytokinin-like activity of synthetic low molecular weight heterocyclic compounds, seeds of muscat pumpkin (*Cucurbita moschata* Duch. et Poir.) cultivar Gilea were sterilized in 1% $KMnO_4$ solution for 3 min and 96% ethanol solution for 1 min and washed three times in the sterilized distilled water. After this procedure

Table 2 Chemical structure of plant hormone cytokinin Kinetin and chemical heterocyclic compounds, derivatives of oxazolopyrimidine and oxazole

№	Chemical structure of compounds	Chemical name and relative molecular mass of compounds
Kinetin	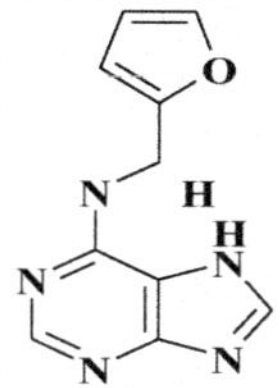	*N*-(2-Furylmethyl)-7*H*-purin-6-amine, MM = 215.22
1	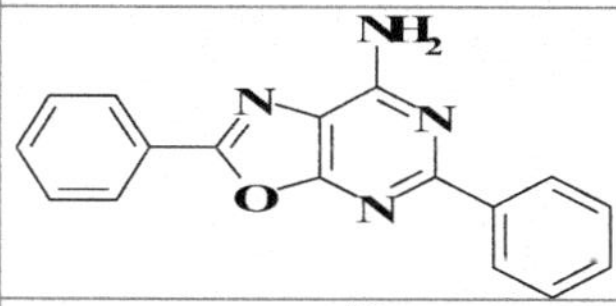	7-Amino-2,5-diphenyl[1,3]oxazolo[5,4-*d*]pyrimidine, MM = 288.31
2	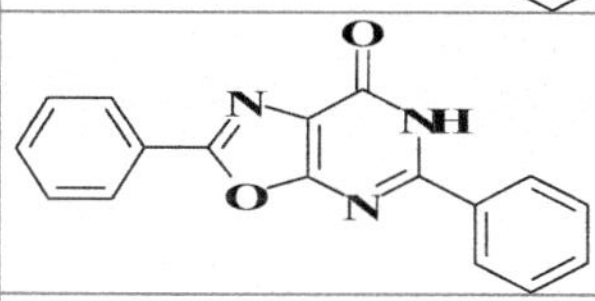	2,5-Diphenyl[1,3]oxazolo[5,4-*d*] pyrimidin-7(6*H*)-one, MM = 289.30
3	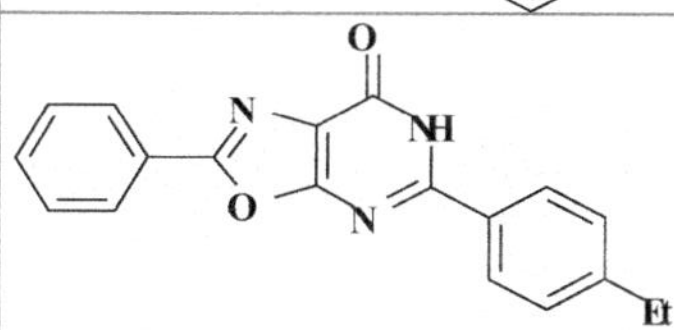	5-(4-Ethylphenyl)-2-phenyl [1,3]oxazolo[5,4-*d*]pyrimidin-7(6*H*)-one, MM = 317.35

(continued)

Table 2 (continued)

№	Chemical structure of compounds	Chemical name and relative molecular mass of compounds
4	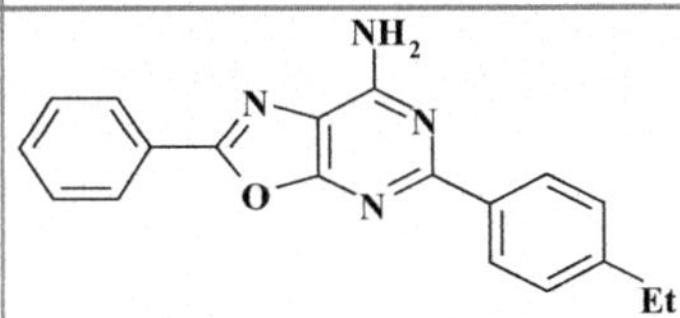	7-Amino-5-(4-ethylphenyl)-2-phenyl [1,3]oxazolo[5,4-*d*]pyrimidine, MM = 316.37
5	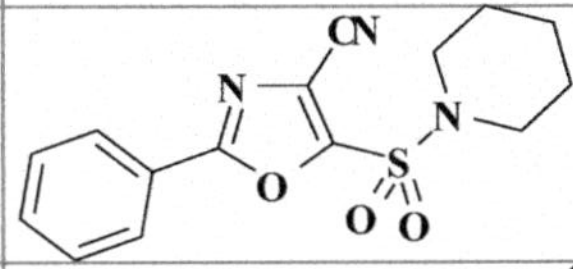	2-Phenyl-5-(piperidin-1-ylsulfonyl)-1,3-oxazole-4-carbonitrile, MM = 317.37
6	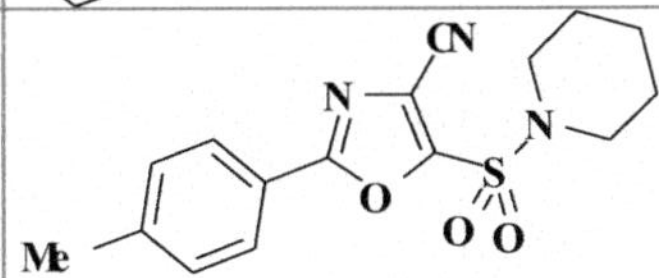	2-Tolyl-5-(piperidin-1-ylsulfonyl)-1,3-oxazole-4-carbonitrile, MM = 331.40

seeds were placed in the cuvettes (each containing 20–25 seeds) on the filter paper moistened with distilled water. Then seeds were placed in the thermostat for their germination in darkness at the temperature 25 °C during 96 h. The 4th-day-old pumpkin seedlings were separated from cotyledons using sterile scalpel. The isolated cotyledons were weighted and placed in the cuvettes (each containing 20 seeds) on the filter paper moistened with distilled water (control) or with water solution of chemical heterocyclic compounds, derivatives of pyridine, pyrimidine, pyrazole, isoflavones used at the concentration 10^{-8} M or with water solution of derivatives of oxazolopyrimidine and oxazole used at the concentration 10^{-9} M, or with water solution of plant hormone cytokinin Kinetin used at the same concentrations 10^{-8} M or 10^{-9} M. Then isolated cotyledons were placed in the plant growth chamber in which they were grown during 16 days or six weeks at above mentioned conditions. To determine indices of growth of biomass (g) of cotyledons isolated from seeds of pumpkin, they were washed with sterilized distilled water and weighted.

3 Statistical Analysis

All experiments were performed in three replicates. Statistical analysis of the data was performed using dispersive Student's-t test with the level of significance at $P \leq 0.05$, the values are mean ± SD (Bang et al. 2010).

4 Results and Discussion

4.1 Study of Auxin-Like Activity of Derivatives of Pyridine, Pyrimidine, Pyrazole and Isoflavones

The conducted researches showed that chemical heterocyclic compounds, derivatives of pyridine, pyrimidine, pyrazole and isoflavones used at the concentration 10^{-8} M revealed expressive auxin-like activity on the formation of adventitious roots on the 14th-day-old leaf petioles isolated from seedlings of haricot bean (*Phaseolus vulgaris* L.) cultivar Belozernaya (Fig. 1).

Conversely, the formation of roots on the control haricot bean leaf petioles treated with distilled water was not observed. Among all heterocyclic compounds, the derivatives of pyrazole and isoflavones, which include compounds № 7, 8, 10–12 showed the greatest stimulating effect on the formation of adventitious roots on the 14th-day-old leaf petioles isolated from haricot bean seedlings (Fig. 1).

The data of the statistical analysis of the indices of average total root number (pcs) and average total root length (mm) calculated per one experimental 14th-day-old haricot bean leaf petiole treated with water solution of chemical heterocyclic compounds at the concentration 10^{-8} M or with water solution of auxins IAA and

Fig. 1 The auxin-like effect of chemical heterocyclic compounds, derivatives of pyrimidine, pyrazole, isoflavones, pyridine, and auxins IAA and NAA on the formation of adventitious roots on the 14th-day-old leaf petioles isolated from seedlings of haricot bean (*Phaseolus vulgaris* L.) cultivar Belozernaya. 1—Compound 6-Methanesulfonyl-imidazo[1,2-*a*]pyrimidine-5-ylamine, 2—Compound 1-Benzyl-5-methanesulfonyl-3-phenyl-1*H*-pyrimidine-2,4-dione, 3—Compound 4-Benzylamino-5-*p*-tolyl-5*H*-pyrrolo-[3,2-*d*]pyrimidin-7-yl)-phosphonic acid diethyl ester, 4—Compound Ethyl 2-(4-oxo-7-methyl-4,7-dihydro-3*H*-pyrazolo[3,4-*d*][1,2,3]triazin-3-yl)acetate, 5—NAA (1-Naphthylacetic acid), 6—IAA (1*H*-Indol-3-ylacetic acid), 7—Compound 5-Hydrazino-1-phenyl-1*H*-pyrazole-4-carbo hydrazide, 8—Compound 3-Ethyl-7-methyl-3,7-dihydro-4*H*-pyrazolo[3,4-*d*][1,2,3]triazin-4-one, 9—Compound (1H-pyrrolo[2,3-*c*]pyridin-3-yl)-acetic acid), 10—Compound 5-Hydroxy-7-methoxy-6-(methoxymethyl)-2-phenyl-4*H*-chromen-4-one, 11—Compound 5-Hydroxy-7-methoxy-6-(methoxymethyl)-3-(2-methoxyphenyl)-4*H*-chromen-4-one, 12—Compound 5-Hydroxy-7-methoxy-8-(methoxymethyl)-3-(4-methoxy phenyl)-4*H*-chromen-4-one

NAA used at the same concentration 10^{-8} M as compared to indices obtained on the control haricot bean leaf petiole treated with distilled water are shown in the Table 3.

It was found that the chemical heterocyclic compound №12 revealed the highest auxin-like stimulating effect on formation of the roots on the 14th-day-old haricot bean leaf petioles, the indices of the total root number increased at the 146% and total root length increased at the 9.18 times as compared to control haricot bean leaf petioles treated with distilled water (Table 3).

The high auxin-like stimulating effect on the formation of roots on the 14th-day-old haricot bean leaf petioles revealed also the chemical heterocyclic compounds: the compound №10, the indices of the total root number increased at the 129% and total root length increased at the 8.45 times as compared to control haricot bean leaf petioles; the compound №11, the indices of the total root number increased at the 117% and total root length increased at the 7.34 times as compared to control haricot bean leaf petioles; the compound №7, the indices of the total root number increased at the 96% and total root length increased at the 5.79 times as compared to control

Table 3 The auxin-like effect of chemical heterocyclic compounds, derivatives of pyrimidine, pyrazole, isoflavones, and pyridine on the average total root number (pcs) and average total root length (mm) formed on the 14th-day-old haricot bean leaf petioles

№ compound	The average total root number per one leaf petiole (pcs)	The average total root length per one leaf petiole (mm)
Control (distilled water)		
1	29 ± 0.76*	138 ± 1.22*
2	43 ± 0.31*	165 ± 1.97*
3	67 ± 1.18*	288 ± 0.35*
4	23 ± 1.48*	34 ± 2.79*
5	79 ± 0.64*	476 ± 2.87*
6	62 ± 0.47*	172 ± 0.39*
7	96 ± 0.62*	579 ± 1.95*
8	83 ± 1.66*	645 ± 1.57*
9	35 ± 0.44*	526 ± 2.13*
10	129 ± 0.32*	845 ± 0.76*
11	117 ± 1.19*	734 ± 2.31*
12	146 ± 1.55*	918 ± 0.53*

Note * Significant differences from control values, $p \leq 0.05$, $n = 3$, (-)—decreasing; (−)—increasing Compound №1—6-Methanesulfonyl-imidazo[1,2-*a*]pyrimidine-5-ylamine, Compound №2—1-Benzyl-5-methanesulfonyl-3-phenyl-1*H*-pyrimidine-2,4-dione, Compound №3—4-Benzylamino-5-*p*-tolyl-5*H*-pyrrolo-[3,2-*d*]pyrimidin-7-yl)-phosphonic acid diethyl ester, Compound №4—Ethyl 2-(4-oxo-7-methyl-4,7-dihydro-3*H*-pyrazolo[3,4-*d*][1,2,3]triazin-3-yl)acetate, Compound №5—NAA (1-Naphthylacetic acid), Compound №6—IAA (1*H*-Indol-3-ylacetic acid), Compound №7—5-Hydrazino-1-phenyl-1*H*-pyrazole-4-carbo hydrazide, Compound №8—3-Ethyl-7-methyl-3,7-dihydro-4*H*-pyrazolo[3,4-*d*][1,2,3]triazin-4-one, Compound №9—(1H-pyrrolo[2,3-*c*]pyridin-3-yl)-acetic acid), Compound №10—5-Hydroxy-7-methoxy-6-(methoxymethyl)-2-phenyl-4*H*-chromen-4-one, Compound №11—5-Hydroxy-7-methoxy-6-(methoxymethyl)-3-(2-methoxyphenyl)-4*H*-chromen-4-one, Compound №12—5-Hydroxy-7-methoxy-8-(methoxymethyl)-3-(4-methoxy phenyl)-4*H*-chromen-4-one

haricot bean leaf petioles; the compound №8, the indices of the total root number increased at the 83% and total root length increased at the 6.45 times as compared to control haricot bean leaf petioles; the compound №3, the indices of the total root number increased at the 67% and total root length increased at the 2 88 times as compared to control haricot bean leaf petioles (Table 3).

The high auxin-like stimulating effect on formation of roots on the 14th-day-old haricot bean leaf petioles revealed also plant hormones auxins: the compound №5 (NAA), the indices of the total root number increased at the 79% and total root length increased at the 4.76 times as compared to control haricot bean leaf petioles, and the compound №6 (IAA), the indices of the total root number increased at the 62% and total root length increased at the 1.72 times as compared to control haricot bean leaf petioles (Table 3).

The lower auxin-like stimulating effect on the formation of the roots on the 14th-day-old haricot bean leaf petioles revealed the chemical heterocyclic compound №2, the indices of the total root number increased at the 43% and total root length increased at the 1.65 times as compared to control haricot bean leaf petioles; the compound №9, the indices of the total root number increased at the 35% and total root length increased at the 5.26 times as compared to control haricot bean leaf petioles; the compound №1, the indices of the total root number increased at the 29% and total root length increased at the 1.38 times as compared to control haricot bean leaf petioles; the compound №4, the indices of the total root number increased at the 23% and total root length increased at the 34% as compared to control haricot bean leaf petioles (Table 3).

Obviously, that the high auxin-like activity of tested chemical heterocyclic compounds, derivatives of pyrimidine, pyrazole, isoflavones, and pyridine may be explained by their inducing auxin-like effect on plant cell elongation, division, and differentiation that are the basic processes of the formation of the adventitious roots on the leaf petioles isolated from seedlings of haricot bean (*Phaseolus vulgaris* L.) cultivar Belozernaya.

4.2 Study of Cytokinin-Like Activity of Derivatives of Pyridine, Pyrimidine, Pyrazole and Isoflavones

The obtained results showed that derivatives of pyridine, pyrimidine, pyrazole and isoflavones used at the concentration 10^{-8} M revealed the expressive cytokinin-like activity on the growth of biomass of cotyledons isolated from seeds of muscat pumpkin (*Cucurbita moschata* Duch. et Poir.) cultivar Gilea during 16 days, which was similar or higher of the activity of plant hormone cytokinin Kinetin used at the same concentration 10^{-8} M (Fig. 2).

It was found also that some chemical heterocyclic compounds used at the concentration 10^{-8} M revealed nonspecific for this bioassay auxin-like activity, which was manifested in formation of the roots on the six-week-old cotyledons isolated from seeds of pumpkin (Fig. 3).

The obtained data of the statistical analysis of indices of average biomass of the 30 cotyledons (g) and average length of one root per 30 cotyledons (cm) of the six-week-old cotyledons isolated from seeds of pumpkin are presented in the Table 4.

The indices of average biomass (g) and average length of one root (cm) per 30 six-week-old cotyledons of pumpkin treated with 10^{-8} M water solution of compound №1 were in average higher of the analogical indices of cotyledons of pumpkin treated either with distilled water (control) or with 10^{-8} M water solution of cytokinin Kinetin (compound №6) as follows: according with average biomass—at the 121% as compared with control and at the 109% as compared with cytokinin Kinetin;

Fig. 2 The cytokinin-like effect of chemical heterocyclic compounds, derivatives of pyridine, pyrimidine, pyrazole, isoflavones, and plant hormone cytokinin Kinetin on the growth of biomass of 16th-day-old cotyledons isolated from seeds of muscat pumpkin (*Cucurbita moschata* Duch. et Poir.) cultivar Gilea. C—Control (distilled water), I—Compound 6-Methanesulfonyl-imidazo[1,2-*a*]pyrimidine-5-ylamine, II—Compound 1-Benzyl-5-methanesulfonyl-3-phenyl-1*H*-pyrimidine-2,4-dione, III—Compound 4-Benzylamino-5-*p*-tolyl-5*H*-pyrrolo-[3,2-*d*]pyrimidin-7-yl)-phosphonic acid diethyl ester, IV—Compound Ethyl 2-(4-oxo-7-methyl-4,7-dihydro-3*H*-pyrazolo[3,4-*d*][1,2,3]triazin-3-yl)acetate, V—Compound 5-Hydrazino-1-phenyl-1*H*-pyrazole-4-carbo hydrazide, VI—Kinetin (*N*-(2-Furylmethyl)-7*H*-purin-6-amine), VII—Compound 3-Ethyl-7-methyl-3,7-dihydro-4*H*-pyrazolo[3,4-*d*][1,2,3]triazin-4-one, VIII—Compound 5-Hydroxy-7-methoxy-6-(methoxymethyl)-2-phenyl-4*H*-chromen-4-one, IX—Compound 5-Hydroxy-7-methoxy-6-(methoxymethyl)-3-(2-methoxyphenyl)-4*H*-chromen-4-one, X—Compound 5-Hydroxy-7-methoxy-8-(methoxymethyl)-3-(4-methoxy phenyl)-4*H*-chromen-4-one, XI—Compound (1H-pyrrolo[2,3-*c*]pyridin-3-yl)-acetic acid)

according with average length of one root—at the 12.7 times as compared with control and at the 4.25 times as compared with cytokinin Kinetin (Table 4).

The indices of average biomass (g) and average length of one root (cm) per 30 six-week-old cotyledons of pumpkin treated with 10^{-8} M water solution of compound №2 were in average higher of the analogical indices of cotyledons of pumpkin treated either with distilled water (control) or with 10^{-8} M water solution of cytokinin Kinetin (compound №6) as follows: according with average biomass—at the 128% as compared with control and at the 116% as compared with cytokinin Kinetin; according with average length of one root—at the 15.9 times as compared with control and at the 5.3 times as compared with cytokinin Kinetin (Table 4).

The indices of average biomass (g) and average length of one root (cm) per 30 six-week-old cotyledons of pumpkin treated with 10^{-8} M water solution of compound №3 were in average higher of the analogical indices of cotyledons of pumpkin treated either with distilled water (control) or with 10^{-8} M water solution of cytokinin Kinetin (compound №6) as follows: according with average biomass—at the 120% as compared with control and at the 108% as compared with cytokinin Kinetin; according with average length of one root—at the 17.5 times as compared with control and at the 5.8 times as compared with cytokinin Kinetin (Table 4).

Fig. 3 The auxin-like effect of chemical heterocyclic compounds, derivatives of pyrimidine, pyrazole, isoflavones, and pyridine on the formation of roots on the six-week-old cotyledons isolated from seeds of muscat pumpkin (*Cucurbita moschata* Duch. et Poir.) cultivar Gilea. I—Compound 6-Methanesulfonyl-imidazo[1,2-*a*]pyrimidine-5-ylamine, II—Compound 1-Benzyl-5-methanesulfonyl-3-phenyl-1*H*-pyrimidine-2,4-dione, III—Compound Ethyl 2-(4-oxo-7-methyl-4,7-dihydro-3*H*-pyrazolo[3,4-*d*][1,2,3]triazin-3-yl)acetate, IV—Compound 5-Hydrazino-1-phenyl-1*H*-pyrazole-4-carbo hydrazide, V—Compound 3-Ethyl-7-methyl-3,7-dihydro-4*H*-pyrazolo[3,4-*d*][1,2,3]triazin-4-one, VI—Compound 5-Hydroxy-7-methoxy-6-(methoxymethyl)-2-phenyl-4*H*-chromen-4-one, VII—Compound 5-Hydroxy-7-methoxy-6-(methoxymethyl)-3-(2-methoxyphenyl)-4*H*-chromen-4-one, VIII—Compound 5-Hydroxy-7-methoxy-8-(methoxymethyl)-3-(4-methoxy phenyl)-4*H*-chromen-4-one, IX—Compound (1H-pyrrolo[2,3-*c*]pyridin-3-yl)-acetic acid)

The indices of average biomass (g) and average length of one root (cm) per 30 six-week-old cotyledons of pumpkin treated with 10^{-8} M water solution of compound №4 were in average higher of the analogical indices of cotyledons of pumpkin treated either with distilled water (control) or with 10^{-8} M water solution of cytokinin Kinetin (compound №6) as follows: according with average biomass—at the 123% as compared with control and at the 112% as compared with cytokinin Kinetin; according with average length of one root—at the 12.9 times as compared with control and at the 4.3 times as compared with cytokinin Kinetin (Table 4).

The indices of average biomass (g) and average length of one root (cm) per 30 six-week-old cotyledons of pumpkin treated with 10^{-8} M water solution of compound №5 were in average higher of the analogical indices of cotyledons of pumpkin treated either with distilled water (control) or with 10^{-8} M water solution of cytokinin Kinetin (compound №6) as follows: according with average biomass—at the 130% as compared with control and at the 118% as compared with cytokinin Kinetin;

Table 4 The cytokinin-like and auxin-like effect of chemical heterocyclic compounds, derivatives of pyrimidine, pyrazole, isoflavones, and pyridine on the growth of biomass (g) and length of roots (cm) formed on the six-week-old cotyledons isolated from seeds of pumpkin

№ compound	The average biomass of the 30 cotyledons (g)	The average length of one root per 30 cotyledons (cm)
Control (distilled water)	45.63 ± 0.29*	1.12 ± 0.45*
1	55.27 ± 0.86**	14.23 ± 1.27**
2	58.64 ± 0.53**	17.78 ± 0.66**
3	54.91 ± 1.64**	19.56 ± 1.14**
4	56.26 ± 1.52**	14.45 ± 0.93**
5	59.67 ± 0.83**	13.21 ± 1.24**
6	50.48 ± 1.18**	3.35 ± 1.57**
7	57.61 ± 0.45**	11.24 ± 1.14**
8	55.22 ± 0.69**	13.15 ± 1.78**
9	61.34 ± 1.94**	10.23 ± 1.44**
10	66.27 ± 1.12**	14.61 ± 1.26**
11	57.49 ± 1.19**	12.45 ± 0.89**

Note ** Significant differences from control values*, $p \leq 0.05$, $n = 3$, (-)—decreasing; (+)—increasing
Compound №1—6-Methanesulfonyl-imidazo[1,2-*a*]pyrimidine-5-ylamine, Compound №2—1-Benzyl-5-methanesulfonyl-3-phenyl-1*H*-pyrimidine-2,4-dione, Compound №3—4-Benzylamino-5-*p*-tolyl-5*H*-pyrrolo-[3,2-*d*]pyrimidin-7-yl)-phosphonic acid diethyl ester, Compound №4—Ethyl 2-(4-oxo-7-methyl-4,7-dihydro-3*H*-pyrazolo[3,4-*d*][1,2,3]triazin-3-yl)acetate, Compound №5—5-Hydrazino-1-phenyl-1*H*-pyrazole-4-carbo hydrazide, Compound №6—Kinetin (*N*-(2-Furylmethyl)-7*H*-purin-6-amine), Compound №7—3-Ethyl-7-methyl-3,7-dihydro-4*H*-pyrazolo[3,4-*d*][1,2,3]triazin-4-one, Compound №8—5-Hydroxy-7-methoxy-6-(methoxymethyl)-2-phenyl-4*H*-chromen-4-one, Compound №9—5-Hydroxy-7-methoxy-6-(methoxymethyl)-3-(2-methoxyphenyl)-4*H*-chromen-4-one, Compound №10—5-Hydroxy-7-methoxy-8-(methoxymethyl)-3-(4-methoxy phenyl)-4*H*-chromen-4-one, Compound №11—(1H-pyrrolo[2,3-*c*]pyridin-3-yl)-acetic acid)

according with average length of one root—at the 11.8 times as compared with control and at the 3.9 times as compared with cytokinin Kinetin (Table 4).

The indices of average biomass (g) and average length of one root (cm) per 30 six-week-old cotyledons of pumpkin treated with 10^{-8} M water solution of compound №7 were in average higher of the analogical indices of cotyledons of pumpkin treated either with distilled water (control) or with 10^{-8} M water solution of cytokinin Kinetin (compound №6) as follows: according with average biomass—at the 126% as compared with control and at the 114% as compared with cytokinin Kinetin; according with average length of one root—at the 10.0 times as compared with control and at the 3.6 times as compared with cytokinin Kinetin (Table 4).

The indices of average biomass (g) and average length of one root (cm) per 30 six-week-old cotyledons of pumpkin treated with 10^{-8} M water solution of compound №8 were in average higher of the analogical indices of cotyledons of pumpkin treated either with distilled water (control) or with 10^{-8} M water solution of cytokinin Kinetin (compound №6) as follows: according with average biomass—at the 121% as compared with control and at the 109% as compared with cytokinin Kinetin; according with average length of one root—at the 11.7 times as compared with control and at the 3.9 times as compared with cytokinin Kinetin (Table 4).

The indices of average biomass (g) and average length of one root (cm) per 30 six-week-old cotyledons of pumpkin treated with 10^{-8} M water solution of compound №9 were in average higher of the analogical indices of cotyledons of pumpkin treated either with distilled water (control) or with 10^{-8} M water solution of cytokinin Kinetin (compound №6) as follows: according with average biomass—at the 134% as compared with control and at the 122% as compared with cytokinin Kinetin; according with average length of one root—at the 9.1 times as compared with control and at the 3.0 times as compared with cytokinin Kinetin (Table 4).

The indices of average biomass (g) and average length of one root (cm) per 30 six-week-old cotyledons of pumpkin treated with 10^{-8} M water solution of compound №10 were in average higher of the analogical indices of cotyledons of pumpkin treated either with distilled water (control) or with 10^{-8} M water solution of cytokinin Kinetin (compound №6) as follows: according with average biomass—at the 145% as compared with control and at the 131% as compared with cytokinin Kinetin; according with average length of one root—at the 13.0 times as compared with control and at the 4.4 times as compared with cytokinin Kinetin (Table 4).

The indices of average biomass (g) and average length of one root (cm) per 30 six-week-old cotyledons of pumpkin treated with 10^{-8} M water solution of compound №11 were in average higher of the analogical indices of cotyledons of pumpkin treated either with distilled water (control) or with 10^{-8} M water solution of cytokinin Kinetin (compound №6) as follows: according with average biomass—at the 125% as compared with control and at the 113% as compared with cytokinin Kinetin; according with average length of one root—at the 11.1 times as compared with control and at the 3.7 times as compared with cytokinin Kinetin (Table 4).

The obtained results suggest that high cytokinin-like and auxin-like activity of chemical heterocyclic compounds, derivatives of pyridine, pyrimidine, pyrazole, isoflavones may be explained by their inducing cytokinin-like and auxin-like effect on plant cell division and elongation resulting in increasing growth of biomass of the cotyledons isolated from seed of muscat pumpkin (*Cucurbita moschata* Duch. et Poir.) cultivar Gilea and formation of the adventitious roots on the isolated cotyledons.

4.3 Study of Cytokinin-Like Activity of Derivatives of Oxazolopyrimidine and Oxazole

The obtained results showed that according to the indices of growth of biomass of cotyledons isolated from seeds of muscat pumpkin (*Cucurbita moschata* Duch. et Poir.) cultivar Gilea during 16 days all tested chemical compounds, derivatives of oxazolopyrimidine and oxazole used at the concentration 10^{-9} M showed the expressive cytokinin-like activity, which was similar or higher of the activity of plant hormone cytokinin Kinetin used at the same concentration 10^{-9} M.

The obtained data of the statistical analysis of indices of growth of biomass of isolated cotyledons of pumpkin showed that the highest cytokinin-like activity revealed the compounds, derivatives of oxazolopyrimidine: the compound №2—2,5-diphenyl[1,3]oxazolo[5,4-*d*]pyrimidin-7(6*H*)-one and compound №4—7-amino-5-(4-ethylphenyl)-2-phenyl[1,3]oxazolo[5,4-*d*]pyrimidine, as well as the compound, derivative of oxazole: the compound №6—2-tolyl-5-(piperidin-1-ylsulfonyl)-1,3-oxazole-4-carbonitrile (Fig. 4).

Among the derivatives of oxazolopyrimidine, the compound №4—7-amino-5-(4-ethylphenyl)-2-phenyl[1,3]oxazolo[5,4-*d*]pyrimidine, which contains amino group at the 7th position of pyrimidine fragment, showed the highest cytokinin-like activity; the indices of growth of biomass of the isolated cotyledons of pumpkin grown on the 10^{-9} M water solution of compound №4 were higher at the 40% and 19% of the indices of growth of biomass of the isolated cotyledons of pumpkin grown either on

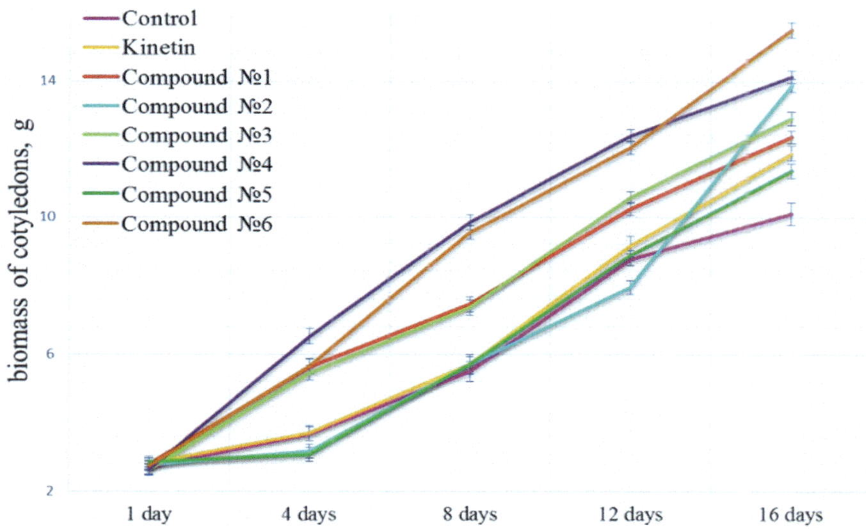

Fig. 4 The cytokinin-like effect of chemical heterocyclic compounds, derivatives of oxazolopyrimidine (compound №1—7-amino-2,5-diphenyl[1,3]oxazolo[5,4-*d*]pyrimidine, compound №2—2,5-diphenyl[1,3]oxazolo[5,4-*d*]pyrimidin-7(6*H*)-one, compound №3—5-(4-ethylphenyl)-2-phenyl[1,3]oxazolo[5,4-*d*]pyrimidin-7(6*H*)-one, compound №4—7-amino-5-(4-ethylphenyl)-2-phenyl[1,3]oxazolo[5,4-*d*]pyrimidine), and derivatives of oxazole (compound №5—2-phenyl-5-(piperidin-1-ylsulfonyl)-1,3-oxazole-4-carbonitrile and compound №6—2-tolyl-5-(piperidin-1-ylsulfonyl)-1,3-oxazole-4-carbonitrile), and plant hormone cytokinin Kinetin (*N*-(2-Furylmethyl)-7*H*-purin-6-amine) on the growth of the biomass of cotyledons isolated from seeds of muscat pumpkin (*Cucurbita moschata* Duch. et Poir.) cultivar Gilea (the biomass was weighted with the interval of each 4 day)

the distilled water (control) or on the 10^{-9} M water solution of cytokinin Kinetin, respectively (Fig. 4).

The high cytokinin-like activity demonstrated also the compound №2—2,5-diphenyl[1,3]oxazolo[5,4-*d*]pyrimidin-7(6*H*)-one, which contains phenyl substituent at the 5th position of pyrimidine fragment; the indices of growth of biomass of the isolated cotyledons of pumpkin grown on the 10^{-9} M water solution of compound №2 were higher at the 38% and 17% of the indices of growth of biomass of the isolated cotyledons of pumpkin grown either on the distilled water (control) or on the 10^{-9} M water solution of cytokinin Kinetin, respectively (Fig. 4).

The lower cytokinin-like activity showed the compound №3—5-(4-ethylphenyl)-2-phenyl[1,3]oxazolo [5,4-*d*]pyrimidin-7(6*H*)-one, which contains 4-ethylphenyl substituent at the 5th position and oxygen at the 7th position of pyrimidine fragment; the indices of growth of biomass of the isolated cotyledons of pumpkin grown on the 10^{-9} M water solution of compound №3 were higher at the 28% and 9% of the indices of growth of biomass of the isolated cotyledons of pumpkin grown either on the distilled water (control) or on the 10^{-9} M water solution of cytokinin Kinetin, respectively (Fig. 4).

The lower cytokinin-like activity showed also the compound №1—7-amino-2,5-diphenyl[1,3] oxazolo[5,4-*d*]pyrimidine, which contains phenyl substituent at the 5th position and amino group at the 7th position of pyrimidine fragment; the indices of growth of biomass of the isolated cotyledons of pumpkin grown on the 10^{-9} M water solution of compound №1 were higher at the 22% and 4% of the indices of growth of biomass of the isolated cotyledons of pumpkin grown either on the distilled water (control) or on the 10^{-9} M water solution of cytokinin Kinetin, respectively (Fig. 4).

Among the compounds, derivatives of oxazole the compound №6—2-tolyl-5-(piperidin-1-ylsulfonyl)-1,3-oxazole-4-carbonitrile, which contains tolyl substituent at the 2th position of oxazole, showed the highest cytokinin-like activity; the indices of growth of biomass of the isolated cotyledons of pumpkin grown on the 10^{-9} M water solution of compound №6 were higher at the 54% and 31% of the indices of growth of biomass of the isolated cotyledons of pumpkin grown either on the distilled water (control) or on the 10^{-9} M water solution of cytokinin Kinetin, respectively (Fig. 4).

At the same time the compound №5—2-phenyl-5-(piperidin-1-ylsulfonyl)-1,3-oxazole-4-carbonitrile that contains phenyl substituent at the 2th position of oxazole revealed lower cytokinin-like activity; the indices of growth of biomass of the isolated cotyledons of pumpkin grown on the 10^{-9} M water solution of compound №5 were higher at the 23% of the indices of growth of biomass of the isolated cotyledons of pumpkin grown on the distilled water (control) (Fig. 4).

The conducted specific bioassay on cytokinin-like activity showed that among heterocyclic compounds, derivatives of oxazolopyrimidine and oxazole the highest activity on the growth of biomass of cotyledons isolated from seed of muscat pumpkin (*Cucurbita moschata* Duch. et Poir.) cultivar Gilea demonstrated the compounds: the compound №2—2,5-diphenyl[1,3]oxazolo[5,4-*d*]pyrimidin-7(6*H*)-one, which contains phenyl substituent at the 5th position of pyrimidine fragment, the compound №4—7-amino-5-(4-ethylphenyl)-2-phenyl[1,3]oxazolo[5,4-*d*]pyrimidine, which contains amino group at the 7th position of pyrimidine fragment, and the compound №6—2-tolyl-5-(piperidin-1-ylsulfonyl)-1,3-oxazole-4-carbonitrile, which contains tolyl substituent at the 2th position of oxazole. It is obvious that cytokinin-like activity on the growth of the biomass of cotyledons isolated from seed of muscat pumpkin (*Cucurbita moschata* Duch. et Poir.) cultivar Gilea of chemical compounds, derivatives of oxazolopyrimidine may depend on substituents at the 5th and 7th positions of pyrimidine fragment, while as activity of compounds, derivatives of oxazole may depend on substituents at the 2th position of oxazole.

Thus, the conducted studies showed that synthetic low molecular weight heterocyclic compounds, derivatives of pyridine, pyrimidine, pyrazole, isoflavones, oxazolopyrimidine and oxazole revealed similar to plant hormones auxins and cytokinins activity (Tsygankova et al. 2018e).

It is possible to assume that the molecular mechanisms of auxin-like and cytokinin-like activity of derivatives of pyridine, pyrimidine, pyrazole, isoflavones, oxazolopyrimidine and oxazole might be associated with their regulatory action (by analogy with plant hormone auxin) on the network of key auxin-binding proteins (ABPs) that may be the auxin receptors involved in auxin signalling and transport, network of auxin response transcription factors (ARFs) that are DNA-binding proteins, which recognize and bind to auxin responsive *cis*-acting promoter elements (AuxREs) in early/primary auxin response genes, and network of transcription factors that bind to promoter elements in genes encoding protein-enzymes responsible for plant cell division and extension (Komaki and Sugimoto 2012; Lavy and Estelle 2016; Leyser 2017; Majda and Robert 2018).

Otherwise, there could be an alternative mode of action related to the inhibitory effect of synthetic low molecular weight heterocyclic compounds, derivatives of pyridine, pyrimidine, pyrazole, isoflavones, oxazolopyrimidine and oxazole on the activity of a key enzyme IAA-oxidase, which is involved in the enzymatic destruction (degradation) of auxin (Hare 1964). As a result, the level of endogenously synthesized auxin IAA is increased in the plant cells, and auxin transport, perception, and signaling are restored leading to improved plant cell division and extension that are the main processes of plant growth and development (Komaki and Sugimoto 2012; Lavy and Estelle 2016; Leyser 2017; Majda and Robert 2018).

Obviously also, that the cytokinin-like effect of synthetic synthetic low molecular weight heterocyclic compounds, derivatives of pyridine, pyrimidine, pyrazole, isoflavones, oxazolopyrimidine and oxazole can be related to their direct influence on the metabolism of endogenous plant hormones or their inhibitory effect on the activity of an enzyme cytokinin oxidase, which is involved in the enzymatic destruction (degradation) of endogenous plant cytokinins (Avalbaev et al. 2012). As a result, the level of endogenously synthesized cytokinins is increased in the plant cells, and cytokinin transport, perception and signal transduction are restored (Kieber and Schaller 2018) leading to improved plant cell division and increased synthesis of photosynthetic pigments such as chlorophylls and carotenoids, and delayed leaf senescence (Zwack and Rashotte 2013).

In support of the bottom concept indicate published works (Tandon and Arya 1982; Liu et al. 1996) which showed the effect of exogenously applied synthetic analogs of auxin on the decrease in the activity of IAA-oxidase and vice versa on the increase in the level of synthesis of endogenous auxin IAA in plant cells.

The authors of the work (Gaspar et al. 1996) also suggested that synthetic auxins might affect the level of synthesis of endogenous auxin modifying directly synthesis of enzyme IAA-oxidase and indirectly through effectors of IAA-oxidase.

Similar studies were conducted in work (Šimonová et al. 2005) that showed that synthetic 2-R substituted benzothiazole derivatives demonstrated dominant auxin-like plant growth promoting activity. Based on obtained results, showing that the plant growth promoting activity of synthetic benzothiazole compounds can be correlated with the activity of IAA synthetase, the authors have proposed that the mode of action of synthetic 2-R substituted benzothiazole derivatives as auxin-like substances is due to their possible regulation of synthesis or degradation of endogenous auxin indole-3-acetic acid (IAA) in plants.

The assumptions discussed in the works (Tandon and Arya 1982; Liu et al. 1996; Gaspar et al. 1996; Šimonová et al. 2005) are consonant with our early published work (Tsygankova et al. 1999), which testified in favor of the indirect, mediated through endogenous phytohormones action of synthetic derivatives of pyridine—lutidine N-oxide (Ivin) and pyrimidine—6-methylthiouracil (Methyur) on plant cell extension, and published works of other authors (Yip and Yang 1986; Murthy et al. 1995, 1998; Hutchinson and Saxena 1996) that showed the effect of exogenously applied synthetic multi-dimensional plant growth regulator Thidiazuron (TDZ; N-phenyl-1,2,3-thidiazole-5ylurea) on increase in concentrations of endogenous cytokinins, auxin, ethylene and ABA in plant cells.

Authors of work (Hutchinson and Saxena 1996) suggested that the powerful cytokinin-like regulatory effect of TDZ on plant growth is associated with its influence on the metabolism of endogenous plant hormones, either directly or indirectly through prevention the breakdown of endogenous purines by inhibiting cytokinin oxidase, due to which plant cell division and regeneration occur.

5 Conclusion

The auxin-like and cytokinin-like activity of synthetic low molecular weight heterocyclic compounds, derivatives of pyridine, pyrimidine, pyrazole, isoflavones, oxazolopyrimidine and oxazole was studied. With this aim the specific bioassay on auxin-like activity conducted on the leaf petioles isolated from seedlings of haricot bean (*Phaseolus vulgaris* L.) cultivar Belozernaya and the specific bioassay on cytokinin-like activity conducted on the cotyledons isolated from seeds of muscat pumpkin (*Cucurbita moschata* Duch. et Poir.) cultivar Gilea were used. It was shown that synthetic low molecular weight heterocyclic compounds used at the concentrations 10^{-8} M and 10^{-9} M demonstrated a high auxin-like and cytokinin-like activity, which was manifested in intensification of growth of isolated plant organs. The obtained results suggested the expressive auxin-like and cytokinin-like inducing effect of synthetic heterocyclic compounds on plant cell division, elongation, and differentiation that are the basic processes of plant growth.

This study confirmed the perspective of practical application of synthetic low molecular weight heterocyclic compounds, derivatives of pyridine, pyrimidine, pyrazole, isoflavones, oxazolopyrimidine and oxazole as a new effective plant growth regulating substances.

References

Avalbaev AM, Somov KA, Yuldashev RA, Shakirova FM (2012) Cytokinin oxidase is a key enzyme of cytokinin degradation. Biochemistry (Mosc) 77(12):1354–1361

Bang H, Zhou XK, van Epps HL, Mazumdar M (2010) Statistical methods in molecular biology. Series: Methods in molecular biology. Humana Press, New York

Basra AS (2000) Plant growth regulators in agriculture and horticulture: their role and commercial uses. Haworth Press Inc., New York, London, Oxford, p 264

Basu RN (1972) Effect of non-auxin chemicals on translocation of auxins in cuttings of *Phaseolus vulgaris* (L.), (kidney beans). J Exp Bot 23:357–365

Baum JS, Chen TM (1987) Plant growth and development modification using benzoxazole derivatives. Patent US 4659360 A.

Cansev A, Gulen H, Zengin MK, Ergin S, Cansev M, Kumral NA (2016) Use of pyrimidines in stimulation of plant growth and development and enhancement of stress tolerance. Patent US 20160000075

Chang JH, Baum JS (1990) Phenylmethyl-4,4-dimethyl-3-isoxazolidinone plant regulators. Patent US 4892578 A

Chen CM, Leisner SM (1985) Cytokinin-modulated gene expression in excised pumpkin. Plant Physiol 77:99–103

Corsi C, Wendeborn SV, Bobbio C, Kessabi J, Schneiter P, Grasso V, Haas UJ (2011) Isothiazole and pyrazole derivatives for use as plant growth regulators. Patent EP 2358699A1

Denancé N, Sánchez-Vallet A, Goffner D, Molina A (2013) Disease resistance or growth: the role of plant hormones in balancing immune responses and fitness costs. Front Plant Sci 4(Article 155):1–12

Enders TA, Strader LC (2015) Auxin activity: past, present, and future. Am J Bot 102(2):180–196

Gaspar T, Kevers C, Penel C, Greppin H, Reid DM, Thorpe TA (1996) Plant hormones and plant growth regulators in plant tissue culture. In Vitro Cell Dev Biol Plant 32(4):272–289

Gyulai G, Heszky LE (1995) Auxin and cytokinin bioassays: a short overview. Acta Agron Hung 43(1/2):185–197

Haggag WM, Abouziena HF, Abd-El-Kreem F, El Habbasha S (2015) Agriculture biotechnology for management of multiple biotic and abiotic environmental stress in crops. J Chem Pharm Res 7(10):882–889

Hare RC (1964) Indoleacetic acid oxidase. Bot Rev 30(1):129–165

Hutchinson MJ, Saxena PK (1996) Role of purine metabolism in thidiazuron-induced somatic embryogenesis of geranium (*Pelargonium hortorum* Bailey) hypocotyl cultures. Physiol Plantarum 98:517–522

Jain KS, Chitre TS, Miniyar PB, Kathiravan MK, Bendre VS, Veer VS, Shahane SR, Shishoo CJ (2006) Biological and medicinal significance of pyrimidines. Curr Sci 90(6):793–803

Jain KS, Arya N, Inamdar NN, Auti PB, Unawane SA, Puranik HH, Sanap MS, Inamke AD, Mahale VJ, Prajapati CS, Shishoo CJ (2016) The chemistry and bio-medicinal significance of pyrimidines & condensed pyrimidines. Curr Top Med Chem 16(28):3133–3174

Kieber JJ, Schaller GE (2018) Cytokinin signaling in plant development. Development 145(dev149344):1–7

Komaki S, Sugimoto K (2012) Control of the plant cell cycle by developmental and environmental cues. Plant Cell Physiol 53(6):953–964

Kumar R, Arora J, Ruhil S, Phougat N, Chhillar AK, Prasad AK (2014) Synthesis and antimicrobial studies of pyrimidine pyrazole heterocycles. Adv Chem. Article ID 329681, 1–12

Lam-Son T, Sikander P (2014) Phytohormones: a window to metabolism, signaling and biotechnological applications. Springer, New York, p 361

Lavy M, Estelle M (2016) Mechanisms of auxin signaling. Development 143(18):3226–3229

Leyser O (2017) Auxin signaling. Plant Physiol 176(1):465–479

Liu ZH, Hsiao IC, Pan YaW (1996) Effect of naphthaleneacetic acid on endogenous indole-3-acetic acid, peroxidase and auxin oxidase in hypocotyl cuttings of soybean during root formation. Bot Bull Acad Sin 37(4):247–253

Majda M, Robert S (2018) The role of auxin in cell wall expansion. Int J Mol Sci 19(4):951

Meena OP (2015) A review: role of plant growth regulators in vegetable production. Int J Agr Sci Res (IJASR) 5(5):71–84

Minn K, Dietrich H, Dittgen J, Feucht D, Häuser-Hahn I, Rosinger CH (2008) Pyrimidine derivatives and their use for controlling undesired plant growth. Patent US 8329717 B2

Mok DWS, Mok MC (2001) Cytokinin metabolism and action. Annu Rev Plant Physiol Plant Mol Biol 52:89–118

Murthy BNS, Murch SJ, Saxena PK (1995) Thidiazuron-induced somatic embryogenesis in intact seedlings of peanut (*Arachis hypogea*): endogenous growth regulator levels and significance of cotyledons. Physiol Plantarum 94:268–276

Murthy BNS, Murch SJ, Saxena PK (1998) Thidiazuron: a potent regulator of in vitro plant morphogenesis. In Vitro Cell Dev Biol Plant 34:267–275

Newton T, Waldeck I (2000) Oxazole carboxamide herbicides. Patent US6096688 A

Nicolopoulou-Stamati P, Maipas S, Kotampasi C, Stamatis P, Hens L (2016) Chemical pesticides and human health: the urgent need for a new concept in agriculture. Front Public Health 4: Article 148.11(3):274–286

Pandey A, Tamta S, Giri D (2011) Role of auxin on adventitious root formation and subsequent growth of cutting raised plantlets of *Ginkgo biloba* L. Int J Biodivers Conserv 3(4):142–146

Pop TI, Pamfil D, Bellini C (2011) Auxin control in the formation of adventitious roots. Not Bot Hort Agrobot Cluj 39(1):307–316

Quin LD, Tyrell JA (2010) Fundamentals of heterocyclic chemistry: importance in nature and in the synthesis of pharmaceuticals. Wiley, Hoboken, New Jersey, 344 pp

Rademacher W (2015) Plant growth regulators: backgrounds and uses in plant production. J Plant Growth Regul 34(4):845–872

Sergiev I, Alexieva V, Ivanov S, Bankova V, Mapelli S (2004) Plant growth regulating activity of some flavonoids. Comptes Rendus de L'academie Bulgare des Sciences 57(4):63–68

Šimonová E, Henselová M, Zahradník P (2005) Benzothiazole derivatives substituted in position two as biologically active substances with plant growth regulation activity. Plant Soil Environ 51:496–505

Takatsuka H, Umeda M (2014) Hormonal control of cell division and elongation along differentiation trajectories in roots. J Exp Bot 65(10):2633–2643

Tandon P, Arya HC (1982) Association of auxin protectors, peroxidase, indoleacetic acid oxidas, and polyphenol oxidase in zizyphus gall and normal stem tissues grown in culture. Biochem Physiol Pflanz 177(2):114–124

Tsygankova V, Andrusevich YaV, Shtompel O, Kopich V, Solomyanny R, Bondarenko O, Brovarets VS (2018a) Phytohormone-like effect of pyrimidine derivatives on regulation of vegetative growth of tomato. Int J Botany Stud 3(2):91–102

Tsygankova V, Andrusevich YaV, Kopich V, Shtompel O, Pilyo S, Kornienko AM, Brovarets V (2018b) Use of oxazole and oxazolopyrimidine to improve oilseed rape growth. Scholars Bull 4(3):301–312

Tsygankova VA, Zayets VN, Galkina LA, Blume YaB (1999) The phytohormone-mediated action of the synthetic regulators on cell extension growth in higher plants. Biopolym Cell 15(5):432–441

Tsygankova V, Andrusevich Ya, Shtompel O, Romaniuk O, Yaikova M, Hurenko A, Solomyanny R, Abdurakhmanova E, Klyuchko S, Holovchenko O, Bondarenko O, Brovarets V (2017a) Application of synthetic low molecular weight heterocyclic compounds derivatives of pyrimidine, pyrazole and oxazole in agricultural biotechnology as a new plant growth regulating substances. Int J Med Biotechnol Genetics S2(002):10–32

Tsygankova VA, Andrusevich YaV, Shtompel OI, Kopich VM, Pilyo SG, Prokopenko VM, Kornienko AM, Brovarets VS (2017b) Intensification of Vegetative Growth of Cucumber by Derivatives of [1,3]oxazolo[5,4-*d*]pyrimidine and N-sulfonyl substituted of 1,3-oxazole. Res J Life Sci Bioinf Pharm Chem Sci (RJLBPCS) 3(4):107–122

Tsygankova VA, Andrusevich YaV, Shtompel OI, Pilyo SG, Kornienko AM, Brovarets VS (2018c) Acceleration of vegetative growth of wheat (*Triticum aestivum* L.) using [1,3]oxazolo[5,4-*d*]pyrimidine and N-sulfonyl substituted 1,3-oxazole. Pharm Chem J 5(2):167–175

Tsygankova VA, Andrusevich YaV, Shtompel OI, Shablykin OV, Hurenko AO, Solomyanny RM, Mrug GP, Frasinyuk MS, Pilyo SG, Kornienko AM, Brovarets VS (2018d) Auxin-like effect of derivatives of pyrimidine, pyrazole, isoflavones, pyridine, oxazolopyrimidine and oxazole on acceleration of vegetative growth of flax. Int J PharmTech Res 11(3):274–286

Tsygankova VA, Andrusevich YaV, Shtompel OI, Solomyanny RM, Hurenko AO, Frasinyuk MS, Mrug GP, Shablykin OV, Pilyo SG, Kornienko AM, Brovarets VS (2018e) Study of auxin-like and cytokinin-like activity of derivatives of pyrimidine, pyrazole, isoflavones, pyridine, oxazolopyrimidine and oxazole on haricot bean and pumpkin plants. Int J ChemTech Res 11(10):174–190

Tsygankova V, Andrusevich Ya, Shtompel O, Kopich V, Solomyanny R, Brovarets V (2019) Study of regulating activity of synthetic low molecular weight heterocyclic compounds, derivatives of pyrimidine on growth of tomato (*Solanum lycopersicum* L.) seedlings. Int J ChemTech Res 12(5):26–38

Wania SH, Kumarb V, Shriramc V, Sah SK (2016) Phytohormones and their metabolic engineering for abiotic stress tolerance in crop plants. Crop J 4:162–176

Whittingham WG, Winn CL, Glithro H, Boussemghoune MA, Aspinall MB (2010) Pyrimidine derivatives and their use as herbicides. WO Patent 2010092339 A1

Yip WK, Yang SF (1986) Effect of thidiazuron, a cytokinin–active urea derivative, in cytokinin-dependent ethylene production system. Plant Physiol 80:515–519

Zhao Yu (2010) Auxin biosynthesis and its role in plant development. Annu Rev Plant Biol 61:49–64

Zwack PJ, Rashotte AM (2013) Cytokinin inhibition of leaf senescence. Plant Signal Behav 8(7):e24737

The manufacturer's authorised representative in the EU is Springer Nature Customer Service Centre GmbH, Europaplatz 3, 69115 Heidelberg, Germany. If you have any concerns regarding our products, please contact ProductSafety@springernature.com

Printed and bound by CPI Group (UK) Ltd, Croydon, CR0 4YY
30/09/2026
02225317-0012